EX VIVO CELL THERAPY

EX VIVO CELL THERAPY

Edited by
Klaus Schindhelm
Robert Nordon
The University of New South Wales
Sydney, Australia

Academic Press
San Diego London Boston New York Sydney Tokyo Toronto

This book is printed on acid-free paper. ∞

Academic Press
a division of Harcourt Brace & Company
525 B Street, Suite 1900, San Diego, California 92101-4495, USA
http://www.apnet.com

Academic Press
24-28 Oval Road, London NW1 7DX, UK
http://www.hbuk.co.uk/ap/

Library of Congress Catalog Card Number: 98-89643

International Standard Book Number: 0-12-624960-1

PRINTED IN THE UNITED STATES OF AMERICA
99 00 01 02 03 04 BB 9 8 7 6 5 4 3 2 1

Contents

3

Cell Signaling by Hematopoietic Growth Factor Receptors

R. Starr and N. A. Nicola

4

Influence of Cytokines and Adhesion Molecules on Hematopoietic Stem Cell Development

P. J. Simmons, D. N. Haylock, and J.-P. Lévesque

5

ALLOGENEIC HEMATOPOIETIC STEM CELL TRANSPLANTATION

P. A. ROWLINGS

6

AUTOLOGOUS STEM CELL TRANSPLANTATION

UWE HAHN AND L. BIK TO

7

CORD BLOOD STEM CELL TRANSPLANTATION

MARCUS R. VOWELS

8

RECONSTITUTION OF IMMUNITY BY ADOPTIVE IMMUNOTHERAPY WITH T CELLS

STANLEY R. RIDDELL, EDUS HOUSTON WARREN, DEBORAH LEWINSOHN, CASSIAN YEE, AND PHILIP D. GREENBERG

9

EXOGENOUS GENE TRANSFER INTO LYMPHOID AND HEMATOPOIETIC PROGENITOR CELLS

LUN QUAN SUN, MICHELLE MILLER, AND GEOFF SYMONDS

10

PROCESS DEVELOPMENT FOR *EX VIVO* CELL THERAPY

CHARLES DOWDING, TOM LEEMHUIS, AUDREY JAKUBOWSKI, AND CHRISTOPHER READING

11

CELL SEPARATION

ROBERT E. NORDON AND KLAUS SCHINDHELM

12

ENVIRONMENTAL REQUIREMENTS OF HEMATOPOIETIC PROGENITOR CELLS IN *EX VIVO* EXPANSION SYSTEMS

PETER W. ZANDSTRA, CONNIE J. EAVES, AND JAMES M. PIRET

13

EX VIVO HEMATOPOIETIC CELL EXPANSION FOR BONE MARROW TRANSPLANTATION

RAMKUMAR MANDALAM, MANFRED KOLLER, AND ALAN SMITH

14

GENE DELIVERY TECHNOLOGY: NONVIRAL AND VIRAL VECTOR SYSTEMS

MITCH RAPONI AND GEOFF SYMONDS

15

SUMMARY AND FUTURE DIRECTIONS

ROBERT E. NORDON AND KLAUS SCHINDHELM

Contributors

Numbers in parentheses indicate the pages on which the author's contributions begin.

C. G. Begley (5) The Walter and Eliza Hall Institute of Medical Research and Rotary Bone Marrow Research Laboratories, P.O. Royal Melbourne Hospital, Melbourne, Victoria 3050, Australia.

Charles Dowding (197) SyStemix Inc., Palo Alto, California 94304.

Connie J. Eaves (245) Department of Medical Genetics, University of British Columbia, Vancouver, British Columbia V6T 1Z3, Canada; and the Terry Fox Laboratory, British Columbia Cancer Agency, Vancouver, British Columbia V5Z 1L3, Canada.

A. G. Elefanty (5) The Walter and Eliza Hall Institute of Medical Research and Rotary Bone Marrow Research Laboratories, P.O. Royal Melbourne Hospital, Melbourne, Victoria 3050, Australia.

Philip D. Greenberg (137) Fred Hutchinson Cancer Research Center, Seattle, Washington 98109.

Uwe Hahn (99) Division of Haematology, Institute of Medical and Veterinary Science, Adelaide, South Australia 5000, Australia.

D. N. Haylock (51) Leukaemia Research Unit, Hanson Centre for Cancer Research, P.O. Box 14, Rundle Mall, Adelaide, South Australia 5000, Australia.

Audrey Jakubowski (197) SyStemix Inc., Palo Alto, California 94304.

Manfred Koller (273) Aastrom Biosciences, Inc., P.O. Box 376, Ann Arbor, Michigan 48106.

Tom Leemhuis (197) SyStemix Inc., Palo Alto, California 94304.

J.-P. Lévesque (51) Leukaemia Research Unit, Hanson Centre for Cancer Research, P.O. Box 14, Rundle Mall, Adelaide, South Australia 5000, Australia.

Deborah Lewinsohn (137) Fred Hutchinson Cancer Research Center, Seattle, Washington 98109.

Ramkumar Mandalam (273) Aastrom Biosciences, Inc., P.O. Box 376, Ann Arbor, Michigan 48106.

Michelle Miller (179) Johnson and Johnson Research, G.P.O. Box 3331, Sydney, New South Wales 2001, Australia.

N. A. Nicola (27) The Walter and Eliza Hall Institute of Medical Research and Cooperative Research Centre for Cellular Growth Factors, P.O. Royal Melbourne Hospital, Melbourne, Victoria, 3050, Australia.

Robert E. Nordon (1, 215, 323) Graduate School of Biomedical Engineering, University of New South Wales, Sydney, New South Wales 2052, Australia.

James M. Piret (245) Biotechnology Laboratory and Department of Chemical and Bio-Resource Engineering, University of British Columbia, Vancouver, British Columbia V6T 1Z3, Canada.

Mitch Raponi (293) Johnson and Johnson Research, Rushcutters Bay, Sydney, New South Wales 2001, Australia.

Christopher Reading (197) SyStemix Inc., Palo Alto, California 94304.

Stanley R. Riddell (137) Fred Hutchinson Cancer Research Center, Seattle, Washington 98109.

L. Robb (5) The Walter and Eliza Hall Institute of Medical Research and Rotary Bone Marrow Research Laboratories, P.O. Royal Melbourne Hospital, Melbourne, Victoria 3050, Australia.

P. A. Rowlings (85) International Bone Marrow Transplant Registry/Autologous Blood and Marrow Transplant Registry Statistical Center, Health Policy Institute, Medical College of Wisconsin, Milwaukee, Wisconsin 53226.

Klaus Schindhlem (1, 215, 323) Graduate School of Biomedical Engineering, University of New South Wales, Sydney, New South Wales 2052, Australia.

P. J. Simmons (51) Leukaemia Research Unit, Hanson Centre for Cancer Research, P.O. Box 14, Rundle Mall, Adelaide, South Australia 5000, Australia.

Alan Smith (273) Aastrom Biosciences, Inc., P.O. Box 376, Ann Arbor, Michigan 48106.

R. Starr (27) The Walter and Eliza Hall Institute of Medical Research and Cooperative Research Centre for Cellular Growth Factors, P.O. Royal Melbourne Hospital, Melbourne, Victoria 3050, Australia.

Lun Quan Sun (179) Johnson and Johnson Research, G.P.O. Box 3331, Sydney, New South Wales 2001, Australia.

Geoff Symonds (179, 293) Johnson and Johnson Research, G.P.O. Box 3331, Sydney, New South Wales 2001, Australia.

L. Bik To (99) Division of Haematology, Institute of Medical and Veterinary Science, Adelaide, South Australia 5000, Australia.

Marcus R. Vowels (127) Haematology/Oncology, Sydney Children's Hospital, Randwick, Sydney, New South Wales 2031, Australia.

Edus Houston Warren (137) Fred Hutchinson Cancer Research Center, Seattle, Washington 98109.

Cassian Yee (137) Fred Hutchinson Cancer Research Center, Seattle, Washington 98109.

Peter W. Zandstra (245) Biotechnology Laboratory and Department of Chemical and Bio-Resource Engineering, University of British Columbia, Vancouver, British Columbia V6T 1Z3, Canada.

FOREWORD

Over the past three decades a steady improvement in the outcome of allogeneic bone marrow transplantation has led to the establishment of marrow transplantation as a standard therapy for selected diseases and stages of disease. Success has been attributed to the development of transfusion support by blood components, to advances in antibacterial and antiviral therapy, and, particularly, to advances in the knowledge of the human leukocyte antigen (HLA) system. These advances have led to prevention of graft rejection and better prevention and control of graft-versus-host disease. More recently, the use of hematopoietic stem cells mobilized and collected from the peripheral blood and from cord blood has expanded the application of hematopoietic cell transplantation.

The field of *ex vivo* cell therapy was born out of the necessity to manipulate the cellular constituents of hematopoietic cell grafts to remove malignant cells and to increase the number of stem cells. *Ex vivo* cell therapy provides the opportunity to manipulate the immune system to reduce the graft-versus-host reaction and autoimmune cells while augmenting the graft-versus-tumor effect and antiviral immunity. The goals of this new generation of cell-based therapeutics will be to increase the safety of hematopoietic cell transplantation and to enable delivery of adoptive cellular immunotherapy and gene therapy. Current research has focused on strategies that require cell selection, expansion, and gene transfer.

Development of these core technologies will facilitate evaluation of the clinical impact of transplanted cells on the patient and on the disease. This field of research will lead to better therapy for malignant diseases, disseminated viral infections, autoimmune diseases, and genetic disorders.

E. Donnall Thomas

1

INTRODUCTION

ROBERT E. NORDON AND KLAUS SCHINDHELM

Graduate School of Biomedical Engineering
University of New South Wales
Sydney, New South Wales 2052, Australia

In recent years, *in vitro* techniques for the generation of cell subsets with functional properties similar to those of unmanipulated cells have provided the key components for development of a new therapy that is based on biological rather than pharmacological intervention. The aim of *ex vivo* cell therapy is to replace, repair, or enhance the biological function of damaged tissue or organs. An *ex vivo* process involves harvesting cells from patients or donors, *in vitro* manipulation to enhance the therapeutic potential of the cell harvest, and subsequent intravenous transfusion. At any stage of the process, cells can be cryopreserved so that therapy can be scheduled according to the patient's requirements. Cell delivery is via the vascular compartment, and, as such, therapy is restricted to those conditions that can be corrected by manipulation of the hematopoietic and immune systems (Table 1.1).

Success of therapy at a clinical level will require the transformation of laboratory-based techniques into individualized, cell-production processes with standards of safety and efficacy similar to those established for pharmaceutical therapeutics. Process development requires a comprehensive overview of scientific and technological advances encompassing *in vitro* methods for cell isolation and growth, devices for cell processing, and the initial clinical evaluation of these *ex vivo* devices and processes. Therefore, in addition to review of technologies for cell selection, cell expansion, and gene transfer, relevant areas of

TABLE 1.1 Diseases That Could Be Treated by *Ex Vivo* Cell Therapy

Malignant
Leukemia
Myelodysplastic syndromes
Lymphoma
Myeloma
Solid tumors
Genetic
Hemoglobinopathies
Immunodeficiency states
Disorders of metabolism
Hemophilia
Infectious
Human immunodeficiency virus
Disseminated cytomegalovirus
Epstein–Barr virus-related lymphoproliferative disorders
Other
Multiple sclerosis
Rheumatoid arthritis
Aplastic anemia

biological and clinical research will be highlighted. These advances are presented by leading researchers and are summarized by the editors in the final chapter.

Often fundamental research provides the rationale or stimulus for development of new therapeutic approaches, and much is to be gained from the study of basic cellular processes. Manipulation of hematopoietic cells *in vitro* requires an understanding of transcriptional control (Chapter 2), cell-signaling pathways (Chapter 3), and the influence of exogenous factors such as cytokines and adhesion molecules (Chapter 4). Likewise, advances in cellular immunology have led to development of *in vitro* systems for selection and expansion of antigen-specific cytotoxic T cells that have the potential to reconstitute immunity against disseminated viral infections or tumor antigens (Chapter 8). Manipulation of cells at a genomic level provides even broader scope to alter cellular function (Chapter 9).

Transplantation medicine is now a complex area of clinical research encompassing treatment toxicity, host/graft/tumor interactions, and the difficulties associated with design of appropriate clinical studies to evaluate treatment regimens. Hematopoietic stem cell transplantation has revolutionized the treatment of cancer by facilitating dose escalation into the myelotoxic range. Despite these clinical advances, patient survival is a balance between treatment response rates and the incidence of fatal sequelae related to the toxicity of therapy. Application of *ex vivo* cell technologies to allogeneic, autologous, and cord blood stem cell transplantation (Chapters 5, 6, and 7, respectively) could increase long-term survival by reducing adverse events associated with high-dose therapy and stem cell transplant.

Enabling technologies should translate laboratory methods into clinically applicable processes, and their development will require collaboration between clinicians, biologists, and technologists. Core technologies for delivery of cell therapy include cell selection (Chapter 11), gene transfer (Chapters 9 and 14), and bioreactor devices (Chapters 12 and 13) for the manipulation of cells in culture.

Quality control and good manufacturing practice developed by blood banks and the pharmaceutical industry have provided an initial template for development of production standards for *ex vivo* cell therapy. As these processes continue to evolve, regulatory authorities, professional organizations, and hospital administrations will be responsible for maintaining adequate standards of safety and efficacy (Chapter 10).

The development of *ex vivo* processes relies on an interactive dialogue between biologists, technologists, and clinicians. Without concepts that are familiar to clinicians and cell biologists, biomedical engineers would have difficulty designing devices for implementing these processes. Likewise, it is important for clinicians to appreciate the physical limitations of various technologies so that they can implement technically and economically feasible processes. In addition to those working in specialist fields, generalists are required to coordinate the specialist areas. The convergence of disciplines required for development of *ex vivo* cell therapy is depicted in Fig. 1.1.

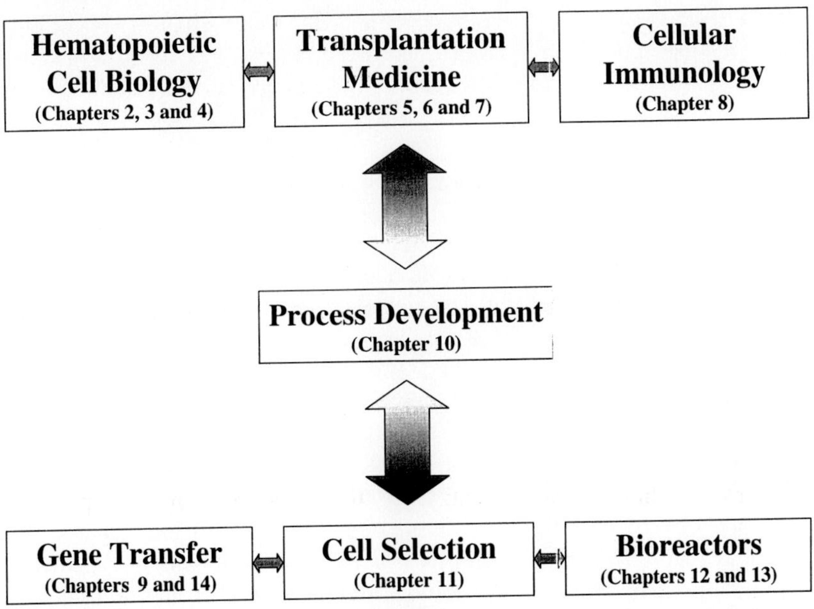

FIGURE 1.1

2

Transcriptional Control of Hematopoiesis

L. Robb, A. G. Elefanty, and C. G. Begley

The Walter and Eliza Hall Institute of Medical Research and Rotary Bone Marrow Research Laboratories
P.O. Royal Melbourne Hospital
Melbourne, Victoria 3050, Australia

The field of experimental hematology has advanced dramatically since the development of the clonal culture assay in the mid-1960s. This technique proved crucial both in allowing the characterization of progenitor cells within the hematopoietic compartment and in stimulating the identification of the extracellular growth factors essential to their development. The subsequent application of molecular cloning techniques allowed the genes encoding these extracellular growth factors, or colony-stimulating factors (CSFs), to be defined and recombinant molecules to be produced. Based on the action of these molecules in *in vitro* and *in vivo* systems, they moved rapidly into the clinic where granulocyte-CSF (G-CSF) and granulocyte-macrophage CSF (GM-CSF) have found widespread application, thus confirming their role as important hematopoietic regulators.

The ability to molecularly clone mammalian genes also provided a new perspective for experimental hematologists. The knowledge that transcription factors were important in determining cell fate in other systems was quickly translated to hematopoietic cells, with identification of key transcriptional regulators. The expression of these regulators in a cell-type-specific manner and the identification of specific target genes provided important evidence that the regulation of, for example, globin genes could begin to be explained at the transcriptional level.

The application of molecular genetic techniques has provided entirely new approaches for dissecting the hematopoietic compartment. The advent of gene targeting or "knockout" technology has allowed the contribution of particular genes and the proteins they encode to be evaluated in the whole animal. In this review, we have selected some of the important transcription factors where new insights into their role in hematopoiesis have been provided by using this approach.

I. SCL: A MASTER HEMATOPOIETIC REGULATOR GENE IDENTIFIED VIA CHROMOSOME TRANSLOCATION

SCL was first identified because of its involvement in a t(1;14) chromosomal translocation in a unique, multipotential leukemia (Begley *et al.,* 1989a, 1989b). The leukemic cells displayed an early T-cell phenotype and differentiated into myeloid cells *in vivo* following treatment with 2-deoxycoformycin (Hershfield *et al.,* 1984). This phenotype was recapitulated *in vitro* when a cell line was established (Kurtzberg *et al.,* 1985). The t(1;14) translocation was molecularly cloned to characterize the gene responsible for the phenotype, and the SCL gene (for stem cell leukemia) was identified (Begley *et al.,* 1989a, 1989b). The majority of leukemias in which this gene has been subsequently implicated are more typical T-cell acute lymphoblastic leukemias (ALL) (thus the alternate name TAL-1) (Brown *et al.,* 1990; Robb and Begley 1996). However, the stem cell nature of the initial leukemia accurately predicted an important aspect of the normal function of this gene.

The protein encoded by the SCL gene is a member of the helix–loop–helix (HLH) family of transcription factors (Murre *et al.,* 1989). This motif serves to allow protein dimerization to occur. The majority of family members have a basic domain that mediates DNA binding to specific sequences (-CANNTG-) that conform to an "E-box" element (Stone *et al.,* 1987; Murre *et al.,* 1989; Weintraub *et al.,* 1991). Protein heterodimers often form between family members that show tissue-specific expression (e.g., SCL) and those that are ubiquitously expressed (e.g., products of the E2A gene) (Hsu *et al.,* 1994).

The HLH motif was first recognized as a common domain in Myc proteins, Myo-D (a "master-regulator" gene of muscle cells), the product of the *Drosophila* daughterless gene and the Ig enhancer binding proteins E12 and E47 (known generically as E proteins) produced by alternate splicing from the E2A gene (Murre *et al.,* 1989). It is noteworthy that, in addition to SCL, other members of the HLH family play critical roles in the development of hematopoietic malignancies. For example, the *c-myc* gene is disrupted in the t(8;14) translocation of Burkitt's lymphoma, and dysregulated expression of *c-myc* initiates lymphomagenesis in transgenic mice harboring a facsimile of the translocation (Cory, 1986). Similarly, the E2A gene is involved in the t(1;19) translocation that is present in approximately 30% of cases of pre-B ALL (Cory, 1986; Kamps *et al.,* 1990; Nourse *et al.,* 1990). LYL-1 and TAL-2 are additional HLH genes implicated in rare T-cell leukemias (Mellentin *et al.,* 1989; Xia *et al.,* 1991). Although SCL can function as an oncogene (Elwood *et al.,* 1993; Condorelli *et al.,* 1996; Kelliher *et al.,* 1996; Larson *et al.,* 1996; Curtis *et al.,* 1997), the mechanism by which it and these related proteins contribute to tumor development is unclear. It is possible that SCL/E protein heterodimers might inappropriately activate new target genes when SCL is aberrantly expressed in a T-cell environment. Conversely, SCL might sequester E proteins (or other HLH family members) and render them unavailable for their normal function.

II. SCL IN NORMAL HEMATOPOIESIS

The first indication that SCL might play a role in hematopoietic differentiation came from studies of SCL gene expression. SCL expression was detected in progenitor/stem cells, mast cells, erythroid cells, and megakaryocytes (Begley *et al.,* 1989b; Green *et al.,* 1991a; Visvader *et al.,* 1991; Mouthon *et al.,* 1993) but was absent from T cells and myeloid cells. This hematopoietic pattern of expression appeared very similar to that of the zinc finger transcription factor GATA-1 (Green *et al.,* 1992). However, SCL expression was also detected in endothelial cells and in developing neural and skeletal tissues (Green *et al.,* 1992; Kallianpur *et al.,* 1994).

Analysis of SCL expression during hematopoietic cell differentiation showed that the level of SCL mRNA increased during erythroid differentiation (Green *et al.*, 1992; Chiba *et al.*, 1993; Cross *et al.*, 1994). Furthermore, gene transfer experiments showed that enforced SCL expression enhanced this process (Aplan *et al.*, 1992). In contrast, myeloid differentiation was associated with decreased SCL expression (Green *et al.*, 1993; Tanigawa *et al.*, 1993, 1995; Hoang *et al.*, 1996) and enforced SCL expression perturbed cytokine-induced macrophage differentiation in M1 cells. Thus SCL is clearly implicated in differentiation events within hematopoietic cells.

SCL also plays a role in modulating hematopoietic cell proliferation. Enforced SCL expression allowed M1 cells to escape from cytokine-induced suppression of clonogenicity (Tanigawa *et al.*, 1993, 1995) and enhanced clonogenicity in a lymphoid cell line (Elwood *et al.*, 1993). In keeping with this, when SCL function was blocked in K562 cells, there was a 50-fold decrease in self-renewal potential (Green *et al.*, 1991a). In addition, it has been suggested that SCL may prevent apoptosis (Leroy-Viard *et al.*, 1995).

Although these experiments documented a role for SCL in hematopoietic proliferation and differentiation, the results of ablation of SCL were even more dramatic. Mice carrying a null mutation of the SCL gene died at day 9.5 of embryonic development, with complete absence of detectable hematopoietic cells (Robb *et al.*, 1995; Shivdasani *et al.*, 1995a). In these animals, the yolk sac was devoid of primitive nucleated erythroid cells or their precursors, and the blood vessels of the embryo, although formed, lacked any hematopoietic cells. Embryonic development appeared otherwise normal. These experiments documented a critical role for SCL in the development of the most primitive embryonic hematopoietic cells ("primitive" hematopoiesis) but did not address the function of SCL in adult hematopoiesis. To examine this, doubly targeted embryonic stem (ES) cells were generated and their behavior was documented *in vitro* and *in vivo*. In *in vitro* culture, the SCL-null ES cells failed to generate hematopoietic colonies and did not express hematopoietic-specific genes (Elefanty *et al.*, 1997). When these ES cells were injected into foster blastocysts, there were no lymphoid, myeloid, or erythroid cells generated from the SCL-null ES cells, although these cells contributed substantially to all other tissues (Porcher *et al.*, 1996; Robb *et al.*, 1996). Thus, these experiments confirmed the absolute requirement for SCL to generate fetal liver and adult hematopoietic cells ("definitive" hematopoiesis) and demonstrated that provision of a normal extracellular environment did not rescue this defect.

Although SCL has a vital role in the development of hematopoietic stem cells in the adult and the embryo, it is still unknown what part SCL plays in erythroid differentiation once a stem cell is formed. This illustrates a limitation of conventional gene targeting, which can only reveal the earliest critical function of a gene. Clearly, alternative approaches are required to dissect out later or nonessential roles.

III. LMO2: ANOTHER PARTNER FOR SCL

As already noted, above, SCL binds DNA as a heterodimer with products of the E2A gene in erythroid and leukemic T cells. Another component of a larger SCL-containing protein complex is the product of the LMO2 (RBTN2/TTG2) gene. LMO2 contains two cysteine-rich, zinc finger-like LIM domains, which appear to be exclusively involved in protein–protein interactions (Sanchez-Garcia and Rabbitts, 1994) despite their structural similarity to the DNA-binding zinc fingers of the GATA transcription factor family (Omichinski *et al.,* 1993; Perez-Alvarado *et al.,* 1994).

LMO2 shows some intriguing functional similarities to SCL. It too was first identified because of its involvement in a t(11;14) chromosomal translocation in T ALL (Boehm *et al.,* 1991; Royer-Pokora *et al.,* 1991). Like SCL, LMO2 is not normally expressed in T lymphocytes, but is abundant in erythroid tissues and putative hematopoietic precursors (Warren *et al.,* 1994). Mice in which the LMO2 gene was ablated showed a very similar phenotype to that of mice lacking SCL. They also died *in utero* with complete absence of erythropoiesis but with an intact vasculature (Warren *et al.,* 1994).

The relationship between SCL and LMO2 extends further. Recent studies have demonstrated their coexpression in T-ALL cells, despite lack of expression of either gene in normal T lymphocytes. In addition, the two proteins cooperate in causing T-cell tumors (Larson *et al.,* 1996). The two proteins have also been shown to form a protein complex in erythroid cells (Valge-Archer *et al.,* 1994; Wadman *et al.,* 1994), where LMO2 may act as a molecular "bridge" between SCL and GATA-1. This results in a multiprotein, DNA-binding protein complex that includes at least SCL, the E2A gene products, LMO2, GATA-1, and the recently identified LIM-motif binding protein Ldb1/NL1 (Osada *et al.,* 1995; Wadman *et al.,* 1997). The similarity between the null phenotype for SCL and LMO2 suggests that these interactions are of physiological significance and that these two proteins play a central role in a common pathway that serves to establish hematopoiesis (Fig. 2.1).

IV. THE GATA FAMILY OF TRANSCRIPTION FACTORS

The GATA nuclear transcription factors comprise an evolutionarily conserved family of proteins that bind the DNA sequence (A/T)GATA(A/G) via a highly conserved zinc finger domain (Evans and Felsenfeld, 1989; Tsai *et al.,* 1989; Orkin, 1992). Through their expression profiles, the GATA proteins may be distinguished as hematopoietic (GATA-1 to GATA-3) and nonhematopoietic (GATA-4 to GATA-6), a functional grouping that mirrors their structural similarities. In this section, only the hematopoietic GATA transcription factors will be considered.

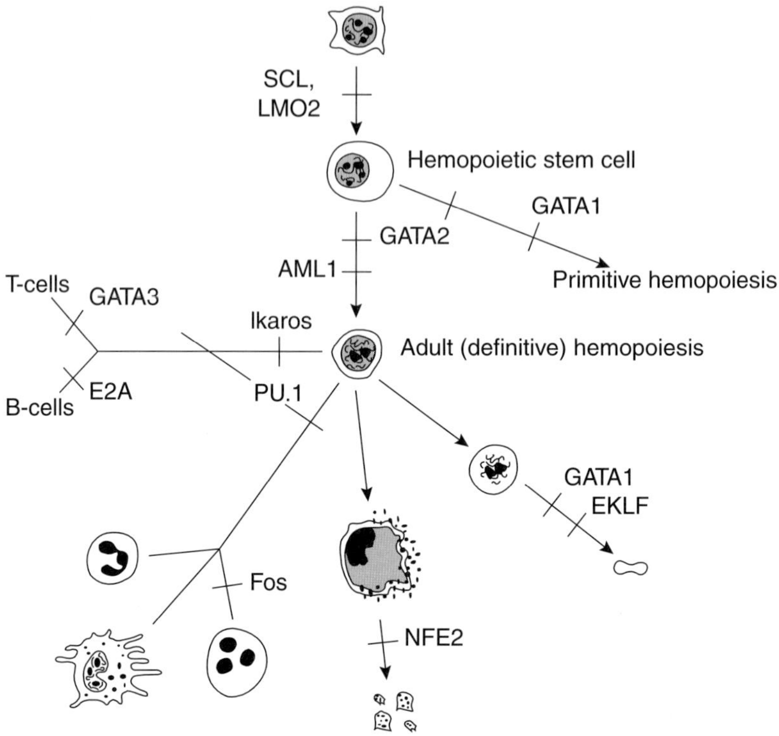

FIGURE 2.1. Site of action of selected transcription factors in hematopoietic development and differentiation. The major site of function, as revealed by gene targeting, is indicated: gene function is required for hematopoietic differentiation to proceed normally beyond this point. Targeted disruption of c-Fos, a widely expressed transcription factor, shows altered osteoclast function, with altered hematopoiesis as a consequence of this effect (Johnson *et al.,* 1992; Wang *et al.,* 1992).

Cells of the erythroid, megakaryocytic, and mast cell lineages are the major sites of GATA-1 expression, but the protein is also expressed in eosinophils and at low levels in multipotential progenitor cells and in the Sertoli cells of the testis (Crotta *et al.,* 1990; Martin *et al.,* 1990; Romeo *et al.,* 1990; Yamamoto *et al.,* 1990; Sposi *et al.,* 1992; Ito *et al.,* 1993; Zon *et al.,* 1993; Weiss and Orkin, 1995). Although GATA-2 shares expression in megakaryocytes and mast cells, it is also found in populations enriched for hematopoietic stem and progenitor cells, endothelial cells, and embryonic brains (Wilson *et al.,* 1990; Yamamoto *et al.,* 1990; Lee *et al.,* 1991; Dorfman *et al.,* 1992; Kornhauser *et al.,* 1994; Nagai *et al.,* 1994). The related molecule GATA-3 is expressed widely during embryogenesis, especially in the developing nervous system. Hematopoietic expression of GATA-3 is restricted, however, to T lymphocytes, and, in the chicken, to erythroid cells (Yamamoto *et al.,* 1990; Ho *et al.,* 1991; Joulin *et al.,* 1991; Ko *et al.,* 1991; Oosterwegel *et al.,* 1992; Kornhauser *et al.,* 1994).

Consistent with these patterns of expression, functionally important GATA-binding sites are present in the *cis*-regulatory elements of many hematopoietically restricted genes in erythroblasts, megakaryocytes, mast cells, and T cells (Weiss and Orkin, 1995). Indeed, GATA-1 has been shown to regulate expression of many such genes, including SCL (Aplan *et al.,* 1992).

The GATA zinc fingers have been shown to mediate protein dimerization as well as DNA binding. Crossley (1995) demonstrated GATA-1 homodimerization *in vitro* requiring the carboxy-terminal zinc finger and showed that the same region mediated heterotypic interactions between GATA-1 and GATA-2 or GATA-3. Also, GATA-1 was shown to functionally interact with the Krüppel family zinc finger proteins, Sp1 and EKLF, in a zinc finger dependent fashion (Merika and Orkin, 1995). In both studies, the authors demonstrated that this interaction was not compromised by GATA-1 mutations that abolished DNA binding. These observations contrasted with the findings of Yang and Evans (1995), who used a chimeric GATA-1 protein—in which the DNA-binding domain of the bacterial repressor protein LexA replaced the carboxy zinc finger—to suggest that GATA-1 homodimerization and GATA-1/GATA-2, but not GATA-1/GATA-5, heterodimerization were possible *in vitro* in the absence of the GATA-1 carboxy-terminal zinc finger.

A. GATA-1

Gene-targeting studies have confirmed the key role played by GATA-1 in erythroid differentiation. GATA-1-deficient embryos died in midgestation from anemia (Fujiwara *et al.,* 1996) and GATA-1-deficient ES cells were unable to form mature erythrocytes in chimeric mice (Pevny *et al.,* 1991). Consistent with this, *in vitro* differentiation of GATA-1-null ES cells resulted in erythroid precursors that were arrested at the proerythroblast stage (Weiss *et al.,* 1994; Pevny *et al.,* 1995) and subsequently underwent apoptosis (Weiss and Orkin, 1995). Although it appears that GATA-1 is required for erythroid survival and differentiation, there are conflicting data concerning the effects of overexpressing GATA-1 in erythroid cells. Expression of a GATA-1–estrogen receptor fusion protein in avian erythroid progenitors led to a hormone-dependent accelerated red blood cell differentiation and suppressed cell proliferation (Briegel *et al.,* 1996). In contrast, overexpression of GATA-1 under the control of the β-globin-locus control region actually inhibited differentiation and enhanced proliferation in murine erythroleukemia cells and in ES cell derived erythroblasts (Whyatt *et al.,* 1997).

Although megakaryocytes and mast cells could develop from GATA-1-null ES cells both *in vitro* and *in vivo* (Pevny *et al.,* 1995), GATA-1 is not necessarily redundant in nonerythroid lineages. Enforced expression of GATA-1 in a promegakaryocyte cell line produced megakaryocytic differentiation (Visvader *et al.,* 1992; Visvader and Adams, 1993) and deletion of an upstream regulatory element from the GATA-1 locus led to mice with selective thrombocytopenia, demonstrating the dependence of normal megakaryopoiesis on GATA-1 (Shivdasani *et al.,* 1997). Retroviral gene delivery experiments in chicken hematopoietic cells also suggested a broader function for GATA-1, with induction of both megakaryocytic and eosinophilic differentiation apparently occurring in a dose-dependent manner (Kulessa *et al.,* 1995).

Recently, a novel protein interacting with the amino-terminal zinc finger of GATA-1 was identified (Tsang *et al.,* 1997). Itself a zinc finger protein, FOG (friend of GATA-1) acted as a GATA-1 cofactor in erythroid and megakaryocytic cells and could physically interact with all the hematopoietic GATA proteins. It has been suggested that FOG may be only one of several related GATA gene cofactors, which may serve to modulate hematopoietic-specific gene expression (Krause and Perkins, 1997). The physiological relevance of the GATA-1/FOG interaction was anticipated by a study in which the phenotypic rescue of a GATA-1-deficient proerythroblast cell line required the expression of a GATA protein incorporating both carboxy-terminal (DNA binding) and amino-terminal (FOG interacting) zinc fingers (Weiss *et al.,* 1997).

B. GATA-2

Consistent with the preferential expression of GATA-2 in immature hematopoietic cells, levels of GATA-2 fall in more differentiated erythroblasts while GATA-1 levels reciprocally increase (Yamamoto *et al.,* 1990; Weiss *et al.,* 1994, 1997). In chicken hematopoietic cells, enforced GATA-2 but not GATA-1 expression promoted proliferation and hindered differentiation (Briegel *et al.,* 1993). These results suggested a role for GATA-2 in the proliferation rather than differentiation of early hematopoietic cells. Ablation of the GATA-2 gene caused a more severe impairment in embryonic hematopoiesis than that seen in GATA-1-deficient mice but the defect was not as complete as the aplasia in SCL- or LMO2-knockout embryos (Tsai *et al.,* 1994). GATA-2-null animals died between embryonic days 10 and 11 (approximately 1 day later than SCL- or LMO2-null embryos), with marked anemia and a 10-fold reduction in colony-forming cells in the yolk sac. Examination of chimeric mice generated by injecting GATA-2-null ES cells into wild-type blastocysts revealed that the GATA-2-deficient ES cells contributed poorly to primitive (yolk sac) hematopoiesis and were unable to give rise to definitive, fetal liver derived blood cells. Consistent with these findings, *in vitro* differentiation of GATA-2-deficient ES cells revealed a lowered frequency of hematopoietic precursors of all lineages and an absolute requirement for GATA-2 for mast cell development (Tsai *et al.,* 1994; Tsai and Orkin, 1997). The rarity and reduced size of mixed lineage or blast cell colonies pointed to an important role for GATA-2 in the expansion of multipotential cells (Tsai and Orkin, 1997). Thus GATA-2 appears to be important at a stage of hematopoietic development intermediate between SCL or LMO2 and GATA-1.

C. GATA-3

The complexity of the phenotype in GATA-3-deficient embryos reflects the widespread expression of GATA-3 during embryogenesis. GATA-3-null mice died between embryonic days 11 and 12, with internal bleeding, growth retardation, and severe deformities of the brain and spinal cord (Pandolfi *et al.*, 1995). Although abnormalities in fetal liver hematopoiesis were noted, it was not clear whether the primary defect lay in the hematopoietic cells or in the fetal liver itself. Examination of the differentiation potential of GATA-3-null ES cells has helped to clarify this point. In chimeric mice, GATA-3-null ES cells contributed to all hematopoietic lineages except for T cells. These studies demonstrated a requirement for GATA-3 at, or before, the $CD4^-/CD8^-$ stage of thymocyte development (Ting *et al.*, 1996).

V. AML-1 AND CBFβ ARE REQUIRED FOR DEFINITIVE HEMATOPOIESIS

The AML (PEBP2/CBF) transcription factors are a family of heterodimeric proteins consisting of a common β-subunit and an α-subunit (Liu *et al.*, 1993; Ogawa *et al.*, 1993a; Wang *et al.*, 1993). There are three mammalian α-subunits, with additional complexity generated by multiple splice variants for each α-subunit transcript (Miyoshi *et al.*, 1991; Bae *et al.*, 1993; Ogawa *et al.*, 1993b; Levanon *et al.*, 1994; Bae *et al.*, 1995). The α-subunits share a DNA-binding region called the Runt domain (defined by its homology to the *Drosophila* Runt protein) that also allows dimerization with the β-subunit (Daga *et al.*, 1992; Kagoshima *et al.*, 1993; Meyers *et al.*, 1993; Ogawa *et al.*, 1993a). PEBP2/CBF DNA-binding sites have been demonstrated in the regulatory regions of several hematopoietic genes, including GM-CSF, myeloperoxidase, elastase, and the receptor for M-CSF (Speck and Stacy, 1995). This observation raised the possibility that CBF would be an important regulator of hematopoietic development. This was confirmed by using gene-targeting approaches.

Mice lacking the AML-1 gene showed normal morphogenesis and normal yolk sac derived (primitive) erythropoiesis but lacked fetal liver (definitive) hematopoiesis, dying at day 12.5 of embryonic development. The extensive regions of hemorrhage observed within the central nervous system of these mice were thought to be secondary to necrosis of endothelial cells in perforating capillaries (Wang *et al.*, 1996a). No definitive hematopoietic progenitor cells of any lineage were detected upon culturing fetal liver or yolk sacs from the mutant animals (Okuda *et al.*, 1996; Wang *et al.*, 1996a).

Experiments using AML-1-null ES cells were consistent with these results. These cells formed embryoid bodies that contained primitive nucleated erythrocytes but failed to generate monocytes or macrophages (Okuda *et al.*, 1996;

Wang *et al.,* 1996). Furthermore, in chimeric mice, the AML-1-null ES cells failed to contribute to bone marrow, spleen, thymus, or peripheral blood, although they contributed substantially to nonhematopoietic tissues (Okuda *et al.,* 1996). These results demonstrated that cells lacking AML-1 were unable to contribute to definitive hematopoiesis and that this defect was intrinsic to the hematopoietic cells themselves. Mice with a null mutation of the CBFβ gene displayed a phenotype identical to that of the AML-1 mutant mice, demonstrating that the two subunits functioned as a heterodimeric complex *in vivo* (Sasaki *et al.,* 1996; Wang *et al.,* 1996b). Although it is clear that the AML-1/CBF transcription factor complex plays a pivotal role in regulating the transcription of genes essential for definitive hematopoiesis, it is uncertain whether the defect lies in the generation of the stem cells required for definitive hematopoiesis or in their survival and/or expansion.

The fusion gene CBFβ/MYH 11 (representing a facsimile of the inv(16) seen in some cases of acute myeloid leukemia) has also been re-created in ES cells by gene targeting (Castilla *et al.,* 1996). These ES cells did not contribute to hematopoietic tissues in chimeric mice, and perhaps because of this, these animals did not develop leukemia. In addition, the phenotype of the CBFβ/MYH 11 mice was virtually indistinguishable from that of the AML-1 and CBFβ mutant mice, supporting the notion that this fusion protein acted in a dominant-negative manner to inhibit the normal function of the CBF complex.

VI. ADDITIONAL EXAMPLES OF DEFECTIVE DEFINITIVE HEMATOPOIESIS: MYB AND PU.1 KNOCKOUTS

The transcription factor c-MYB is the cellular homolog of v-MYB, the transforming element encoded by the avian myeloblastosis virus that causes myelomonocytic leukemia (Moscovici, 1975; Klempnauer *et al.,* 1982). A truncated MYB gene fused with an activated ETS gene forms the oncogenic component in another avian virus, E26, which is capable of transforming both myeloid and erythroid lineages (Radke *et al.,* 1982; Moscovici *et al.,* 1983).

Consistent with the ability of its oncogenic forms to transform hematopoietic cells, c-Myb expression was abundant in immature hematopoietic cells and decreased as these cells differentiated (Gonda and Metcalf, 1984; Kastan *et al.,* 1989; Westin *et al.,* 1996). Expression was not restricted to these cell types, however, and c-MYB was present in a number of other tissues, including the gut and nervous system (Torelli *et al.,* 1987; Thiele *et al.,* 1988; Sitzmann *et al.,* 1995). Enforced expression of c-MYB in murine erythroleukemia cells inhibited their erythroid differentiation (Clarke *et al.,* 1988) whereas MYB-antisense oligonucleotides caused growth arrest and inhibited hematopoietic colony formation from human bone marrow mononuclear cells (Gewirtz and Calabretta, 1988).

Inactivation of the c-MYB gene resulted in embryos that showed normal primitive (yolk sac) hematopoiesis but had a severe defect in definitive (fetal liver) hematopoiesis (Mucenski *et al.,* 1991). The major difference between the defect in definitive hematopoiesis in the AML-1 compared to c-MYB mutant animals was the presence of hematopoietic progenitors in c-MYB-deficient mice, although at a markedly reduced level. Also, in contrast to AML-1-null mice, c-MYB mutant mice had normal numbers of fetal liver megakaryocytes.

It remains unclear which cells within the hematopoietic compartment are compromised as a result of deficiency of c-MYB. A defect in the proliferation of definitive hematopoietic stem cells or multipotential progenitor cells would be analogous to the role postulated for GATA-2 in the expansion of primitive and definitive hematopoietic stem/multipotential cells. Alternatively, the sparing of the megakaryocyte lineage may be consistent with a role for c-MYB in the regulation of other lineage-committed progenitors. Either scenario would predict that c-MYB is responsible for a more restricted subset of genes than those regulated by AML-1.

The transcription factor PU.1 is a hematopoietic-specific member of the ETS gene family, expressed at high levels in monocytes, granulocytes, B lymphocytes, and, to a lesser extent, erythroid cells (Moreau-Gachelin *et al.,* 1989; Klemsz *et al.,* 1990; Galson *et al.,* 1993; Pahl *et al.,* 1993; Pongubala *et al.,* 1993; Shin and Koshland, 1993). Consistent with this, PU.1 appears more important for myeloid than erythroid differentiation of human $CD34^+$ progenitor–stem cells *in vitro* (Voso *et al.,* 1994).

Using different mutational strategies, two groups have generated mice lacking a functional PU.1 gene (Scott *et al.,* 1994; McKercher *et al.,* 1996). Although impaired lymphoid and myeloid differentiation was evident in both cases, there were some unexplained disparities in the reported phenotypes. The mice produced by Scott *et al.* (1994) died between embryonic days 16 and 18, whereas the PU.1-deficient animals generated by McKercher *et al.* (1996) succumbed from septicemia within 48 h of birth and survived for up to 2 weeks if they were treated with antibiotics. Secondly, the defect in early lympho- and myelopoiesis appeared more complete in the study by Scott *et al.* (1994), raising the possibility that the PU.1 mutant produced by McKercher *et al.* (1996) represented a "knockdown" rather than a "knockout." In both studies, macrophages, granulocytes, and lymphoid lineages were the major cell types affected whereas megakaryocytes and erythrocytes were relatively spared. Consistent with these data, there were no multilineage or myeloid colony-forming cells in the fetal livers of PU.1-deficient embryos, although erythroid and megakaryocytic colonies were present in normal numbers (Scott *et al.,* 1997).

In follow-up studies using chimeric mice to evaluate the hematopoietic potential of PU.1-deficient ES cells, Scott *et al.* (1997) confirmed that PU.1 expression was required for the formation of granulocytes, macrophages, T cells, and B cells. Although definitive erythrocytes derived from PU.1-deficient ES cells were detected in day 16.5 chimeric embryos, no ES cell derived erythrocytes were detected in adult chimeras. This finding may indicate that bone marrow and fetal liver erythroid progenitors differ in their requirements for PU.1. In an elegant study which exploited the fact that their PU.1-deficient mice were born alive, workers from the laboratory of Richard Maki (Tondravi *et al.,* 1997) demonstrated that PU.1 was required for osteoclast differentiation, presumably acting "upstream" of a common macrophage–osteoclast progenitor. PU.1-null mice showed evidence of osteopetrosis at birth and their bone marrow cavities were devoid of osteoclasts as well as macrophages. The phenotype was reversed following transplantation of bone marrow cells from PU.1-wild-type animals, confirming the cell-autonomous nature of the defect.

The detection of early myeloid gene expression (such as the GM-CSF receptor and myeloperoxidase) in differentiating PU.1-deficient embryoid bodies by reverse transcriptase–polymerase chain reaction demonstrated that myelopoiesis was initiated in the absence of PU.1 even though mature myeloid cells could not be produced (Olson *et al.,* 1995).

As with the c-MYB-null mice, it is uncertain at what level within the hematopoietic compartment PU.1 deficiency is manifest. Whether there are defects in multiple distinct lineages or the involvement of a common myelolymphoid progenitor remains to be determined.

VII. LINEAGE-SPECIFIC DEFECTS: EKLF AND NF-E2 KNOCKOUTS

In addition to the aforementioned phenotypes that display features affecting multiple hematopoietic lineages, the gene-targeting approach has served to identify genes whose normal function is critical to a specific stage within a hematopoietic lineage. EKLF (erythroid Krüppel-like factor) was identified as an erythroid-specific protein that bound to the nucleotide sequence -CACCC-, a *cis*-regulatory element critical for the transcription of many erythroid expressed genes (Orkin, 1990; Miller and Bieleer, 1993). EKLF binds preferentially to the -CACCC- site in the β-globin promoter and fails to activate promoters that are mutated in some patients with β-thalassemia (Feng *et al.,* 1994). Employing a gene-targeting approach in which a lacZ reporter gene was used to disrupt the EKLF locus, Nuez *et al.* (1995) demonstrated that high-level EKLF expression was restricted to the fetal liver, consistent with its requirement for transcription of adult, but not embryonic, β-globin genes. However, Perkins *et al.* (1995) reported that EKLF was equally abundant in both yolk sac and fetal liver derived erythroblasts, although embryonic day 11 yolk sac blood only expressed very low levels of the EKLF target gene, adult β-globin. Nevertheless, it was not surprising that mice carrying a null mutation of the EKLF gene died with a β-thalassemia-like syndrome between embryonic day 14 and embryonic day 16 (Nuez *et al.,* 1995; Perkins *et al.,* 1995). Normal numbers of colony-forming cells were present in the fetal liver in EKLF-deficient embryos but the failure to efficiently activate β-globin transcription resulted in imbalanced globin chain synthesis and hence lethality due to ineffective definitive erythropoiesis. Since embryonic erythroblasts do not express large amounts of adult β-globin, yolk sac erythropoiesis was ostensibly normal in EKLF-deficient embryos. Interestingly, other CACCC-box containing erythroid-specific genes, including embryonic βh1 globin, the erythropoietin receptor, GATA-1, and genes of the heme biosynthetic pathway, were expressed at normal or near-normal levels, suggesting the involvement of additional EKLF-like proteins in their regulation.

In vitro differentiation of EKLF-deficient ES cells confirmed their ability to give rise to normal numbers of definitive erythroid colonies but with a marked impairment of β-globin expression and poor hemoglobinization (Lim *et al.,* 1997). Analysis of chimeric mice demonstrated that EKLF-deficient ES cells did not give rise to adult erythrocytes, consistent with the analyses of EKLF-deficient embryos.

In addition to -CACCC- sites and -GATA- sequences, other critical regulatory motifs have been identified in genes expressed in erythroid cells. One such is the AP-1-like motif recognized by NF-E2 (-[T/C]TGCTGA[C/G]TCA[T/C]-) (Mignotte *et al.,* 1989; Ney *et al.,* 1990; Talbot & Grosveld, 1991). NF-E2 is a heterodimeric basic leucine zipper transcription factor composed of a hematopoietic-specific subunit (p45) and a ubiquitous subunit (p18), related to the chicken *v-maf* oncogene (Andrews *et al.,* 1993a, 1993b).

Expression of p45 NF-E2 is restricted to progenitor cells, erythroid cells, mast cells, and megakaryocytes. On the basis of numerous previous studies implicating NF-E2 as a major regulator of globin gene expression (Andrews *et al.,* 1993a), it was anticipated that NF-E2-deficient mice would display marked abnormalities in erythropoiesis. Surprisingly, mice with a null mutation of p45 were not markedly anemic but died within the first week of life due to thrombocytopenia and consequent hemorrhage (Shivdasani *et al.,* 1995b). Megakaryocytes were present and although early phases of differentiation proceeded normally, ultrastructural analysis revealed a failure of platelet development in NF-E2-deficient megakaryocytes. This defect in megakaryocyte differentiation was not overcome by administration of the platelet-specific growth factor thrombopoietin.

Detailed analysis of erythropoiesis in NF-E2-deficient mice revealed a mild hypochromic anemia, dysmorphic changes in erythrocytes, and a moderate degree of splenomegaly, reflecting increased extramedullary hematopoiesis (Shivdasani and Orkin, 1995). In keeping with the mild phenotype, decreases in hematopoietic colony numbers were not detected in the bone marrow of NF-E2-deficient mice and RNase protection did not reveal reduced synthesis of globin genes in their fetal or adult erythrocytes. These studies raised the possibility that NF-E2 deficiency could be compensated for by other proteins in erythroid cells or that the NF-E2 binding sites within the globin locus control region were dispensable *in vivo.* Recently, targeted disruption of the p18 NF-E2 subunit was described (Kotkow and Orkin, 1996). The p18 NF-E2-deficient mice were indistinguishable from wild-type littermates, and NF-E2 DNA-binding activity was unaffected, indicating that other members of the *v-maf* family functionally substituted for p18 NF-E2.

VIII. IKAROS AND E2A: EXAMPLES OF KEY TRANSCRIPTION FACTORS IN LYMPHOPOIESIS

The principles illustrated in the previous section also apply to the lymphoid system, where some of the genes responsible for development and differentiation in the lymphoid lineages have been defined. For example, mice deficient in the IKAROS gene, which encodes a family of six lymphoid-restricted zinc finger proteins, display hematopoietic abnormalities in B and T lymphocytes, NK (natural killer) cells, and thymic dendritic cells. The initial IKAROS mutant animals were created by targeting the DNA-binding domain (Georgopoulos *et al.,* 1994). Homozygous IKAROS mutant mice lacked B cells, T cells, and NK cells, were growth retarded, and died with evidence of sepsis during the first 3 weeks of life. The rather surprising observation that mice heterozygous for this mutation succumbed to T-cell leukemias (Winandy *et al.,* 1995) was followed by the realization that the IKAROS mutation actually created a dominant-negative protein. The severe defects in these mice were thus hypothesized to be due to the combined loss of Ikaros activity plus the effects of dominant-negative interference with other Ikaros-interacting factors such as the related gene AIOLOS (Wang *et al.,* 1996).

The same investigators reported a true IKAROS-deficient mouse created through deletion of a critical carboxy-terminal dimerization domain (IKAROS C−/−) (Wang *et al.,* 1996). These mice were not growth retarded and did not die from sepsis in early postnatal life. Although these IKAROS C−/− mice lacked B cells, NK cells, and thymic dendritic cells, the defect in T lymphopoiesis was more complex. Fetal thymic immigration and expansion were absent, but the thymus was colonized shortly after birth (presumably from bone marrow derived stem cells), eventually leading to thymic cellularity only several-fold lower than that of wild-type littermates. Abnormalities in T lymphoid differentiation and proliferation persisted, however, with an increased proportion of $CD4^+$ thymocytes and the emergence of a clonal population of thymocytes in older IKAROS C−/− mice, suggesting a role for IKAROS as a tumor suppressor gene.

A null mutation of the HLH gene E2A, which encodes the SCL dimerization partners E12 and E47, also resulted in a complex phenotype manifesting as high perinatal mortality and severe postnatal growth retardation in surviving homozygous mutants. As with the IKAROS mutant mice, these animals displayed defective lymphopoiesis. In this case, the defect was restricted to early B-cell development, with a failure of mutant cells to initiate immunoglobulin gene rearrangement (Zhuang *et al.,* 1992; Bain *et al.,* 1994; Zhuang *et al.,* 1994).

IX. CONCLUSIONS

Although gene-targeting approaches have defined important functions for some genes, many questions remain to be answered. As illustrated by SCL/LMO2, the role of cooperating/compensating proteins is complex. For many of these transcription factors, the target genes that they regulate are unknown. Equally, the noncritical or later actions of these molecules have not been defined and will depend on the development of alternative strategies for their elucidation, such as conditional gene targeting, where the gene of interest is deleted in a tissue-specific manner or at a given time during development or adult life. This may be accomplished by tagging the gene of interest with recognition sites for site-specific recombinases (such as Cre recombinase) which are then expressed from tissue-restricted or developmentally restricted promoters. Alternatively, a similar end may be achieved through germline deletion of tissue-restricted or developmentally restricted gene regulatory sequences. In either case, genetic approaches will continue to be vital in furthering our initial insights into the role of transcription factors in the regulation of hematopoiesis.

REFERENCES

Andrews, N. C., Erdjument-Bromage, H., Davidson, M. B., *et al.* (1993a). Erythroid transcription factor NF-E2 is a haematopoietic-specific basic-leucine zipper protein. *Nature* **362,** 722–728.

Andrews, N. C., Kotkow, K. J., Ney, P. A., *et al.* (1993b). The ubiquitous subunit of erythroid transcription factor NF-E2 is a small basic-leucine zipper protein related to the v-maf oncogene. *Proc. Natl. Acad. Sci. USA* **90,** 11488–11492.

Aplan, P. D., Nakahara, K., Orkin, S. H., *et al.* (1992). The SCL gene product: A positive regulator of erythroid differentiation. *EMBO J.* **11,** 4073–4081.

Bae, S. C., Takahashi, E., Zhang, Y. W., *et al.* (1995). Cloning, mapping and expression of PEBP2 alpha C, a third gene encoding the mammalian Runt domain. *Gene* **159,** 245–248.

Bae, S. C., Yamaguchi, I. Y., Ogawa, E., *et al.* (1993). Isolation of PEBP2 alpha B cDNA representing the mouse homolog of human acute myeloid leukemia gene, AML1. *Oncogene* **8,** 809–814.

Bain, G., Maandag, E. C., Izon, D. J., *et al.* (1994). E2A proteins are required for proper B cell development and initiation of immunoglobulin gene rearrangements. *Cell* **79,** 885–892.

Begley, C. G., Aplan, P. D., Davey, M. P., *et al.* (1989a). Chromosomal translocation in a human leukemic stem-cell line disrupts the T-cell antigen receptor delta-chain diversity region and results in a previously unreported fusion transcript. *Proc. Natl. Acad. Sci. USA* **86,** 2031–2035.

Begley, C. G., Aplan, P. D., Denning, S. M., *et al.* (1989b). The gene SCL is expressed during early hematopoiesis and encodes a differentiation-related DNA-binding motif. *Proc. Natl. Acad. Sci. USA* **86,** 10128–10132.

Boehm, T., Foroni, L., Kaneko, Y., *et al.* (1991). The rhombotin family of cysteine-rich LIM-domain oncogenes: Distinct members are involved in T-cell translocations to human chromosomes 11p15 and 11p13. *Proc. Natl. Acad. Sci. USA* **88,** 4367–4371.

Briegel, K., Bartunek, P., Stengl, G., *et al.* (1996). Regulation and function of transcription factor GATA-1 during red blood cell differentiation. *Development* **122,** 3839–3850.

Briegel, K., Lim, K. C., Plank, C., *et al.* (1993). Ectopic expression of a conditional GATA-2/estrogen receptor chimera arrests erythroid differentiation in a hormone-dependent manner. *Genes Dev.* **7,** 1097–1109.

Brown, L., Cheng, J. T., Chen, Q., *et al.* (1990). Site-specific recombination of the tal-1 gene is a common occurrence in human T cell leukemia. *EMBO J.* **9,** 3343–3351.

Castilla, L., Wijmenga, C., Wang, Q., *et al.* (1996). Failure of embryonic hematopoiesis and lethal hemorrhages in mouse embryos heterozygous for a knocked-in leukemia gene CBFB-MYH11. *Cell* **87,** 687–696.

Chiba, T., Nagata, Y., Kishi, A., *et al.* (1993). Induction of erythroid-specific gene expression in lymphoid cells. *Proc. Natl. Acad. Sci. USA* **90,** 11593–11597.

Clarke, M. F., Kukowska-Latallo, J. F., Westin, E., *et al.* (1988). Constitutive expression of a c-myb cDNA blocks Friend murine erythroleukemia cell differentiation. *Mol. Cell. Biol.* **8,** 884–892.

Condorelli, G. L., Facchiano, F., Valtieri, M., *et al.* (1996). T-cell-directed TAL-1 expression induces T-cell malignancies in transgenic mice. *Cancer Res.* **56,** 5113–5119.

Cory, S. (1986). Activation of cellular oncogenes in hemopoietic cells by chromosome translocation. *Adv. Cancer Res.* **47,** 189–234.

Cross, M. A., Heyworth, C. M., Murrell, A. M., *et al.* (1994). Expression of lineage restricted transcription factors precedes lineage specific differentiation in a multipotent haemopoietic progenitor cell line. *Oncogene* **9,** 3013–3016.

Crossley, M., Merika, M., and Orkin, S. H. (1995). Self-association of the erythroid transcription factor GATA-1 mediated by its zinc finger domains. *Mol. Cell. Biol.* **15,** 2448–2456.

Crotta, S., Nicolis, S., Ronchi, A., *et al.* (1990). Progressive inactivation of an erythroid transcriptional factor in GM- and G-CSF dependent myeloid cell lines. *Nucleic Acids Res.* **18,** 6863–6869.

Curtis, D., Robb, L., Strasser, A., *et al.* (1997). The CD2-SCL transgene alters the phenotype and the frequency of T-lymphomas in N-ras transgenic or p53 deficient mice. *Oncogene* **15,** 2975–2983.

Daga, A., Tighe, J. E., and Calabi, F. (1992). Leukaemia/*Drosophila* homology. *Nature* **356,** 484.

Dorfman, D. M., Wilson, D. B., Bruns, G. A. P., *et al.* (1992). Human transcription factor GATA-2. Evidence for regulation of preproendothelin-1 gene expression in endothelial cells. *J. Biol. Chem.* **267,** 1279–1285.

Elefanty, A., Robb, L., Birner, R., *et al.* (1997). Hematopoietic-specific genes are not induced during *in-vitro* differentiation of SCL-null embryonic stem cells. *Blood* **90,** 1435–1447.

Elwood, N. J., Cook, W. D., Metcalf, D., *et al.* (1993). SCL, the gene implicated in human T-cell leukaemia, is oncogenic in a murine T-lymphocyte cell line. *Oncogene* **8,** 3093–3101.

Evans, T., and Felsenfeld, G. (1989). The erythroid-specific transcription factor Eryf1: A new finger protein. *Cell* **58,** 877–885.

Feng, W. C., Southwood, C. M., and Bieker, J. J. (1994). Analyses of beta-thalassemia mutant DNA interactions with erythroid Kruppel-like factor (EKLF), an erythroid cell-specific transcription factor. *J. Biol. Chem.* **269,** 1493–1500.

Fujiwara, Y., Browne, C. P., Cunniff, K., *et al.* (1996). Arrested development of embryonic red cell precursors in mouse embryos lacking transcription factor GATA-1. *Proc. Natl. Acad. Sci. USA* **93,** 12355–12358.

Galson, D. L., Hensold, J. O., Bishop, T. R., *et al.* (1993). Mouse beta-globin DNA-binding protein B1 is identical to a proto-oncogene, the transcription factor Spi-1/PU.1, and is restricted in expression to hematopoietic cells and the testis. *Mol. Cell. Biol.* **13,** 2929–2941.

Georgopoulos, K., Bigby, M., Wang, J. H., *et al.* (1994). The Ikaros gene is required for the development of all lymphoid lineages. *Cell* **79,** 143–156.

Gewirtz, A. M., and Calabretta, B. (1988). A c-myb antisense oligodeoxynucleotide inhibits normal human hematopoiesis *in vitro*. *Science* **242,** 1303–1306.

Gonda, T. J., and Metcalf, D. (1984). Expression of myb, myc and fos proto-oncogenes during the differentiation of a murine myeloid leukaemia. *Nature* **310,** 249–251.

Green, A. R., DeLuca, E., and Begley, C. G. (1991b). Antisense SCL suppresses self-renewal and enhances spontaneous erythroid differentiation of the human leukaemic cell line K562. *EMBO J.* **10,** 4153–4158.

Green, A. R., Lints, T., Visvader, J., *et al.* (1992). SCL is coexpressed with GATA-1 in hemopoietic cells but is also expressed in developing brain. *Oncogene* **7,** 653–660.

Green, A. R., Rockman, S., DeLuca, E., *et al.* (1993). Induced myeloid differentiation of K562 cells with downregulation of erythroid and megakaryocytic transcription factors: A novel experimental model for hemopoietic lineage restriction. *Exp. Hematol.* **21,** 525–531.

Green, A. R., Salvaris, E., and Begley, C. G. (1991a). Erythroid expression of the "helix–loop–helix" gene, SCL. *Oncogene* **6,** 475–479.

Hershfield, M. S., Kurtzberg, J., Harden, E., *et al.* (1984). Conversion of a stem cell leukemia from a T-lymphoid to a myeloid phenotype induced by the adenosine deaminase inhibitor 2′-deoxycoformycin. *Proc. Natl. Acad. Sci. USA* **81,** 253–257.

Ho, I. C., Vorhees, P., Marin, N., *et al.* (1991). Human GATA-3: A lineage-restricted transcription factor that regulates the expression of the T cell receptor α gene. *EMBO J.* **10,** 1187–1192.

Hoang, T., Paradis, E., Brady, G., *et al.* (1996). Opposing effects of the basic helix–loop–helix transcription factor SCL on erythroid and monocytic differentiation. *Blood* **87,** 102–111.

Hsu, H. L., Wadman, I., and Baer, R. (1994). Formation of *in vivo* complexes between the TAL1 and E2A polypeptides of leukemic T cells. *Proc. Natl. Acad. Sci. USA* **91,** 3181–3185.

Ito, E., Toki, T., Ishihara, H., *et al.* (1993). Erythroid transcription factor GATA-1 is abundantly transcribed in mouse testis. *Nature* **362,** 466–468.

Johnson, R. S., Spiegelman, B. M., and Papaioannou, V. (1992). Pleiotropic effects of a null mutation in the c-fos proto-oncogene. *Cell* **71,** 577–586.

Joulin, V., Bories, D., Eleouet, J. F., *et al.* (1991). A T cell-specific TCR δ DNA binding protein is a member of the human GATA family. *EMBO J.* **10,** 1809–1816.

Kagoshima, H., Shigesada, K., Satake, M., *et al.* (1993). The Runt domain identifies a new family of heteromeric transcriptional regulators. *Trends Genet.* **9,** 338–341.

Kallianpur, A. R., Jordan, J. E., and Brandt, S. J. (1994). The SCL/TAL-1 gene is expressed in progenitors of both the hematopoietic and vascular systems during embryogenesis. *Blood* **83,** 1200–1208.

Kamps, M. P., Murre, C., Sun, X. H., *et al.* (1990). A new homeobox gene contributes the DNA binding domain of the t(1;19) translocation protein in pre-B ALL. *Cell* **60,** 547–555.

Kastan, M. B., Stone, K. D., and Civin, C. I. (1989). Nuclear oncoprotein expression as a function of lineage, differentiation stage, and proliferative status of normal human hematopoietic cells. *Blood* **74,** 1517–1524.

Kelliher, M. A., Seldin, D. C., and Leder, P. (1996). Tal-1 induces T cell acute lymphoblastic leukemia accelerated by casein kinase II alpha. *EMBO J.* **15,** 5160–5166.

Klempnauer, K. H., Gonda, T. D., and Bishop, J. M. (1982). Nucleotide sequence of the retroviral leukemia gene v-myb and its cellular progenitor c-myb: The architecture of a transduced oncogene. *Cell* **31,** 453–463.

Klemsz, M. J., McKercher, S. R., Celada, A., *et al.* (1990). The macrophage and B cell-specific transcription factor PU.1 is related to the *ets* oncogene. *Cell* **61,** 113–124.

Ko, L. J., Yamamoto, M., Leonard, M. W., *et al.* (1991). Murine and human T-lymphocyte GATA-3 factors mediate transcription through a cis-regulatory element within the human T-cell receptor δ gene enhancer. *Mol. Cell. Biol.* **11,** 2778–2784.

Kornhauser, J. M., Leonard, M. W., Yamamoto, M., *et al.* (1994). Temporal and spatial changes in GATA transcription factor expression are coincident with development of the chicken optic tectum. *Mol. Brain Res.* **23,** 100–110.

Kotkow, K. J., and Orkin, S. H. (1996). Complexity of the erythroid transcription factor NF-E2 as revealed by gene targeting of the mouse p18 NF-E2 locus. *Proc. Natl. Acad. Sci. USA* **93,** 3514–3518.

Krause, D. S., and Perkins, A. S. (1997). Gotta find GATA a friend. *Nat. Med.* **3,** 960–961.

Kulessa, H., Frampton, J., and Graf, T. (1995). GATA-1 reprograms avian myelomonocytic cell lines into eosinophils, thromboblasts, and erythroblasts. *Genes Dev.* **9,** 1250–1262.

Kurtzberg, J., Bigner, S. H., and Hershfield, M. S. (1985). Establishment of the DU.528 human lymphohemopoietic stem cell line. *J. Exp. Med.* **162,** 1561–1578.

Larson, R. C., Lavenir, I., Larson, T. A., *et al.* (1996). Protein dimerization between Lmo2 (Rbtn2) and Tall alters thymocyte development and potentiates T cell tumorigenesis in transgenic mice. *EMBO J.* **15,** 1021–1027.

Lee, M. E., Temizer, D. H., Clifford, J. A., *et al.* (1991). Cloning of the GATA-binding protein that regulates endothelin-1 gene expression in endothelial cells. *J. Biol. Chem.* **266,** 16188–16192.

Leroy-Viard, K., Vinit, M. A., Lecointe, N., *et al.* (1995). Loss of TAL-1 protein activity induces premature apoptosis of Jurkat leukemic T cells upon medium depletion. *EMBO J.* **14,** 2341–2349.

Levanon, D., Negreanu, V., Bernstein, Y., *et al.* (1994). AML1, AML2, and AML3, the human members of the runt domain gene-family: cDNA structure, expression, and chromosomal localization. *Genomics* **23,** 425–432.

Lim, S. K., Bieker, J. J., Lin, C. S., *et al.* (1997). A shortened life span of EKLF−/− adult erythrocytes, due to a deficiency of β-globin chains, is ameliorated by human γ-globin chains. *Blood* **90,** 1291–1299.

Liu, P., Tarle, S. A., Hajra, A., *et al.* (1993). Fusion between transcription factor CBF beta/PEBP2 beta and a myosin heavy chain in acute myeloid leukemia. *Science* **261,** 1041–1044.

Martin, D. I. K., Zon, L. I., Mutter, G., *et al.* (1990). Expression of an erythroid transcription factor in megakaryocytic and mast cell lineages. *Nature* **344,** 444–447.

McKercher, S. R., Torbett, B. E., Anderson, K. L., *et al.* (1996). Targeted disruption of the PU.1 gene results in multiple hematopoietic abnormalities. *EMBO J.* **15,** 5647–5658.

Mellentin, J. D., Smith, S. D., and Cleary, M. L. (1989). Lyl-1, a novel gene altered by chromosomal translocation in T cell leukemia, codes for a protein with a helix–loop–helix DNA binding motif. *Cell* **58,** 77–83.

Merika, M., and Orkin, S. H. (1995). Functional synergy and physical interactions of the erythroid transcription factor GATA-1 with the Kruppel family proteins Sp1 and EKLF. *Mol. Cell. Biol.* **15,** 2437–2447.

Meyers, S., Downing, J. R., and Hiebert, S. W. (1993). Identification of AML-1 and the (8;21) translocation protein (AML-1/ETO) as sequence-specific DNA-binding proteins: The runt homology domain is required for DNA binding and protein–protein interactions. *Mol. Cell. Biol.* **13,** 6336–6345.

Mignotte, V., Wall, L., deBoer, E., *et al.* (1989). Two tissue-specific factors bind the erythroid promoter of the human porphobilinogen deaminase gene. *Nucleic Acids Res.* **17,** 37–54.

Miller, I. J., and Bieker, J. J. (1993). A novel, erythroid cell-specific murine transcription factor that binds to the CACCC element and is related to the Kruppel family of nuclear proteins. *Mol. Cell. Biol.* **13,** 2776–2786.

Miyoshi, H., Shimizi, K., Kozu, T., *et al.* (1991). t(8;21) breakpoints on chromosome 21 in acute myeloid leukemia are clustered within a limited region of a single gene, AML1. *Proc. Natl. Acad. Sci. USA* **88,** 10431–10434.

Moreau-Gachelin, F., Ray, D., Mattei, M. G., *et al.* (1989). The putative oncogene Spi-1: Murine chromosomal localization and transcriptional activation in murine acute erythroleukemias. *Oncogene* **4,** 1449–1456.

Moscovici, C. (1975). Leukemic transformation with avian myeloblastosis virus: Present status. *Curr. Top. Microbiol. Immunol.* **71,** 79–101.

Moscovici, M. G., Jurdic, P., Samarut, J., *et al.* (1983). Characterization of the hematopoietic target cells of the avian leukemia virus E26. *Virology* **129,** 65–78.

Mouthon, M. A., Bernard, O., Mitjavila, M. T., *et al.* (1993). Expression of tal-1 and GATA-binding proteins during human hematopoiesis. *Blood* **81,** 647–655.

Mucenski, M. L., McLain, K., Kier, A. B., *et al.* (1991). A functional c-MYB gene is required for normal murine fetal hepatic hematopoiesis. *Cell* **65,** 677–689.

Murre, C., McCaw, P. S., and Baltimore, D. (1989). A new DNA binding and dimerization motif in immunoglobulin enhancer binding, daughterless, Myo-D, and Myc proteins. *Cell* **56,** 777–783.

Nagai, T., Harigae, H., Ishihara, H, *et al.* (1994). Transcription factor GATA-2 is expressed in erythroid, early myeloid, and CD34$^+$ human leukemia-derived cell lines. *Blood* **84,** 1074–1084.

Ney, P. A., Sorrentino, B. P., Lowrey, C. H., *et al.* (1990). Inducibility of the HS II enhancer depends on binding of an erythroid specific nuclear protein. *Nucleic Acids Res.* **18,** 6011–6017.

Nourse, J., Mellentin, J. D., Galili, N., *et al.* (1990). Chromosomal translocation t(1;19) results in synthesis of a homeobox fusion mRNA that codes for a potential chimeric transcription factor. *Cell* **60,** 535–545.

Nuez, B., Michalovich, D., Bygrave, A., *et al.* (1995). Defective haematopoiesis in fetal liver resulting from inactivation of the EKLF gene. *Nature* **375,** 316–318.

Ogawa, E., Inuzuka, M., Maruyama, M., *et al.* (1993a). Molecular cloning and characterization of PEBP2 beta, the heterodimeric partner of a novel *Drosophila* runt-related DNA binding protein PEBP2 alpha. *Virology* **194,** 314–331.

Ogawa, E., Maruyama, M., Kagoshima, H., *et al.* (1993b). PEBP2/PEA2 represents a family of transcription factors homologous to the products of the *Drosophila* runt gene and the human AML 1 gene. *Proc. Natl. Acad. Sci. USA* **90,** 6859–6863.

Okuda, T., van Deursen, J., Hiebert, S. W., *et al.* (1996). AML1, the target of multiple chromosomal translocations in human leukemia, is essential for normal fetal liver hematopoiesis. *Cell* **84,** 321–330.

Olson, M. C., Scott, E. W., Hack, A. A., *et al.* (1995). PU.1 is not essential for early myeloid gene expression but is required for terminal myeloid differentiation. *Immunity* **3,** 703–714.

Omichinski, J. G., Clore, G. M., Schaad, O., *et al.* (1993). NMR structure of a specific DNA complex of Zn-containing DNA binding domain of GATA-1. *Science* **261,** 438–446.

Oosterwegel, M., Timmerman, J., Leiden, J., *et al.* (1992). Expression of GATA-3 during lymphocyte differentiation and mouse embryogenesis. *Dev. Immunol.* **3,** 1–11.

Orkin, S. H. (1990). Globin gene regulation and switching: Circa 1990. *Cell* **63,** 665–672.

Orkin, S. H. (1992). GATA-binding transcription factors in hematopoietic cells. *Blood* **80,** 575–581.

Osada, H., Grutz, G., Axelson, H., *et al.* (1995). Association of erythroid transcription factors: Complexes involving the LIM protein RBTN2 and the zinc-finger protein GATA1. *Proc. Natl. Acad. Sci. USA* **92,** 9585–9589.

Pahl, H. L., Scheibe, R. J., Zhang, D. E., *et al.* (1993). The proto-oncogene PU.1 regulates expression of the myeloid-specific CD11b promoter. *J. Biol. Chem.* **268,** 5014–5020.

Pandolfi, P. P., Roth, M. E., Karis, A., *et al.* (1995). Targeted disruption of the GATA3 gene causes severe abnormalities in the nervous system and in fetal liver haematopoiesis [see comments]. *Nat. Genet.* **11,** 40–44.

Perez-Alvarado, G. C., Miles, C., Michelsen, J. W., *et al.* (1994). Structure of the carboxy-terminal LIM domain from the cysteine rich protein CRP. *Nat. Struct. Biol.* **1,** 388–398.

Perkins, A. C., Sharpe, A. H., and Orkin, S. H. (1995). Lethal β-thalassaemia in mice lacking the erythroid CACCC-transcription factor EKLF. *Nature* **375,** 318–322.

Pevny, L., Lin, C. S., D'Agati, V., *et al.* (1995). Development of hematopoietic cells lacking transcription factor GATA-1. *Development* **121,** 163–172.

Pevny, L., Simon, M. C., Robertson, E., *et al.* (1991). Erythroid differentiation in chimaeric mice blocked by a targeted mutation in the gene for transcription factor GATA-1. *Nature* **349,** 257–260.

Pongubala, J. M., Van Beveren, C., Nagulapalli, S., *et al.* (1993). Effect of PU.1 phosphorylation on interaction with NF-EM5 and transcriptional activation. *Science* **259,** 1622–1625.

Porcher, C., Swat, W., Rockwell, K., *et al.* (1996). The T cell leukemia oncoprotein SCL/Tal-1 is essential for development of all hematopoietic lineages. *Cell* **86,** 47–57.

Radke, K., Beug, H., Kornfeld, S., *et al.* (1982). Transformation of both erythroid and myeloid cells by E26, an avian leukemia virus that contains the myb gene. *Cell* **31,** 643–653.

Robb, L., and Begley, C. G. (1996). The helix–loop–helix gene SCL: Implicated in T-cell acute lymphoblastic leukaemia and in normal haematopoietic development. *Int. J. Biochem.* **28,** 609–618.

Robb, L., Elwood, N. J., *et al.* (1996). The *scl* gene product is required for the generation of all hematopoietic lineages in the adult mouse. *EMBO J* **155:** 4123–4129.

Robb, L., Lyons, I., Li, R., *et al.* (1995). Absence of yolk sac hematopoiesis from mice with a targeted disruption of the scl gene. *Proc. Natl. Acad. Sci. USA* **92,** 7075–7079.

Romeo, P. H., Prandini, M. H., Joulin, V., *et al.* (1990). Megakaryocytic and erythrocytic lineages share specific transcription factors. *Nature* **344,** 447–449.

Royer-Pokora, B., Loos, U., and Ludwig, W. D. (1991). TTG-2, a new gene encoding a cysteine-rich protein with the LIM motif, is overexpressed in acute T-cell leukaemia with the t(11; 14)(p13]]1). *Oncogene* **6,** 1887–1893.

Sanchez-Garcia, I., and Rabbitts, T. H. (1994). The LIM domain: A new structural motif found in zinc-finger-like proteins. *Trends Genet.* **10,** 315–320.

Sasaki, K., Yagi, H., Bronson, R. T., *et al.* (1996). Absence of fetal liver hematopoiesis in mice deficient in transcriptional coactivator core binding factor β. *Proc. Natl. Acad. Sci. USA* **93,** 12359–12363.

Scott, E. W., Fisher, R. C., Olson, M. C., *et al.* (1997). PU.1 functions in a cell-autonomous manner to control the differentiation of multipotential lymphoid–myeloid progenitors. *Immunity* **6,** 437–447.

Scott, E. W., Simon, M. C., Anastasi, J., *et al.* (1994). Requirement of transcription factor PU.1 in the development of multiple hematopoietic lineages. *Science* **265,** 1573–1577.

Shin, M. K., and Koshland, M. E. (1993). Ets-related protein PU.1 regulates expression of the immunoglobulin J-chain gene through a novel Ets-binding element. *Genes Dev.* **7,** 2006–2015.

Shivdasani, R. A., Fujiwara, Y., McDevitt, M. A., *et al.* (1997). A lineage-selective knockout establishes the critical role of transcription factor GATA-1 in megakaryocyte growth and platelet development. *EMBO J.,* **16,** 3965–3973.

Shivdasani, R. A., Mayer, E. L., and Orkin, S. H. (1995a). Absence of blood formation in mice lacking the T-cell leukaemia oncoprotein tal-1/SCL. *Nature* **373,** 432–434.

Shivdasani, R. A., and Orkin, S. H. (1995). Erythropoiesis and globin gene expression in mice lacking the transcription factor NF-E2. *Proc. Natl. Acad. Sci. USA* **92,** 8690–8694.

Shivdasani, R. A., Rosenblatt, M. F., Zucker-Franklin, D., *et al.* (1995b). Transcription factor NF-E2 is required for platelet formation independent of the actions of thrombopoietin/MGDF in megakaryocyte development. *Cell* **81,** 695–704.

Sitzmann, J., Noben-Trauth, K., and Klempnauer, K. H. (1995). Expression of mouse c-myb during embryonic development. *Oncogene* **11,** 2273–2279.

Speck, N. A., and Stacy, T. (1995). A new transcription factor family associated with human leukemias. *In* "Critical Reviews in Eukaryotic Gene Expression" (G. S. Stein, J. L. Stein, and J. B. Lian, Eds.), pp. 337–364. Begell House, New York.

Sposi, N. M., Zon, L. I., Care, A., *et al.* (1992). Cycle-dependent initiation and lineage-dependent abrogation of GATA-1 expression in pure differentiating hematopoietic progenitors. *Proc. Natl. Acad. Sci. USA* **89,** 6353–6357.

Stone, J., de Lange, T., Ramsay, G., *et al.* (1987). Definition of regions in human c-myc that are involved in transformation and nuclear localization. *Mol. Cell. Biol.* **7,** 1697–1709.

Talbot, D., and Grosveld, F. (1991). The 5′HS2 of the globin locus control region enhances transcription through the interaction of a multimeric complex binding at two functionally distinct NF-E2 binding sites. *EMBO J.* **10,** 1391–1398.

Tanigawa, T., Elwood, N., Metcalf, D., *et al.* (1993). The SCL gene product is regulated by and differentially regulates cytokine responses during myeloid leukemic cell differentiation. *Proc. Natl. Acad. Sci. USA* **90,** 7864–7868.

Tanigawa, T., Nicola, N., McArthur, G. A., *et al.* (1995). Differential regulation of macrophage differentiation in response to leukemia inhibitory factor/oncostatin-M/interleukin-6: The effect of enforced expression of the SCL transcription factor. *Blood* **85,** 379–390.

Thiele, C. J., Cohen, P. S., and Israel, M. A. (1988). Regulation of c-myb expression in human neuroblastoma cells during retinoic acid-induced differentiation. *Mol. Cell. Biol.* **8,** 1677–1683.

Ting, C. N., Olson, M. C., Barton, K. B., *et al.* (1996). Transcription factor GATA-3 is required for development of the T-cell lineage. *Nature* **384,** 474–478.

Tondravi, M. M., McKercher, S. R., Anderson, K., *et al.* (1997). Osteopetrosis in mice lacking haematopoietic transcription factor PU.1. *Nature* **386,** 81–84.

Torelli, G., Venturelli, D., Colo, A., *et al.* (1987). Expression of c-myb protooncogene and other cell cycle-related genes in normal and neoplastic colonic mucosa. *Cancer Res.* **47,** 5266–5269.

Tsai, F. Y., Keller, G., Kuo, F. C., *et al.* (1994). An early haematopoietic defect in mice lacking the transcription factor GATA-2. *Nature* **371,** 221–226.

Tsai, F. Y., and Orkin, S. H. (1997). Transcription factor GATA-2 is required for proliferation/survival of early hematopoietic cells and mast cell formation, but not for erythroid and myeloid terminal differentiation. *Blood* **89,** 3636–3643.

Tsai, S. F., Martin, D. I., Zon, L. I., *et al.* (1989). Cloning of cDNA for the major DNA-binding protein of the erythroid lineage through expression in mammalian cells. *Nature* **339,** 446–451.

Tsang, A. P., Visvader, J. E., Turner, C. A., *et al.* (1997). FOG, a multitype zinc finger protein, acts as a cofactor for transcription factor GATA-1 in erythroid and megakaryocyte differentiation. *Cell* **90,** 109–119.

Valge-Archer, V. E., Osada, H., Warren, A. J., *et al.* (1994). The LIM protein RBTN2 and the basic helix–loop–helix protein TAL1 are present in a complex in erythroid cells. *Proc. Natl. Acad. Sci. USA* **91,** 8617–8621.

Visvader, J., and Adams, J. M. (1993). Megakaryocytic differentiation induced in 416B myeloid cells by GATA-2 and GATA-3 transgenes or 5-azacytidine is tightly coupled to GATA-1 expression. *Blood* **82,** 1493–1501.

Visvader, J., Begley, C. G., and Adams, J. M. (1991). Differential expression of the LYL, SCL and E2A helix–loop–helix genes within the hemopoietic system. *Oncogene* **6,** 187–194.

Visvader, J. E., Elefanty, A. G., Strasser, A., *et al.* (1992). GATA-1 but not SCL induces megakaryocytic differentiation in an early myeloid line. *EMBO J.* **11,** 4557–4564.

Voso, M. T., Burn, T. C., Wulf, G., *et al.* (1994). Inhibition of hematopoiesis by competitive binding of transcription factor PU.1. *Proc. Natl. Acad. Sci. USA* **91,** 7932–7936.

Wadman, I., Li, J., Bash, R. O., *et al.* (1994). Specific *in vivo* association between the bHLH and LIM proteins implicated in human T cell leukemia. *EMBO J.* **13,** 4831–4839.

Wadman, I. A., Osada, H., Grutz, G. G., *et al.* (1997). The LIM-only protein Lmo2 is a bridging molecule assembling an erythroid, DNA-binding complex which includes the TAL1, E47, GATA-1 and Ldb1/NLI proteins. *EMBO J.* **16,** 3145–3157.

Wang, J. H., Nichogiannopoulou, A., Wu, L., *et al.* (1996). Selective defects in the development of the fetal and adult lymphoid system in mice with an Ikaros null mutation. *Immunity* **5,** 537–549.

Wang, Q., Stacy, T., Binder, M., *et al.* (1996a). Disruption of the Cbfa2 gene causes necrosis and hemorrhaging in the central nervous system and blocks definitive hematopoiesis. *Proc. Natl. Acad. Sci. USA* **93,** 3444–3449.

Wang, Q., Stacy, T., Miller, J. D., *et al.* (1996b). The CBFβ subunit is essential for CBFα2 (AML1) function *in vivo*. *Cell* **87,** 697–708.

Wang, S., Wang, Q., Crute, B. E., *et al.* (1993). Cloning and characterization of subunits of the T-cell receptor and murine leukemia virus enhancer core-binding factor. *Mol. Cell. Biol.* **13,** 3324–3339.

Wang, Z. Q., Ovitt, C., Grigoriadis, A. E., *et al.* (1992). Bone and haematopoietic defects in mice lacking c-fos. *Nature* **360,** 741–745.

Warren, A. J., Colledge, W. H., Carlton, M. B., *et al.* (1994). The oncogenic cysteine-rich LIM domain protein rbtn2 is essential for erythroid development. *Cell* **78,** 45–57.

Weintraub, H., Davis, R., Tapscott, S., *et al.* (1991). The myoD gene family: Nodal point during specification of the muscle cell lineage. *Science* **251,** 761–766.

Weiss, M. J., Keller, G., and Orkin, S. H. (1994). Novel insights into erythroid development revealed through *in vitro* differentiation of GATA-1 embryonic stem cells. *Genes Dev.* **8,** 1184–1197.

Weiss, M. J., and Orkin, S. H. (1995). GATA transcription factors: Key regulators of hematopoiesis. *Exp. Hematol.* **23,** 99–107.

Weiss, M. J., Yu, C., and Orkin, S. H. (1997). Erythroid-cell-specific properties of transcription factor GATA-1 revealed by phenotypic rescue of a gene-targeted cell line. *Mol. Cell. Biol.* **17,** 1642–1651.

Westin, E. H., Gallo, R. C., Arya, S. K., *et al.* (1996). Differential expression of the amv gene in human hematopoietic cells. *Proc. Natl. Acad. Sci. USA* **93,** 12359–12363.

Whyatt, D., Karis, A., Harkes, I., *et al.* (1997). The level of the tissue-specific factor GATA-1 affects the cell-cycle machinery. *Genes Funct.* **1,** 11–24.

Wilson, D. B., Dorfman, D. M., and Orkin, S. H. (1990). A nonerythroid GATA-binding protein is required for function of the human preproendothelin-1 promoter in endothelial cells. *Mol. Cell. Biol.* **10,** 4854–4862.

Winandy, S., Wu, P., and Georgopoulos, K. (1995). A dominant mutation in the Ikaros gene leads to rapid development of leukemia and lymphoma. *Cell* **83,** 289–299.

Xia, Y., Brown, L., Yang, C. Y., *et al.* (1991). TAL2, a helix–loop–helix gene activated by the (7; 9)(q34;q32) translocation in human T-cell leukemia. *Proc. Natl. Acad. Sci. USA* **88,** 11416–11420.

Yamamoto, M., Ko, L. J., Leonard, M. W., *et al.* (1990). Activity and tissue-specific expression of the transcription factor NF-E1 multigene family. *Genes Dev.* **4,** 1650–1662.

Yang, H. Y., and Evans, T. (1995). Homotypic interactions of chicken GATA-1 can mediate transcriptional activation. *Mol. Cell. Biol.* **15,** 1353–1363.

Zhuang, Y., Kim, C. G., Bartelmez, S., *et al.* (1992). Helix–loop–helix transcription factors E12 and E47 are not essential for skeletal or cardiac myogenesis, erythropoiesis, chondrogenesis, or neurogenesis. *Proc. Natl. Acad. Sci. USA* **89,** 12132–12136.

Zhuang, Y., Soriano, P., and Weintraub, H. (1994). The helix–loop–helix gene E2A is required for B cell formation. *Cell* **79,** 875–884.

Zon, L. I., Yamaguchi, Y., Yee, K., *et al.* (1993). Expression of mRNA for the GATA-binding proteins in human eosinophils and basophils: Potential role in gene transcription. *Blood* **81,** 3234–3241.

3

CELL SIGNALING BY HEMATOPOIETIC GROWTH FACTOR RECEPTORS

R. STARR AND N. A. NICOLA

The Walter and Eliza Hall Institute of Medical Research
and Cooperative Research Centre for Cellular Growth Factors
P.O. Royal Melbourne Hospital
Melboure, Victoria 3050, Australia

I. INTRODUCTION

The hematopoietic growth factor system exemplifies many of the problems that need to be addressed to understand how extracellular events (growth factor binding to its receptor) result in intracellular changes that alter the behavior of cells.

TABLE 3.1 Activation of JAKs and STATs by Various Growth Factors and Cytokines

Ligand	JAK activated	STAT activated
IFNα/β	1, Tyk2	1, 2
IFNγ	1, 2	1
IL-2	1, 3	3, 5
IL-4	1, 3	6
IL-7	1, 3	1, 5
IL-9	1, 3	1, 3, 5
IL-13	2	6
IL-15	1, 3	3, 5
IL-3	2	5
IL-5	2	1, 5
GM-CSF	2	5
IL-6	1, 2, Tyk2	1, 3
LIF	1, 2, Tyk2	1, 3
OSM	1, 2, Tyk2	1, 3
CNTF	1, 2, Tyk2	1, 3
G-CSF	1, 2	3
EPO	2	5
GH	2	1, 3, 5
PRL	2	1, 3, 5
IL-12	2, Tyk2	3, 4
EGF	1	1, 3, 5
PDGF	1, 2, Tyk2	1, 3
CSF-1	1, 2, Tyk2	1, 3

First, the binding of a single growth factor to a single type of receptor complex can result in a multitude of biological responses in a single cell type. Most hematopoietic cytokines influence cell survival (usually they are antiapoptotic), cell proliferation, cell differentiation, and cell activation (Metcalf and Nicola, 1994). Do these events occur as a result of distinct or common intracellular signals and, if the former, are the pathways independent or interdependent (i.e., are they parallel or branching pathways)?

Second, the biological consequences of growth factor binding to its receptor can be cell-specific. Some extreme examples are the induction or suppression of differentiation by leukemia inhibitory factor (LIF) in myeloid versus embryonic stem cells, respectively (Nicola and Hilton, 1997), and the induction or suppression of growth by transforming growth factor-β (TGF-β) (Lawrence, 1996). Do such cells mount intrinsically different cell signaling components as a result of receptor activation or is the cell's transcriptional machinery simply different so that the same signal is interpreted differently?

Third, the hematopoietic literature abounds with examples of synergistic activity of growth factors. The best documented of these is the inability of primitive hematopoietic cells to respond to any single cytokine but a profound cellular expansion and differentiation in the presence of multiple cytokines (Li and Johnson, 1995). How is this synergy mediated at the level of intracellular signaling pathways? Does it involve some unique form of cross-activation of distinct signaling molecules?

There have been very great advances in the last decade in delineating specific intracellular biochemical pathways that are activated in response to hematopoietic growth factor activation of receptors. However, there has been much less success in definitively assigning specific biochemical pathways to biological responses in cells. The lesson from lower organisms such as *Drosophila* or *C. elegans* is that a genetic analysis of essential signaling components is perhaps the most powerful strategy for correlating pathways with biological function, and since methods for doing this in mammalian cells now exist (Darnell *et al.*, 1994), it can be expected that major advances will occur in the next several years. The definition of such essential pathways will provide the targets for the next generation of therapeutics that can be expected to be far more selective in their actions than those currently available.

In this brief review we will describe a few of the best defined stimulatory and inhibitory pathways involved in intracellular signaling by growth factors in hematopoietic cells. However, it should be emphasized that, in general, there is little evidence for a causal relationship between activation of these pathways and biological responses such as cell proliferation, survival, and differentiation. Where possible, we reference more detailed reviews on particular pathways and key signaling molecules.

II. THE JAK/STAT SIGNAL TRANSDUCTION PATHWAYS

A new pathway of signal transduction, which connects events at the cell surface directly to the nucleus, has been described in recent years, initially through studies of interferon-α (IFNα)- and interferon-γ (IFNγ)-induced signal transduction (Darnell *et al.*, 1994). Ligand–receptor interactions lead to the activation of members of the Janus kinase (JAK) family, which then phosphorylate latent cytoplasmic transcription factors called STATs (signal transducers and activators of transcription). Phosphorylated STAT proteins dimerize and translocate to the nucleus, where they bind to specific DNA elements and activate gene transcription. Although IFNα and IFNγ were the first ligands described to activate this pathway, the JAK/STAT pathway is now known to be a general mechanism for connecting receptor activation to changes in gene transcription and is utilized by many different cytokine receptors, in addition to certain members of the receptor tyrosine kinase family (Table 3.1 and Fig. 3.1). The importance of this pathway is illustrated by its conservation through evolution, which extends to *Drosophila* (Binari and Perrimon, 1994; Hou *et al.*, 1996; Yan *et al.*, 1996).

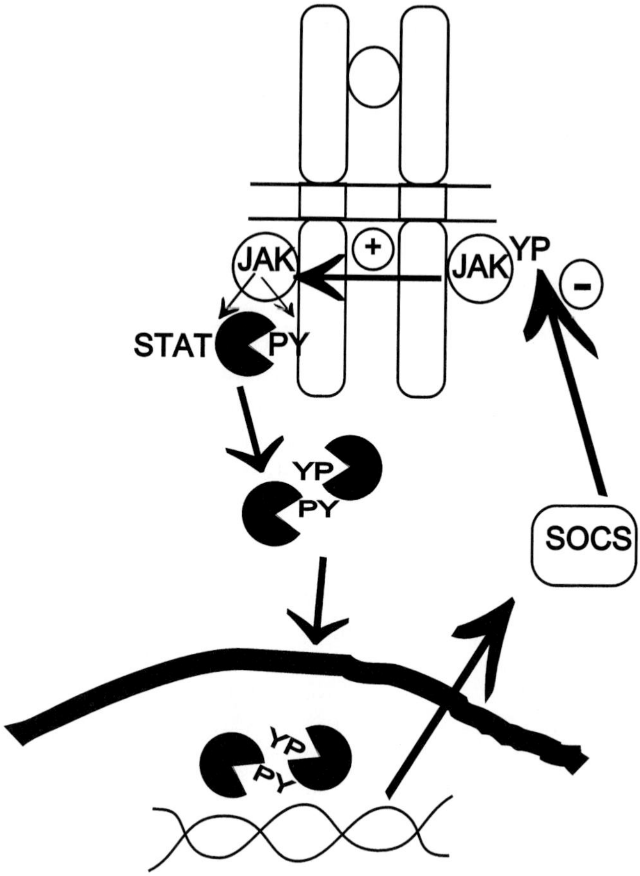

FIGURE 3.1 Outline of the JAK/STAT pathway of transcriptional activation by cytokine receptors. Cytokine binding results in receptor dimerization and cross-activation of constitutively associated JAK kinases by tyrosine phosphorylation. Active JAKs phosphorylate the receptors on tyrosine, creating docking sites for STATs, which are in turn phosphorylated. Phosphorylated STATs form dimers, translocate to the nucleus, and activate transcription of effector genes as well as SOCS, which then bind to activated JAKs and inhibit their activity, completing a negative feedback loop.

The JAK family currently consists of four members, Tyk2, JAK1, JAK2, and JAK3, which range in molecular mass from 125 to 135 kDa. With the exception of JAK3, which is mainly expressed in myelocytic and lymphocytic lineages, all are widely expressed. The overall structure of these kinases is similar, with seven conserved domains, the most striking of which are the two kinase domains, JH1 and JH2. The more carboxy-terminal kinase domain, JH1, conforms to all the consensus sequences associated with tyrosine kinases (Hanks *et al.*, 1988) and is predicted to be catalytically active, whereas the JH2 domain lacks several critical residues of the consensus and is unlikely to have kinase activity. However, the JH2 domain appears to fulfill a critical structural role, as mutant Tyk2 lacking this domain is inactive (Velazquez *et al.*, 1995). The functions of the five additional regions of homology have yet to be defined. Notably, the JAKs lack *src*-homology-2 (SH2) and SH3 domains, specific sequence modules common to a variety of cytoplasmic protein tyrosine kinases, which are involved in mediating protein–protein interactions. SH2 domains recognize short peptide sequences that contain phosphotyrosine residues, whereas SH3 domains bind to sequences containing one or more proline residues (Pawson, 1995).

STAT proteins are latent transcription factors, which reside in the cytoplasm of resting cells. Six STATs have been described (STAT1 through STAT6) which share several conserved structural and functional domains. In particular, the STATs contain a conserved SH2 domain, which is critical for the interaction of STATs with activated receptor complexes and for STAT dimerization. A critical tyrosine residue located immediately carboxy-terminal to the SH2 domain is phosphorylated in response to receptor activation (Shuai *et al.*, 1992; Gouilleux *et al.*, 1994; Improta *et al.*, 1994). STAT function is strictly dependent on phosphorylation of this residue, as mutation of this tyrosine to phenylalanine blocks IFNγ-induced STAT phosphorylation, DNA binding activity, and gene activation function (Shuai *et al.*, 1993). The central region of STATs contains a DNA binding domain, which is highly conserved between STAT family members but is distinct from that found in other DNA-binding proteins.

Dimerization of STAT molecules is essential for DNA binding activity and is thought to occur through intermolecular phosphotyrosine–SH2 domain interactions (Shuai *et al.*, 1994). In resting cells, JAK kinases are catalytically inactive and are constitutively associated with the cytoplasmic domains of receptor chains. Each receptor selectively associates with a distinct subset of JAK kinases (Table 3.1). Certain receptor chains, such as the βc chain common to the IL-3, IL-5, and GM-CSF receptors, bind a single JAK family member (JAK2), in contrast to gp130, which can bind Tyk2, JAK1, or JAK2 (Lutticken *et al.*, 1994; Quelle *et al.*, 1994).

Ligand-induced receptor dimerization brings two JAK kinases into close proximity, resulting in their activation by cross-phosphorylation of critical tyrosine residues (Gauzzi *et al.*, 1996; Feng *et al.*, 1997). Activated JAKs then tyrosine phosphorylate several substrates, including the receptors themselves. Specific phosphotyrosine residues on the receptor serve as docking sites for a variety of SH2-containing signaling molecules, including SHC, Vav, SHP-1, SHP-2, phospholipase C-γ (PLC-γ), the 85-kDa subunit of phosphatidylinositol (PI) 3-kinase, and STATs. The STATs are then phosphorylated on tyrosine, dissociate from the receptor, and form either homodimers or heterodimers. STAT dimers then migrate to the nucleus and activate transcription.

In addition to the foregoing, there are clearly other mechanisms of activation of these pathways. For instance, full activation of STATs appears to require phosphorylation on serine and threonine residues, in addition to tyrosine, possibly by MAP kinase (David *et al.,* 1995a; Wen *et al.,* 1995). Furthermore, mutant cytokine receptors, lacking any tyrosine residues, have been shown to activate STATs, suggesting that STAT molecules may interact with other proteins in the cell (Wang *et al.,* 1995; Nicholson *et al.,* 1996). One of these substrates appears to be the JAK kinases themselves. A recent study has used a yeast two-hybrid screening system to show that the JH2 domain of JAK1, JAK2, or JAK3 specifically associates with STAT5 (Fujitani *et al.,* 1997). Neither STAT1 nor STAT3 was able to interact with the JAK JH2 domain, and unlike STAT5, activation of STAT3 was dependent on receptor tyrosine phosphorylation. This suggests that STAT activation may occur by two distinct mechanisms, one which depends on the presence of docking sites on the receptor and the other which requires a direct interaction between JAKs and STATs.

The JAK/STAT pathway has been defined in most detail in the context of signaling through cytokine receptors, which lack intrinsic tyrosine kinase activity. In this system, the need for cytoplasmic kinases such as JAKs is clear, to couple ligand binding to tyrosine phosphorylation using noncovalently associated kinases. However, the JAK/STAT pathway has also been shown to be activated by receptor tyrosine kinases such as the receptors for epidermal growth factor (EGF), platelet-derived growth factor (PDGF), and colony-stimulating factor (CSF-1). PDGF and CSF-1 activate JAK1, JAK2, and Tyk2 (Novak *et al.,* 1995; Vignais *et al.,* 1996), whereas EGF specifically phosphorylates JAK1 (Leaman *et al.,* 1996). Studies using mutant cell lines have shown that activation of STATs by EGF does not require JAK kinases but is dependent on the activity of the intrinsic receptor tyrosine kinase (David *et al.,* 1996; Leaman *et al.,* 1996). This suggests first that STATs can be activated by kinases other than JAKs, possibly including receptor tyrosine kinases, and second that JAK kinase activation may be required for functions distinct from STAT phosphorylation.

Activation of JAK kinases has been shown to be essential for induction of *c-myc* expression, and this appears to be independent of STAT activation (Kawahara *et al.,* 1995; Watanabe *et al.,* 1996). A study aiming to isolate molecules acting downstream of JAKs in this pathway has identified a novel signal transducing adaptor molecule (STAM) which is tyrosine phosphorylated by JAK2 and JAK3 in response to a variety of cytokines, including IL-2, IL-3, GM-CSF, and PDGF (Takeshita *et al.,* 1996). STAM contains an SH3 domain and an immunoreceptor tyrosine-based activation motif (ITAM) through which it associates with JAKs (Takeshita *et al.,* 1997). Expression of wild type, but not mutant forms of STAM lacking either the SH3 or ITAM domains, enhanced *c-myc* induction mediated by IL-2 and GM-CSF in BAF-BO3 cells (Takeshita *et al.,* 1997). Molecules involved downstream of STAM in this pathway have yet to be defined but are likely to interact with the SH3 domain of STAM.

How do different cytokines generate distinct biological outcomes when they appear to activate similar or overlapping combinations of JAKs and STATs? There appear to be several levels of specificity operating. First, some JAKs and

STATs have a restricted pattern of expression in different tissues. Second, specific STATs are recruited to the receptors, depending on the particular sequence motifs flanking the docking sites. For instance, STAT3 specifically binds to a YXXQ motif in the cytoplasmic region of gp130 and LIFRα (Stahl *et al.,* 1995). Furthermore, the sequence of the STAT SH2 domains contributes to the specificity of signaling since exchanging these domains between STATs can change the receptor complex with which they bind (Heim *et al.,* 1995). Additional specificity can be achieved by selective heterodimerization of different STATs. Finally, different STAT dimers have demonstrated distinct affinities for different nucleotide sequences (Rothman *et al.,* 1994; Schindler *et al.,* 1994). In contrast, JAKs do not appear to contribute to the specificity of these pathways to the same extent, as JAKs have been shown to be functionally interchangeable in signaling by IFNγ (Kotenko *et al.,* 1996).

In contrast to *in vitro* studies in which each STAT has been shown to signal a variety of responses, examination of mice lacking expression of individual STAT proteins has shown a nonredundant biological role for each STAT tested. Mice deficient in STAT1 were completely unresponsive to stimulation by IFNα and IFNγ but showed a normal response to other cytokines that activate STAT1 *in vitro.* Furthermore, these mice were highly susceptible to infection by viruses and microbial pathogens, demonstrating that STAT1 is critical for mediating IFN-induced responses which provide innate immunity (Durbin *et al.,* 1996; Meraz *et al.,* 1996). Similar studies have shown that STAT4 is specific for IL-12-mediated biological responses, whereas STAT6 is essential for IL-4 and IL-13 biological activities (Takeda *et al.,* 1996; Thierfelder *et al.,* 1996). Thus, despite common activation mechanisms, some degree of cytokine specificity appears to be inherent to the JAK/STAT signaling pathway.

III. RAS ACTIVATION PATHWAY

Activated in response to a variety of cytokines, the Ras pathway is a highly conserved process that mediates multiple aspects of cellular growth and differentiation. The Ras superfamily consists of small 20- to 25-kDa GTP-binding proteins, which act as critical control points by linking cytokine receptor activation to a cascade of phosphorylation events that lead to changes in gene transcription (Lim *et al.,* 1996; Wittinghofer and Nassar, 1996). Ras proteins cycle between an active GTP-bound state and an inactive GDP-bound form to which they are converted by an intrinsic GTPase activity. Regulation of Ras activity is achieved by interaction with guanine-nucleotide exchange proteins such as Sos, which promote GTP binding and thus activate Ras, and GTPase-activating proteins (GAP), which inactivate Ras by stimulating the intrinsic GTPase activity. Constitutively active forms of Ras that lack GTPase activity have been found in many human tumors, indicating the importance of the Ras pathway to the regulation of mitogenesis and differentiation.

Ras is located on the cytoplasmic surface of the plasma membrane and mediates signaling from both cytokine receptors and receptor tyrosine kinases (Duronio *et al.,* 1992; Torti *et al.,* 1992), in addition to G protein-coupled receptors (van Corven *et al.,* 1993). Ligand-induced phosphorylation of receptor tyrosine kinases creates docking sites for SH2-containing adaptor molecules such as Grb2. Grb2 is constitutively associated with Sos through the SH3 domain of Grb2 (Chardin *et al.,* 1993; Egan *et al.,* 1993). The main role of Grb2/Sos recruitment to the receptor is to position Sos closer to Ras, since the nucleotide exchange activity of Sos is not dependent on ligand stimulation (Buday and Downward, 1993; Chardin *et al.,* 1993), and furthermore, Sos membrane translocation alone appears sufficient to activate the Ras pathway (Aronheim *et al.,* 1994; Quilliam *et al.,* 1994).

Cytokine receptors lacking intrinsic tyrosine kinase activity commonly use an alternate pathway to activate Ras. This signal is mediated by binding of the SH2-containing protein, SHC, to phosphorylated tyrosine residues of the receptor. SHC is then itself tyrosine phosphorylated, possibly by JAK kinases (He *et al.,* 1995; VanderKuur *et al.,* 1995), which allows the recruitment of Grb2/Sos to the receptor through the Grb2 SH2 domain (Rozakis-Adcock *et al.,* 1992) (Fig. 3.2).

A series of phosphorylation events are triggered by the activation of Ras. The serine/threonine kinase Raf-1 is recruited to the membrane by binding with high affinity to GTP-bound Ras (Leevers *et al.,* 1994; Stockoe *et al.,* 1994) where Raf-1 is activated by phosphorylation, possibly by Src kinase or protein kinase C (Carroll and May, 1994; Marais *et al.,* 1995). Activated Raf-1 then phosphorylates MEK (Kyriakis *et al.,* 1992), which in turn phosphorylates and activates MAP kinase (Johnson and Vaillancourt, 1994). Finally, MAP kinase activation results in the induction of immediate-early genes such as *c-fos* and *c-jun.*

These basic pathways of Ras activation mask a number of complexities. First, although receptor tyrosine kinases can interact directly with Grb2, other adaptor proteins are also used to link activated receptors with the Grb2/Sos complex. For example, SHC binds to the activated EGF receptor (Pelicci *et al.,* 1992; Rozakis-Adcock *et al.,* 1992), and the protein tyrosine phosphatase SHP-2 links PDGF receptor activation to the Ras pathway (Bennett *et al.,* 1994; Li *et al.,* 1994). Other adaptor proteins, such as Grap, are implicated in coupling signals from both cytokine receptors and tyrosine kinase receptors to Ras (Feng *et al.,* 1996). Furthermore, MAP kinase can be activated by a Raf-1-independent pathway (Czar *et al.,* 1997). Thus, the Ras pathway is best considered to be a complex network of interactions, the physiological significance of which is not clear at present, rather than a simple linear pathway from activated receptor to the nucleus.

The Ras pathway has traditionally been associated with the induction of mitogenesis. However, several lines of evidence suggest that the Ras pathway may stimulate multiple biological responses. First, Ras activation requires the

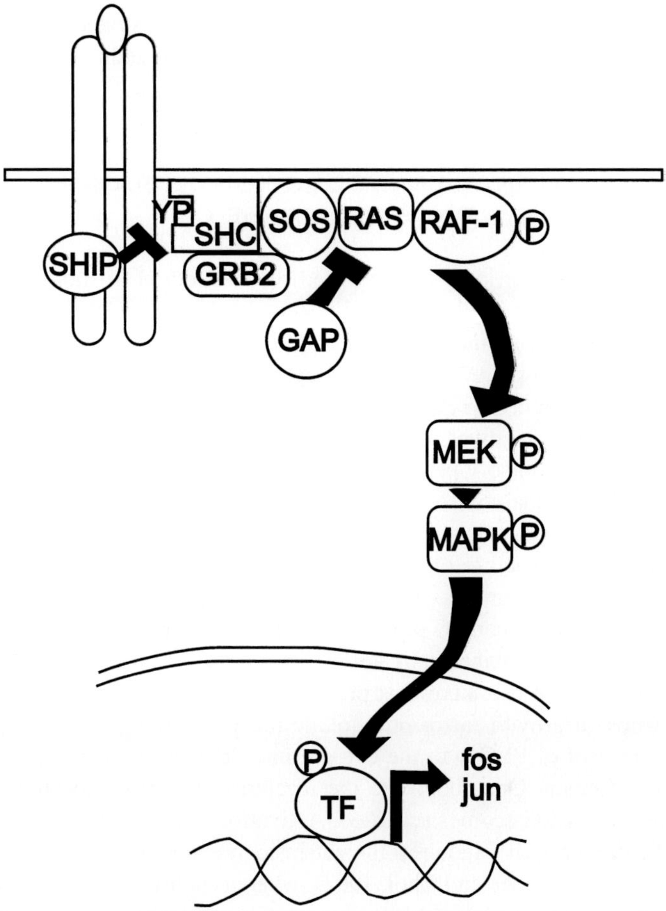

FIGURE 3.2 Ras-mediated activation of the MAP kinase pathway of transcriptional activation. Activated tyrosine phosphorylated cytokine receptors bind and activate the adapter protein SHC while activated tyrosine kinase receptors bind to the adapter molecule Grb2. In either case the complexes recruit Sos to the membrane where it can activate the GTP exchange activity of Ras and activate Raf-1 kinase. This initiates a serine/threonine kinase activation cascade leading to the activation of transcription factors in the nucleus. SHIP, an SH2 domain-containing inositol phosphatase, inhibits SHC activation while GAP inhibits GTP exchange in Ras.

membrane distal region of cytokine receptors (Miura *et al.,* 1994), whereas the proximal region of the receptor is necessary for inducing mitogenic responses (Quelle *et al.,* 1994). Furthermore, in contrast to most cytokines, IL-4 does not activate the Ras pathway (Satoh *et al.,* 1991; Welham, 1994), and proliferation of IL-4-dependent cell lines does not require the activation of Ras or MAP kinase (Welham *et al.*, 1994). Other studies have demonstrated different roles for Ras pathway intermediates, including involvement in cellular differentiation (Alexander *et al.*, 1996).

IV. SIGNAL TRANSDUCTION THROUGH PHOSPHOLIPASE C

A signal transduction pathway activated by a variety of membrane receptors involves the hydrolysis of phosphatidylinositol 4,5-bisphosphate (PIP_2) by phospholipase C (PLC) and the subsequent generation of the second messengers inositol 1,4,5-trisphosphate (IP_3) and diacylglycerol (DAG). The activation of PLC is an early event in response to receptor occupancy and is common to receptors that regulate diverse cellular processes ranging from differentiation and proliferation to sensory perception.

To date, 10 mammalian isoforms of PLC have been identified, which can be divided into three types (β, γ, δ) on the basis of distinct structural features (Rhee *et al.*, 1989). The structure of PLC-γ is distinguished by the presence of one SH3 and two SH2 domains. In general, PLC-β isozymes are activated by stimulation of heterotrimeric G protein-coupled receptors (Exton, 1996), whereas growth factors and cytokines activate PLC-γ (Noh *et al.*, 1995). The activation mechanism of PLC-δ is unknown at present.

Stimulation of growth factor or cytokine receptors by ligand binding results in the recruitment of PLC-γ to the cytoplasmic domain of the receptor, via the PLC-γ SH2 domain. Once bound to the receptor, PLC-γ is phosphorylated on tyrosine residues and becomes activated. Activation of PLC-γ by growth factors, such as PDGF or EGF, requires the intrinsic tyrosine kinase activity of the associated receptor (Rhee and Choi, 1992), whereas nonreceptor protein tyrosine kinases such as JAKs are most likely involved in cytokine-induced activation of PLC-γ. The SH3 domain of PLC-γ interacts with components of the cytoskeleton (Bar-Sagi *et al.*, 1993), which is thought to position the enzyme in close proximity to its substrates at the cell membrane.

The specific role of PLC-γ in hematopoiesis has yet to be defined. Evidence suggests that PLC-γ activation may initiate mitogenesis, as PDGF-induced cell proliferation can be blocked by monoclonal antibodies to PIP_2, the substrate of PLC-γ (Matuoka *et al.*, 1988). Furthermore, an analysis of mutant PDGF receptors has shown that PDGF-induced activation of PLC-γ is sufficient to induce a mitogenic response (Valius and Kazlauckas, 1993). However, PLC-γ activation is not always required for cellular proliferation, as treatment of cells with certain growth factors such as CSF-1 stimulates proliferation but does not result in phosphoinositide turnover (Whetton *et al.*, 1986).

Activated PLC-γ hydrolyzes PIP_2 to generate IP_3 and DAG. IP_3 acts as a second messenger by inducing the release of calcium from intracellular stores. This affects a variety of cellular processes but also contributes to the second messenger role of DAG, which is the activation of the serine/threonine-specific protein kinase C family (PKC). DAG stimulates the activity of PKC by greatly increasing the affinity of the kinase for calcium ions. Activated PKC is translocated from the cytosol to the cell membrane where it phosphorylates downstream targets (Fig. 3.3). Raf-1 appears to be one target of PKC (Carroll and May, 1994), suggesting a role for PKC in the regulation of Ras pathways.

PKC is the intracellular receptor for the potent tumor promoters, phorbol esters. Phorbol esters interact with PKC at the same site as DAG and the two are thought to activate the enzyme in a similar way, although the former are considerably stronger activators of PKC. This suggests that PKC may contribute to cell survival or proliferation, or modulate differentiation processes.

Several lines of evidence suggest that PKC plays a role in suppressing apoptosis, possibly by regulating Bcl-2 (Gomez *et al.,* 1994; Rinaudo *et al.,*

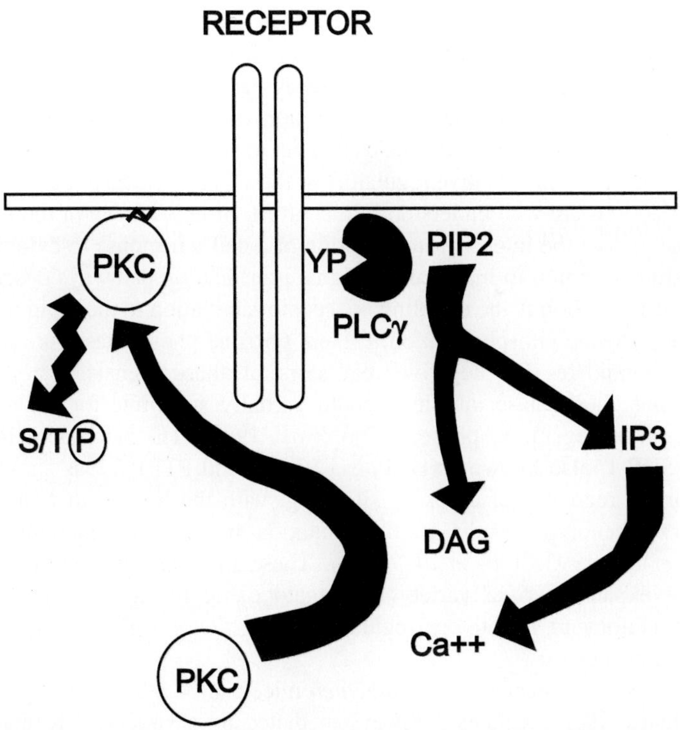

FIGURE 3.3 The phospholipase C-γ (PLC-γ) and inositol lipid pathway of activation of protein kinase C (PKC) isoforms. Activated receptors contain tyrosine phosphorylated docking sites for PLC-γ that initiate phospholipid hydrolysis to diacylglycerol (DAG) and inositol trisphosphate (IP_3). IP_3 leads to Ca^{2+} release from the endoplasmic reticulum, which, with DAG, binds to cytoplasmic PKC and induces its translocation to the membrane. This activates its serine/threonine kinase activity and phosphorylation of downstream targets.

1995). PKC also appears to modulate cell proliferation, since induction of the immediate-early genes *c-jun* and *c-fos* by hematopoietic cytokines is blocked by inhibitors of PKC (Adnuyah *et al.,* 1991; Spangler and Sytkowski, 1992). In addition, PKC has been associated with lineage determination. Inhibition of PKC activity in HEL cells blocked phorbol ester-induced megakaryocyte differentiation, instead promoting differentiation along the erythroid lineage (Hong *et al.,* 1996). Furthermore, increased PKC activity has been correlated with monocyte differentiation of myeloid cells, whereas down-regulation of PKC activity was associated with granulocytic differentiation (Devalia *et al.,* 1992; Whetton *et al.,* 1994; Rossi *et al.,* 1996). It should be noted, however, that many of the foregoing studies relied on the use of PKC activators or inhibitors, rather than directly assessing the biological response of cells to PKC activation by cytokines or growth factors.

V. INHIBITORY PATHWAYS

Hematopoietic differentiation occurs through the activation of cytoplasmic signal transduction pathways. These pathways are finely tuned processes that require a delicate balance between the action of molecules that activate and amplify the signal and the mechanisms that act to suppress the signal. As already described, many of the positive regulators of these pathways have been identified and their actions are well understood. In contrast, little is known of the feedback loops that restrict the intensity and duration of a cell's response to cytokines.

A feature common to many of the signal transduction pathways described in the preceding section is the coupling of receptor activation to the rapid induction of protein tyrosine phosphorylation. Protein tyrosine phosphatases are therefore potential candidates for negative regulators of these signals. Early studies showed that phosphatase inhibitors could partially substitute for the action of cytokines in mitogenic responses (Tojo *et al.,* 1987). The hematopoietic phosphatase SHP-1, also known as HCP, SH-PTP-1, and PTP1C, was shown to be an important regulator of cytokine signaling, with the finding that the murine *motheaten* phenotype results from a mutation in the gene encoding SHP-1 (Schulz *et al.,* 1993; Tsui *et al.,* 1993). These mice are characterized by an abnormal expansion of a variety of hematopoietic lineages, suggesting that SHP-1 functions to negatively regulate growth factor signal transduction in hematopoietic cells.

The pleiotropic phenotype of *motheaten* mice can be attributed in part to the finding that SHP-1 regulates pathways mediated by a variety of receptors, including both cytokine receptors and receptor tyrosine kinases (Yi *et al.,* 1993; Klingmuller *et al.,* 1995; Chen *et al.,* 1996; Paulson *et al.,* 1996). SHP-1 has been most closely studied in the context of Epo signaling and in this system has been shown to bind via its SH2 domain to the tyrosine phosphorylated Epo receptor (Klingmuller *et al.,* 1995; Yi *et al.,* 1995). This interaction then mediates the dephosphorylation and inactivation of JAK2, thereby down-modulating signals generated by the activated Epo receptor.

It is of particular interest that SHP-1 can selectively regulate distinct components of JAK/STAT signal transduction pathways *in vivo.* Comparison of IFNα/β signaling in macrophages from normal and *motheaten* mice has shown that negative regulation by SHP-1 is specific to JAK1 but not Tyk2, as well as STAT1α but not STAT2 (David *et al.,* 1995b). This observation demonstrates that phosphatases can be highly specific regulatory molecules that can distinguish between closely related substrates and also suggests that other as yet unidentified phosphatases may be present which may exhibit different specificity, for example, for Tyk2.

A functional approach to clone inhibitors of IL-6-induced signal transduction has recently identified a second group of regulators of cytokine signaling which appears to act as a classic negative feedback system (Starr *et al.,* 1997). The SOCS (suppressor of cytokine signaling) family of proteins, also known as JAB (Endo *et al.,* 1997) and SSI-1 (Naka *et al.,* 1997), show some similarities to SHP-1. Notably, they are SH2-containing proteins and appear to function by inhibiting the catalytic activity of JAKs (Endo *et al.,* 1997). Also, like SHP-1, SOCS-1 suppresses signals induced by several cytokines, including IL-6, LIF, IFNγ, and TPO (Starr *et al.,* 1997). Despite these similarities, the mode of action of SHP-1 and SOCS is quite different. SOCS expression is not constitutive in the cell but is induced by cytokine stimulation, probably through the activation of STATs (Matsumoto *et al.,* 1997; Naka *et al.,* 1997). Furthermore, SOCS proteins do not appear to have any catalytic activity and most likely bind directly to JAK kinases (Endo *et al.,* 1997; Naka *et al.,* 1997), whereas SHP-1 is thought to associate with the activated receptor itself (Klingmuller *et al.,* 1995). An exception to this is CIS (cytokine-inducible SH2-containing protein), also a member of the SOCS family, which appears to negatively regulate signaling by binding to sites on cytokine receptors which are essential for STAT5 activation, thereby inhibiting the docking of STAT5 to the receptor (Yoshimura *et al.,* 1995; Naka *et al.,* 1997).

In addition to a conserved SH2 domain, the SOCS proteins share a carboxy-terminal region of sequence similarity termed the SOCS box (Starr *et al.,* 1997). This sequence resembles a critical autoregulatory region of the JAK kinases (Feng *et al.,* 1997), which suggests that one mechanism by which the SOCS proteins might inhibit signal transduction is by binding to phosphorylated JAK kinases through their SH2 domain and suppressing the catalytic activity of the kinase via the SOCS box (Starr *et al.,* 1997) (Fig. 3.1). Expression of SOCS genes is induced by multiple cytokines. However, both the kinetics of induction and the spectrum of cytokines able to induce expression differ for each SOCS gene, implying that each SOCS family member may play specific roles in inhibiting different signal transduction pathways (Starr *et al.,* 1997).

A novel family of proteins which inhibit signaling through tyrosine kinase receptors has recently been described (Kharitonenkov *et al.,* 1997). The SIRP (signal regulatory proteins) family comprises at least 15 members and was initially identified by purification of a tyrosine-phosphorylated glycoprotein that associates with the activating phosphatase SHP-2 and appears to act as its substrate. Two subtypes of SIRPs can be distinguished. SIRPα proteins contain a putative transmembrane domain, an extracellular region containing three immunoglobulin-like structures, and a cytoplasmic domain that contains four potential tyrosine phosphorylation sites and a proline-rich sequence. SIRPβ proteins have a similar extracellular domain and a putative membrane-spanning region but lack the cytoplasmic domain. The SIRPα proteins act as substrates of activated receptor tyrosine kinases such as receptors for EGF, PDGF, and insulin and, when phosphorylated, bind to SHP-2 through SH2 interactions. Although the exact mechanism by which SIRPs negatively regulate signaling is unclear at present, they may function to relocate SHP-2 molecules in the cell away from activated receptor tyrosine kinases, thereby preventing SHP-2 from linking Grb2/Sos complexes to the receptor, as described earlier.

VI. PATHWAYS LEADING TO CELL DEATH (APOPTOSIS) OR SURVIVAL

Many cytokine-dependent cells undergo an orderly process of cell suicide (apoptosis) upon withdrawal of cytokine. Other cytokines—notably members of the tumor necrosis factor or FAS ligand family—directly induce apoptosis in target cells via receptor-mediated processes. Both cell survival and cell death are important physiological processes, with the latter being required for several developmental processes, for the control of steady-state cell population size, and for the removal of autoreactive or infected cells (Jacobson *et al.,* 1997).

The effector proteins of the cell suicide program are a family of at least 10 cysteinyl aspartate-specific proteinases (caspases) that recognize the approximate sequence Asp-Glu-X-Asp ↓ , which is present in DNA repair enzymes, cell cycle progression proteins, and cytoskeletal proteins, among others, and leads to their degradation. The action of caspases also activates the proenzyme form of other caspases as well as enzymes involved in DNA fragmentation and sterol biosynthesis that leads to engulfment of the cell by macrophages (Nicholson and Thornberry, 1997).

The procaspases at the top of the caspase activation hierarchy are not known, nor is the signal for their activation. However, in the nematode *C. elegans,* the single caspase CED-3 requires a protein CED-4 to bind to it and activate it. Recently, a mammalian homolog of CED-4—apoptosis activating factor-1 (Apaf-1)—has been described that, under appropriate conditions (cytochrome *c,* ATP, and Apaf-3), can bind to and activate procaspase-3 (Zou *et al.,* 1997).

In *C. elegans,* an antiapoptotic protein, CED-9, prevents activation of CED-3, possibly by binding to CED-4. Again, a large number of mammalian homologs of CED-9 have been described. Surprisingly, some of these, such as CED-9, are antiapoptotic (Bcl-2, Bcl-X_L, Bcl-W) whereas others, such as Bax, Bak, Bcl-X3, Bad, and Bid, are proapoptotic. These proteins form a variety of hetero- and homodimers, with the outcome—cell death or survival—dependent on the relative concentrations of each component. Heterodimers of Bad with Bcl-2 or Bcl-X_L inactivate the antiapoptotic activity of Bcl proteins, homodimers of Bax are directly proapoptotic, and heterodimers of Bax with Bcl-2 or Bcl-X_L are antiapoptotic (Jacobson, 1997; Kroemer, 1997). It is not known how these proteins affect caspase activation. However, mitochondria are essential for the process, and the influence on mitochondrial permeability by Bcl-like proteins may be a critical process. For example, efflux of cytochrome *c* from the mitochondrion appears to provide an essential component for Apaf-1 activation of procaspase-3.

Prosurvival signals initiated by cytokine receptors include the phosphorylation of Bad on serine embedded in consensus 14–3–3 protein-binding sites (Zha *et al.,* 1996). Phosphorylation appears to cause dissociation of Bad from Bad.Bcl-2 or Bad/Bcl-X_L complexes and sequestration in cytosolic complexes with the 14–3–3 protein. This leaves Bcl proteins free to associate with Bax and to be targeted to the mitochondria where they exert their effects on cytochrome *c* retention.

One mechanism of cytokine-induced phosphorylation of Bad is through the serine/threonine kinase Akt or protein kinase B. Receptor activation leads to activation of PI-3 kinase (possibly through the Ras pathway), which leads to the production of phosphatidylinositol 3,4-bisphosphate (PtdIns-3, 4-P2). This lipid binds to the Akt pleckstrin homology domain and induces Akt dimerization and kinase activation (Franke *et al.,* 1997). A second mechanism of phosphorylation of Bad is through cytokine-activated Raf-1 kinase, which can also be targeted to the mitochondrial membrane by Bcl-2 (Wang *et al.,* 1996a). The Bcl-2-interacting protein BAG-1 interacts with Raf-1 and increases its kinase activity, possibly accounting for its prosurvival activity (Wang *et al.,* 1996b).

Apoptosis induced by FAS or TNF receptors appears to involve specific death adaptor proteins (Nicholson and Thornberry, 1997). For example, FAS interacts with the adaptor protein FADD/MORT 1 through sequence elements in each called death domains (DD). In turn, FADD/MORT 1 interacts with procaspases through homologous death effector domains (DED). A similar situation occurs for TNF receptors except that additional adaptor molecules (TRADD, RAIDD, and RIP kinase) are involved. The association of procaspases at the cytoplasmic face of these receptor complexes may lead to intermolecular activation of the protease activity and, to some extent, bypass the Bcl pathway (Fig. 3.4). Some evidence suggests that Bcl-2 is a poor inhibitor of FAS or TNF receptor-mediated cell death whereas it is an effective inhibitor of cell death induced by cytokine withdrawal (Strasser *et al.,* 1995).

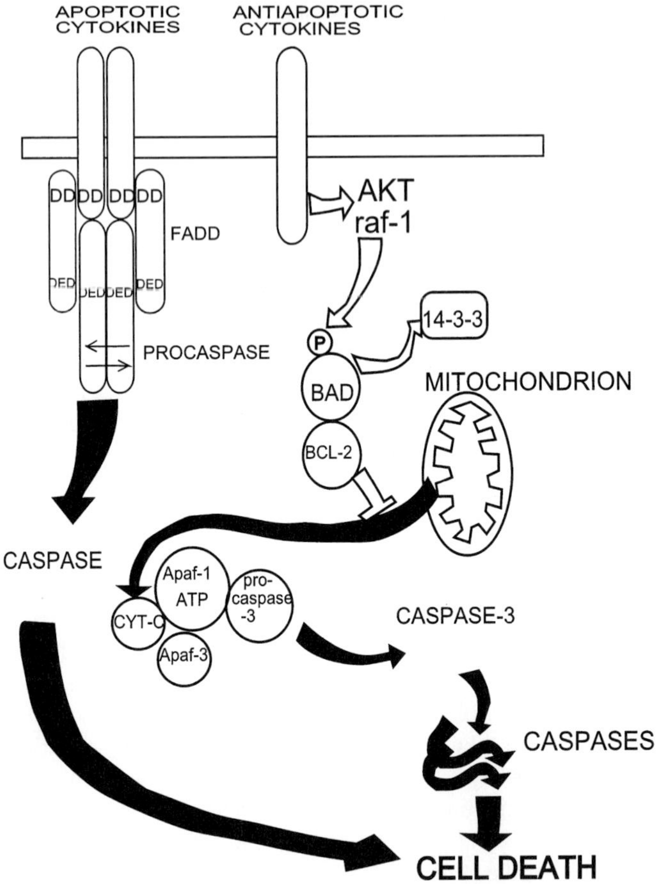

FIGURE 3.4 Prevention of cell death (apoptosis) by survival cytokines and induction of cell death by apoptotic cytokines. Prosurvival cytokines activate Akt or Raf-1 kinase (through PI-3 kinase) which phosphorylate Bad on serine/threonine. This causes dissociation of Bad from heterodimers with Bcl-2 and binding to 14–3–3 proteins. Bcl-2 is then targeted to the mitochondrial membrane where it prevents cytochrome *c* release and therefore prevents activation of caspases. Apoptotic cytokine receptors, upon cytokine binding, aggregate, bringing together associated procaspases, which cross-activate each other. Survival pathways are shown with white arrows and death pathways are shown in shaded arrows.

VII. TGF-β FAMILY SIGNALING

Members of the TGF-β family of growth and differentiation factors—TGF-β, activin, inhibin, and bone morphogenetic proteins (BMPs)—have essential roles in embryonic patterning as well as inhibiting hematopoietic cell growth (Roberts and Spern, 1990).

TGF-β isoforms bind to a heterodimeric receptor consisting of type I and type II subunits, each of which displays intrinsic serine/threonine kinase activity. Mutation data suggest that the ligand-induced transphosphorylation of the type I receptor activates it for signal transduction. It is thought that TGF-β binds first to the type II receptor, which has constitutive kinase activity, and that this then induces heterodimerization with the type I receptor. The type I receptor is then transphosphorylated in a glycine/serine-rich region adjacent to the kinase domain which activates its own kinase activity. The events after this have a strong parallel with that of the JAK/STAT signaling pathway used by other cytokines.

Genetic evidence in *Drosophila* and *C. elegans* has shown that intracellular proteins called Mad or Smads, respectively, are required for downstream signaling from TGF-β-like receptors. In mammalian cells there are at least five homologs of these proteins (Smad 1 through Smad 5). Smads interact directly or indirectly with TGF-β receptors and are phosphorylated on serine or threonine. Smads then form homo- or heterocomplexes and are translocated to the nucleus where they act as transcriptional activators. They do not appear to bind directly to DNA but rather form complexes with DNA-binding proteins such as those of the forkhead family to induce transcriptional activation of cell-cycle regulatory genes (Fig. 3.5).

As for the JAK/STAT pathway, there appears to be some specificity in Smad binding and activation by different members of the TGF-β-receptor family. For example, Smads 1 and 3 appear to be selectively activated by BMPs whereas Smads 2 and 5 are selectively activated by activin/TGF-β receptors.

Smad 4, on the other hand, does not associate with TGF-β receptors and is not phosphorylated, but is synergistic with the other Smads by forming heterodimers. A number of Smads appear to be constitutively transcriptionally active and localized in the nucleus when the N-terminal domain is deleted. This suggests that the N-terminal domain acts to repress the activation domain of Smads and that this repression is relieved by receptor activation, possibly by phosphorylation of Smads and heteromeric complex formation (Baker and Harland, 1997).

VIII. SOME OTHER PATHWAYS

There are many other signaling pathways activated by receptors on hematopoietic cells that we have not been able to cover due to lack of space. Fortunately, there are excellent recent reviews on signaling by T-cell (Cantrell, 1996; Alberola-Ila *et al.,* 1997) and B-cell (DeFranco, 1997) antigen receptors, G-protein-coupled receptors (Post and Brown, 1996; Gudermann *et al.,* 1997), activation of the stress-activated kinase pathways (Kyriakis and Avruch, 1996; Woodgett *et al.,* 1996), activation of NF-κB (Baldwin, 1996), and the important signaling molecules Vav (Bonnefoy-Berard *et al.,* 1996), the negative regulatory inositol phosphatase SHIP (Liu *et al.,* 1997), and the activating phosphatase SHP-2 (Frearson and Alexander, 1997).

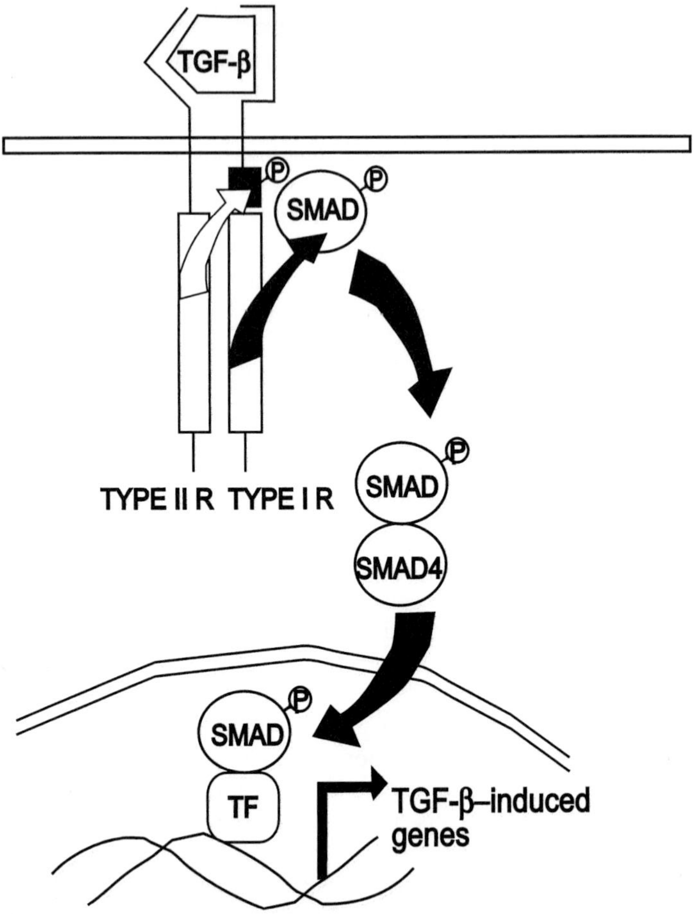

FIGURE 3.5 TGF-β pathway of transcriptional activation. TGF-β isoforms bind to type II receptors with intrinsic serine/threonine kinase activity and induce association with type I receptors. The type I receptor is then phosphorylated, thus activating its own kinase activity. Smads bind to the type I receptor, are phosphorylated, and are then released to form heterodimers with other Smads that are translocated to the nucleus. Smads then bind to DNA-binding proteins and activate their transcriptional activity.

IX. SUMMARY

Although several biochemical pathways involved in signal transduction by hematopoietic growth factors have been described in some detail, a major challenge for the future is to link biological function to specific mediators. The only pathway where the link from receptor activation to biological effector is beginning to become apparent is the pathway leading to cell death through the activation of caspases by TNF-like receptors. The critical effectors for initiation of entry into the cell cycle, inhibition of cell death, induction of cell differentiation, and activation of mature cell functions mediated by hematopoietic growth factors are still unknown. The roles of the JAK/STAT and MAPK pathways in cell proliferation versus differentiation are still not clearly understood and are somewhat controversial.

The pathways described in this chapter are to some extent generic to many different types of receptor systems, and the generation of signaling specificity by different growth factors is only partially understood. Individual cells appear to be primed to respond in different ways (e.g., the type of cellular differentiation that occurs) to the same apparent signal, but the nature of this priming is essentially unknown. Similarly, there is accumulating evidence that few biological functions are mediated by a single, linear signaling pathway, but rather by multiple parallel and branching pathways. The integration of these pathways by the cell is achieved by unknown mechanisms.

Clearly, genetic manipulation of signaling pathways will be an important tool in forging the links between signaling pathways and function and in coming to terms with signaling specificity and complexity.

REFERENCES

Adnuyah, S., Unlap, T., Wagner, F., *et al.* (1991). Regulation of c-jun expression and AP-1 enhancer activity by granulocyte-macrophage colony stimulating factor. *J. Biol. Chem.* **266,** 5670–5675.

Alberola-Ila, J., Takaki, S., Kerner, J. D., *et al.* (1997). Differential signaling by lymphocyte antigen receptors. *Annu. Rev. Immunol.* **15,** 125–154.

Alexander, W. S., Maurer, A. B., Novak, U., *et al.* (1996). Tyrosine-599 of the c-Mpl receptor is required for Shc phosphorylation and the induction of cellular differentiation. *EMBO J.* **15,** 6531–6540.

Aronheim, A., Engelberg, D., Li, N., *et al.* (1994). Membrane targeting of the nucleotide exchange factor Sos is sufficient for activating the ras signaling pathway. *Cell* **78,** 949–961.

Baker, J. C., and Harland, R. M. (1997). From receptor to nucleus: The Smad pathway. *Curr. Opin. Genet. Dev.* **7,** 467–473.

Baldwin, A. S. (1996). The NF-kappa B and I kappa B proteins: New discoveries and insights. *Annu. Rev. Immunol.* **14,** 649–683.

Bar-Sagi, D., Rotin, D., Batzer, A., *et al.* (1993). SH3 domains direct cellular localization of signaling molecules. *Cell* **74,** 83–91.

Bennett, A. M., Tang, T. L., Sugimoto, S., *et al.* (1994). Protein-tyrosine-phosphatase SHPTP2 couples platelet-derived growth factor receptor β to Ras. *Proc. Natl. Acad. Sci. USA* **91,** 7335–7339.

Binari, R., and Perrimon, N. (1994). Stripe-specific regulation of pair-rule genes by hopscotch, a putative Jak family tyrosine kinase in *Drosophila. Genes Dev.* **8,** 300–312.

Bonnefoy-Berard, N., Munshi, A., Yron, I., *et al.* (1996). Vav: Function and regulation in hematopoietic cell signaling. *Stem Cells* **14,** 250–268.

Buday, L., and Downward, J. (1993). Epidermal growth factor regulates p21ras through the formation of a complex of receptor, Grb2 adaptor protein, and Sos nucleotide exchange factor. *Cell* **73,** 611–620.

Cantrell, D. (1996). T cell antigen receptor signal transduction pathways. *Annu. Rev. Immunol.* **14,** 259–274.

Carroll, M. P., and May, W. S. (1994). Protein kinase C mediated serine phosphorylation directly activates raf-1 in murine hematopoietic cells. *J. Biol. Chem.* **269,** 1249–1256.

Chardin, P., Camonis, J. H., Gale, N., *et al.* (1993). Human Sos1: A guanine nucleotide exchange factor for Ras that binds to GRB2. *Science* **260,** 1338–1343.

Chen, H. E., Chang, S., Trub, T., *et al.* (1996). Regulation of colony-stimulating factor 1 receptor signaling by the SH2 domain-containing tyrosine phosphatase SHPTP1. *Mol. Cell. Biol.* **16,** 3685–3697.

Czar, X. F., Ward, A. C., Hoffman, B. W., *et al.* (1997). cAMP suppresses p21ras and Raf-1 responses but not the Erk-1 response to granulocyte-colony-stimulating factor: Possible Raf-1-independent activation of Erk-1. *Biochem. J.* **322,** 79–87.

Darnell, J., Kerr, I. M., and Stark, G. R. (1994). Jak-STAT pathways and transcriptional activation in response to IFNs and other extracellular signaling proteins. *Science* **264,** 821–825.

David, M., Petricoin, E., Benjamin, C., *et al.* (1995a). Requirement for MAP kinase (ERK2) activity in interferon-alpha- and interferon-beta-stimulated gene expression through STAT proteins. *Science* **269,** 1721–1723.

David, M., Chen, H. E., Goelz, S., *et al.* (1995b). Differential regulation of the alpha/beta interferon-stimulated Jak/Stat pathway by the SH2 domain-containing tyrosine phosphatase SHPTP1. *Mol. Cell. Biol.* **15,** 7050–7058.

David, M., Wong, L., Flavell, R., *et al.* (1996). STAT activation by epidermal growth factor (EGF) and amphiregulin. Requirement for the EGF receptor kinase but not for tyrosine phosphorylation sites or JAK1. *J. Biol. Chem.* **271,** 9185–9188.

DeFranco, A. L. (1997). The complexity of signaling pathways activated by the BCR. *Curr. Opin. Immunol.* **9,** 296–308.

Devalia, V., Thomas, N. S. B., Roberts, P. J., *et al.* (1992). Down-regulation of human protein kinase C α is associated with terminal neutrophil differentiation. *Blood* **80,** 68–76.

Durbin, J. E., Hackenmiller, R., Simon, M. C., *et al.* (1996). Targeted disruption of the mouse Stat1 gene results in compromised immunity to viral disease. *Cell* **84,** 443–450.

Duronio, V., Welham, M. J., Abraham, S., *et al.* (1992). p21ras activation via hemopoietin receptors and c-kit requires tyrosine kinase activity but not tyrosine phosphorylation of p21ras-GTPase-activating protein. *Proc. Natl. Acad. Sci. USA* **89,** 1587–1591.

Egan, S. E., Giddings, B. W., Brooks, M. W., *et al.* (1993). Association of Sos Ras exchange protein with Grb2 is implicated in tyrosine kinase signal transduction and transformation. *Nature* **363,** 45–51.

Endo, T. A., Masuhara, M., Yokouchi, M., *et al.* (1997). A new protein containing an SH2 domain that inhibits JAK kinases. *Nature* **387,** 921–924.

Exton, J. H. (1996). Regulation of phosphoinositide phospholipase by hormones, neurotransmitters and other agonists linked to G-proteins. *Annu. Rev. Pharmacol. Toxicol.* **36,** 481–509.

Feng, G. S., Ouyang, Y. B., Hu, D. P., *et al.* (1996). Grap is a novel SH3-SH2-SH3 adaptor protein that couples tyrosine kinases to the Ras pathway. *J. Biol. Chem.* **271,** 12129–12132.

Feng, J., Witthuhn, B. A., Matsuda, T., *et al.* (1997). Activation of Jak2 catalytic activity requires phosphorylation of Y1007 in the kinase activation loop. *Mol. Cell. Biol.* **17,** 2497–2501.

Franke, T. F., Kaplan, D. R., Cantley, L. C., *et al.* (1997). Direct regulation of the Akt proto-oncogene product by phosphatidylinositol-3,4-bisphosphate. *Science* **275,** 665–668.

Frearson, K. A., and Alexander, D. R. (1997). The role of phosphotyrosine phosphatases in haematopoietic cell signal transduction. *Bioessays* **19,** 417–427.

Fujitani, Y., Hibi, M., Fukada, T., *et al.* (1997). An alternative pathway for STAT activation that is mediated by the direct interaction between JAK and STAT. *Oncogene* **14,** 751–761.

Gauzzi, M. C., Velazquez, L., McKendry, R., *et al.* (1996). Interferon-alpha-dependent activation of Tyk2 requires phosphorylation of positive regulatory tyrosines by another kinase. *J. Biol. Chem.* **271,** 20494–20500.

Gomez, J., Heva, A., Silva, A., *et al.* (1994). Implication of protein kinase C in IL-2 mediated proliferation and apoptosis in a murine T-cell clone. *Exp. Cell Res.* **213,** 178–182.

Gouilleux, F., Wakao, H., Mundt, M., *et al.* (1994). Prolactin induces phosphorylation of Tyr 694 of Stat5 (MGF), a prerequisite for DNA binding and induction of transcription. *EMBO J.* **13,** 4361–4369.

Gudermann, T., Schoneberg, T., and Schultz, G. (1997). Functional and structural complexity of signal transduction via G-protein-coupled receptors. *Annu. Rev. Neurosci.* **20,** 399–427.

Hanks, S. K., Quinn, A. M., and Hunter, T. (1988). The protein tyrosine kinase family: Conserved features and deduced phylogeny of the catalytic domains. *Science* **241,** 42–52.

He, T. E., Jiang, N., Zhuang, H., *et al.* (1995). Erythropoietin-induced recruitment of Shc via a receptor phosphotyrosine-independent, Jak2-associated pathway. *J. Biol. Chem.* **270,** 11055–11061.

Heim, M. H., Kerr, I. M., Stark, G. R., *et al.* (1995). Contribution of STAT SH2 groups to specific interferon signaling by the Jak-STAT pathway. *Science* **267,** 1347–1349.

Hong, Y., Martin, J. F., Vainchenker, W., *et al.* (1996). Inhibition of protein kinase C suppresses megakaryocytic differentiation and stimulates erythroid differentiation in HEL cells. *Blood* **87,** 123–131.

Hou, X. S., Melnick, M. B., and Perrimon, N. (1996). Marelle acts downstream of the *Drosophila* HOP/JAK kinase and encodes a protein similar to the mammalian STATs. *Cell* **84,** 411–419.

Improta, T., Schindler, C., Horvath, C. M., *et al.* (1994). Transcription factor ISGF-3 formation requires phosphorylated Stat91 protein, but Stat113 protein is phosphorylated independently of Stat91 protein. *Proc. Natl. Acad. Sci. USA* **91,** 4776–4780.

Jacobson, M. D. (1997). Apoptosis: Bcl-2-related proteins get connected. *Curr. Biol.* **7,** R277–R281.

Jacobson, M. D., Weil, M., and Raff, M. C. (1997). Programmed cell death in animal development. *Cell* **88,** 347–354.

Johnson, G. L., and Vaillancourt, R. R. (1994). Sequential protein kinase reactions controlling cell growth and differentiation. *Curr. Opin. Cell Biol.* **6,** 230–238.

Kawahara, A., Minami, Y., Miyazaki, T., *et al.* (1995). Critical role for the interleukin 2 (IL-2) receptor g-chain-associated Jak3 in the IL-2-induced c-fos and c-myc, but not bcl-2, gene induction. *Proc. Natl. Acad. Sci. USA* **92,** 8724–8728.

Kharitonenkov, A., Chen, Z., Sures, I., *et al.* (1997). A family of proteins that inhibit signaling through tyrosine kinase receptors. *Nature* **386,** 181–186.

Klingmuller, U., Lorenz, U., Cantley, L. C., *et al.* (1995). Specific recruitment of SH-PTP1 to the erythropoietin receptor causes inactivation of JAK2 and termination of proliferative signals. *Cell* **80,** 729–738.

Kotenko, S. V., Izotova, L. S., Pollack, B. P., *et al.* (1996). Other kinases can substitute for Jak2 in signal transduction by interferon-gamma. *J. Biol. Chem.* **271,** 17174–17182.

Kroemer, G. (1997). The proto-oncogene Bcl-2 and its role in regulating apoptosis. *Nat. Med.* **3,** 614–620.

Kyriakis, J. M., App, H., Zhang, X. F., *et al.* (1992). Raf-1 activates MAP kinase-kinase. *Nature* **358,** 417–421.

Kyriakis, J. M., and Avruch, J. (1996). Protein kinase cascades activated by stress and inflammatory cytokines. *Bioessays* **18,** 567–577.

Lawrence, D. A. (1996). Transforming growth factor-beta: A general review. *Eur. Cytokine Netw.* **7,** 363–374.

Leaman, D. W., Pisharody, S., Flickinger, T. W., *et al.* (1996). Roles of JAKs in activation of STATs and stimulation of c-fos gene expression by epidermal growth factor. *Mol. Cell. Biol.* **16,** 369–375.

Leevers, S., Paterson, H., and Marshall, C. J. (1994). Requirement for ras in raf activation is overcome by targeting raf to the plasma membrane. *Nature* **369,** 411–414.

Li, C. L., and Johnson, G. R. (1995). Murine hematopoietic stem and progenitor cells: Enrichment and biologic characterization. *Blood* **85,** 1472–1479.

Li, W., Nishimura, R., Kashishian, A., *et al.* (1994). A new function for a phosphotyrosine phosphatase: Linking GRB2-Sos to a receptor tyrosine kinase. *Mol. Cell. Biol.* **14,** 509–517.

Lim, L., Manser, E., Leung, T., *et al.* (1996). Regulation of phosphorylation pathways by p21 GTPases. *Eur. J. Biochem.* **242,** 171–185.

Liu, L., Damen, J. E., Ware, M., *et al.* (1997). SHIP, a new player in cytokine-induced signaling. *Leukemia* **11,** 181–184.

Lutticken, C., Wegenka, U. M., Yuan, J., *et al.* (1994). Association of transcription factor APRF and protein kinase Jak1 with the interleukin-6 signal transducer gp130. *Science* **263,** 89–92.

Marais, R., Light, Y., Paterson, H. F., *et al.* (1995). Ras recruits Raf-1 to the plasma membrane for activation by tyrosine phosphorylation. *EMBO J.* **14,** 3136–3145.

Matsumoto, A., Masuhara, M., Mitsui, K., *et al.* (1997). A novel cytokine-inducible gene CIS encodes an SH2-containing protein that binds to tyrosine-phosphorylated interleukin 3 and erythropoietin receptors. *Blood* **89,** 3148–3154.

Matuoka, K., Fukami, K., Natanishi, O., *et al.* (1988). Mitogenesis in response to PDGF and bombesin abolished by microinjection of antibody to PIP2. *Science* **239,** 640–643.

Meraz, M. A., White, J. M., Sheehan, K. C., *et al.* (1996). Targeted disruption of the Stat1 gene in mice reveals unexpected physiological specificity in the JAK-STAT signaling pathway. *Cell* **84,** 431–442.

Metcalf, D., and Nicola, N. A. (1994). "The Hemopoietic Colony-Stimulating Factors: From Biology to Clinical Applications." Cambridge Univ. Press, Cambridge, UK.

Miura, Y., Miura, O., Ihle, J. N., *et al.* (1994). Activation of the mitogen-activated protein kinase pathway by the erythropoietin receptor. *J. Biol. Chem.* **269,** 29962–29969.

Naka, T., Narazaki, M., Hirata, M., *et al.* (1997). Structure and function of a new STATs- induced STATs inhibitor-1 (SSI-1). *Nature* **387,** 924–929.

Nicholson, D. W., and Thornberry, N. A. (1997). Caspases: Killer proteases. *Trends Biochem. Sci.* **22,** 299–306.

Nicholson, S. E., Starr, R., Novak, U., *et al.* (1996). Tyrosine residues in the granulocyte colony-stimulating factor (G-CSF) receptor mediate G-CSF-induced differentiation of murine myeloid leukemic (M1) cells. *J. Biol. Chem.* **271,** 26947–26953.

Nicola, N. A., and Hilton, D. J. (1997). Leukemia inhibitory factor and its receptor. *In* "Growth Factors and Cytokines in Health and Disease" D. Le Roith D. and C. Bandy, Eds.), pp. 613–668. JAI Press, New York.

Noh, D. Y., and Shin, S. H., and Rhee, S. G. (1995). Phosphoinositide-specific phospholipase C and mitogenic signaling. *Biochim. Biophys. Acta* **1242,** 99–113.

Novak, U., Harpur, A. G., Paradiso, L., *et al.* (1995). Colony-stimulating factor 1-induced STAT1 and STAT3 activation is accompanied by phosphorylation of Tyk2 in macrophages and Tyk2 and JAK1 in fibroblasts. *Blood* **86,** 2948–2956.

Paulson, R. F., Vesely, S., Siminovitch, K. A., *et al.* (1996). Signaling by the W/Kit receptor tyrosine kinase is negatively regulated *in vivo* by the protein tyrosine phosphatase Shp1. *Nat. Genet.* **13,** 309–315.

Pawson, T. (1995). Protein modules and signaling networks. *Nature* **373,** 573–579.

Pelicci, G., Lanfrancone, L., Grignani, F., *et al.* (1992). A novel transforming protein (SHC) with an SH2 domain is implicated in mitogenic signal transduction. *Cell* **70,** 93–104.

Post, G. R., and Brown, J. H. (1996). G protein-coupled receptors and signaling pathways regulating growth factors. *FASEB J.* **10,** 741–749.

Quelle, F. W., Sato, N., Witthuhn, B. A., *et al.* (1994). JAK2 associates with the bc chain of the receptor for granulocyte-macrophage colony-stimulating factor, and its activation requires the membrane proximal region. *Mol. Cell. Biol.* **14,** 4335–4341.

Quilliam, L. A., Huff, S. Y., Rabun, K. M., *et al.* (1994). Membrane-targeting potentiates guanine nucleotide exchange factor CDC25 and SOS1 activation of Ras transforming activity. *Proc. Natl. Acad. Sci. USA* **91,** 8512–8516.

Rhee, S. G., and Choi, K. D. (1992). Multiple forms of phospholipase C isozymes and their activation mechanism. *Adv. Second Messenger Phosphoprotein Res.* **26,** 35–61.

Rhee, S. G., Suh, P. G., Ryu, S. H., *et al.* (1989). Studies of inositol phospholipid-specific phospholipase C. *Science* **244,** 546–550.

Rinaudo, M., Su, K., Falk, L. A., *et al.* (1995). Human interleukin-3 receptor modulates bcl-2 mRNA and protein levels through protein kinase C in TF-1 cells. *Blood* **86,** 80–88.

Roberts, A. B., and Sporn, M. B. (1990). The transforming growth factors. *In* "Peptide Growth

Factors and Their Receptors" (M. B. Sporn and A. C. Roberts, Eds.), pp. 419–472. Springer, Heidelberg.

Rossi, F., McNagny, K. M., Smith, G., *et al.* (1996). Lineage committment of transformed haematopoietic progenitors is determined by the level of PKC activity. *EMBO J.* **15,** 1894–1901.

Rothman, P., Kreider, B., Azam, M., *et al.* (1994). Cytokines and growth factors signal through tyrosine phosphorylation of a family of related transcription factors. *Immunity,* **1,** 457–468.

Rozakis-Adcock, M., McGlade, J., Mbamalu, G., *et al.* (1992). Association of the Shc and Grb2/Sem5 SH2-containing proteins is implicated in activation of the Ras pathway by tyrosine kinases. *Nature* **360,** 689–692.

Satoh, T., Nakafuku, M., Miyajima, A., *et al.* (1991). Involvement of ras p21 protein in signal-transduction pathways from interleukin 2, interleukin 3, and granulocyte/macrophage colony-stimulating factor, but not from interleukin 4. *Proc. Natl. Acad. Sci. USA* **88,** 3314–3318.

Schindler, C., Kashleva, H., Pernis, A., *et al.* (1994). STF-IL-4: A novel IL-4-induced signal transducing factor. *EMBO J.* **13,** 1350–1356.

Schulz, L. D., Schweitzer, P. A., Rajan, T. V., *et al.* (1993). Mutations at the murine motheaten locus are within the hematopoietic cell protein tyrosine phosphatase (Hcph) gene. *Cell* **73,** 1445–1454.

Shuai, K., Horvath, C. M., Huang, L. H., *et al.* (1994). Interferon activation of the transcription factor Stat91 involves dimerization through SH2-phosphotyrosyl peptide interactions. *Cell* **76,** 821–828.

Shuai, K., Schindler, C., Prezioso, V. R., *et al.* (1992). Activation of transcription by IFN-gamma: Tyrosine phosphorylation of a 91-kD DNA binding protein. *Science* **258,** 1808–1812.

Shuai, K., Stark, G. R., Kerr, I. M., *et al.* (1993). A single phosphotyrosine residue of Stat91 required for gene activation by interferon-gamma. *Science* **261,** 1744–1746.

Spangler, R., and Sytkowski, A. (1992). C-myc is an early response gene in normal erythroid cells: Evidence for a protein kinase C mediated signal. *Blood* **179,** 52–57.

Stahl, N., Farruggella, T. J., Boulton, T. G., *et al.* (1995). Choice of STATs and other substrates specified by modular tyrosine-based motifs in cytokine receptors. *Science* **267,** 1349–1353.

Starr, R., Willson, T. A., Viney, E. M., *et al.* (1997). A family of cytokine-inducible inhibitors of signaling. *Nature* **387,** 917–921.

Stockoe, D., MacDonald, S., Cadwallader, K., *et al.* (1994). Activation of ras as a result of recruitment to the plasma membrane. *Science* **264,** 1463–1467.

Strasser, A., Harris, A. W., Huang, D. C., *et al.* (1995). Bcl-2 and FAS/APO-1 regulate distinct pathways to lymphocyte apoptosis. *EMBO J.* **14,** 6136–6147.

Takeda, K., Tanaka, T., Shi, W., *et al.* (1996). Essential role of Stat6 in IL-4 signaling. *Nature* **380,** 627–630.

Takeshita, T., Arita, T., Asao, H., *et al.* (1996). Cloning of a novel signal-transducing adaptor molecule containing an SH3 domain and ITAM. *Biochem. Biophys. Res. Commun.* **225,** 1035–1039.

Takeshita, T., Arita, T., Higuchi, M., *et al.* (1997). STAM, signal transducing adaptor molecule, is associated with Janus kinases and involved in signaling for cell growth and *c-myc* induction. *Immunity* **6,** 449–457.

Thierfelder, W. E., van Deursen, J. M., Yamamoto, K., *et al.* (1996). Requirement for Stat4 in interleukin-12-mediated responses of natural killer and T cells. *Nature* **382,** 171–174.

Tojo, A., Kasuga, M., Urabe, A., *et al.* (1987). Vanadate can replace interleukin 3 for transient growth of factor-dependent cells. *Exp. Cell Res.* **171,** 16–23.

Torti, M., Marti, K. B., Altschuler, D., *et al.* (1992). Erythropoeitin induces p21ras activation and p120GAP tyrosine phosphorylation in human erythroleukemia cells. *J. Biol. Chem.* **267,** 8293–8298.

Tsui, H. W., Siminovitch, K. A., de Souza, L., *et al.*(1993). Motheaten and viable motheaten mice have mutations in the haematopoietic cell phosphatase gene. *Nat. Genet.* **4,** 124–129.

Valius, M., and Kazlauskas, A. (1993). Phospholipase C-gamma 1 and phosphatidylinositol 3 kinase are the downstream mediators of the PDGF receptor's mitogenic signal. *Cell* **73,** 321–334.

van Corven, E. J., Hordijk, P. L., Medema, R. H., *et al.* (1993). Pertussis toxin-sensitive activation of p21ras by G protein-coupled receptor agonists in fibroblasts. *Proc. Natl. Acad. Sci. USA* **90,** 1257–1261.

VanderKuur, J., Allevato, G., Billestrup, N., *et al.* (1995). Growth hormone-promoted tyrosyl phosphorylation of SHC proteins and SHC association with Grb2. *J. Biol. Chem.* **270,** 7587–7593.

Velazquez, L., Mogensen, K. E., Barbieri, G., *et al.* (1995). Distinct domains of the protein tyrosine kinase tyk2 required for binding of interferon-alpha/beta and for signal transduction. *J. Biol. Chem.* **270,** 3327–3334.

Vignais, M. L., Sadowski, H. B., Watling, D., *et al.* (1996). Platelet-derived growth factor induces phosphorylation of multiple JAK family kinases and STAT proteins. *Mol. Cell. Biol.* **16,** 1759–1769.

Wang, H. G., Rapp, U. R., and Reed, J. C. (1996a). Bcl-2 targets the protein kinase Raf-1 to mitochondria. *Cell* **87,** 629–638.

Wang, H. G., Takayama, S., Rapp, U. R., *et al.* (1996b). Bcl-2 interacting protein, BAG-1, binds to and activates the kinase Raf-1. *Proc. Natl. Acad. Sci. USA* **93,** 7063–7068.

Wang, Y. D., Wong, K., and Wood, W. I. (1995). Intracellular tyrosine residues of the human growth hormone receptor are not required for the signaling of proliferation or Jak-STAT activation. *J. Biol. Chem.* **270,** 7021–7024.

Watanabe, S., Itoh, T., and Arai, K. (1996). JAK2 is essential for activation of c-fos and c-myc promoters and cell proliferation through the human granulocyte-macrophage colony-stimulating factor receptor in BA/F3 cells. *J. Biol. Chem.* **271,** 12681–12686.

Welham, M. J. (1994). Multiple hemopoietins, with the exception of interleukin-4, induce modification of Shc and mSos1, but not their translocation. *J. Biol. Chem.* **269,** 21165–21176.

Welham, M. J., Duronio, V., and Schrader, J. W. (1994). Interleukin-4-dependent proliferation dissociates p44erk-1, p42erk-2, and p21ras activation from cell growth. *J. Biol. Chem.* **269,** 5865–5873.

Wen, Z., Zhong, Z., and Darnell, J. (1995). Maximal activation of transcription by Stat1 and Stat3 requires both tyrosine and serine phosphorylation. *Cell* **82,** 241–250.

Whetton, A. D., Heyworth, C. M., Nicholls, S. E., *et al.* (1994). Cytokine-mediated protein kinase C activation is a signal for lineage determination in bipotential granulocyte macrophage colony-forming cells. *J. Cell Biol.* **125,** 651–659.

Whetton, A. D., Monk, P. N., Consalvey, S. D., *et al.* (1986). The haemopoietic growth factors interleukin 3 and colony stimulating factor-1 stimulate proliferation but do not induce inositol lipid breakdown in murine bone-marrow-derived macrophages. *EMBO J.* **5,** 3281–3286.

Wittinghofer, A., and Nassar, N. (1996). How Ras-related proteins talk to their effectors. *Trends Biochem. Sci.* **21,** 488–491.

Woodgett, J. R., Avruch, J., and Kyriakis, J. (1996). The stress activated protein kinase pathway. *Cancer Surv.* **27,** 127–138.

Yan, R., Small, S., Desplan, C., *et al.* (1996). Identification of a Stat gene that functions in Drosophila development. *Cell* **84,** 421–430.

Yi, T., Mui, A.-F., Krystal, G., *et al.* (1993). Hematopoietic cell phosphatase associates with the interleukin-3 (IL-3) receptor β chain and down-regulates IL-3-induced tyrosine phosphorylation and mitogenesis. *Mol. Cell. Biol.* **13,** 7577–7586.

Yi, T., Zhang, J., Miura, O., *et al.* (1995). Hematopoietic cell phosphatase associates with erythropoeitin (Epo) receptor after Epo-induced receptor tyrosine phosphorylation. *Blood* **85,** 87–95.

Yoshimura, A., Ohkubo, T., Kiguchi, T., *et al.* (1995). A novel cytokine-inducible gene CIS encodes an SH2-containing protein that binds to tyrosine-phosphorylated interleukin 3 and erythropoietin receptors. *EMBO J.* **14,** 2816–2826.

Zha, J., Harada, H., Yang, E., *et al.* (1996). Serine phosphorylation of death agonist BAD in response to survival factor results in binding to 14-3-3 not BCL-X. *Cell* **87,** 619–628.

Zou, H., Henzel, W. J., Liu, X., *et al.* (1997). Apaf-1, a human protein homologous to *C. elegans* CED-4, participates in cytochrome *c*-dependent activation of caspase-3. *Cell* **90,** 405–413.

4

Influence of Cytokines and Adhesion Molecules on Hematopoietic Stem Cell Development

P. J. Simmons, D. N. Haylock, and J.-P. Lévesque

Leukaemia Research Unit
Hanson Centre for Cancer Research
P.O. Box 14 Rundle Mall
Adelaide, South Australia 5000, Australia

I. INTRODUCTION

In the adult mammal, hematopoiesis is restricted to the extravascular compartment of the bone marrow (BM) where hematopoietic stem cells (HSCs) and their progeny develop in intimate contact with a heterogeneous population of stromal cells that comprise the hematopoietic microenvironment (HM). It is well established that cellular interactions between primitive hematopoietic progenitor cells (HPCs) and BM stromal cells play a critical role in regulating hematopoiesis, although the molecular mechanisms responsible for this control remain to be defined. Current data suggest that at least two classes of molecules including hematopoietic growth factors (HGFs) and members of several cell adhesion molecule (CAM) superfamilies contribute to the regulation of hematopoiesis, although the exact contribution made by each class of molecule remains to be determined. There are abundant data documenting HGFs as potent regulators of HPC survival, growth, and differentiation. Emerging evidence suggests that in addition to their well-documented role in initiating and maintaining contact between HPCs and stromal cells, CAMs, as bona fide signaling molecules, also participate more directly in the growth and development of primitive HPCs. Hematopoiesis can therefore be considered as a process regulated by signals provided to developing hematopoietic cells by their surrounding microenvironment both by stromal cell–HPC interactions mediated by various CAMs and also through the action of specific HGFs following binding to their cognate cell surface receptors. Recent studies document considerable functional overlap between the HGFs and CAM families as demonstrated by the capacity of HGFs to regulate the functional properties of CAMs on primitive HPCs. This review will discuss the contribution of both classes of molecule to the regulation of HPC growth and development and will examine the mechanisms that may be responsible for the functional interdependence between HGFs and CAMs.

II. PRIMITIVE HEMATOPOIETIC PROGENITOR CELLS REQUIRE MULTIPLE CYTOKINES FOR PROLIFERATION

The concept of HGFs interacting additively and/or synergistically to regulate the survival and proliferation of HPCs is now well established. The first indication that HGFs could effect such potent influences on HPCs, at least *in vitro,* came from early studies utilizing colony-forming assays of murine and human bone marrow cells. These assays, performed in semisolid media, have been widely used to investigate the function of hematopoietic regulators and the hematopoietic potential of cell populations isolated from various hematopoietic tissues (Bradley and Metcalf, 1966; Ichikawa *et al.,* 1966; Metcalf and Nicola, 1983; Metcalf, 1984; Ema *et al.,* 1990). A major advantage of this system is that the progeny of individual precursor cells remain physically localized during the process of colony formation, thus facilitating identification of colony type and also providing a direct means of describing how individual HGFs or combinations of HGFs influence colony-forming cell (CFC) growth. A number of important aspects of HPC and HGF interactions have been determined from clonogenic assay. As reviewed in detail by Metcalf (1993), the colony-stimulating factors (CSFs) in particular and other HGFs were found to have pleiotropic biological activities, including the ability to prevent HPC apoptosis (Metcalf, 1982), stimulation of cell proliferation, and activation of mature cell function (Stanley and Burgess, 1983; Gasson *et al.,* 1984; Weisbart *et al.,* 1985). In addition, the CSFs display a high degree of functional overlap, at least in *in vitro* clonogenic assays. For example, each of the four colony-stimulating factors, granulocyte-CSF (G-CSF), granulocyte-macrophage-CSF (GM-CSF), macrophage-CSF (M-CSF), and interleukin-3 (IL-3), were found to stimulate the formation of distinctive types of colonies (Metcalf and Nicola, 1983; Metcalf, 1984; Metcalf *et al.,* 1986a; Metcalf and Nicola, 1992). However, it was noted that more than one factor could stimulate the formation of what appeared to be the same type of colony. This was best illustrated by generation of small neutrophilic colonies by G-CSF, GM-CSF, IL-3, stem cell factor (SCF), and IL-6. A similar situation also exists with regard to factors that can stimulate or potentiate the formation of megakaryocyte and eosinophil colonies (Rennick *et al.,* 1989).

The contribution of single CSFs or HGFs to steady-state hematopoiesis remains unclear, but valuable insights have been gained through studies of naturally occurring mutations (Ruscetti *et al.,* 1976; Russell, 1979) and gene-targeted mice lacking either hematopoietic growth factors (Lieschke *et al.,* 1994; Stanley *et al.,* 1994) or their receptors (Nishinakamura *et al.,* 1995; Robb *et al.,* 1995; Nicola *et al.,* 1996; Nishinakamura *et al.,* 1996). It is evident that particular HGFs have critical roles in maintenance of steady-state hematopoiesis. Notable in this regard is SCF, the ligand for the tyrosine kinase receptor c-kit. Mutations in the steel (*Sl*) locus, which encodes SCF, form the basis for the hematopoietic defects observed in *Sl/S*^{d} mice (Williams *et al.,* 1990; Zsebo *et al.,* 1990). Similarly, mutations in the *Op* gene result in a lack of active M-CSF and are the basis for the hematopoietic defects in *op/op* mice (Begg *et al.,* 1993). Gene knockout studies also confirm that some HGFs, despite having potent *in vitro* hematopoietic capacity, do not contribute to steady-state hematopoiesis. A good example is GM-CSF, where mice lacking this factor have no detectable deficiency in steady-state hematopoiesis (Stanley *et al.,* 1994). In contrast, G-CSF-deficient mice have reduced hematopoietic progenitors in the bone marrow and spleen and neutropenia (Lieschke *et al.,* 1994). Interestingly, in adult mice lacking both G-CSF and GM-CSF (Seymour *et al.,* 1997), the reduction in both marrow and splenic progenitor numbers, the percentage of bone marrow myeloid precursors, and the degree of neutropenia were no greater than in G-CSF-deficient mice, suggesting that GM-CSF is not involved in the residual hematopoiesis of G-CSF-deficient mice.

Clonogenic assays have also provided a particularly valuable means for investigating interactions between hematopoietic regulators because two distinct parameters can be distinguished: alterations in colony size and alteration in colony number. The term synergy is used to define a process where two or more regulators acting on the same precursor cell induce a greater number of progeny. The second event, where two or more regulators promote increased numbers of precursors to proliferate, is termed recruitment. Increased recruitment is considered to be an indication that some progenitors require simultaneous stimulation by two or more factors before being able to respond. Many studies have provided evidence of both synergistic interactions and increased recruitment of both mouse and human HPCs when cultured with combinations of CSFs or HGFs (Metcalf *et al.,* 1986b; Williams *et al.,* 1987; McNeice *et al.,* 1988; Migliaccio *et al.,* 1988; Sonoda *et al.,* 1988; Bartelmez *et al.,* 1989; Sieff *et al.,* 1989). Metcalf *et al.* (1986b) described how twofold to threefold more erythroid bursts formed in the presence of GM-CSF and EPO as compared to cultures stimulated with GM-CSF alone. Agar culture of human BM also demonstrated a potent synergy between IL-3 and GM-CSF (McNeice *et al.,* 1989) and also M-CSF and GM-CSF (Falk and Vogel, 1988), which was evident by the presence of high proliferative potential (HPP) colonies. Subsequent studies demonstrated that HPP colony forming cells (HPP-CFC) could be further divided according to their growth factor requirements (McNeice *et al.,* 1986) and for the first time suggested that primitive HPCs required multiple HGFs to induce proliferation (McNeice *et al.,* 1990).

With the identification of antigens present on mouse and human HPCs (Civin *et al.,* 1984; Spangrude *et al.,* 1988; Li and Johnson, 1992) and the use of multiparameter fluorescent activated cell sorting (FACS), more sophisticated analysis of the responses of immunophenotypically defined HPC subpopulations to HGFs became feasible. These advances, together with the development of serum-free or serum-depleted media (Lansdorp and Dragowska, 1992; Rebel *et al.,* 1994), allowed investigators to define how purified recombinant hematopoietic regulators influenced the survival, recruitment, and proliferation of HPCs. In this respect, groups who developed culture systems for identifying precursors of HPCs made significant contributions (Clark and Kamen, 1987; Moore *et al.,* 1987; Muller-Sieburg *et al.,* 1988; Bartelmez *et al.,* 1989; Iscove *et al.,* 1989). Collectively, these investigators demonstrated that early hematopoietic precursors proliferate and differentiate in response to IL-1 or IL-6 in synergy with IL-3 while progressively committed populations responded preferentially to GM-

CSF, G-CSF, and M-CSF. Similar investigations, using so-called delta or pre-CFU (colony-forming units) assay, were performed with human bone marrow CD34$^+$ cells (Smith *et al.,* 1991). In this assay system the generation of nascent CFU-GM by HGFs, which act upon primitive HPCs, serves as an index of precursors to CFU-GM. Hierarchically primitive progenitor cells identified as pre-CFU were CD34$^+$ and found to lack detectable markers for T cell, B cell, natural killer cell, and myeloid lineages. The generation of nascent CFU-GM from 4-hydroperoxycyclophosphamide-resistant (4HC) CD34$^+$ cells was highly dependent on the cytokine(s) used for culture: IL-3 when used alone was consistently better than either IL-1 or IL-6 whereas the combination of IL-1 and IL-3 was better than any single cytokine or any other two-cytokine combination tested (Smith *et al.,* 1991). A combined approach of HPC isolation by FACS followed by liquid culture demonstrated that candidate multipotent hematopoietic stem cells were restricted to the CD34$^+$CD38$^-$ cells within normal human adult bone marrow. Approximately 25% of single cells of this phenotype when cultured for 14 days in media containing 1% bovine serum albumin without cytokines were able to form blast colonies within 14–21 days following addition of a combination of IL-3, IL-6, GM-CSF, G-CSF, and erythropoietin (EPO) (Terstappen *et al.,* 1991). These data together with data from a number of other studies performed with either murine or human BM HPCs isolated by FACS have shown convincingly that stimulation with multiple HGFs including SCF, IL-3, and IL-11 is required to initiate division of primitive HPCs (Bodine *et al.,* 1991; Musashi *et al.,* 1991; Muench *et al.,* 1992; Williams *et al.,* 1992; Mayani *et al.,* 1993).

Few studies have systematically analyzed the HGF requirements of primitive (pre-CFU) hematopoietic progenitors in stromal cell-free suspension cultures. Two notable exceptions are the studies of Muench *et al.* and Haylock *et al.,* where various combinations of IL-1, IL-3, IL-6, GM-CSF, G-CSF, and SCF were used to stimulate post-5-fluorouracil (5FU) BM and mobilized CD34$^+$ cells, respectively (Haylock *et al.,* 1992; Muench *et al.,* 1992). These studies, based on pre-CFU assays, substantiated the roles of IL-1, IL-6, and SCF as regulators of primitive HPCs and demonstrated that maximal generation of myeloid cells from CD34$^+$ cells required stimulation by both early acting cytokines and multiple CSFs. Moreover, the report by Haylock *et al.* (1992) introduced the concept of treating posttransplant cytopenia by infusion of *ex vivo* generated neutrophil and megakaryocyte precursors. This rationale has subsequently been adopted and tested clinically (Brugger *et al.,* 1995; Williams *et al.,* 1996). In later reports, SCF, G-CSF, and, to a lesser extent, IL-3 were found to be the most critical factors for expansion of CFU-GM and for generation of nucleated myeloid cells from mobilized blood CD34$^+$ cells (Makino *et al.,* 1997). Collectively, these studies suggest that primitive HPCs, defined as pre-CFU, can be induced to proliferate and differentiate *in vitro* provided appropriate combinations of HGFs are supplied.

Verfaillie (1993, 1994) has shown that maintenance of long-term, culture-initiating cells (LTC-IC) in human long-term marrow cultures can occur both in the absence of direct physical contact with marrow stromal cells and, moreover, when cells are cultured either in conditioned medium from marrow stromal cells or in a combination of recombinant cytokines added at the concentrations typically measured in the stromal-cell conditioned medium. What then for stromal cell–stem cell interactions in regulating responses of primitive HPCs: are such interactions redundant in cytokine-dependent, stroma-free culture systems? This issue was highlighted by Rowley *et al.* (1993), who showed that generation of nascent CFC from a 4HC-resistant subpopulation of $CD34^{+}$ $lineage^{-}$ cells required interaction with an irradiated allogeneic stromal layer in addition to IL-1, IL-3, IL-6, G-CSF, GM-CSF, and SCF. This study thus suggests that despite use of a combination of six HGFs, very primitive hematopoietic cell populations *in vitro* still require additional signals provided by the stromal layer to stimulate optimal production of nascent CFC.

This consideration formed the basis of a method for large-scale bioreactor expansion of human HPCs from BM mononuclear cells (MNCs) (described in detail in Chapter 13). Growth of HPCs in this culture system was dependent on the generation of an autologous stromal layer. Bioreactors were inoculated and then incubated for 1 day without perfusion to facilitate stromal layer attachment (Koller *et al.,* 1993a). In a subsequent study performed with cord blood MNCs, expansion of LTC-IC was only achieved when exogenous IL-3, IL-6, and SCF were added to the perfusion bioreactor culture (Koller *et al.,* 1993b), indicating that despite contact with stromal cells, primitive HPCs required additional cytokines to either allow survival or promote division. Stromal-based culture of HPCs may represent a system that is too complex for comprehensive analysis of the regulation of HPC development by CAMs and HGFs. Clearly, a well-defined culture system where the constituents of the culture media and the chemistry of the adhesive substrate are well characterized will permit more precise analysis of interactions.

We have developed a pre-CFU culture system in Terasaki plates that is ideal for monitoring survival, apoptosis, recruitment, proliferation, and differentiation of target cells (Haylock *et al.,* 1997a). Single cells are cultured in stroma-free, serum-free conditions, and their growth is absolutely dependent on provision of exogenous cytokines. Moreover, wells can also be coated with specific antibodies (as surrogate ligands), purified extracellular matrix proteins, or recombinant adhesive ligands to investigate how adhesion modulates growth of HPCs. This system has proven to be extremely versatile and has allowed us to investigate in detail the contribution of several cytokines to the survival and proliferation of pre-CFU.

III. CYTOKINES WHICH ACT ON PRIMITIVE HPC: FLT3-LIGAND AND THROMBOPOIETIN

The ligand for the flt3/flk2 receptor tyrosine kinase (Flt3-ligand, Flt3L) was recently demonstrated to interact synergistically with other cytokines to stimulate proliferation of $CD34^+$ cells (Lyman *et al.,* 1993; Broxmeyer *et al.,* 1995). The action of Flt3L is restricted to primitive HPCs and is able to increase the proportion of primitive HPCs that divide and, moreover, provide a potent proliferative stimulus to recruited cells (Haylock *et al.,* 1997b). These effects were observed on single $CD34^+CD38^-$ cells cultured under pre-CFU conditions: only 35% of $CD34^+CD38^-$ cells isolated from adult bone marrow divided during 14 days when cultured in a HGF combination of IL-3, IL-6, G-CSF, and SCF (36GS). The addition of Flt3L to this four-HGF combination resulted in a significant increase in the proportion of $CD34^+CD38^-$ cells recruited into division. This effect was evident from day 3 of culture and maintained to day 14 when approximately twofold more $CD34^+CD38^-$ cells divided when Flt3L was added to 36GS. The combination of 36GS and Flt3L (36GSF) was also highly effective at initiating cell division in a population of hierarchically more primitive progenitor cells that were $CD34^+CD38^-$, rhodamine 123 dull and resistant to 4HC.

The notion that Flt3L has potent effects on the growth and development of primitive HPCs is supported by the studies of Shah *et al.* (1996), who demonstrated similar results with single $CD34^+CD38^-$ cells cultured on preformed, irradiated-BM stromal cells supplemented with HGFs. Additional evidence for the action of Flt3L on primitive HPCs within the $CD34^+CD38^-$ compartment comes from a study reported by Petzer *et al.* (1996a) where Flt3L alone or in combination with other HGFs including SCF, IL-3, IL-6, G-CSF, and β-nerve growth factor was found to expand LTC-IC within the $CD34^+CD38^-$ cell population. In a further study, Petzer *et al.* (1996b) reported that aside from thrombopoietin, Flt3L was the only factor that on its own was able to stimulate expansion of LTC-IC numbers in serum-free, stroma-free cultures of $CD34^+CD38^-$ cells.

Additional studies have since described a key role for thrombopoietin (TPO) in maintaining survival and inducing division of primitive HPCs. In a report by Borge *et al.* (1997), TPO, when used as a single factor, was found to support survival of 22% of single $CD34^+CD38^-$ cells when cultured for 5 days under stroma-free, serum-deplete conditions, which was significantly better than with IL-3, SCF, or Flt3L. This effect of TPO was attributed to its ability to suppress apoptosis and is consistent with an earlier report by Ritchie *et al.* (1996), who described TPO as being able to suppress apoptosis and promote survival of the factor-dependent cell line MO7e. Additional evidence for the role of TPO as a survival factor for $CD34^+CD38^-$ adult BM cells was observed by Haylock *et al.,* who incubated cells in single cytokines for 14 days, and then cultured these cells with a combination of six cytokines to identify residual viable cells (Haylock *et al.,* 1997b). In these experiments, TPO was able to support survival of 21% of $CD34^+CD38^-$ cells (Fig. 4.1). In more recent studies, Ramsfjell *et al.* reported that thrombopoietin has synergy with SCF, Flt3L, or IL-3 to potently enhance clonogenic growth of $CD34^+CD38^-$ cells (Ramsfjell *et al.,* 1997). These data are supported by our own studies investigating the ability of different HGF combinations to induce division of single adult BM $CD34^+CD38^-$ cells.

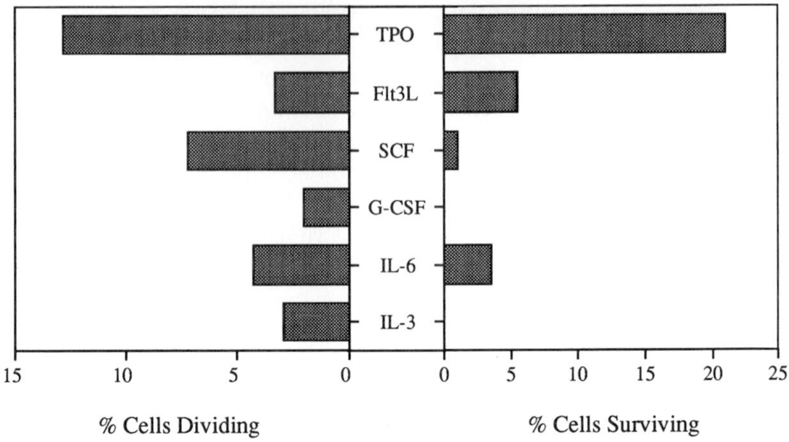

FIGURE 4.1 Recruitment and survival of single adult bone marrow CD34⁺CD38⁻ cells cultured with single HGFs. CD34⁺CD38⁻ cells were deposited by fluorescent activated cell sorting into each well of two Terasaki plates containing 10 μL of pre-CFU medium without HGFs or supplemented with either single HGFs (IL-3 and IL-6 at 10 ng/ml, G-CSF, SCF, Flt3L, and TPO at 100 ng/ml) or a combination of the six HGFs. Plates were incubated at 37°C, 5% CO_2, and the proportion of dividing cells on each day was recorded for 14 days. On day 14, 5 μL of media was removed from each well and then replaced with fresh media supplemented with six HGFs. Growth was monitored for a further 7 days and the proportion of cells undergoing division was used as an index of survival over the initial 14 days of culture. Under these conditions, 92% of single CD34⁺CD38⁻ cells divided within 14 days when cultured with six HGFs (IL-3 + IL-6 + G-CSF + SCF + Flt3L + TPO). The left-hand panel shows the proportion (%) of single CD34⁺CD38⁻ cells dividing after 7 days in the single HGFs (mean of two separate experiments). The right-hand panel indicates the proportion (%) of single CD34⁺CD38⁻ cells surviving after culture in an individual HGF for 14 days.

As depicted in Fig. 4.2, the addition of TPO to a combination of 36GSF results in division of 92% of these cells during 14 days of culture. Remarkably, when cultured in 36GSFT, 34% of dividing $CD34^+CD38^-$ cells were capable of at least 12 divisions, producing more than 4096 cells, and 2.6% of cells produced 15,000 cells or more. In subsequent cultures of 1000 $CD34^+CD38^-$ cells followed for 10 weeks, we observed a (64–271) $\times$ 10^6-fold increase in total nucleated cells and a 270,000-fold increase in $CD34^+$ cells. Despite the proportion of $CD34^+CD38^-$ cells decreasing, there was a 570-fold increase in $CD34^+CD38^-$ cells at 4 weeks.

Collectively our data, together with the data of Borge and Ramsfjell and the impaired hematopoiesis observed in the *c-mpl* (TPO receptor) knockout mouse (Alexander *et al.,* 1996), provide convincing evidence that TPO in addition to its well-documented role as a regulator of platelet mass also has a critical role in supporting the survival and proliferation of primitive HPCs. Additional evidence that signaling through *c-mpl* may have an important role in amplification

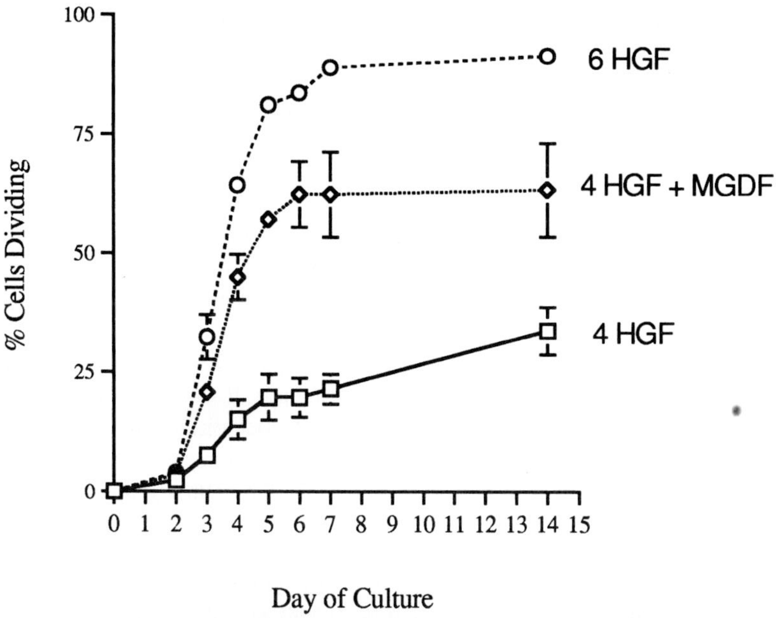

FIGURE 4.2 Recruitment of single adult human bone marrow CD34⁺CD38⁻ cells. Single CD34⁺CD38⁻ cells were deposited into wells of a Terasaki plate containing pre-CFU media supplemented with different combinations of HGFs including four HGFs (IL-3 + IL-6 + G-CSF + SCF), four HGFs + TPO, or six HGFs (IL-3 + IL-6 + G-CSF + SCF + Flt3L + TPO). Cells were examined each day for 7 days and then again after 14 days of culture. The figure shows the proportion (%) of single cells that had undergone division at each time point. Results are the mean (SEM) from experiments performed with 11 different sources of adult human bone marrow, with at least 100 CD34⁺CD38⁻ cells examined in each experiment.

of the primitive HSC pool comes from the work of Goncalves (1997), who enforced expression of *c-mpl* in murine stem cells and observed proliferation and differentiation of progenitors of several lineages but without preferential differentiation toward megakaryopoiesis. The notion of TPO acting as a potent stem cell stimulator and acting synergistically with other cytokines is also strongly supported by recent studies performed with umbilical cord blood CD34⁺ cells (Ohmizono *et al.*, 1997; Piacibello *et al.*, 1997). Piacibello reported that the combination of TPO and Flt3L in the absence of stromal cells resulted in a considerable amplification in the total number of HPCs and a concordant increase in the absolute numbers of cells exhibiting a CD34⁺CD38⁻ phenotype, characteristic of candidate HSCs in adult human BM. Moreover, cell generation was sustained for more than 6 months under these culture conditions. As cited by these authors, the clinical applications of this study are potentially very exciting: transplanting of adults would become possible if CB HPCs could be readily expanded. Exciting though these observations are, caution should nevertheless be exercised in interpreting these data since expansion of a primitive cell phenotype does not necessarily equate with expansion of transplantable HSC number as clearly demonstrated by Lansdorp and colleagues in the mouse (Rebel *et al.,* 1994).

IV. WHAT THEN FOR *EX VIVO* MANIPULATION OF HEMATOPOIETIC STEM CELLS?

Accumulating data indicate that it may now be possible with cytokines alone to induce division of most, if not all, cells with a putative HSC phenotype. The key growth factors required for this process include IL-3, IL-6, G-CSF, SCF, Flt3L, and TPO (36GSFT), though it remains to be determined which, if any, of these HGFs are redundant with respect to stimulating stem cell division. The use of expanded HSC populations for clinical applications raises a number of fundamental biological questions.

Perhaps the most significant question is whether it will be possible to expand the absolute numbers of stem cells in any given source of hematopoietic tissue while retaining their full biological potential and transplantability. This issue may be addressed in part by analysis of the outcome of transplantation studies in immunocompromised animal models such as the NOD-SCID mouse system (Pflumio *et al.,* 1996; Conneally *et al.,* 1997). What is required to induce division of the 36GSFT nonresponsive cells? What role, if any, does signaling through cell adhesion molecules play in modulating the response of primitive HPCs to HGFs? Will cooperation between particular CAMs together with cytokines promote or inhibit recruitment of stem cells? We now discuss these questions in part.

Of concern is the notion that exposure of HSCs to cytokines may alter the ability of primitive hematopoietic cells to engraft. Does culture of HSCs with HGFs or cytokines change their ability to either home to the bone marrow or remain within the BM microenvironment or both? The most comprehensive data supporting this concept come from Quesenberry and colleagues (Stewart *et al.,* 1993; Peters *et al.,* 1995; Ramshaw *et al.,* 1995; Peters *et al.,* 1996; Kittler *et al.,* 1997). In an initial study reported by Stewart *et al.* (1993), bone marrow collected from 5FU-treated mice showed markedly defective engraftment when transplanted into nonmyeloablated hosts. These findings were confirmed by Ramshaw *et al.* (1995) and Haylock *et al.* (1997b) and were somewhat surprising given that they and others (Lerner and Harrison, 1990; Harrison and Lerner, 1991) demonstrated that post 5FU BM competed effectively with normal marrow in irradiated-myeloablated hosts and contributed to long-term hematopoiesis. These results were in part explained by the different microenvironment of the two models. The irradiated hosts were more likely to have localized damage within the microenvironment, which would facilitate entry of HSCs, whereas nonablated mice would have an intact marrow stromal–endothelial cell interface.

The data also suggested that the defect might be related to stem cell cycle status, with actively cycling stem cells displaying impaired engraftment in nonmyeloablated hosts. A subsequent study showed that murine BM cells expanded in culture with IL-3, IL-6, IL-11, and SCF also have an engraftment defect (Peters *et al.,* 1995). Additional evidence that preincubation of HPCs with HGFs reduces seeding to the BM was provided by van der Loo (1995). In this study, a 2- to 3-h incubation with IL-3 or a combination of IL-3, IL-12, and SCF led to a substantial decrease in seeding of all hematopoietic subsets measured, in both the spleen and bone marrow. More recently, retrovirally transduced murine HPCs were found to be less effective than nontransduced HPCs in contributing to long-term stable engraftment (Kittler *et al.,* 1997). It was proposed that incubation of HPCs with IL-3, IL-6, IL-11, and SCF, although stimulating progression through S phase and facilitating retroviral transduction, also impaired homing or lodgment of these cells to the BM.

In contrast to these reports, the studies of Bodine *et al.* (1992) and Neben *et al.* (1994) suggest that exposure of primitive HPCs to growth factors may enhance engraftment. Bodine showed that *in vitro* exposure of marrow cells to IL-3, IL-6, and SCF for 6 days augmented *in vivo* repopulation in W/W^V animals several fold. Similarly, Neben reported that culture of murine bone marrow with the same combination of cytokines (IL-3, IL-6, IL-11, and SCF) for 6 days improved repopulating ability in myeloablated hosts by fourfold.

Collectively, these studies raise concerns about exposure of HPCs to cytokines and their subsequent capacity for engraftment. There is a need for more detailed study of this phenomenon and of the underlying mechanisms involved. Changes in the homing/lodgment and retention of primitive HPCs within the BM microenvironment could include alteration in the function of particular cell adhesion molecules as a direct consequence of signaling through cytokine receptors.

We have recently described activation of the two β_1-integrins VLA-4 and VLA-5 following exposure to HGFs and subsequent increased adhesion to fibronectin (Lévesque *et al.,* 1995, 1996) and VCAM-1 (Lévesque and Haylock, unpublished data; described in detail in the following section). In addition, Schofield (1997) reported that low levels of IL-3 produce significant reduction in adhesion of $CD34^+$ cells to the alternatively spliced IIICS region of fibronectin, which can be attributed to change in the activation state of VLA-4. They also reported that migration of $CD34^+$ cells was slightly enhanced by G-CSF, SCF, and, to a greater extent, IL-3. Despite these concerns about exposure of HPCs to HGFs adversely affecting transplantation, BM HSCs, which are mobilized into the circulation by HGFs and chemotherapy, give rise to the most rapid rate of hematopoietic reconstitution in humans. Can *ex vivo* manipulation of primitive HPCs be further improved by utilizing specific adhesive interactions between HPCs and adhesive ligands present within the BM microenvironment to facilitate transplantation? The remaining section of this review focuses on cell adhesion mediated by two classes of cell adhesion molecules expressed by HPCs, namely, integrins and sialomucins. These data indicate that both have the potential to modulate HPC survival and proliferation.

V. ADHESIVE INTERACTIONS AND THEIR ROLE IN HEMATOPOIETIC REGULATION

The development of the hematopoietic system during ontogeny is characterized by an ordered pattern of migration of aorta-associated primordial hematopoietic stem cells (Charbord *et al.,* 1996; Tavian *et al.,* 1996) that ultimately come to lodge in the bone marrow, a tissue which in the adult displays a unique capacity to support hematopoiesis. Although the precise mechanisms that govern this sequential migration from one hematopoietic microenvironment to the next remain largely unknown, a considerable body of evidence suggests that the precise localization of hematopoiesis both during ontogeny and in the adult involves developmentally regulated adhesive interactions between primitive HSCs and the various cellular and extracellular matrix (ECM) components that collectively constitute the stromal tissue of the hematopoietic organs. Growth and development of HSCs are restricted to the extravascular compartment of the bone marrow by a single layer of specialized vascular endothelial cells, which emerge from the capillary bed along the endosteum as a series of arborized sinuses. Thus entry of primitive HPCs into the bone marrow ("homing") requires specific recognition of the luminal surface of the vascular endothelium lining the venous sinuses and subsequent transmigration to gain entry into the BM. Exit of such cells from the marrow, as, for example, occurs during cytokine-induced mobilization of blood stem cells, is presumed to represent the reverse of these events (Chapter 6). A full description of the stromal tissue of the bone marrow is beyond the scope of this article and interested readers are referred to several recent reviews (Trentin, 1970; Lichtman, 1981; Bentley, 1982; Tavassoli *et al.,* 1983; Torok-storb, 1988).

Bone marrow HPCs encounter a diversity of adhesive ligands (cell surface molecules, membrane or ECM-anchored HGFs, ECM components) that are presented by the marrow stroma (Kincade *et al.,* 1989; Clark *et al.,* 1992; Long, 1992; Simmons *et al.,* 1994, 1997), which, in addition to modulating growth and survival, appear to contribute to the lodgment of HPCs within the BM microenvironment. This multiplicity of potential adhesive mechanisms presents a major technical challenge to the identification and characterization of those molecules responsible for mediating these various adhesive interactions between HPCs and the vascular and nonvascular (stromal) components of the BM. Normal HPCs express at their surface a variety of cell adhesion molecules representing at least five superfamilies of CAMs, including integrins, immunoglobulins (Ig), selectins, sialomucins, and the CD44 family of adhesion molecules (Kincade *et al.,* 1989; Clark *et al.,* 1992; Long, 1992; Simmons *et al.,* 1994, 1997). Why do primitive hematopoietic cells exhibit such a broad range of CAMs and what are their roles in engraftment, HSC mobilization, and HPC development? The answers to these questions are at present unknown, though a number of potential mechanisms have been postulated. Individual CAMs may perform specific roles in association with "homing" of HPCs whereas the contribution of other adhesion molecules may be more evident in the later lodgment phase and in the retention of these primitive cells within the bone marrow. Hematopoietic progenitor cells express a very similar cohort of CAMs to circulating leukocytes, raising the possibility that the trafficking of primitive HPCs between the BM and the peripheral circulation involves a multistep paradigm similar to that recently described for leukocyte emigration and lymphocyte recirculation (Springer, 1994).

There is abundant evidence in nonhematopoietic tissues that CAMs participate in a large variety of signal transduction events important not only for regulating cell adhesion and motility but also for cell growth (Hansen *et al.,* 1994; Symington, 1995; Fang *et al.,* 1996; Zhu *et al.,* 1996), apoptosis (Zhang *et al.,* 1995), and specific gene regulation (Yurochko *et al.,* 1992). Although the signaling function of CAMs has not been studied extensively in HPCs, extrapolation from other systems would suggest that signals generated locally by CAMs interact with cytokine signal transduction pathways to help control HPC growth and differentiation.

Given that the specific role and signaling function of the many CAMs expressed on HPCs remains poorly documented, we will present some recent investigations which suggest that there are significant functional interactions (cross-talk) between HGFs and CAM signaling pathways. We will initially focus on members of the β_1-integrin superfamily, in particular CD49d/CD29 (VLA-4 or $\alpha_4\beta_1$-integrin) (Miyake *et al.,* 1991; Williams *et al.,* 1991; Simmons *et al.,* 1992; Hirsch *et al.,* 1996), in mediating HPC–stromal cell adhesive interactions both *in vitro* and *in vivo.* Secondly, we will briefly examine the sialomucins, an additional family of CAMs expressed in abundance by primitive HPCs, about which relatively little is known regarding their potential contribution to HPC–stromal cell interactions.

VI. THE CROSS-TALK BETWEEN INTEGRINS AND CYTOKINE RECEPTORS

Integrins are a superfamily of receptors mediating adhesive cell to cell and cell to extracellular matrix interactions, which has been extremely well conserved during animal evolution (Hynes, 1992). Integrins are heterodimers of one α-chain noncovalently linked to a β-chain. Among 17 identified α-chains and eight β-chains, the specific pairing of a particular α-chain with a β-chain, defines a unique integrin receptor with a unique repertoire of ligand specificity, cell distribution, and regulation (Hynes, 1992). Integrins have been subclassified into eight different groups corresponding to the eight β-chains so far identified (Hynes, 1992). Hematopoietic cells express different integrins during the various stages of their development. For instance, β_3-integrins (CD61) are expressed on megakaryocytic progeny and mature platelets, integrin $\alpha_{IIb}\beta_3$ being the platelet fibrinogen receptor responsible for platelet aggregation. β_2-Integrins (CD18) are found on committed myeloid progenitors, mature granulocytes, monocytes, and lymphocytes but are absent from primitive HPCs. They are involved in various phenomena such as opsonization, phagocytosis, and lymphocyte activation. The integrin repertoire in primitive HPCs and HSCs is restricted to the β_1-integrins (CD29), namely, VLA-2 ($\alpha_2\beta_1$, CD49b/CD29), VLA-4 ($\alpha_4\beta_1$, CD49d/CD29), VLA-5 ($\alpha_5\beta_1$, CD49e/CD29), and VLA-6 ($\alpha_6\beta_1$, CD49e/CD29) (Simmons *et al.,* 1994; Lévesque *et al.,* 1995), which are receptors for collagen, the stromal cell surface protein VCAM-1, and the extracellular matrix proteins fibronectin and laminin, respectively.

The dominant role of β_1-integrins, particularly VLA-4 and VLA-5, as mediators of adhesive interactions between HPCs, the ECM, and cellular components of the hematopoietic microenvironment has recently been demonstrated *in vivo.* Administration of anti-VLA-4 antibodies to primates induces HPC mobilization (Papayannopoulou and Nakamoto, 1993), whereas injection of anti-β_1-integrin antibodies into mice has been shown to reduce medullar hematopoiesis with its relocation to spleen (Williams *et al.,* 1991). In addition, β_1-integrins also appear to fulfill an essential role in the establishment of hematopoiesis during ontogeny as shown by experiments performed using chimetic mice generated from β_1-integrin-deficient and β_1-integrin$^{+/+}$ murine strains. In these chimeras, β_1-integrin$^{-/-}$ HPCs were produced in the yolk sac but failed to home to the liver, spleen, or bone marrow, resulting in an absolute lack of β_1-integrin$^{-/-}$ blood cells in adult chimeras (Hirsch *et al.,* 1996). Similar studies performed using chimeras derived from α_4-integrin$^{-/-}$ stem cells resulted in a less severe phenotype as demonstrated by normal monocyte and NK development but impairment of T- and B-cell development during adult life (Arroyo *et al.,* 1996). Therefore, it is likely that the VLA-4/VCAM-1 interaction may not be the sole integrin-mediated interaction controlling homing of HPCs in BM but that other β_1-integrins such as VLA-5 (Traycoff *et al.,* 1997: van der Loo *et al.,* 1997) contribute to various stages of the hematopoietic system development, including its final location in the bone marrow. Is the role of integrins on primitive HPCs limited to that of a simple homing signal, or do they transduce specific integrin-mediated signals that contribute to HPC development? Experiments performed using β_1-integrin null chimeras do not provide insight into this question since, although β_1-integrin$^{-/-}$ HPCs derived from the yolk sac were unable to home to the hematopoietic organs and to sustain hematopoiesis, these cells were still able to differentiate normally *in vitro* (Hirsch *et al.,* 1996). These data therefore provide very convincing evidence that the lack of β_1-integrin expression does not impair the growth of clonogenic HPCs *in vitro,* but importantly, these observations do not exclude a direct role of β_1-integrins in development and differentiation of HPCs *in vivo.* This is suggested by the fact that integrins are transducing molecules in many cellular systems and that the signals they transduce regulate many aspects of cellular physiology (Schwartz *et al.,* 1995; Gumbinert, 1996). Moreover, the discovery of a cross-talk between integrins and cytokine receptors (Lévesque *et al.,* 1995, 1996) suggests a means by which integrins may contribute directly to the control of HPC development.

β_1-Integrins are expressed on quiescent HSCs as nonactive, nonligand-binding receptors. However, exposure of HSCs to cytokines which stimulate their growth, such as IL-3, GM-CSF, G-CSF, IL-6, and SCF (Lévesque *et al.,* 1995, 1996), activates transiently and very selectively both VLA-4 and VLA-5, promoting HSC attachment to fibronectin and VCAM-1 (Fig. 4.3). Interestingly, the affinity of the other β_1-integrins expressed by HPCs and HSCs, VLA-2 and VLA-6, is unaffected (Lévesque *et al.,* 1995), and we have yet to identify cytokines with the capacity to stimulate the adhesive properties of these integrins. This effect of cytokines on VLA-4 and VLA-5 function is mediated by an activation of these two integrins since their expression at the cell surface is not altered during the period of cytokine exposure (Lévesque *et al.,* 1995) and, additionally, is accompanied by the specific expression of β_1-integrin activation-related epitopes (Takamatsu *et al.,* 1998). Following these findings, we have proposed a two-step model with (1) integrin activation by "inside-out" signaling generated by cytokine receptor ligation and (2) generation of a secondary signal, or "outside-in" signal, resulting from the ligation of VLA-4 and VLA-5 to their adhesive ligands (Fig. 4.4) (Lévesque *et al.,* 1996).

The inside-out signaling, which leads to the activation of VLA-4 and VLA-5 in hematopoietic cells, is not fully understood. Although integrins themselves are not phosphorylated during cell activation by cytokines, some kinases undoubtedly participate in the inside-out signaling. This is illustrated by the facts that staurosporin, a nonspecific protein kinase inhibitor, abolishes β_1-integrin activation in hematopoietic cells and that genistein, a tyrosine kinase inhibitor, blocks VLA-4 and VLA-5 activation by stem cell factor (c-kit ligand) (Lévesque *et al.,* 1995). Although an abundant literature has shown the importance of the small GTPase rhoA to activate β_1-integrins in fibroblasts (Ridley and Hall, 1992; Barry and Crithley, 1994; Hotchin and Hall, 1995), there is no evidence of the involvement of this protein in inside-out signaling in hematopoietic cells. Instead, several reports, including our data, suggest a major role of phosphatidyl inositol 3′ (OH) kinase (PI3K). In mast cells, activation of VLA-4 and VLA-5 by SCF is blocked by a combination of protein kinase Cγ and PI3K inhibitors (Kinashi *et al.,* 1995; Vosseller *et al.,* 1997). In contrast, the activation of VLA-4 and VLA-5 by SCF, IL-3, and GM-CSF in cytokine-dependent myeloid cell lines (Takahira *et al.,* 1997) and CD34$^+$ HPCs (J. P. Lévesque, unpublished observations) can be completely blocked by the sole addition of PI3K inhibitors. The affinity of integrins is believed to be regulated by two NPXY motifs present in the cytoplasmic tail of several β-integrin chains including β_1-, β_2-, and β_3-

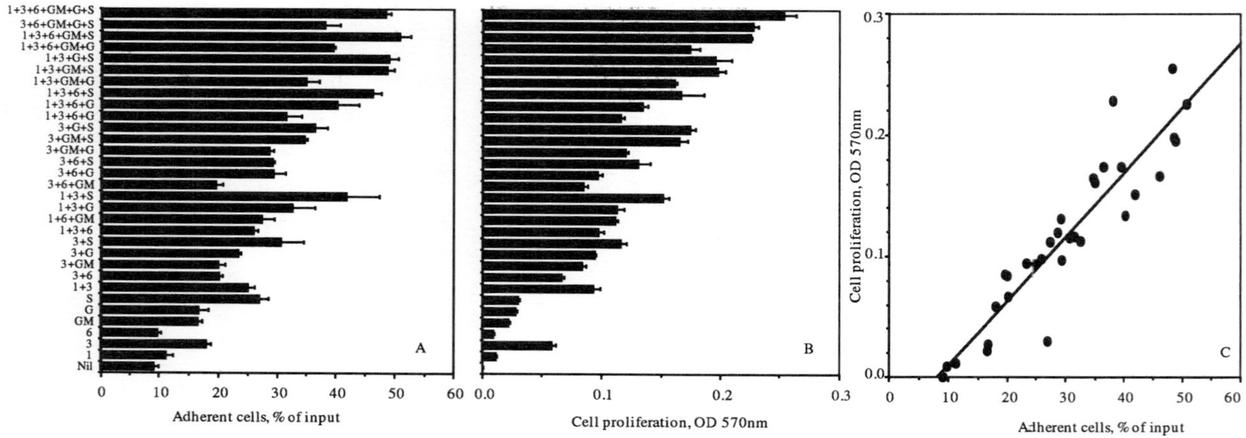

FIGURE 4.3 The cytokine activation of proliferation and VLA-4/VLA-5-mediated cell adhesion are correlated in human bone marrow CD34$^+$ hematopoietic progenitors. Panel A shows the percentage adhesion of human bone marrow CD34$^+$ HPC to immobilized fibronectin after a 30-min stimulation by 30 different combinations of IL-1β (1), IL-3 (3), IL-6 (6), GM-CSF (GM), G-CSF (G), and SCF (S). Panel B shows the proliferative response of the same cell population after 3 days of culture in the presence of the same combinations of cytokines as measured by MTT assay. The correlation between cell adhesion and cell proliferation in response to these 30 combinations of cytokines is plotted in panel C.

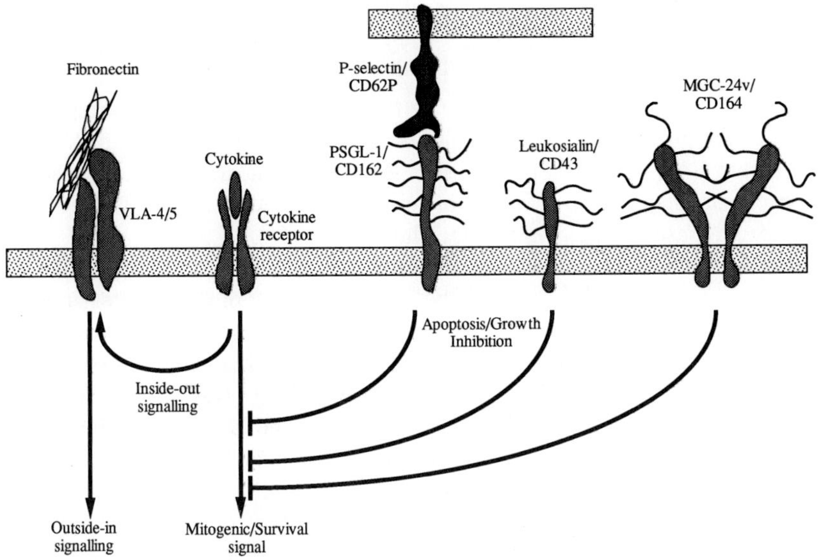

FIGURE 4.4 Model of cross-talk between adhesive and cytokine receptors in CD34$^+$ hematopoietic progenitors.

chains (O'Toole *et al.,* 1995). Therefore PI3K or phosphoinositide lipids, once phosphorylated by PI3K, may control the activity of cytoplasmic proteins binding to the NPXY motifs contained within the cytoplasmic tail of the β_1-integrin chains and lead to their activation. The α_4- and α_5-integrins, however, must also play a role in the specificity of inside-out signaling in HPCs since, as already noted, cytokines activate VLA-4 and VLA-5 but neither VLA-2 nor VLA-6.

Whether the outside-in signal generated by ligated VLA-4 and VLA-5 cooperates with or antagonizes the mitogenic signal generated by cytokine receptors is still ambiguous. Some authors have reported that enforcing cell attachment to fibronectin with the function-activating anti-human β_1-integrin monoclonal antibody 8A2 results in an inhibition of HPC cycling (Hurley *et al.,* 1995). These results should be interpreted with caution since they may not reflect the physiological situation. Indeed, unlike physiological inside-out activation of integrins, which is transient (Lévesque *et al.,* 1995), enforced adhesion by function-activating anti-human β_1-integrin monoclonal antibody 8A2 is stable and cannot be regulated during cell cycle (J. P. Lévesque, unpublished observations) and may therefore inhibit cytokinesis, resulting in cell death. Secondly, studies in different cell systems and in hematopoietic cell lines have clearly demonstrated a spatial and functional convergence of the transduction pathways activated by mitogenic cytokine receptors and by integrin-mediated outside-in signaling (Miyamoto *et al.,* 1995; Plopper, 1995; Assoian, 1997). For example, both classes of receptor result in the activation of similar transducers such as MAP kinases (Chen *et al.,* 1994; Morino *et al.,* 1995; Renshaw *et al.,* 1997), the proto-oncogene *vav* (Gotoh *et al.,* 1997), the adapter Grb2 (Schlaepfer *et al.,* 1994), the Na^+/H^+ antiporter (Schwartz *et al.,* 1991), cytoskeleton-regulating proteins such as paxillin and tensin (Bockholt and Burridge, 1993), and the focal adhesion kinase FAK (Takahira *et al.,* 1997). Moreover, integrin-mediated adhesion is a necessary step for the synthesis of some cyclins and the activity of cell cycle dependent kinases (Fang *et al.,* 1996; Zhu *et al.,* 1996).

How do β_1-integrins transduce these outside-in signals? Although they are devoid of any enzymatic activities, the cytoplasmic tail of the β_1 integrin chain is known to be able to interact in nonhematopoietic cells with a variety of proteins regulating the cytoskeleton such as talin and α-actinin (Otey *et al.,* 1993) and the focal adhesion tyrosine kinase $p125^{FAK}$ (Lewis and Schwartz, 1995). Talin via vinculin (Gilmore *et al.,* 1996) and α-actinin (Mimura and Asano, 1986) can bind to and regulate actin stress fiber formation via multiple binding sites. Once autophosphorylated and activated, the tyrosine kinase FAK can recruit, phosphorylate, and modulate the activity of a large number of proteins involved in the regulation of both cytoskeletal organization and proliferation. These regulatory proteins include paxillin (Tachibana *et al.,* 1995) and talin (Chen *et al.,* 1995), other tyrosine kinases such as $p60^{c\text{-}src}$ and $p59^{c\text{-}fyn}$ (Cobb *et al.,* 1994), adapters and docking proteins such as Grb2 (Schlaepfer *et al.,* 1994), and the proto-oncogene $p120^{c\text{-}cbl}$ (Manié *et al.,* 1997; Ojaniemi *et al.,* 1997). This cascade of interactions leads to the formation in fibroblasts and endothelial cells of large multimolecular complexes (Miyamoto *et al.*, 1995) localized within adhesion plaques or focal contacts, linking the extracellular matrix substratum to the nucleus via actin stress fibers (Wang *et al.,* 1993; Burridge and Chrzanowska-Wodnicka, 1996).

How is this model of integrin-mediated outside-in signaling relevant to that occurring in primitive human HPCs? $CD34^+$ HPCs and cytokine-dependent myeloid cells form adhesion plaques on ECM proteins in which integrins colocalize with cytoskeletal proteins talin, α-actinin, and vinculin (Lévesque and Simmons, 1999) in which specific tyrosine phosphorylations occur (Gotoh *et al.,* 1997). However, these changes are not associated with cytoskeletal reorganization such as the formation of radial F-actin stress fibers (Lévesque and Simmons, 1999).

VII. INTEGRINS AND HEMATOPOIETIC LEUKEMIAS: THE EXAMPLE OF CHRONIC MYELOGENOUS LEUKEMIA (CML)

In murine models, it has been well documented that integrins are involved in the metastasis of solid tumors (Fujita *et al.,* 1992; Hardan *et al.,* 1993). In hematological neoplasms, abnormal integrin function has also been reported in chronic myelogenous leukemia (CML). This malignant disorder of hematopoietic stem cells is characterized by a translocation between chromosomes 9 and 22 resulting in the formation of the Philadelphia (Ph) chromosome. Clinically, CML Ph^+ HPCs expand abnormally and leave the bone marrow prematurely to circulate in peripheral blood. It has been shown *in vitro* that circulating primitive Ph^+ HPCs have reduced attachment to bone marrow stromal cells (Gordon *et al.,* 1987) and fibronectin (Verfaillie *et al.,* 1992; Moore *et al.,* 1998). At the molecular level, the Philadelphia translocation gives rise to the juxtaposition of the coding sequence of *bcr* in frame with the 3′ end of the coding sequence of the *c-abl* proto-oncogene (De Klein *et al.,* 1982). The translation products of this gene are either the $p185^{bcr\text{-}abl}$ or $p210^{bcr\text{-}abl}$ oncoproteins, which are constitutively active tyrosine kinases (Lugo *et al.,* 1990). It has been shown *in vitro* that the bcr-abl oncoproteins target both inside-out and outside-in integrin-mediated signaling. For instance, using thermosensitive mutants of $p210^{bcr\text{-}abl}$ in cytokine-dependent myeloid cell lines, Bazzoni *et al.* (1996) have shown that the stimulation of bcr-abl tyrosine kinase induces a rapid and transient activation of both VLA-4 and VLA-5 followed by an inhibition of these two β_1-integrins, mimicking the effect of mitogenic cytokines (Lévesque *et al.,* 1995). Stable transduction of bcr-abl in cytokine-dependent myeloid cell lines also induces the tyrosine phosphorylation and activation of the focal adhesion kinase $p125^{FAK}$, rendering these cells cytokine-independent for proliferation (Gotoh *et al.,* 1995). A number of other cytoskeletal elements, including actin, appear to interact with or be phosphorylated by $p210^{bcr\text{-}abl}$ (Salgia *et al.,* 1995), contributing to the abnormal motility and cytoskeleton of CML blasts (Salgia *et al.,* 1997). Interestingly, interferon-α, which can restore normal hematopoiesis in 20% of CML cases, can reverse the abnormal β_1-integrin function (Bhatia *et al.,* 1994, 1995) and cytoskeleton (Salgia *et al.,* 1997) in CML blasts. Therefore, there is growing evidence that one of the causes of the abnormal behavior of CML cells is the profound alteration in the integration of integrin and cytokine-mediated signaling induced by the bcr-abl oncoprotein.

VIII. MUCIN-LIKE MOLECULES: AN EMERGING FAMILY OF CELL ADHESION MOLECULES EXPRESSED BY HEMATOPOIETIC PROGENITOR CELLS

Mucin-like molecules are a family of glycoproteins expressed by tissues of the hematopoietic and immune systems (Baumheter *et al.,* 1993; Shimizu and Shaw, 1993), which exhibit the common structural feature within their extracellular domains of regions rich in serine and threonine residues that act as sites for the attachment of O-linked glycans. The presence of this dense array of O-linked carbohydrates confers upon this family of molecules an extended threadlike structure providing an ideal platform for the presentation of multiple terminal carbohydrate residues (Fig. 4.4). Members of this family include GlyCAM-1 and CD34 (Lasky, 1992; Baumheter *et al.,* 1993) (ligands for L-selectin), P-selectin glycoprotein ligand-1 (PSGL-1/CD162) (Sako *et al.,* 1993), and MAdCAM-1, a counterreceptor on high endothelial venules in mucosal lymph nodes for L-selectin and integrin $\alpha_4\beta_7$ (Briskin *et al.,* 1993). Other CAMs which possess O-linked carbohydrate moieties include the ubiquitously expressed leukosialin (CD43), CD45RA, and MGC-24 (multiglycosylated core of 24 kDa) (Masuzawa *et al.,* 1992), which has recently been identified as CD164 (Zannettino *et al.,* 1998). Of these mucin-like molecules, five have been shown to be expressed by primitive human HPCs: CD34, CD43, CD45RA, CD162 (PSGL-1), and CD164 (MGC-24).

CD34 is a sialomucin which is selectively expressed at high levels by primitive HPCs (Krause *et al.,* 1996) and has long been speculated to be involved in mediating adhesive interactions between HPCs and bone marrow stromal cells. Widely distributed on vascular endothelium in the mouse and in humans (Fina *et al.,* 1990), CD34 has been shown to function as an additional endothelial ligand for L-selectin (CD62L) (Baumheter *et al.,* 1993). However, the CD34 glycoform expressed on HPCs does not appear to bind L-selectin, and the putative ligand(s) recognized by the hematopoietic form of the molecule has yet to be identified. In a recent study murine thymocytes expressing the human form of CD34 were shown to selectively bind to human, but not murine, marrow stromal cells *in vitro* (Healy *et al.,* 1995). While suggestive of a specific counter receptor for CD34 on marrow stromal tissue, the possibility that this is an indirect effect mediated by activation or up-regulation of other CAMs on the thymocytes cannot be excluded. In this regard, antibody engagement of CD34 on human hematopoietic cell lines has been shown to activate both β_1-integrin-mediated adhesion to fibronectin and VCAM-1 (Edwards and Sun, 1995) and β_2-integrin-mediated homotypic aggregation (Madjic *et al.,* 1994).

CD43 (leukosialin, sialophorin) is expressed ubiquitously on mature leukocytes with the exception of a population of B cells (Shelley *et al.,* 1986) and is also expressed at high copy number by candidate human HSCs defined by the $CD34^+$ $Thy\text{-}1^{Lo}$ $Lineage^-$ phenotype (Bazil *et al.,* 1996). Despite reports demonstrating that CD43 can transduce an activating signal to peripheral T cells, monocytes, and neutrophils (Pallant *et al.,* 1989; Piller *et al.,* 1991), the function of this molecule remains largely unknown. Nevertheless, a role in cell adhesion is suggested by reports demonstrating that cross-linking of CD43 on monocytes and neutrophils results in integrin-mediated homotypic aggregation (Nong *et al.,* 1989; Kuijpers *et al.,* 1992). CD43 has also been suggested to function as an alternate receptor for intercellular adhesion molecule-1 (ICAM-1; CD54) (Rosenstein *et al.,* 1991), although this has not been confirmed in subsequent reports (Ardman *et al.,* 1992; de Fougerolles *et al.,* 1993). Evidence for a proadhesive role of CD43 on HPCs has not been demonstrated, although antibody-mediated cross-linking of CD43 on $CD34^+$ HPCs results in the induction of apoptosis (see later) Bazil *et al.,* 1995, 1996).

CD 162 (PSGL-1) is a 240-kDa disulfide-linked homodimeric sialomucin which was shown to function as a specific counterreceptor on myeloid cells for P-selectin (CD62P) (Sako *et al.,* 1993). In a previous study, human CD34$^+$ cells including primitive (pre-CFU) and clonogenic progenitors were also shown to exhibit cation-dependent binding to P-selectin (Zannettino *et al.,* 1995). In accord with this adhesive property, CD34$^+$ cells were shown to express PSGL-1 as demonstrated initially by RT-PCR analysis (Zannettino *et al.,* 1995) and subsequently confirmed by others using flow cytometric analyses with the availability of anti-PSGL-1 antibody reagents (Laszik *et al.,* 1996; Tracey and Rinder, 1996). Our own recent studies demonstrate that PSGL-1 on human HPCs is the sole receptor for CD62P.

MGC-24 (CD164) has only recently been identified and is expressed both by HPCs and marrow stromal cells (Zannettino *et al.,* 1998). Antibodies to CD164 bind to a minor subpopulation of human BM mononuclear cells that contain the majority of CD34$^+$ HPCs and pre-CFU, and from representative cell lines antibodies to CD164 immunoprecipitate a 160-kDa antigen comprising two 80-kDa monomers. An adhesive function for CD164 (MGC-24) is demonstrated by the capacity of antibodies directed against certain epitopes of MGC-24 to partially block the adhesion of CD34$^+$ HPCs to allogeneic marrow stromal cells. The nature of the stromal cell ligand for CD164 is currently unknown.

IX. OUTSIDE-IN SIGNALING THROUGH MUCIN-LIKE MOLECULES ON HEMATOPOIETIC PROGENITOR CELLS: NEGATIVE REGULATORS OF HEMATOPOIESIS?

A series of recent reports suggest that mucin-like molecules can also directly mediate outside-in signaling in HPCs. Constitutive expression of the full-length splice variant of CD34 in murine M1 cells was shown to inhibit cytokine-induced differentiation (Fackler *et al.,* 1995). Antibody cross-linking of CD43 resulted in the induction of apoptosis in human CD34$^+$ HPCs.

Multipotential progenitors (CFU-GEMM) and erythroid progenitors (BFU-E) were substantially more sensitive to CD43-induced apoptosis than myeloid progenitors (CFU-GM) (Bazil *et al.,* 1995). Although expressing CD43 at high level, candidate HSCs were apparently refractory to the apoptosis-inducing effects of the anti-CD43 antibody (Bazil *et al.,* 1996). Similarly, recent studies from this laboratory demonstrate that adhesion of CD34$^+$ cells to P-selectin markedly inhibits hematopoiesis in stromal cell-free, cytokine-supported culture. This response appears to be due both to the induction of apoptosis in a subpopulation of primitive CD34$^+$CD38$^-$ cells and to a slowing of the proliferation rate of the more mature CD34$^+$CD38$^+$ subpopulation (Lévesque *et al.,* manuscript submitted). These studies therefore imply a key role for the PSGL-1 sialomucin as a signaling molecule on primitive human HPCs. Addition of anti-CD164 (MGC-24) antibody to human HPC clonogenic assays was found to induce a dose-dependent inhibition of both CFU-GM and BFU-E colony formation. Similarly, addition of antibody to CD164 at optimal concentrations markedly suppressed hematopoiesis in cytokine-dependent, stroma-free suspension culture of $CD34^+$ cells (Zannettino *et al.,* 1998). The mechanism responsible for the inhibition of hematopoiesis by CD164 antibody has yet to be determined.

Thus mucin-like molecules on HPCs appear to be negative regulators of hematopoiesis (Fig. 4.4). The question of whether the growth inhibitory property of these molecules occurs *in vivo* requires further study. In the case of PSGL-1, some clues are provided by the phenotype of mice deficient in CD62P (P-selectin) and CD62E (E-selectin). $CD62P^{-/-}$ mice demonstrate increased numbers of megakaryocyte progenitors in the bone marrow (Banu *et al.,* 1995). Moreover, mice doubly deficient in CD62P and CD62E exhibit an extreme leukocytosis and abnormally elevated HPC numbers (Frenette *et al.,* 1996). A similar phenotype was recently reported for mice deficient in $\alpha(1,3)$-fucosyltransferase Fuc-TVII, an enzyme required for selectin ligand biosynthesis (Maly *et al.,* 1996). Given the susceptibility of the $CD62P/E^{-/-}$ mice to opportunistic bacterial infections, the raised leukocyte counts may in part be driven by infection. However, even neonatal animals exhibited a leukocytosis, suggesting that this is a direct consequence of the absence of both selectins. This raises the intriguing possibility that the unanticipated role of the two endothelial selectins in regulating leukocyte homeostasis may be due to negative regulation of HPCs by interactions between selectins and their ligands.

One possible role of these mucin-mediated interactions *in vivo* might be as a powerful negative regulatory mechanism to dampen excessive expansion of HPCs. Alternatively, mucin-like molecules may deliver a positive proliferative stimulus to HPCs if combined with an appropriate, as yet unknown, stimulus. This hypothesis is suggested by studies with CD43 antibody, which provides a costimulatory proliferative stimulus to T cells (Park *et al.,* 1991), in contrast to its ability to induce apoptosis of HPCs. Thus in the absence of this additional stimulus, CD43-mediated signaling in HPCs could result in the elimination of these improperly activated cells through the induction of apoptosis. Other scenarios are clearly possible but for the present, additional studies will be required to elucidate both the physiological significance of these observations and the signal transduction pathways involved.

X. SUMMARY

In this article we have attempted to review the contribution of hematopoietic growth factors and cell adhesion molecules to the regulation of the growth and development of primitive hematopoietic progenitors. We have shown that such cells exhibit responses to a wide range of HGFs but that one of the key features that distinguishes hierarchically primitive HPCs from their lineage-restricted clonogenic progeny is the apparent obligatory requirement for multiple HGFs in order to elicit entry into cell cycle and proliferation. Although the molecular mechanisms underlying this multifactor requirement remain to be fully elucidated, the necessity for multiple cytokines is of great importance. *Ex vivo* manipulations of hematopoietic tissues which seek, for example, to expand the number of transplantable stem cells in tissues such as umbilical cord blood or induce division of such cells as a prerequisite for retroviral-mediated gene therapy would require multiple HGFs for proliferation of primitive HPCs.

An additional feature of primitive hematopoietic cells is their expression of an extensive array of CAMs with diverse ligand specificity. We have chosen to focus on the function of two families only, the integrins and the sialomucins, but the important contribution of other CAMs such as CD44 and CD31 must also be considered (Kincade *et al.,* 1989; Watt *et al.,* 1993). Why primitive hematopoietic cells exhibit such a repertoire of CAMs is essentially unknown, but answers to this question are likely to emerge from recent burgeoning evidence documenting the participation of CAMs in an increasingly large variety of signal transduction events. Thus, in addition to their likely roles as mediators of physical interactions between HPCs and stromal elements, the diverse CAM–ligand interactions in part reviewed here may also participate more directly in controlling growth and development of hematopoietic cells.

Given the large array of cytokine receptors on primitive HPCs dedicated to signal transduction, why is there an additional need for cross-talk between cytokine and CAM signaling? The control of cell adhesion by cytokines, as we have described, represents an obvious example, but others include the possibility of integrating aspects of cell growth and differentiation with physical cell adhesion events and promoting synergy between growth factor-triggered and cell adhesion-mediated signals. Much of the data in support of this derives from studies performed in nonhematopoietic tissues. A major objective for the future will therefore be to understand the molecular basis for the coupling between CAM and growth factor receptor-mediated signaling in primitive human hematopoietic cells. From a practical standpoint, however, it is also important to appreciate the implications of this functional overlap between CAM and cytokine receptor families. In doing so, this may yield significant improvements in strategies for the *ex vivo* growth of hematopoietic tissues, particularly in regard to the control of cycling and maintenance of primitive hematopoietic cells, including stem cells.

REFERENCES

Alexander, W. S., Roberts, A. W., Nicola, N. A., *et al.* (1996). Deficiencies in progenitor cells of multiple hematopoietic lineages and defective megakaryocytopoiesis in mice lacking the thrombopoietic receptor c-Mpl. *Blood* **87,** 2162–2170.

Ardman, B., Sikorski, M. A., and Staunton, T. A. (1992). CD43 interferes with T-lymphocyte adhesion. *Proc. Natl. Acad. Sci. USA* **89,** 5001–5005.

Arroyo, A. G., Yang, J. T., Rayburn, H., *et al.* (1996). Differential requirements for alpha4 integrins during fetal and adult hematopoiesis. *Cell* **85,** 997–1008.

Assoian, R. (1997). Anchorage-dependent cell cycle progression. *J. Cell Biol.* **136,** 1–4.

Banu, N., Wang, J. F., Deng, B., *et al.* (1995). Modulation of megakaryocytopoiesis by thrombopoietin: The c-Mpl ligand. *Blood* **86,** 1331–1338.

Barry, S. T., and Crithley, D. R. (1994). The rhoA-dependent assembly of focal adhesions in Swiss 3T3 is associated with increased tyrosine phosphorylation and the recruitment of both pp125FAK and protein kinase C-d to focal adhesions. *J. Cell Sci.* **107,** 2033–2045.

Bartelmez, S. H., Bradley, T. R., Bertoncello, I., *et al.* (1989). Interleukin 1 plus interleukin 3 plus colony stimulating factor are essential for clonal proliferation of primitive myeloid bone marrow cells. *Exp. Hematol.* **17,** 240–245.

Baumheter, S., Singer, M. S., Henzel, W., *et al.* (1993). Binding of L-selectin to the vascular sialomucin CD34. *Science* **262,** 436–438.

Bazil, V., Brandt, J., Chen, S., *et al.* (1996). A monoclonal antibody recognizing CD43 (leukosialin) initiates apoptosis of human hematopoietic progenitor cells but not stem cells. *Blood* **87,** 1272–1281.

Bazil, V., Brandt, J., Tsukamoto, A., *et al.* (1995). Apoptosis of human hematopoietic progenitor cells induced by crosslinking of surface CD43, the major sialoglycoprotein of leukocytes. *Blood* **86,** 502–511.

Bazzoni, G., Carlesso, N., Griffin, J. D., *et al.* (1996). Bcr/Abl expression stimulates integrin function in hemopoietic cell lines. *J. Clin. Invest.* **98,** 521–528.

Begg, S. K., Radley, J. M., Pollard, J. W., *et al.* (1993). Delayed hematopoietic development in osteopetrotic (op/op) mice. *J. Exp. Med.* **177,** 237–242.

Bentley, S. A. (1982). Bone marrow connective tissue and the haemopoietic microenvironment. *Br. J. Haematol.* **50,** 1–6.

Bhatia, R., McGlave, P. B., and Verfaillie, C. M. (1995). Treatment of marrow stroma with interferon-alpha restores normal beta 1 integrin-dependent adhesion of chronic myelogenous leukemia hematopoietic progenitors. Role of MIP-1 alpha. *J. Clin. Invest.* **96,** 931–939.

Bhatia, R., Wayner, E. A., McGlave, P. B., *et al.* (1994). Interferon-alpha restores normal adhesion of chronic myelogenous leukemia hematopoietic progenitors to bone marrow stroma by correcting impaired beta 1 integrin receptor function. *J. Clin. Invest.* **94,** 384–391.

Bockholt, S. M., and Burridge, K. (1993). Cell spreading on extracellular matrix proteins induces tyrosine phosphorylation of tensin. *J. Biol. Chem.* **268,** 14565–14567.

Bodine, D. M., Crosier, P. S., and Clark, S. C. (1991). Effects of hematopoietic growth factors on the survival of primitive stem cells in liquid suspension culture. *Blood* **78,** 914–920.

Bodine, D. M., Orlic, D., Birkett, N. C., *et al.* (1992). Stem cell factor increases colony-forming unit-spleen number *in vitro* in synergy with interleukin-6, and *in vivo* in Sl/sld mice as a single factor. *Blood* **79,** 913–919.

Borge, O. J., Ramsfjell, V., Cui, L., *et al.* (1997). Ability of early acting cytokines to directly promote survival and suppress apoptosis of human primitive $CD34^+CD38^-$ bone marrow cells with multilineage potential at the single-cell level: Key role of thrombopoietin. *Blood* **90,** 2282–2292.

Bradley, T. R., and Metcalf, D. (1966). The growth of mouse bone marrow cells *in vitro. Aust. J. Exp. Biol. Med. Sci.* **44,** 287–299.

Briskin, M. J., McEvoy, L. M., and Butcher, E. C. (1993). MAdCAM-1 has homology to immunoglobulin and mucin-like adhesion receptors and to IgA1. *Nature* **363,** 461–464.

Broxmeyer, H. E., Lu, L., Cooper, S., *et al.* (1995). Flt3 ligand stimulates/costimulates the growth of myeloid stem/progenitor cells. *Exp. Hematol.* **23,** 1121–9.

Brugger, W., Heimfeld, S., Berenson, R. J., *et al.* (1995). Reconstitution of hematopoiesis after high-dose chemotherapy by autologous progenitor cells generated *ex vivo. N. Engl. J. Med.* **333,** 283–287.

Burridge, K., and Chrzanowska-Wodnicka, M. (1996). Focal adhesions, contractility and signaling. *Annu. Rev. Cell. Dev. Biol.* **12,** 463–513.

Charbord, P., Tavian, M., Humeau, L., *et al.* (1996). Early ontogeny of the human marrow from long bones: An immunohistochemical study of hematopoiesis and its micro-environment. *Blood* **87,** 4109–4119.

Chen, H. C., Appeddu, P. A., Parsons, J. T., *et al.* (1995). Interaction of focal adhesion kinase with cytoskeletal protein talin. *J. Biol. Chem.* **270,** 16995–16999.

Chen, Q., Kinch, M. S., Lin, T. H., *et al.* (1994). Integrin-mediated cell adhesion activates mitogen-activated protein kinases. *J. Biol. Chem.* **269,** 26602–26605.

Civin, C. I., Strauss, L. C., Brovall, C., *et al.* (1984). Antigenic analysis of hematopoiesis. III. A hematopoietic progenitor cell surface antigen defined by a monoclonal antibody raised against KG-1a cells. *J. Immunol.* **133,** 157–165.

Clark, B. R., Gallager, J. T., and Dexter, T. M. (1992). Cell adhesion in the stromal regulation of adhesion. *Baillère's Clin. Hematol.* **10,** 21–53.

Clark, S. C., and Kamen, R. (1987). The human colony stimulating factors. *Science* **236,** 1229–1237.

Cobb, B. S., Schaller, M. D., Leu, T. H., *et al.* (1994). Stable association of pp60src and pp59fyn with the focal adhesion-associated protein tyrosine kinase, pp125FAK. *Mol. Cell. Biol.* **14,** 147–155.

Conneally, E., Cashman, J., Petzer, A., *et al.* (1997). Expansion *in vitro* of transplantable human cord blood stem cells demonstrated using a quantitative assay of their lympho-myeloid repopulating ability in nonobese diabetic-scid/scid mice. *Proc. Natl. Acad. Sci. USA* **94,** 9836–9841.

de Fougerolles, A. R., Klickstein, L. B., and Springer, T. A. (1993). Cloning and expression of intercellular adhesion molecule 3 reveals strong homology to other immunoglobulin family counter-receptors for lymphocyte function-associated antigen 1. *J. Exp. Med.* **177,** 1187–1192.

De Klein, A., Van Kessel, A. G., Grosveld, G., *et al.* (1982). A cellular oncogene is translocated to the Philadelphia chromosome in chronic myelocytic leukemia. *Nature* **300,** 765–767.

Edwards, J. R., and Sun, M. F. (1995). CD34 binding enhances β_1 integrin mediated adhesion of protein kinase C (PKC) activated KG1 and is blocked by G-CSF. *Blood* **86,** (suppl 1), 306a (abstract).

Ema, H., Suda, T., Miura, Y., *et al.* (1990). Colony formation of clone-sorted human hematopoietic progenitors. *Blood* **75,** 1941–1946.

Fackler, M., Krause, D., Smith, O., *et al.* (1995). Full length but not truncated CD34 inhibits hematopoietic cell differentaition of M1 cells. *Blood* **85,** 3040–3047.

Falk, L. A., and Vogel, S. N. (1988). Granulocyte-macrophage colony stimulating factor (GM-CSF) and macrophage colony stimulating factor (CSF-1) synergise to stimulate progenitor cells with high proliferative potential. *J. Leuk. Biol.* **44,** 455–464.

Fang, F., Orend, G., Watanabe, N., *et al.* (1996). Dependence of cyclinE-cdk2 kinase activity on cell anchorage. *Science* **271,** 499–502.

Fina, L., Molgaard, H. V., Robertson, D., *et al.* (1990). Expression of the CD34 gene in vascular endothelial cells. *Blood* **75,** 2417–2426.

Frenette, P. S., Mayadas, T. N., Rayburn, H., *et al.* (1996). Susceptibility to infection and altered hematopoiesis in mice deficient in both P- and E-selectins. *Cell* **84,** 563–574.

Fujita, S., Suzuki, H., Kinoshita, M., *et al.* (1992). Inhibition of cell attachment, invasion and metastasis of human carcinoma cells by anti-integrin beta 1 subunit antibody. *Jpn. J. Cancer Res.* **83,** 1317–1326.

Gasson, J. C., Weisbart, R. H., Kaufman, S. E., *et al.* (1984). Purified human granulocyte-macrophage colony-stimulating factor: Direct action on neutrophils. *Science* **226,** 1339–1342.

Gilmore, A. P., and Burridge, K. (1996). Regulation of vinculin binding to talin and actin by phosphatidyl-inositol-4-5-bisphosphate. *Nature* **381,** 531–535.

Goncalves, F., Lacout, C., Villeval, J. L., *et al.* (1997). Thrombopoietin does not induce lineage-restricted commitment of Mpl-r expressing pluripotent progenitors but permits their complete erythroid and megakaryocytic differentiation. *Blood* **89,** 3544–3553.

Gordon, M., Dowding, C., Riley, G., *et al.* (1987). Altered adhesive interactions with marrow stroma of hematopoietic progenitor cells in chronic myeloid leukaemia. *Nature* **328,** 342–344.

Gotoh, A., Miyazawa, K., Ohyashiki, K., *et al.* (1995). Tyrosine phosphorylation and activation of focal adhesion kinase (p125FAK) by Bcr-Abl oncoprotein. *Exp. Hematol.* **23,** 1153–1159.

Gotoh, A., Takahira, H., Gaehlen, R. L., *et al.* (1997). Cross-linking of integrins induces tyrosine phosphorylation of the proto-oncogene product vav and the protein tyrosine kinase syk in human factor-dependent myeloid cells. *Cell Growth Differ.* **8,** 721–729.

Gumbiner, B. M. (1996). The molecular basis of tissue architecture and morphogenesis. *Cell* **84,** 345–357.

Hansen, L. K., Mooney, D. J., Vacanti, J. P., *et al.* (1994). Integrin binding and cell spreading on extracellular matrix act at different points in the cell cycle to promote hepatocyte growth. *Mol. Biol. Cell* **5,** 967–975.

Hardan, I., Weiss, L., Hershkoviz, R., *et al.* (1993). Inhibition of metastatic cell colonization in murine lungs and tumor-induced morbidity by non-peptidic Arg-Gly-Asp mimetics. *Int. J. Cancer* **55,** 1023–1028.

Harrison, D. E., and Lerner, C. P. (1991). Most primitive hematopoietic stem cells are stimulated to cycle rapidly after treatment with 5-fluorouracil. *Blood* **78,** 1237–1240.

Haylock, D. N., Horsfall, M. J., Dowse, T. L., *et al.* (1997a). Increased recruitment of hematopoietic progenitor cells underlies the *ex vivo* expansion potential of FLT3 ligand. *Blood* **90,** 2260–2272.

Haylock, D. N., Nuitta, S., Lévesque, J. P., *et al.* (1997b). Megakaryocyte growth and development factor (MGDF) preferentially promotes survival and is essential for maximal recruitment of adult bone marrow $CD34^+CD38^-$ cells into division. *Blood* **90,** 474a.

Haylock, D. N., To, L. B., Dowse, T. L., *et al.* (1992). *Ex vivo* expansion and maturation of peripheral blood $CD34^+$ cells into the myeloid lineage. *Blood* **80,** 1405–1412.

Healy, L., May, G., Gale, K., *et al.* (1995). The stem cell antigen CD34 functions as a regulator of hemopoietic cell adhesion. *Proc. Natl. Acad. Sci. USA* **92,** 12240–12244.

Hirsch, E., Iglesias, A., Potocnik, A. J., *et al.* (1996). Impaired migration but not differentiation of hemopoietic stem cells in the absence of β_1 integrins. *Nature* **380,** 171–175.

Hotchin, N. A., and Hall, A. (1995). The assembly of integrin adhesion complexes requires both extracellular matrix and intracellular rho/rac GTPases. *J. Cell Biol.* **131,** 1857–1865.

Hurley, R. W., McCarthy, J. B., and Verfaillie, C. M. (1995). Direct adhesion to bone marrow stroma via fibronectin receptors inhibits hematopoietic progenitor proliferation. *J. Clin. Invest.* **96,** 511–519.

Hynes, R. O. (1992). Integrins: Versatility, modulation, and signaling in cell adhesion. *Cell* **69,** 11–25.

Ichikawa, Y., Pluznik, D. H., and Sachs, L. (1966). *In vitro* control of the development of macrophage and granulocyte colonies. *Proc. Natl. Acad. Sci. USA* **56,** 488–495.

Iscove, N. N., Shaw, A. R., and Keller, G. (1989). Net increase of pluripotential hematopoietic precursors in suspension culture to IL-1 and IL-3. *J. Immunol.* **142,** 2332–2337.

Kinashi, T., Escobedo, J. A., Williams, L. T., *et al.* (1995). Receptor tyrosine kinase stimulates cell-matrix adhesion by phosphatidylinositol 3 kinase and phospholipase C-γ_1 pathways. *Blood* **86,** 2086–2090.

Kincade, P. W., Lee, G., Pietrangeli, C. E., *et al.* (1989). Cells and molecules that regulate B-lymphopoiesis in the bone marrow. *Annu. Rev. Immunol.* **7,** 111–143.

Kittler, E. L. W., Peters, S. O., Crittenden, R. B., *et al.* (1997). Cytokine-facilitated transduction leads to low-level engraftment in non-ablated hosts. *Blood* **90,** 865–872.

Koller, M. R., Bender, J. G., Miller, W. M., *et al.* (1993a). Expansion of primitive human hematopoietic progenitors in a perfusion bioreactor system with IL-3, IL-6, and stem cell factor. *Biotechnology* **11,** 358–363.

Koller, M. R., Emerson, S. G., and Palsson, B. O. (1993b). Large-scale expansion of human stem cell and progenitor cells from bone marrow mononuclear cells in continuous perfusion cultures. *Blood* **82,** 378–384.

Krause, D. S., Fackler, M. J., Civin, C. I., *et al.* (1996). CD34: Structure, biology and clinical utility. *Blood* **87,** 1–13.

Kuijpers, T. W., Hoogerwerf, M., Kuijpers, K. C., *et al.* (1992). Cross-linking of sialophorin (CD43) induces neutrophil aggregation in a CD18-dependent and a CD18-independent way. *J. Immunol.* **149,** 998–1003.

Lansdorp, P. M., and Dragowska, W. (1992). Long-term erythropoiesis from constant numbers of $CD34^+$ cells in serum-free cultures initiated with highly purified progenitor cells from human bone marrow. *J. Exp. Med.* **175,** 1501–1509.

Lasky, L. A. (1992). Selectins: Interpreters of cell-specific carbohydrate information during inflammation. *Science* **258,** 964–969.

Laszik, Z., Jansen, P. J., Cumming, R. D., *et al.* (1996). P-Selectin glycoprotein ligand-1 is broadly expressed in cells of myeloid, lymphoid and dendritic lineage and in some non- hematopoietic cells. *Blood* **88,** 3010–3021.

Lerner, C., and Harrison, D. E. (1990). 5-Fluorouracil spares hemopoietic stem cells responsible for long-term repopulation. *Exp. Hematol.* **18,** 114–118.

Lévesque, J. P., Haylock, D., and Simmons, P. J. (1996). Cytokine regulation of proliferation and cell adhesion are correlated events in human $CD34^+$ hemopoietic progenitors. *Blood* **88,** 1168–1176.

Lévesque, J. P., Leavesley, D. I., Niutta, S, *et al.* (1995). Cytokines increase human hemopoietic cell adhesiveness by activation of very late antigen (VLA)-4 and (VLA)-5 integrins. *J. Exp. Med.* **181,** 1805–1815.

Lévesque, J. P., and Simmons, P. J. (1999). Cytoskeleton and integrin-mediated adhesion signaling in human $CD34^+$ hemopoietic progenitor cells. *Exp. Hematol.* **27,** in press.

Lewis, J. M., and Schwartz, M. A. (1995). Mapping *in vivo* associations of cytoplasmic proteins with integrin beta 1 cytoplasmic domain mutants. *Mol. Biol. Cell* **6,** 151–160.

Li, C. L., and Johnson, G. R. (1992). Long-term hemopoietic repopulation by $Thy\text{-}1^{lo}$, lin^-,$Ly6A/E^+$ cells. *Exp. Hematol.* **20,** 1309–1315.

Lichtman, M. A. (1981). The ultrastructure of the hemopoietic environment of the marrow: A review. *Exp. Hematol.* **9,** 391–410.

Lieschke, G. J., Grail, D., Hodgson, G., *et al.* (1994). Mice lacking granulocyte colony-stimulating factor have chronic neutropenia, granulocyte and macrophage progenitor cell deficiency, and impaired neutrophil mobilisation. *Blood* **84,** 1737–1746.

Long, M. W. (1992). Blood cell cytoadhesion molecules. *Exp. Hematol.* **20,** 288–301.

Lugo, T. G., Pendergast, A. M., Muller, A. J., *et al.* (1990). Tyrosine kinase activity and transformation potency of bcr-abl oncogene product. *Science* **247,** 1079–1082.

Lyman, S. D., James, L., Vanden Bos, T., *et al.* (1993). Molecular cloning of a ligand for the flt3/flk2 tyrosine kinase receptor: A proliferative factor for primitive hematopoietic cells. *Cell* **74,** 1157–1167.

Madjic, O., Stockl, J., Pickl, W., *et al.* (1994). Signaling and induction of enhanced adhesiveness via the hematopoietic progenitor cell surface molecule CD34. *Blood* **83,** 1226–1234.

Makino, S., Haylock, D. N., Dowse, T. L., *et al.* (1997). *Ex-vivo* culture of peripheral blood $CD34^+$ cells: Effects of hemopoietic growth factors on the production of neutrophilic precursors. *J. Hematother.* **6,** 475–489.

Maly, P., Thall, A., Petryniak, B., *et al.* (1996). The alpha(1,3) fucosyltransferase Fuc-TVII controls leukocyte trafficking through an essential role in L-, E- and P-selectin ligand biosynthesis. *Cell* **86,** 643–653.

Manié, S. N., Sattler, M., Astier, A., *et al.* (1997). Tyrosine phosphorylation of the product of c-cbl protooncogene induced after integrin stimulation. *Exp. Hematol.* **25,** 45–50.

Masuzawa, Y., Miyauchi, T., Hamanoue, M., *et al.* (1992). A novel core protein as well as polymorphic epithelial mucin carry peanut agglutinin binding sites in human gastric carcinoma cells: Sequence analysis and examination of gene expression. *J. Biochem.* **112,** 609–615.

Mayani, H., Dragowska, W., and Lansdorp, P. M. (1993). Cytokine-induced selective expansion and maturation of erythroid versus myeloid progenitors from purified cord blood precursor cells. *Blood* **81,** 3252–3258.

McNeice, I. K., Bertoncello, I., Breigler, A. B., *et al.* (1990). Colony-forming cells with high proliferative potential (HPP). *Int. J. Cell Cloning* **8,** 146–160.

McNeice, I. K., Bradley, T. R., Kreigler, A. B., *et al.* (1986). Subpopulations of mouse bone marrow high proliferative potential colony forming cells (HPP-CFC). *Exp. Hematol.* **14,** 856–860.
McNeice, I. K., Robinson, B. E., and Quesenberry, P. J. (1988). Stimulation of murine colony-forming cells with high proliferative potential by the combination of GM-CSF and CSF-1. *Blood* **72,** 191–195.
McNeice, I. K., Stewart, F. M., Deacon, D. M, *et al.* (1989). Detection of human CFC with a proliferative potential. *Blood* **74,** 609–612.
Metcalf, D. (1982). Effects of GM-CSF deprivation on precursors of granulocytes and macrophages. *J. Cell Physiol.* **112,** 411–418.
Metcalf, D. (1984). "The Hemopoietic Colony Stimulating Factors." Elsevier, Amsterdam.
Metcalf, D. (1993). Hematopoietic regulators: Redundancy or subtlety? *Blood* **82,** 3515–3523.
Metcalf, D., Begley, C. G., Johnson, G. R., *et al.* (1986a). Biological properties *in vitro* of a recombinant human granulocyte-macrophage colony stimulating factor. *Blood* **67,** 37–45.
Metcalf, D., Burgess, A. W., Johnson, G. R., *et al.* (1986b). *In vitro* actions on hemopoietic cells of recombinant murine GM-CSF purified after production in *Escherichia coli:* Comparison with purified native GM-CSF. *J. Cell Physiol.* **128,** 421–431.
Metcalf, D., and Nicola, N. A. (1983). Proliferative effects of purified granulocyte colony-stimulating factor (G-CSF) on normal mouse hematopoietic cells. *J. Cell Physiol.* **116,** 198–206.
Metcalf, D., and Nicola, N. A. (1992). The clonal proliferation of normal mouse hemopoietic cells: Enhancement and suppression by CSF combinations. *Blood* **79,** 2861–2866.
Migliaccio, G., Migliaccio, A. R., and Visser, J. W. M. (1988). Synergism between erythropoietin and interleukin-3 in the induction of hematopoietic stem cell proliferation and erythroid burst colony formation. *Blood* **72,** 944–951.
Mimura, N., and Asano, A. (1986). Isolation and characterization of a conserved actin-binding domain from rat hepatic actinogelin, rat skeletal muscle, and chicken gizzard α-actinins. *J. Biol. Chem.* **261,** 10680–10687.
Miyake, K., Weissman, I. L., Greenberger, J. S., *et al.* (1991). Evidence for a role of the integrin VLA-4 in lympho-hemopoiesis. *J. Exp. Med.* **173,** 599–607.
Miyamoto, S., Teramoto, H., Coso, O. A., *et al.* (1995). Integrin function: Molecular hierarchies of cytoskeletal and signaling molecules. *J. Cell Biol.* **131,** 791–805.
Moore, M. A. S., Warren, D. J., and Souza, L. M. (1987). Synergistic interactions between interleukin 1 and CSF's in hematopoiesis. *In* "UCLA Symposium on Leukemia: Recent Advances in Leukemia and Lymphoma" (R. P. Gale and D. W. Golde, Eds.) p. 445. A. R. Liss, New York.
Moore, S., Haylock, D. N., Lévesque, J. P., *et al.* (1998). Stem cell factor as a single agent induces selective proliferation of the Philadelphia chromosome positive fraction of chronic myeloid leukemia $CD34^+$ cells. *Blood* **92,** 2461–2470.
Morino, N., Mimura, T., Hamasaki, K., *et al.* (1995). Matrix/integrin interaction activates the mitogen-activated protein kinase, $p44^{erk-1}$ and $p42^{erk-2}$. *J. Biol. Chem.* **270,** 269–273.
Muench, M. O., Schneider, J. G., and Moore, M. A. S. (1992). Interactions among colony-stimulating factors, IL-1β, IL-6, and kit-ligand in the regulation of primitive murine hematopoietic cells. *Exp. Hematol.* **20,** 339–349.
Muller-Sieburg, C. E., Townsend, K., Weismann, I. L., *et al.* (1988). Proliferation and differentiation of highly enriched mouse hematopoietic stem and progenitor cells in response to defined growth factors. *J. Exp. Med.* **167,** 1825–1840.
Musashi, M., Yang, Y. C., Paul, S. R., *et al.* (1991). Direct and synergistic effects of interleukin 11 on murine hemopoiesis in culture. *Proc. Natl. Acad. Sci. USA* **88,** 765–769.
Neben, S., Donaldson, D., Sieff, C., *et al.* (1994). Synergistic effects of interleukin 11 with other growth factors on the expansion of murine hematopoietic progenitors and maintenance of stem cells in liquid cultures. *Exp. Hematol.* **22,** 353–359.
Nicola, N. A., Robb, L., Metcalf, D., *et al.* (1996). Functional inactivation in mice of the gene for the interleukin-3 (IL-3)-specific receptor β-chain: Implications for the IL-3 function and the mechanism of receptor transmodulation in hematopoietic cells. *Blood* **87,** 2665–2674.
Nishinakamura, R., Miyajima, A., Mee, P. J., *et al.* (1996). Hematopoiesis in mice lacking the entire granulocyte-macrophage colony-stimulating factor/interleukin-3/interleukin-5 functions. *Blood* **88,** 2458–2464.

Nishinakamura, R., Nakayama, N., Hirabayashi, Y., *et al.* (1995). Mice deficient for the IL3/GM-CSF/IL5 βc receptor exhibit lung pathology and impaired immune response while βIL3 receptor-deficient mice are normal. *Immunity* **2,** 211–222.

Nong, Y. H., Remold-O'Donnell, E., and LeBien, T. W. (1989). A monoclonal antibody to siolophorin (CD43) induces homotypic adhesion and activation of human monocytes. *J. Exp. Med.* **170,** 259–267.

Ohmizono, Y., Sakabe, H., Kimura, T., *et al.* (1997). Thrombopoietin augments *ex vivo* expansion of human cord blood-derived hematopoietic progenitors in combination with stem cell factor and flt3 ligand. *Leukemia* **11,** 524–530.

Ojaniemi, M., Martin, S. S., Dolfi, F., *et al.* (1997). The proto-oncogene product p120(cbl) links c-src and phosphatidylinositol 3′-kinase to the integrin signaling pathway. *J. Biol. Chem.* **272,** 3780–3787.

Otey, C. A., Vasquez, G. B., Burridge, K., *et al.* (1993). Mapping of the α-actinin binding site within the beta 1 integrin cytoplasmic domain. *J. Biol. Chem.* **268,** 21193–21197.

O'Toole, T. E., Ylanne, J., and Culley, B. M. (1995). Regulation of integrin affinity states through an NPXY motif in the β subunit cytoplasmic domain. *J. Biol. Chem.* **270,** 8553–8558.

Pallant, A., Eskenazi, A., Mattei, M. G., *et al.* (1989). Characterization of cDNAs encoding leukosialin and localization of the leukosialin gene to chromosome 16. *Proc. Natl. Acad. Sci. USA* **86,** 1328–1332.

Papayannopoulou, T., and Nakamoto, B. (1993). Peripheralization of hemopoietic progenitors in primates treated with anti-VLA4 integrin. *Proc. Natl. Acad. Sci. USA* **90,** 9374–9378.

Park, J. K., Rosenstein, Y. J., Remold, O'Donnell, E., *et al.* (1991). Enhancement of T-cell activation by the CD43 molecule whose expression is defective in Wiskott–Aldrich syndrome. *Nature* **350,** 706–709.

Peters, S. O., Kittler, E. L. W., Ramshaw, H. S., *et al.* (1996). *Ex vivo* expansion of murine marrow cells with interleukin-3 (IL-3), IL-6, IL-11, and stem cell factor leads to impaired engraftment in irradiated hosts. *Blood* **87,** 30–37.

Peters, S. O., Kittler, H. S., Ramshaw, P. J., *et al.* (1995). Murine marrow cells expanded in culture with IL-3, IL-6, IL-11, and SCF acquire an engraftment defect in normal hosts. *Exp. Hematol.* **23,** 461–469.

Petzer, A. L., Hogge, D. E., Lansdorp, P. M., *et al.* (1996a). Self-renewal of primitive human hematopoietic cells (long-term-culture-initiating cells) *in vitro* and their expansion in defined medium. *Proc. Natl. Acad. Sci. USA* **93,** 1470–1474.

Petzer, A. L., Zandstra, P. W., Piret, J. M., *et al.* (1996b). Differential cytokine effects on primitive ($CD34^+CD38^-$) human hematopoietic cells: Novel responses to Flt3-ligand and thrombopoietin. *J. Exp. Med.* **183,** 2551–2558.

Pflumio, F., Izac, B., Katz, A., *et al.* (1996). Phenotype and function of human hematopoietic cells after engrafting immune-deficient CB 17-severe combined immunodeficiency mice and nonobese diabetic-severe combined immunodeficiency mice after transplantation of human cord blood mononuclear cells. *Blood* **88,** 3731–3740.

Piacibello, W., Sanavio, F., Severino, A., *et al.* (1997). Extensive amplification and self-renewal of human primitive hematopoietic stem cells from cord blood. *Blood* **89,** 2644–2653.

Piller, F., Le Deist, F., Weinberg, K. I., *et al.* (1991). Altered *O*-glycan synthesis in lymphocytes from patients with Wiskott–Aldrich syndome. *J. Exp. Med.* **173,** 1501–1510.

Plopper, G. (1995). Convergence of integrin and growth factor receptor signaling pathways with the focal adhesion complex. *Mol. Biol. Cell* **6,** 1349–1365.

Ramsfjell, V., Borge, O. J., Cui, L., *et al.* (1997). Thrombopoietin directly and potently stimulates multilineage growth and progenitor expansion from primitive ($CD34^+CD38^-$) human bone marrow progenitor cells: Distinct and key interactions with the ligands for c-kit and flt3, and inhibitory effects of TGF-beta and TNF-alpha. *J. Immunol.* **158,** 5169–5177.

Ramshaw, H. S., Rao, S. S., Crittenden, R. B., *et al.* (1995). Engraftment of bone marrow cells into normal unprepared hosts: Effects of 5-fluorouracil and cell cycle status. *Blood* **86,** 924–929.

Rebel, V. I., Dragowska, W., Eaves, C. J., *et al.* (1994). Amplification of Sca-1^{+}Lin-WGA^{+} cells in serum-free cultures containing steel factor, interleukin-6, and erythropoietin with maintenance of cells with long-term *in vivo* reconstituting potential. *Blood* **83,** 128–136.

Rennick, D., Jackson, J., Yang, G., *et al.* (1989). Interleukin-6 interacts with interleukin-4 and other hematopoietic growth factors to selectively enhance the growth of megakaryocytic, erythroid, myeloid and multipotential progenitor cells. *Blood* **73,** 1828–1835.

Renshaw, M. W., Ren, X. D., and Schwartz, M. A. (1997). Growth factor activation of MAP kinase requires cell adhesion. *EMBO J.* **16,** 5592–5599.

Ridley, A. J., and Hall, A. (1992). The small GTP-binding protein rho regulates the assembly of focal adhesions and actin stress fibers in response to growth factors. *Cell* **70,** 389–399.

Ritchie, A., Vadhan-Raj, S., and Broxmeyer, H. E. (1996). Thrombopoietin suppresses apoptosis and behaves as a survival factor for the human growth factor-dependent cell line MO7e. *Stem Cells* **14,** 330–336.

Robb, L., Drinkwater, C., Metcalf, D., *et al.* (1995). Hematopoietic and lung abnormalities in mice with a null mutation of the common β subunit of the receptors for granulocyte-macrophage colony-stimulating factor and interleukins 3 and 5. *Proc. Natl. Acad. Sci. USA* **92,** 9565–9569.

Rosenstein, Y., Park, J. K., Hahn, W. C., *et al.* (1991). CD43, a molecule defective in Wiskott–Aldrich syndrome, binds ICAM-1. *Nature* **354,** 233–235.

Rowley, S. D., Brashem-Stein, C., Andrews, R., *et al.* (1993). Hematopoietic precursors resistant to treatment with 4-hydroperoxycyclophosphamide: Requirement for an interaction with marrow stroma in addition to hematopoietic growth factors for maximal generation of colony-forming activity. *Blood* **82,** 60–65.

Ruscetti, F. N., Boggs, D. R., Torok, B. J., *et al.* (1976). Reduced blood and bone marrow neutrophils and granulocyte colony forming cells in Sl/Sld mice. *Proc. Soc. Exp. Biol. Med.* **152,** 398–402.

Russell, E. S. (1979). Hereditary anemias of the mouse: A review for geneticists. Adv. *Genet.* **20,** 357–459.

Sako, D., Chang, X. J., Barone, K. M., *et al.* (1993). Expression cloning of a functional glycoprotein ligand for P-selectin. *Cell* **75,** 1179–1186.

Salgia, R., Brunkhorst, B., Pisick, E., *et al.* (1995). Increased tyrosine phosphorylation of focal adhesion proteins in myeloid cell lines expressing p210bcr/abl. *Oncogene* **11,** 1149–1155.

Salgia R., Li, J. L., Ewaniuk, D. S., *et al.* (1997). BCR/ABL induces multiple abnormalities of cytoskeletal function. *J. Clin. Invest.* **100,** 46–57.

Schlaepfer, D. D., Hanks, S. K., Hunter, T., *et al.* (1994). Integrin-mediated signal transduction linked to ras pathway by GRB2 binding to focal adhesion kinase. *Nature* **372,** 786–791.

Schofield, K. P., Rushton, G., Humphries, M. J., *et al.* (1997). Influence of interleukin-3 and other growth factors on $\alpha 4\beta 1$ integrin-mediated adhesion and migration of human hematopoietic progenitor cells. *Blood* **90,** 1858–1866.

Schwartz, M. A., Lechene, C., and Ingber, D. E. (1991). Insoluble fibronectin activates the Na^{+}/H^{+} antiporter by clustering and immobilizing integrin $\alpha 5\beta 1$, independent of cell shape. *Proc. Natl. Acad. Sci. USA* **88,** 7849–7853.

Schwartz, M. A., Schaller, M. D., and Ginsberg, M. H. (1995). Integrins: Emerging paradigms of signal transduction. *Annu. Rev. Cell. Dev. Biol.* **11,** 549–599.

Seymour, J. F., Lieschke, G. J., Grail, D., *et al.*, (1997). Mice lacking both granulocyte colony-stimulating factor (CSF) and granulocyte-macrophage CSF have impaired reproductive capacity, perturbed neonatal granulopoiesis, lung disease, amyloidosis and reduced long-term survival. *Blood* **90,** 3037–3049.

Shah, A. J., Smogorzewska, E. M., Hannum, C., *et al.* (1996). Flt3 ligand induces proliferation of quiescent human bone marrow CD34^{+}CD38^{-}cells and maintains progenitor cells *in vitro. Blood* **87,** 3563–3570.

Shelley, C. S., Remold-O'Donnell, E., Davis, A. E., *et al.* (1986). Signaling and induction of enhanced adhesiveness via the hematopoietic progenitor cell surface molecule CD34. *Proc. Natl. Acad. Sci. USA* **86,** 2819–2823.

Shimizu, Y., and Shaw, S. (1993). Mucins in the mainstream. *Nature* **366,** 630–631.

Sieff, C. A., Ekern, S. C., Nathan, D. G., *et al.* (1989). Combinations of recombinant colony-stimulating factors are required for optimal hematopoietic differentiation in serum-deprived culture. *Blood* **73,** 688–693.

Simmons, P. J., Lévesque, J. P., and Zannettino, A. C. W. (1997). Cell adhesion molecules and their role in regulating hemopoiesis. *Baillère's Clin. Hematol.* **10,** 485–505.

Simmons, P. J., Mesinosky, B., Longenecker, B. M., *et al.* (1992). Vascular-cell adhesion molecule-1 expressed by bone marrow stromal cells mediated the binding of hematopoietic progenitor cells. *Blood* **80,** 388–395.

Simmons, P. J., Zannettino, A., Gronthos, A., *et al.* (1994). Potential adhesion mechanisms for localisation of haematopoietic progenitors to bone marrow stroma. *Leuk. Lymph.* **12,** 353–363.

Smith, C., Gasparetto, C., Collins, N., *et al.* (1991). Purification and partial characterization of a human hematopoietic precursor population. *Blood* **77,** 2122–2128.

Sonoda, Y., Yang, Y. C., Wong, G. G., *et al.* (1988). Analysis in serum-free culture of the targets of recombinant human hemopoietic growth factors: Interleukin-3 and granulocyte/macrophage colony stimulating factors are specific for early developmental stages. *Proc. Natl. Acad. Sci. USA* **85,** 4360–4364.

Spangrude, G. J., Heimfeld, S., and Weissman, I. L. (1988). Purification and characterization of mouse hematopoietic stem cells. *Science* **241,** 58–62.

Springer, T. (1994). Traffic signals for lymphocyte recirculation and leukocyte emigration: The multistep paradigm. *Cell* **76,** 301–314.

Stanley, E., Lieschke, G. J., Grail, D., *et al.* (1994). Granulocyte/macrophage colony-stimulating factor-deficient mice show no major pertubation of hematopoiesis but develop a characteristic pulmonary pathology. *Proc. Natl. Acad. Sci. USA* **91,** 5592–5596.

Stanley, I. R., and Burgess, A. W. (1983). GM-CSF stimulates the synthesis of membrane and nuclear proteins in murine neutrophils. *J. Cell. Biochem.* **23,** 241–258.

Stewart, F. M., Crittenden, R. B., Lowry, P. A., *et al.* (1993). Long-term engraftment of normal and post-5-fluorouracil murine bone marrow into normal nonmyeloablated mice. *Blood* **81,** 2473–2474.

Symington, B. E. (1995). Growth signaling through the alpha5beta1 fibronectin receptor. *Biochem. Biophys. Res. Commun.* **208,** 136–134.

Tachibana, K., Sato, T., D'Avirro, N., *et al.* (1995). Direct association of pp125FAK with paxillin, the focal adhesion-targeting mechanism of pp125FAK. *J. Exp. Med.* **182,** 1089–1099.

Takahira, H., Gotoh, A., Ritchie, A., *et al.* (1997). Steel factor enhances integrin-mediated tyrosine phosphorylation of focal adhesion kinase (pp125FAK) and paxillin. *Blood* **89,** 1574–1584.

Takamatsu, Y., Simmons, P. J., and Lévesque, J. P. (1998). Dual control by divalent cations and mitogenic cytokines of $\alpha 4\beta 1$ and $\alpha 5\beta 1$ integrin affinity on human hemopoietic cells. *Cell Adhes. Commun.* **5,** 349–366.

Tavassoli, M., and Friedenstein, A. (1983). Hemopoietic stromal micro-environment. *Ann. J. Hematol.* **15,** 195–203.

Tavian, M., Coulombel, L., Luton, D., *et al.* (1996). Aorta-associated $CD34^+$ hematopoietic cells in the early human embryo. *Blood* **87,** 67–72.

Terstappen, L. W., Huang, S. M., Safford, M., *et al.* (1991). Sequential generations of hematopoietic colonies derived from single nonlineage-committed $CD34^+CD38^-$progenitor cells. *Blood* **77,** 1218–1227.

Torok-storb, B. (1988). Cellular interactions. *Blood* **73,** 373–385.

Tracey, J. B., and Rinder, H. M. (1996). Characterization of the P-selectin ligand on human hematopoietic progenitors. *Exp. Hematol.* **24,** 1494–1500.

Traycoff, C. M., Yoder, M. C., Hiatt, K., *et al.* (1997). Functional association between expression of adhesion molecules and marrow repopulating potential of primitive murine hematopoietic progenitor cells (HPC). *Exp. Hematol.* **25,** 736a.

Trentin, J. J. (1970). Influence of hematopoietic organ stroma (hematopoietic inductive microenvironments) on stem cell differentiation. *In* "Regulation of Hematopoiesis" (A. S. Gordon, Ed.), pp. 616–686. Appleton-Century-Crofts, New York.

van der Loo, J. C. M., and Ploemacher, R. E. (1995). Marrow- and spleen-seeding efficiencies of all murine hematopoietic stem cell subsets are decreased by preincubation with hematopoietic growth factors. *Blood* **85,** 2598–2606.

van der Loo, J. C. M., Xiao, X. L., McMillin, D., *et al.* (1997). Functional VLA-5 on hematopoietic stem cells. *Exp. Hematol.* **25,** 743a.

Verfaillie, C. M. (1993). Soluble factor(s) produced by human bone marow stroma increase cytokine-induced proliferation and maturation of primitive hematopoietic progenitors while preventing their terminal differentiation. *Blood* **82,** 2045–2053.

Verfaillie, C. M., Catanzarro, P. M., and Li, W. N. (1994). Macrophage inflammatory protein 1α, interleukin 3 and diffusible marrow stromal factors maintain human hematopoietic stem cells for at least eight weeks *in vitro*. *J. Exp. Med.* **179,** 643–649.

Verfaillie, C., McCarthy, J., and McGlave, P. (1992). Mechanisms underlying abnormal trafficking of maligant progenitors in chronic myelogenous leukemia: Decreased adhesion to stroma and fibronectin but increased adhesion to the membrane components laminin and collagen I. *J. Clin. Invest.* **90,** 1232–1241.

Vosseller, K., Stella, G., Yee, N. S., *et al.* (1997). c-kit receptor signaling through its phosphatidylinositide-3′-kinase-binding site and protein kinase C: Role in mast cell enhancement of degranulation, adhesion and membrane ruffling. *Mol. Biol. Cell* **8,** 909–922.

Wang, N., Butler, J., and Ingber, D. (1993). Mechanotransduction across the cell surface through the cytoskeleton. *Science* **260,** 1124–1127.

Watt, S. M., Williamson, J., Genevier, H., *et al.* (1993). The heparin binding PECAM-1 adhesion molecule is expressed by CD34$^+$ hematopoietic precursor cells with early myeloid and B-lymphoid cell phenotypes. *Blood* **82,** 2649–2663.

Weisbart, R. H., Golde, D. W., Clarke, S. C., *et al.* (1985). Human granulocyte-macrophage colony-stimulating factor is a neutrophil activator. *Nature* **314,** 361–363.

Williams, D. A., Rios, M., Stephens, C., *et al.* (1991). Fibronectin and VLA-4 in haemopoietic stem cell-micro-environment. *Nature* **352,** 438–441.

Williams, D. E., Eisenman, J., Baird, A., *et al.* (1990). Identification of a ligand for the c-kit proto-oncogene. *Cell* **63,** 167–174.

Williams, D. E., Hangoc, G., Cooper, S., *et al.* (1987). The effects of purified recombinant murine interleukin-3 and/or purified natural murine CSF-1 *in vivo* on the proliferation of murine high- and low-proliferative potential colony-forming cells: Demonstration of *in vivo* synergism. *Blood* **70,** 401–403.

Williams, N., Bertoncello, I., Kavnoudias, H., *et al.* (1992). Recombinant rat stem cell factor stimulates the amplification and differentiation of fractionated mouse stem cell populations. *Blood* **79,** 58–64.

Williams, S. F., Lee, W. J., Bender, J. G., *et al.* (1996). Selection and expansion of peripheral blood CD34$^+$ cells in autologous stem cell transplantation for breast cancer. *Blood* **87,** 1687–1691.

Yurochko, A. D., Liu, D. Y., Eierman, D., *et al.* (1992). Integrins as a primary signal transduction molecule regulating monocyte immediate-early gene induction. *Proc. Natl. Acad. Sci. USA* **89,** 9034–9038.

Zannettino, A. C. W., Berndt, M. C., Butcher, C., *et al.* (1995). Primitive human hematopoietic progenitors adhere to P-selectin (CD62P). *Blood* **85,** 3466–3477.

Zannettino, A. C. W., Buhring, H. J., Simmons, P. J., *et al.* (1998). The sialomucin CD164 (MGC-24v) is an adhesive glycoprotein expressed by human hematopoietic progenitors and bone marrow stromal cells that serves as a potent negative regulator of hematopoiesis. *Blood* **92,** 2613–2628.

Zannettino, A. C. W., Rappold, I., Buhring, H. J., *et al.* (1996). "Cluster Report: MGC-24 (CD164). VIth International Workshop and Conference on Human Leukocyte Differentiation Antigens, Japan, 1996". Graland Publishing Inc., London.

Zhang, Z., Vuori, K., Reed, J. C., *et al.* (1995). The $\alpha5\beta1$ integrin supports survival of cells on fibronectin and up-regulates Bcl-2 expression. *Proc. Natl. Acad. Sci. USA* **92,** 6161–6165.

Zhu, X., Ohtsubo, M., Bohmer, R. M., *et al.* (1996). Adhesion-dependent cell cycle progression linked to the expression of cyclin D1, activation of cyclin E-cdk2, and phosphorylation of the retinoblastoma protein. *J. Cell Biol.* **133,** 391–403.

Zsebo, K. M., Williams, D. A., Geissler, E. N., *et al.* (1990). Stem cell factor is encoded at the Sl locus of the mouse and is the ligand for the c-kit tyrosine kinase receptor. *Cell* **63,** 213–224.

5

Allogeneic Hematopoietic Stem Cell Transplantation

P. A. Rowlings[1]

International Bone Marrow Transplant Registry/Autologous Blood and Marrow Transplant Registry Statistical Center
Health Policy Institute
Medical College of Wisconsin
Milwaukee, Wisconsin 53226

I. INTRODUCTION

Allogeneic hematopoietic transplants have been performed for over 30 years, with now over 10,000 performed each year for a variety of malignant and

[1]Current address: Haematology Department, Prince of Wales Hospital, Randwick, Sydney, NSW 2031, Australia.

nonmalignant diseases. Over the past 25 years the International Bone Marrow Transplant Registry (IBMTR), an international research organization, has collected and analyzed data on recipients of allogeneic hematopoietic stem cell transplants. The IBMTR collects data from over 300 institutions in 47 countries. Since 1991 the Autologous Blood and Marrow Transplant Registry (ABMTR) has collected data from over 150 institutions, predominantly located in North and South America, using autologous hematopoietic cells to reconstitute marrow function following high-dose chemotherapy and/or radiotherapy. The IBMTR/ABMTR database has information for greater than 75,000 transplant recipients, about 40% of allogeneic transplants performed since 1964 and about half of autotransplants in North and South America since 1989. In this chapter current allogeneic hematopoietic stem cell transplant activity, as reported to the IBMTR, is summarized as well as key findings from recently published studies.

II. INDICATIONS FOR ALLOGENEIC HEMATOPOIETIC STEM CELL TRANSPLANTS

Hematopoietic stem cell transplants are used to treat a wide range of malignant diseases as well as congenital and acquired cytopenias, disorders of immune function, and metabolic storage diseases. The distribution of diseases for almost 46,000 patients transplanted from 1964 to 1997 reported to the IBMTR is presented in Table 5.1.

TABLE 5.1 Indications for Allogeneic Hematopoietic Stem Cell Transplants As Reported to the IBMTR between 1964 and 1997

Disease	Number of cases
Acute myelogenous leukemia	11,085
Acute lymphoblastic leukemia	8,721
Chronic myelogenous leukemia	11,020
Myelodysplastic syndromes	2,393
Non-Hodgkin lymphoma	2,056
Hodgkin disease	322
Multiple myeloma	1,119
Other malignancies[a]	1,234
Aplastic anemia	3,844
Hemoglobinopathies and related disorders	1,781
Immunodeficiency states	1,416
Inherited disorders of metabolism	550
Histiocytic disorders	127
Other	265
Total	45,984

[a] Includes undifferentiated and other leukemias.

III. DONOR AND TISSUE SOURCES OF HEMATOPOIETIC STEM CELLS

Most allogeneic hematopoietic stem cell transplants use sibling donors matched with recipients for human leukocyte antigens (HLA). Since only about 30% of transplant candidates have an HLA-identical sibling, there is considerable interest in alternative allogeneic donors. With establishment of large international panels of volunteer bone marrow donors—about 3 million donors available worldwide—use of HLA-matched unrelated donor transplants is increasing. About 25% of allogeneic transplants now use unrelated donors compared to <10% in 1990 according to data reported to the IBMTR. Another 5–10% use related donors mismatched for one or more HLA antigens. The IBMTR recently compared outcomes of transplants for leukemia using HLA-identical sibling, HLA-mismatched related, and HLA-matched and mismatched-unrelated donors performed between 1985 and 1991 (Szydlo *et al.,* 1997). Serologic techniques were used to define HLA types. Transplant-related mortality was two to three times higher after alternative donor than after HLA-identical sibling transplants. Data from this study indicate that, despite higher transplant-related mortality, alternative donor transplants are effective in some patients with leukemia. The outcome of such transplants depends on leukemia state and donor–recipient histocompatibility. It is possible that current technology that better defines histocompatibility using DNA-based methods for HLA typing may lead to better results by allowing selection of more closely HLA-matched donors (Petersdorf *et al.,* 1995, 1996).

Hematopoietic stem cells for transplantation were traditionally obtained from bone marrow. However, in the 1980s it was shown that autologous cells harvested from peripheral blood could restore hematopoiesis and that large numbers of these cells could be obtained in the recovery phase following chemotherapy and after administration of hematopoietic growth factors (To *et al.,* 1984; Kessinger *et al.,* 1986; Körbling *et al.,* 1986; Reiffers *et al.,* 1986; Sheridan *et al.,* 1992). These blood-derived cells provide more rapid hematopoietic recovery than bone marrow. More than 60% of autologous hematopoietic stem cell transplants now use blood cells (Fig. 5.1). Most allogeneic transplants still use bone marrow because of concerns about administering growth factors to normal donors and the potential for increased graft-versus-host disease (GVHD) from the large numbers of T lymphocytes in blood-derived grafts. However, in 1995 three centers reported rapid hematopoietic recovery and acceptable acute GVHD after blood cell allogeneic transplants from HLA-identical relatives (Bensinger *et al.,* 1995; Körbling *et al.,* 1995; Schmitz *et al.,* 1995). As reported to the IBMTR, there have been over 2000 blood allogeneic transplants since 1992. For 1996 over 20% of all allogeneic hematopoietic transplants reported to the IBMTR used cells obtained from the blood.

As seen in Fig. 5.1, umbilical cord blood is now also used as a source of hematopoietic stem cells from either related or unrelated donors. Recent studies

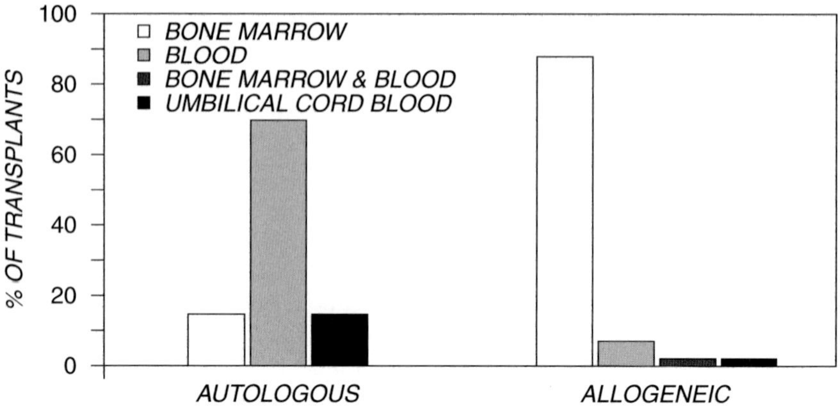

FIGURE 5.1 Sources of hematopoietic stem cells for autologous and allogeneic transplant in 1995 (Rowlings, 1996).

suggest that this may be a feasible source of allogeneic stem cells for persons without a suitable related bone marrow donor (Wagner *et al.,* 1995; Kurtzburg *et al.,* 1996; Gluckman *et al.,* 1997). Hematopoietic recovery is slow and predictable engraftment in adults remains to be demonstrated in large numbers of patients. For this approach to be useful a large supply of readily available cord blood units is required. Several cord blood banks are now established, with about 10,000 units worldwide, and plans for rapid expansion are being implemented. About 500 cord blood transplants have been done, most from unrelated donors and most in children.

IV. RESULTS OF ALLOGENEIC HEMATOPOIETIC STEM CELL TRANSPLANTS

A. CHRONIC MYELOGENOUS LEUKEMIA (CML)

Allogeneic transplantation remains the treatment of choice for young patients with CML with an HLA-identical sibling donor available. Among 3409 recipients of HLA-identical sibling transplants done between 1989 and 1995, reported to the IBMTR, 3-year actuarial probabilities of relapse (95% confidence interval) are 16% (14–18%) for 2753 patients transplanted in first chronic phase, 36% (30–42%) for 490 in accelerated phase, and 61% (50–72%) for 166 in blast phase (Rowlings, 1996). Three-year probabilities of survival are 66% (64–68%), 44% (39–49%), and 19% (12–26%), respectively (Fig. 5.2). Survival after HLA-identical sibling transplants is higher when transplants are done within 1

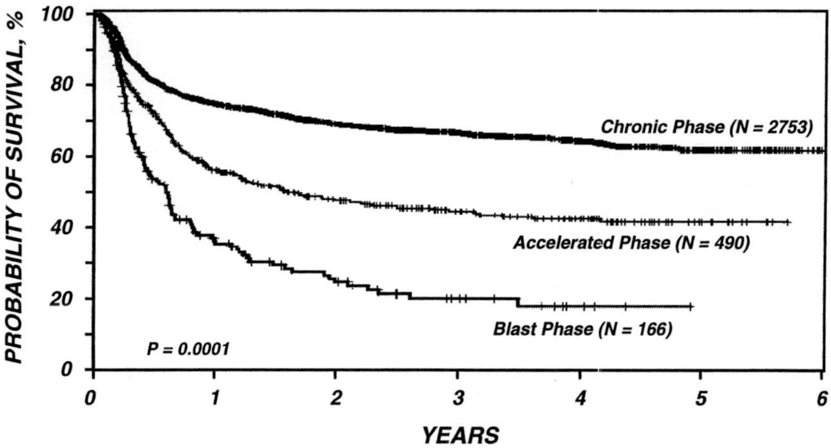

FIGURE 5.2 Probability of survival after HLA-identical sibling transplant for chronic myelogenous leukemia. 1989–1995 (Rowlings, 1996).

year of diagnosis and when busulfan is not used for pretransplant therapy (Thomas *et al.,* 1986; Goldman *et al.,* 1993).

Only about 30% of persons with CML have an HLA-identical sibling donor. Unrelated-donor transplants can cure CML but have higher transplant-related mortality (Szydlo *et al.,* 1997). Unrelated-donor transplants are often delayed because of time required to identify a donor and reluctance to risk high transplant-related mortality. The median interval between diagnosis and transplantation for CML is 10 months for HLA-identical sibling and 22 months for unrelated-donor transplants. Three-year leukemia-free survival (LFS) is 64% (61–67%) after 1623 HLA-identical sibling transplants done less than 1 year after diagnosis of CML, 51% (48–54%) after 1127 HLA-identical sibling transplants done greater than 1 year after diagnosis, 47% (34–60%) after 122 unrelated-donor transplants done less than 1 year after diagnosis, and 35% (30–40%) after 497 unrelated-donor transplants done later (Rowlings, 1996).

B. ACUTE LYMPHOBLASTIC LEUKEMIA (ALL)

Most patients with ALL are cured with conventional chemotherapy. Therefore, transplants are generally reserved for patients failing chemotherapy or having prognostic factors predicting a high risk of leukemia recurrence. Such factors are older age, high leukocyte count at diagnosis, Philadelphia chromosome, and difficulty obtaining first remission. Among 2497 recipients of HLA-identical sibling transplants between 1989 and 1995, 3-year probabilities of relapse are 25% (21–29%) for 1005 transplants in first remission, 46% (42–

50%) for 1074 in greater than or second remission, and 68% (61–75%) for 418 not in remission (Rowlings, 1996). Three-year probabilities of LFS are 54% (50–58%), 40% (37–43%), and 20% (15–25%), respectively (Fig. 5.3).

Unrelated-donor transplants are also used for patients with ALL unlikely to be cured with chemotherapy. Among 402 recipients of unrelated-donor transplants for ALL, 3-year LFS is 37% (23–51%) after 102 transplants done in first remission and 36% (30–42%) after 300 in second or subsequent remission (Rowlings, 1996).

C. ACUTE MYELOGENOUS LEUKEMIA (AML)

As in ALL, outcome of transplants for AML correlates with remission state. Among 3503 recipients of HLA-identical sibling transplants done between 1989 and 1995, 3-year probabilities of relapse are 24% (22–26%) for 2247 transplants done in first remission, 45% (37–53%) for 459 in greater than or second remission, and 57% (52–62%) for 797 not in remission (Rowlings, 1996). Three-year probabilities of LFS are 59% (57–61%), 35% (30–40%), and 26% (22–30%), respectively (Fig. 5.4).

As in ALL, unrelated-donor transplants are used in some patients with AML at high risk of failing conventional therapy. Among 208 patients receiving an unrelated-donor transplant for AML between 1989 and 1995, 3-year probabilities of LFS are 57% (44–70%) after 87 transplants done in first remission and 25% (13–37%) for 121 in or greater than second remission (Rowlings, 1996).

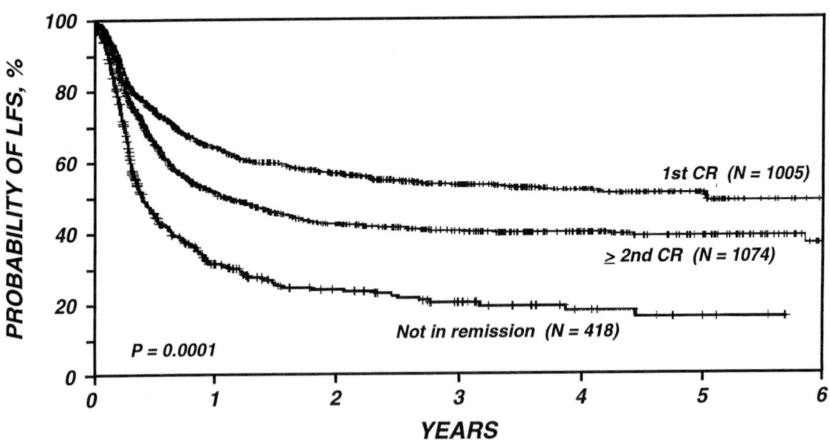

FIGURE 5.3 Probability of leukemia-free survival after HLA-identical sibling transplant for acute lymphoblastic leukemia, 1989–1995 (Rowlings, 1996).

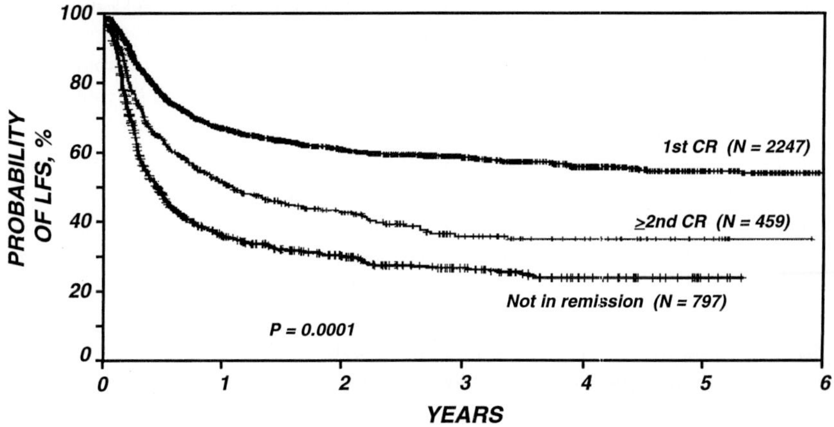

FIGURE 5.4 Probability of leukemia-free survival after HLA-identical sibling transplant for acute myelogenous leukemia, 1989–1995 (Rowlings, 1996).

Autologous hematopoietic stem cell transplants are sometimes utilized to ensure long-term LFS following remission when HLA-identical sibling donors are not available. Comparative studies indicate a higher transplant-related mortality but lower probability of relapse and subsequently higher leukemia-free survival when allogeneic stem cells are used (Hermans *et al.,* 1989; Ferrant *et al.,* 1991; Burnett *et al.,* 1994; Keating *et al.,* 1995; Zittoun *et al.,* 1995; Reiffers *et al.,* 1996). Data indicate that *in vitro* treatment of the autologous graft with 4-hydroperoxycyclophosphamide or mafosfamide to remove leukemia cells leads to higher leukemia-free survival (Sharkis *et al.,* 1980; Kaizer *et al.,* 1985; Yeager *et al.,* 1986; Gorin *et al.,* 1990; Lenarsky *et al.,* 1990; Cassileth *et al.,* 1993; Chao *et al.,* 1993; Linker *et al.,* 1993; Woods *et al.,* 1993; Miller *et al.,* 1996).

D. SEVERE APLASTIC ANEMIA

Improved results of allogeneic transplants for aplastic anemia since the early 1980s were recently demonstrated in a study from the IBMTR (Passweg *et al.,* 1997). Higher survival results primarily from decreased transplant-related mortality in the first 3 months posttransplant. This study suggested that changes in transplant strategies accounted for most of the improved outcome, especially use of cyclosporine to prevent GVHD. Transplantation is the treatment of choice for aplastic anemia in young patients with an HLA-identical sibling. Three-year probabilities of survival after HLA-identical sibling transplants between 1989 and 1995 are 73% (69–77%) for 868 patients less than 20 years of age and 61% (56–66%) for 591 older patients. Results are not as good with unrelated-donor transplants: 41% (31–51%) in 136 patients younger than 20 years and 40% (27–53%) in 64 older patients (Rowlings, 1996).

The IBMTR recently analyzed data from 40 identical twin transplants for aplastic anemia (Hinterberger *et al.,* 1997). Seven of 23 recipients recovered normal hematopoiesis after a single transplant with no conditioning. One of the 16 who did not recover hematopoiesis initially recovered following a second transplant without conditioning. The remaining 15 received one or more subsequent transplants with pretransplant conditioning: 13 recovered normal hematopoiesis. Seventeen other patients received conditioning before their first twin transplant: 12 recovered normal bone marrow function. Failure to recover hematopoiesis without conditioning supports an immune-mediated etiology for aplastic anemia in most of the patients in this study. Actuarial 10-year survival for the 40 patients was 78% (59–92%).

V. ALLOGENEIC DONOR LYMPHOCYTE INFUSIONS

A. TREATMENT OF RELAPSED LEUKEMIA

A major therapeutic advance in the management of leukemia this decade has been the harnessing of the graft-versus-leukemia effect of donor lymphocytes. Following relapse of leukemia, post-allogeneic transplant infusion of lymphocytes from the same donor can result in remission in the absence of cytotoxic therapy (Kolb *et al.,* 1990; Drobyski *et al.,* 1993; Collins *et al.,* 1997). This antileukemic effect is most predictable in patients who relapse into the chronic phase of CML; however, responses in other diseases and disease state occur. Difficulties with this approach are development of aplasia and graft-versus-host disease. Aplasia can be avoided by infusion of the donor lymphocytes in early relapse before the leukemic repopulation of the marrow has replaced the normal hematopoiesis achieved posttransplant. The severity of graft-versus-host disease can be reduced in some patients by administering specific T cell doses, titrated to be low enough to achieve graft-versus-leukemia without severely damaging the host through graft-versus-host disease. Even more elegant is the transfer of the herpes simplex virus thymidine kinase “suicide” gene into donor lymphocytes (Bonini *et al.,* 1997). Patients are able to benefit from the antileukemic effect, but if graft-versus-host disease develops, it can be controlled by ganciclovir-induced elimination of the transduced cells.

B. TREATMENT OF EPSTEIN–BARR VIRUS-ASSOCIATED POSTTRANSPLANT LYMPHOPROLIFERATIVE DISORDERS

Lymphoma associated with Epstein–Barr virus is a complication of allogeneic bone marrow transplant that responds poorly to standard forms of therapy.

These lymphomas are due to unregulated proliferation of donor B lymphocytes. Investigators showed that infusions of low doses of cytotoxic T lymphocytes were able to cause involution of these lymphomas (Papadopoulos *et al.,* 1994). However, graft-versus-host disease can also result from infusion of these non-specific-T lymphocytes. This can be avoided either by the transfer of the "suicide" gene referred to earlier (Bonini *et al.,* 1997) or by preparing Epstein–Barr virus-specific clones (Rooney *et al.,* 1995). The antigen-specific nature of these T lymphocytes avoids the problem of graft-versus-host disease. These infused cells have been demonstrated to continue to exist and respond *in vivo* for as long as 18 months (Heslop *et al.,* 1996).

C. RECONSTITUTION OF CELLULAR IMMUNITY AGAINST CYTOMEGALOVIRUS

Cytomegalovirus disease post-allogeneic transplant is a major cause of morbidity and mortality. Although antiviral therapy has reduced the impact of this virus, toxic effects of therapy, cost, and emerging viral resistance make this a suboptimal approach. The problem of cytomegalovirus disease arises because of reduced numbers of T lymphocytes directed at the virus. Infusion of large numbers of donor T lymphocytes would again have the risk of graft-versus-host disease. However, clones of cytotoxic T lymphocytes specific for cytomegalovirus can be developed if the donor has previously been exposed to the virus (Walter *et al.,* 1995). Infusion of these cells is a safe and effective way to reconstitute cellular immunity against cytomegalovirus after allogeneic stem cell transplantation.

VI. ADVERSE EVENTS IN LONG-TERM SURVIVORS

Over 20,000 persons now survive 5 or more years after transplant and that number will increase rapidly. Although most 5-year survivors are generally well, transplant recipients remain at risk for complications long after transplant. These include late infections, cataracts, abnormalities of growth and development, thyroid disorders, chronic lung disease, and avascular necrosis. Additionally, data from the IBMTR and others suggest that allograft recipients have increased risks of death compared to the general population as long as 8 years after transplantation (Socié *et al.,* 1996; Duell *et al.,* 1997). There is also an increased incidence of leukemias, myelodysplasias, and solid tumors in transplant recipients compared to the general population. A collaborative study of the IBMTR, the Fred Hutchinson Cancer Research Center, and the Cancer Epidemiology Branch of the United States National Cancer Institute found a 2.7-fold increased risk of new solid cancers in allogeneic transplant recipients compared to the general population (Curtis *et al.,* 1997). The risk was 8.3 times as high as expected among patients surviving 10 or more years after transplantation. Cumulative incidence rates were 2.2% (1.5–3.0%) at 10 years and 6.7% (3.7–9.6%) at 15 years. Significantly increased risks were detected for malignant melanoma and cancers of the buccal cavity, liver, central nervous system, thyroid, bone, and connective tissue. The risk was higher for recipients who were children at the time of transplantation. Many of these second cancers may not be due to the transplant but to chemotherapy and/or radiation given as treatment earlier in the management of the primary disease for which the patient is eventually transplanted. The need for lifelong surveillance of transplant recipients is necessary.

VII. CONCLUSION

Allogeneic hematopoietic stem cell transplant is an effective therapy for many malignant and nonmalignant conditions. It is an exciting area of cellular immune therapy incorporating many of the latest techniques of molecular biology to improve safety and efficacy. There are now many long-term survivors who, although would not have survived without their transplant, require careful observation for late adverse events such as secondary malignancies.

ACKNOWLEDGMENTS

We thank Melodee Nugent, Kim Hyler, Lisa Lehrmann, and Linda Schneider for assistance in manuscript preparation. We also thank Diane J. Knutson, Barbara McGary, Sharon Nell, Hemant Patani, and Jane Rebro for data collection and management. Dr. Rowlings research is supported by the Michael Foundation, NSW, Australia. IBMTR/ABMTR is supported by Public Health Service Grant PO1-CA-40053 from the National Cancer Institute, the National Institute of Allergy and Infectious Diseases, and the National Heart, Lung and Blood Institute; Contract No. CP-21161 from the National Cancer Institute of the U.S. Department of Health and Human Services; Grant No. DAMD17-95-I-5002 from the Department of the U.S. Army Medical Research and Development Command; and grants from Activated Cell Therapy; Alpha Therapeutic Corp.; American Oncology Resources; Amgen, Inc.; Anonymous; Astra Pharmaceutical; Baxter Healthcare Corp.; Bayer Corp.; Biogen; BioWhittaker, Inc.; BIS Laboratories; Blue Cross and Blue Shield Association; Lynde and Harry Bradley Foundation; Bristol-Myers Squibb Co.; Frank G. Brotz Family Foundation; Caremark, Inc.; CellPro, Inc.; Cell Therapeutics; Centeon; Center for Advanced Studies in Leukemia; Chimeric Therapies; Chiron Therapeutics; Cigna HealthCare; COBE BCT Inc.; Coulter Corp.; Coram Healthcare; Charles E. Culpeper Foundation; Eleanor Naylor Dana Charitable Trust; Deborah J. Dearholt Memorial Fund; Eppley Foundation for Research: Fujisawa USA, Inc.; Genentech, Inc.; Glaxo Wellcome Co.; Hewlett-Packard Co.; Hoechst Marion Roussel, Inc.; Immunex Corp.; Janssen Pharmaceutica; Kettering Family Foundation; Kirin Brewery Co.; Robert J. Kleberg, Jr. and Helen C. Kleberg Foundation; Herbert H. Kohl Charities; Lederle Laboratories; Life Technologies, Inc.; Eli Lilly Company Foundation; The Liposome Co.; Nada and Herbert P. Mahler Charities; Medical SafeTEC; MGI Pharma, Inc.; Milstein Family Foundation; Milwaukee Foundation/Elsa Schoeneich Research Fund; NCSG and Associates, Inc.; NeXstar Pharmaceuticals, Inc; Samuel Roberts Noble Foundation; Novartis Pharmaceuticals; Ortho Biotech Corp.; John Oster Family Foundation; Elsa U. Pardee Foundation; Jane and Lloyd Pettit Foundation; Alirio Pfiffer Bone Marrow Transplant Support Association; Pfizer, Inc.; Pharmacia and Upjohn; QLT PhotoTherapeutics; Quantum Health

Resources; RGK Foundation; Roche Laboratories; RPR GenCell; SangStat Medical Corp.; Schering-Plough International; Walter Schroeder Foundation; Searle; SEQUUS Pharmaceuticals Inc.; Stackner Family Foundation; Starr Foundation; StemCell Technologies; Joan and Jack Stein Charities; Sy-Stemix; Therakos; TS Scientific and Planer Products; Wyeth-Ayerst Laboratories; and Xoma Corp.

REFERENCES

Bensinger, W. I., Weaver, C. H., Appelbaum, F. R., et al. (1995). Transplantation of allogeneic peripheral blood stem cells mobilized by recombinant human granulocyte colony-stimulating factor. *Blood* **85,** 1655–1658.

Bonini, C., Ferrari, G., Verzeletti, S., et al. (1997). HSV-TK gene transfer into donor lymphocytes for control of allogeneic graft-versus-leukemia. *Science* **276,** 1719–1723.

Burnett, A. K., Goldstone, A. H., Stevens, R. F., et al. (1994). The role of bone marrow transplantation in addition to intensive chemotherapy in AML in first CR: Results of the MRC AML-10 trial. *Blood* **84,** 252a.

Cassileth, P. A., Anderson, J., Lazarus, H. M., et al. (1993). Autologous bone marrow transplant in acute myeloid leukemia in first remission. *J. Clin. Onco.* **11,** 314–319.

Chao, N. J., Stein, A. S., Long, G. D., et al. (1993). Busulfan/etoposide: Initial experience with a new preparatory regimen for autologous bone marrow transplantation in patients with acute nonlymphoblastic leukemia. *Blood* **81,** 319–323.

Collins, R. H., Jr., Shpilberg, O., Drobyski, W. R., et al. (1997). Donor leukocyte infusions in 140 patients with relapsed malignancy after allogeneic bone marrow transplantation. *J. Clin. Oncol.* **15,** 433–444.

Curtis, R. E., Rowlings, P. A., Deeg, H. J., et al. (1997). Solid tumors developing after allogeneic bone marrow transplantation. *N. Engl. J. Med.* **336,** 897–904.

Drobyski, W. R., Keever, C. A., Roth, M. S., et al. (1993). Salvage immunotherapy using donor leukocyte infusions as treatment for relapsed chronic myelogenous leukemia after allogeneic bone marrow transplantation: Efficacy and toxicity of a defined T-cell dose. *Blood* **82,** 2310–2318.

Duell, T., van Lint, M. T., Ljungman, P., et al. (1997). Health and functional status of long-term survivors of bone marrow transplantation. EBMT Working Party on Late Effects and EULEP Study Group on Late Effects, European Group for Blood and Marrow Transplantation. *Ann. Intern. Med.* **126,** 184–192.

Ferrant, A., Doyen, C., Delannoy, A., et al. (1991). Allogeneic or autologous bone marrow transplantation for acute non-lymphocytic leukemia in first remission. *Bone Marrow Transplant* **7,** 303–309.

Gluckman, E., Rocha, V., Boyer-Chammard, A., et al. (1997). Outcome of cord-blood transplantation from related and unrelated donors. *N. Eng. J. Med.* **337,** 373–381.

Goldman, J. M., Szydlo, R., Horowitz, M. M., et al. (1993). Choice of pretransplant treatment and timing of transplants for chronic myelogenous leukemia in chronic phase. *Blood* **82,** 2235–2238.

Gorin, N. C., Aegerter, P., Auvert, B., et al. (1990). Autmlogous bone marrow transplantation for acute myelocytic leukemia in first remission: A European survey of the role of marrow purging. *Blood* **75,** 1606–1614.

Hermans, J., Suciu, S., Stijnen, T., et al. (1989). Treatment of acute myelogenous leuiemia: An EBMT–EORTC retrospective analysis of chemotherapy versus allogeneic or autologous bone marrow transplantation. *Eur. J. Cancer Clin. Oncol.* **25,** 545–550.

Heslop, H. E., Ng, C. Y. C., Li, C., et al. (1996). Long-term restoration of immunity against Epstein–Barr virus infection by adoptive transfer of gene-modified virus-specific T lymphocytes. *Nat. Med.* **2,** 551–555.

Hinterberger, W., Rowlings, P. A., Hinterberger-Fischer, M., et al. (1997). Results of bone marrow transplants from genetically-identical twins in persons with aplastic anemia. *Ann. Intern. Med.* **126,** 116–122.

Kaizer, H., Stuart, R. K., Brookmeyer, R., et al. (1985). Autologous bone marrow transplantation in acute leukemia: A phase I study of *in vitro* treatment of marrow with 4-hydroperoxycyclophosphamide to purge tumor cells. *Blood* **65,** 1504–1510.

Keating, A., Rowlings, P. A., Zhang, M. J., et al. (1995). Autologous versus HLA-identical sibling bone marrow transplants for aucte myelogenous leukemia. *Proc. Annu. Meet. Am. Soc. Clin. Oncol.* **14,** 325.

Kessinger, A., Armitage, J. O., Landmark, J. D., et al. (1986). Reconstitution of human hematopoietic function with autologous cryopreserved circulating stem cells. *Exp. Hematol.* **14,** 192–196.

Kolb, H. J., Mittermuller, J., Clemm, C., et al. (1990). Donor leukocyte transfusions for treatment of recurrent chronic myelogenous leukemia in marrow transplant patients. *Blood* **76,** 2462–2465.

Körbling, M., Dorken, B., Ho, A. D., et al. (1986). Autologous transplantation of blood-derived hemopoietic stem cells after myeloablative therapy in a patient with Burkitt's lymphoma. *Blood* **67,** 529–532.

Körbling, M., Przepiorka, D., Huh, Y. O., et al. (1995). Allogeneic blood stem cell transplantation for refractory leukemia and lymphoma: Potential advantage of blood over marrow allografts. *Blood* **85,** 1659–1665.

Kurtzberg, J., Laughlin, M., Graham, M. L., et al. (1996). Placental blood as a source of hematopoietic stem cells for transplantation into unrelated recipients. *N. Engl. J. Med.* **335,** 157–166.

Lenarsky, C., Weinberg, K., Petersen, J., et al. (1990). Autologous bone marrow transplantation with 4-hydroperoxycyclophosphamide purged marrows for children with acute non-lymphoblastic leukemia in second remission. *Bone Marrow Transplant* **6,** 425–429.

Linker, C. A., Ries, C. A., Damon, L. E., et al. (1993). Autologous bone marrow transplantation for acute myeloid leukemia using busulfan plus etoposide as a preparative regimen. *Blood* **81,** 311–318.

Miller, C. B., Rowlings, P. A., Jones, R. J., et al. (1996). Autotransplants for acute myelogenous leukemia (AML): Effect of purging with 4-hydroxyperoxycyclophosphamide (4HC). *Proc. Annu. Meet. Am. Soc. Clin. Oncol.* **15,** A976.

Papadopoulos, E. B., Ladanyi, M., Emanuel, D., et al. (1994). Infusions of donor leukocytes to treat Epstein–Barr virus-associated lymphoproliferative disorders after allogeneic bone marrow transplantation. *N. Engl. J. Med.* **330,** 1185–1191.

Passweg, J. R., Socié, G., Hinterberger, W., et al. (1997). Bone marrow transplantation for severe aplastic anemia: Has outcome improved? *Blood* **90,** 858–864.

Petersdorf, E. W., Longton, G. M., Anasetti, C., et al. (1995). The significance of HLA-DRBI matching on clinical outcome after HLA-A, B, DR identical unrelated donor marrow transplantation. *Blood* **86,** 1606–1613.

Petersdorf, E. W., Longton, G. M., Anasetti, C., et al. (1996). Definition of HLA-DQ as a transplantation antigen. *Proc. Natl. Acad. Sci. USA* **93,** 15358–15363.

Reiffers, J., Bernard, P., David, B., et al. (1986). Successful autologous transplantation with peripheral blood hemopoietic cells in a patient with acute leukemia. *Exp. Hematol.* **14,** 312–315.

Reiffers, J., Stoppa, A. M., Attal, M., et al. (1996). Allogeneic vs autologous stem cell transplantation vs chemotherapy in patients with acute myeloid leukemia in first remission: The BGMT 87 study. *Leukemia* **10,** 1874–1882.

Rooney, C. M., Smith, C. A., Ng, C. Y. C., et al. (1995). Use of gene-modified virus-specific T lymphocytes to control Epstein–Barr-virus-related lymphoproliferation. *Lancet* **345,** 9–13.

Rowlings, P. A. (1996). Current use and outcome of blood and marrow transplantation. *ABMTR Newsl.* **3,** 6–12.

Schmitz, N., Dreger, P., Suttorp, M., et al. (1995). Primary transplantation of allogeneic peripheral blood progenitor cells mobilized by filgrastim (granulocyte colony-stimulating factor). *Blood* **85,** 1666–1672.

Sharkis, S. J., Santos, G. W., and Colvin, M. (1980). Elimination of acute myelogenous leukemic cells from marrow and tumor suspensions in the rat with 4-hydroxyperoxycyclophosphamide. *Blood* **55,** 521–523.

Sheridan, W. P., Begley, C. G., Juttner, C. A., et al. (1992). Effect of peripheral blood progenitor cells mobilized by filgrastim (G-CSF) on platelet recovery after high dose chemotherapy. *Lancet* **339,** 640-644.

Socié, G., Sobocinski, K. A., Veum-Stone, J., et al. (1996). Long-term survival and analysis of late causes of death after allogeneic bone marrow transplantation. *Blood* **88,** 643a.

Szydlo, R., Goldman, J. M., Klein, J. P., et al. (1997). Results of allogeneic bone marrow transplants for leukemia using donors other than HLA-identical siblings. *J. Clin. Oncol.* **15,** 1767–1777.

Thomas, E. D., Clift, R. A., Fefer, A., et al. (1986). Marrow transplantation for the treatment of chronic myelogenous leukemia. *Ann. Intern. Med.* **104,** 155–163.

To, L. B., Haylock, D. N., Kimber, R. J., *et al.* (1984). High levels of circulating haemopoietic stem cells in very early remission from acute non-lymphoblastic leukaemia and their collection and cryopreservation. *Br. J. Haematol.* **58,** 399–410.

Wagner, J. E., Kernan, N. A., Steinbuch, M., et al. (1995). Allogeneic sibling umbilical-cord-blood transplantation in children with malignant and non-malignant disease. *Lancet* **346,** 214–219.

Walter, E. A., Greenberg, P. D., Gilbert, M. J., et al. (1995). Reconstitution of cellular immunity against cytomegalovirus in recipients of allogeneic bone marrow by transfer of T-cell clones from the donor. *N. Eng. J. Med.* **333,** 1038–1044.

Woods, W. G., Kobrinsky, N., and Buckley, J. (1993). Intensively timed induction therapy followed by autologous or allogeneic bone marrow transplantation for children with acute myeloid leukemia or myelodysplastic syndrome: A Childrens Cancer Group pilot study. *J. Clin. Oncol.* **11,** 1448–1457.

Yeager, A. M., Kaizer, H., Santos, G. W., et al. (1986). Autologous bone marrow transplantation in patients with acute nonlymphocytic leukemia, using *ex vivo* marrow treatment with 4-hydroxyperoxycyclophosphamide. *N. Engl. J. Med.* **315,** 141–148.

Zittoun, R. A., Mandelli, F., Willemze, R., et al. (1995). Autologous or allogeneic bone marrow transplantation compared with intensive chemotherapy in acute myelogenous leukemia. *N. Engl. J. Med.* **332,** 217–223.

6

Autologous Stem Cell Transplantation

Uwe Hahn and L. Bik To

Division of Haematology
Institute of Medical and Veterinary Science
Adelaide, South Australia 5000, Australia

I. Introduction and Historical Background

The indications for intensive chemoradiotherapy followed by stem cell transplantation (SCT) have greatly expanded in recent years (Thomas *et al.,* 1975; Frei *et al.,* 1980). The theoretical basis for high-dose chemotherapy or the combination of chemotherapy and total body irradiation (TBI) is the steep dose–response relationship (Hryniuk and Bush, 1984; Van Hoff *et al.,* 1986). As profound myelosuppression is inevitable, hematological rescue using hematopoietic progenital cells is mandatory. Autologous stem cell transplants circumvent the need for matched—related or unrelated—bone marrow donors (Chapter 5) and thus allow a more widespread utilization of the procedure. It is noteworthy that only 30% of potential recipients will finally have an acceptable family or unrelated donor (Ferrara and Deeg, 1991). The virtually complete absence of graft-versus-host disease (GVHD), the absence of engraftment failure, and the significantly lower incidence of cytomegalovirus (CMV) infection make autologous SCT a much safer procedure (Wingard *et al.,* 1988; Storb *et al.,* 1989). In contrast, the possibility of malignant contamination of the autograft and the lack of graft-versus-tumor (GVT) effect comprise the most important drawbacks of autotransplantation (Weiden *et al.,* 1981; Horowitz *et al.,* 1990). According to recent European Bone Marrow Transplant (EBMT) Registry data, the number of autologous transplants for lymphoma and breast cancer in Europe from 1990 through 1994 increased fivefold (Gratwohl *et al.,* 1996). In this period, the percentage of peripheral blood stem cell (PBSC) transplants rose from 15 to 75%, reflecting a considerable shift in the preferred source of hematopoietic progenitor cells.

A milestone in the history of autologous SCT was the successive reports of clinical studies demonstrating mobilization of hematopoietic precursors into the peripheral blood (Goldman *et al.,* 1978; Körbling *et al.,* 1978). Mobilization of progenitor cells was initially detected during hematopoietic recovery following myelosuppressive chemotherapy (Stiff *et al.,* 1983; To *et al.,* 1989). The emergence of recombinant human hematopoietic growth factors improved mobilization by enhancing yield and ameliorating cytopenia when used with chemotherapy (Socinski *et al.,* 1988; Gianni *et al.,* 1989). Consequently, an advantage in earlier hematopoietic reconstitution as compared with bone marrow transplantation (BMT) was shown (Haas *et al.,* 1994a). Enhanced granulocyte and platelet recovery translated to less infection and hemorrhage and a shortened hospital stay with lower antibiotic and transfusion requirements (Henon *et al.,* 1992; To *et al.,* 1992; Peters *et al.,* 1993a). Furthermore, immunological studies suggest an earlier reconstitution of T-cell-mediated immunity (Roberts *et al.,* 1993). The improved margin of safety enabled a widening of the indications of high-dose chemotherapy. Therefore, patients with high-risk solid tumors such as breast cancer as well as older patients up to 65 years of age are now enrolled in high-dose chemotherapy protocols.

This chapter reviews the potential indications for high-dose chemotherapy and autologous SCT for a range of hematological malignancies and solid tumors. A number of other topics relevant to the development of autologous SCT will be reviewed. These include recent advances in peripheral stem cell mobilization techniques, the clinical significance of tumor cell contamination of autologous grafts, and potential therapies that would involve the *ex vivo* modification of autologous bone marrow stem cells.

II. HEMATOLOGICAL MALIGNANCIES

A. ACUTE MYELOID LEUKEMIA

In patients with *de novo* acute myeloid leukemia (AML), complete remission (CR) can be achieved in 65–70% of patients using induction chemotherapy with Ara-C and anthracyclines (Stone and Mayer, 1993). This results in a long-term, disease-free survival (DFS) of 20–40%. Age at diagnosis and chromosomal aberiations are the two most important indicators of prognosis (Curtis *et al.,* 1984; Schiffer *et al.,* 1989). Chromosomal aberrations have been identified in 60–70% of patients at diagnosis even without refined molecular techniques such as fluorescence *in situ* hybridization (FISH) and polymerase chain reaction (PCR) (Bloomfield and de la Chapelle, 1987). Based on their cytogenetics, patients with AML can be subdivided into three risk categories (Table 6.1) differing in the response rate and the duration of response (Bloomfield and de la Chapelle, 1987; Bloomfield, 1992).

Patients with inv(16), t(8;21), and t(15;17) have a high CR rate and long DFS, and thus high-dose chemotherapy is not indicated in CR1 (first complete remission) and should be reserved for patients in CR2. In contrast, patients with intermediate to poor prognostic markers such as t(9;11) and t(9;22) seem to have a fatal outcome in 90% of cases, and high-dose chemotherapy is preferable in CR1 (Keating *et al.,* 1988; Schiffer *et al.,* 1989). Cure rates are higher with allogeneic SCT (including grafts from unrelated donors) than autologous SCT,

TABLE 6.1 Prognostic Subgroups for Acute Myeloid Leukemia

Prognostic subgroup	Chromosomal aberration	Response to primary chemotherapy	Risk-adjusted consolidation therapy
Favorable	t(8;21), t(15;17), inv(16)	High CR rate, prolonged DFS	Conventional consolidation chemotherapy (two cycles)
Intermediate	t(9;11), normal karyotype	High CR rate, short DFS	Allogeneic sibling transplant in CR1; if no match or too old, autologous SCT with PBSC or purged marrow can be considered
Poor	All other aberrations, particularly t(9;22) and very complex chromosomal abnormalities	Low CR rate, early relapses, primary refractory disease	Allogeneic transplant including MUD in CR1; if no match or too old, autologous SCT with PBSC or purged marrow can be considered

probably due to a GVT effect and lack of leukemic contamination (Clift *et al.*, 1987). The intermediate-risk group with normal karyotype or t(9;11) often achieves remission, but remission duration is usually short. A European Organization for Research and Treatment of Cancer (EORTC) and Gruppo Italiano Malattie Ematologiche Maligne dell'Adulto (GIMEMA) Leukemia Cooperative Group study demonstrated a superior probability of DFS 5 years post transplant for both allogeneic and autologous SCT as compared to conventional intensive chemotherapy (55 and 48% versus 30%, respectively) (Zittoun *et al.*, 1995). Results for autologous SCT in CR2 are less favorable, but this still remains an option for patients without a related or unrelated donor or patients in whom older age precludes allogeneic SCT (Koerbling *et al.*, 1989). Purging of autologous marrow seems to result in improved clinical outcome as shown by Gorin *et al.* (1991), which will be discussed later in this chapter.

B. ACUTE LYMPHATIC LEUKEMIA

Intensive and prolonged conventional chemoradiotherapy provides a 70% probability of cure in children (Goekbuget and Hoelzer, 1995). Unfortunately, adult acute lymphatic leukemia (ALL) has a less favorable prognosis, with an overall long-term DFS of 40% (Hoelzer *et al.*, 1993). In the standard-risk group, treatment consists of induction, reinduction, and consolidation chemotherapy including prophylactic radiotherapy directed to the central nervous system (CNS) and intrathecal chemotherapy (Hoelzer *et al.*, 1993). Long-term maintenance therapy, i.e., for 24 months, is mandatory in B-lineage ALL, but probably not in T-lineage ALL (Goekbuget and Hoelzer 1995). Risk factors defining poor prognosis include age over 40 years at diagnosis, white cell count greater than 30×10^9/L, slow response to treatment, B- or T-lineage, and specific chromosomal aberrations such as t(4;11) and t(9;22) (Gaynor *et al.*, 1988; Hoelzer *et al.*, 1988; Boucheix *et al.*, 1994).

These risk factors seem more complex relative to AML, but future studies may show that, once again, age and chromosomal aberrations are the most important factors. The Philadelphia (Ph) translocation t(9;22) is detected in 55% of adult patients with ALL, especially pre-B-ALL, and is associated with a very poor prognosis, i.e., a long-term DFS of 10–15%. In contrast, T-ALL is rarely associated with t (9;22), which may explain the favorable prognosis of this subset (Maurer *et al.*, 1991; Rieder *et al.*, 1993). Pre-pre-B or pro-B-ALL also has a distinct chromosomal marker, t(4;11), as does B-ALL, t(8;14). Both are associated with a less favorable outcome with standard ALL treatment (Gaynor *et al.*, 1988; Hoelzer *et al.*, 1988; Fenaux *et al.*, 1989).

Current strategies for high-risk patients include high-dose chemotherapy and autologous or allogeneic SCT as consolidation in CR1, resulting in improved DFS (34–61%) (Hoelzer, 1994). In Ph-positive disease, allogeneic transplants are preferred (Barrett *et al.*, 1992), and the patient should be screened for t(9; 22) before autologous SCT. Experimental approaches based on tumor purging (e.g., antisense oligonucleotides) and posttransplant interferon are under investigation.

B-ALL requires treatment with rapidly alternating aggressive polychemotherapy, and high-dose tumor chemotherapy is not generally indicated in CR1 (Fenaux *et al.,* 1989). In CR2, prognosis with conventional chemotherapy is poor (DFS $< 10\%$), and allogeneic or autologous transplantation achieves superior results (DFS 40%) (Freund *et al.,* 1994).

C. CHRONIC MYELOGENOUS LEUKEMIA

At present, allogeneic stem cell transplantation provides the only curative therapy for patients with chronic myelogenous leukemia (CML) (McGlave, 1991). For chronic-, accelerated-, and blast-phase disease, the 5-year leukemia-free survival rates are 45, 30, and 6%, respectively (Gratwohl *et al.,* 1991). Donor T lymphocytes have an outstanding antileukemic effect as shown by the high relapse rate after syngeneic and T-cell-depleted allogeneic SCT (Goldman *et al.,* 1992). Furthermore, in patients who relapsed after allogeneic SCT, infusion of donor T cells produced clinical and even cytogenetic CR (Kolb *et al.,* 1990; Porter *et al.,* 1994). Consequently, at this stage autologous SCT is not indicated for CML because of a lack of GVT effect. Besides, t(9;22)-positive cells frequently contaminate autografts.

In cases where a suitable donor is not available, tumor-purging strategies may be required. These include purging with cyclophosphamide derivatives, purging with γ-interferon (McGlave *et al.,* 1990; Carlo-Stella *et al.,* 1991), or long-term culture of bone marrow, where culture purging leads to t(9;22) negativity in 30% of the patients (Turhan *et al.,* 1990; Udomsakdi *et al.,* 1992). Expansion of primitive hematopoietic precursors ($CD34^+CD38^-$ cells) may also result in Ph-negative grafts (Lewis *et al.,* manuscript in preparation).

In vivo purging is also possible as shown by Carella (1991) where aggressive induction chemotherapy with idarubicin, etoposide, and cytarabine resulted in Ph-negative harvests in 50% of the patients. Of the nine patients autografted, five were cytogenetically Ph-negative up to 19 months posttransplant. Similar results were published by Lo Coco (1990). Thus, autologous transplantation with blood mobilized progenitor cells is a reasonable approach in chronic-phase patients lacking an allogeneic donor. Although cure may not be achievable with such a regimen, it may establish transient or even long-lasting Ph-negative hematopoiesis and prolong the chronic phase.

D. HODGKIN'S DISEASE

Hodgkin's disease is one of the most chemosensitive neoplasias, and cure is achieved in 60–70% of the patients even in advanced disease (Bonadonna *et al.,* 1986; Canellos *et al.,* 1992). In relapse after cyclophosphamide/mechlorethamine, vincristine, procarbazine, and prednisone (COPP/MOPP) or doxorubicin, bleomycin, vinblastine, and dacarbazine (ABVD), 60% of relapsed patients are responsive to methyl-GAG, ifosfamide, methotrexate, and etoposide (MIME) or dexamethasone, BCNU, etoposide, ara-C, and melphalan (Dexa-BEAM) (Hagemeister *et al.,* 1987; Pfreundschuh *et al.,* 1994).

Autologous SCT provides long-term DFS in 40–50% of relapsed patients (Reece *et al.,* 1991; Schmitz *et al.,* 1993). Blood stem cell transplantation seems to achieve superior results as compared with autologous BMT (Kořbling *et al.,* 1990; Kessinger *et al.,* 1991). Most patients have undergone radiotherapy during their first-line treatment, and cyclophosphamide, BCNU, and etoposide (CBV) or BEAM are frequently used high-dose protocols. The role of allogeneic transplantation remains uncertain as previous studies failed to show a benefit in terms of improved long-term DFS (Anderson *et al.,* 1993). The results for autologous SCT are sufficiently encouraging to be recommended to patients with refractory disease, provided that the patient has no more than two failed regimens and a normal performance status (Bierman *et al.,* 1993). Allogeneic SCT should also be considered, although no clear GVT effect has been demonstrated.

E. NON-HODGKIN'S LYMPHOMA

Non-Hodgkin's lymphoma (NHL) comprises a heterogenous group of malignancies. Burkitt's lymphoma and lymphoblastic lymphoma are highly proliferating tumors which, like B-ALL, respond well to treatment with alternating aggressive polychemotherapy (Soussain *et al.,* 1995; Magrath *et al.,* 1996). High-dose chemotherapy would be considered for relapsed patients or patients initially presenting with factors indicating poor prognosis, i.e., bone marrow or CNS involvement and/or lactate dehydrogenase levels greater than 300 units at diagnosis (Coleman *et al.,* 1986; Troussard *et al.,* 1990). At present, early intensification with high-dose therapy and autologous or allogeneic stem cell transplantation for high-risk patients is under investigation (Troussard *et al.,* 1990; Baro *et al.,* 1992).

For patients with high- and intermediate-grade NHL who relapse, high-dose chemotherapy provides the only option to achieve long-term survival. The efficacy of high-dose chemotherapy and autologous SCT is related to tumor sensitivity to prior chemotherapy regimens. Relapsed patients with disease which had responded to prior chemotherapy regimens such as dexamethasone, cisplatin, and cytarabine (DHAP) had a 36% 3-year DFS, whereas DFS for patients with lymphoma resistant to salvage chemotherapy was only 14% (Philip *et al.,* 1987). A prospective randomized trial showed that the 5-year overall survival for chemotherapy- sensitive, non-Hodgkin's lymphoma in relapse treated with high-dose chemotherapy and autologous BMT was 53% as compared to 32% in the conventional-treatment arm (Philip *et al.,* 1995).

Indicators of a poor prognosis have been defined according to the International NHL Prognostic Factors Group as age greater than 60, disease stage greater than stage II, extranodal involvement, and elevated lactate dehydrogenase (Atkins, 1993). Up-front high-dose therapy in patients presenting with high-risk features is still a matter of debate. At present, the retrospective analysis of a large, multicenter French trial indicated an improved DFS in the high-dose arm. Gianni *et al.* obtained a similar result, showing improved DFS and overall survival (OS) in the SCT arm (Haioun *et al.,* 1994; Gianni *et al.,* 1996). The question whether sequential submyeloablative regimens with consecutive PBSC rescue might result in higher CR rates has been raised. Obviously, randomized trials to answer some of these questions are necessary.

Low-grade NHL in advanced stages is incurable by conventional chemotherapy. High-dose chemotherapy is capable of achieving complete remissions, but the probability of cure remains unclear (Haas *et al.,* 1994c). One factor hampering randomized studies is the slowly progressing nature of these malignancies, requiring follow-up in excess of 5 or even 10 years. Thus, in low-grade lymphomas, randomized clinical studies are even more critical to define the role of high-dose chemotherapy, taking into account the heterogeneity of the diseases and various stages.

F. MULTIPLE MYELOMA

Multiple myeloma (MM) is an incurable disease by chemotherapy in standard doses, and only less than 10% of patients with a high tumor burden (stage III) will survive longer than 5 years (Boccadoro *et al.,* 1991). Following the introduction of high-dose chemotherapy and autologous SCT for the treatment of MM, high response rates (CR $>$ 50%) were achieved even in patients with advanced disease, resulting in improved DFS after 5 years (Barlogie, 1991; Cunningham *et al.,* 1994). Attal (1996) reported the results of a randomized trial comparing up-front high-dose therapy consisting of melphalan and TBI with conventional chemotherapy. A significantly improved CR rate, DFS, and OS in the transplant arm was shown. In a subsequent trial, the same investigators did not demonstrate superior clinical outcome following double transplants using higher doses of melphalan.

Recent data suggest that B cells harboring the same *VDJ* gene rearrangement as the malignant plasma cells are frequently detected in the peripheral blood of myeloma patients, and they seem to be only marginally affected by current high-dose regimens (Pilarski and Belch, 1994; Bergsagel *et al.,* 1995). Immunophenotyping suggested that myeloma cells coexpress CD34, which was confirmed by the detection of mRNA in single cells using rt-PCR (reverse transcriptase-polymerase chain reaction) (Szczepek *et al.,* 1997). Thus, one of the most important issues in myeloma research lies in the characterization of the clonogenic precursor cell because it may impact on autologous SCT and CD34$^+$ cell selection.

At present, melphalan is the most effective single agent and is therefore administered alone or in combination with TBI as a conditioning regimen for

SCT (Jagannath *et al.,* 1990). High mortality due to infection and/or severe GVHD (40–45%) has cast some doubt on allogeneic BMT (Tricot *et al.,* 1996). This is at least in part attributable to patients with MM being usually older, more heavily pretreated, and having preexisting organ dysfunction secondary to MM. However, a graft-versus-myeloma effect was demonstrated by Barlogic and other investigators (Bensinger *et al.,* 1996). Bensinger *et al.* (1996) reported a relapse-free survival for patients achieving CR following allogeneic SCT of nearly 40% after 5 years, with four patients being in continuous complete remission more than 6 years posttransplant. This clearly suggests that future trials should incorporate allogeneic SCT earlier in the course of disease and aim to use less toxic preparative regimens (Tricot *et al.,* 1996). Introduction of allogeneic peripheral blood SCT in the treatment of MM may further reduce infectious complications due to the earlier neutrophil recovery (Grigg *et al.,* 1995; Russell *et al.,* 1996).

III. SOLID TUMORS

A. BREAST CANCER

Breast cancer is a common disease in western countries, and the yearly mortality in Europe and the United States is approximately 80 per 100,000 per year (Miller, 1991). Adjuvant high-dose chemotherapy has been used in women with primary breast cancer in the high-risk setting, i.e., in patients with 10 or more involved lymph nodes in the ipsilateral axilla (N10+). The probability of DFS after 10 years in patients with N10+ is poor (10–20%) despite conventional adjuvant chemotherapy (Bonadonna *et al.,* 1995). Thus, the introduction of a dose-escalated chemotherapy regimen is clearly attractive.

Numerous nonrandomized clinical trials have suggested a superior outcome with high-dose chemotherapy. Compared with historical controls, the probability of event-free survival was 80% versus 52% after 3 years and 93% versus 43% after 21 months, respectively (Gianni *et al.,* 1992; Peters *et al.,* 1993b). These encouraging results have led to the initiation of a number of randomized trials in the United States, Europe, and Australia, which should ultimately define the role of high-dose therapy in the adjuvant setting.

Two randomized studies suggest superior results following high-dose chemotherapy. Bezwoda reported a higher response rate (95% versus 53%) and longer DFS in the high-dose arm. Patients achieving CR before high-dose therapy had a particularly good prognosis (Bezwoda *et al.,* 1995). This finding was confirmed in recent data from the International Bone Marrow Transplant Registry (IBMTR) showing a significantly better long-term DFS for this patient population even 5 or more years after transplantation. Peters *et al.* (1988) showed improved DFS in patients receiving high-dose consolidation after induction chemotherapy with doxorubicin, cyclophosphamide, and 5-fluorouracil. The conventional treatment arm of Bezwoda's study (cyclophosphamide, mitoxantrone, and vincristine) did not receive adequate standard chemotherapy as reflected in no 2-year overall survival and was not directly comparable to the high-dose arm, which received a different combination of agents (cyclophosphamide, mitoxantrone, and etoposide). Nevertheless, these data indicate that high-dose consolidation may improve long-term DFS in women with metastatic breast cancer who do not have more than two involved sites and whose disease is responsive to conventional induction chemotherapy.

B. OVARIAN CANCER

There are increasing reports of high remission rates and possible DFS in patients with primary ovarian cancer who were enrolled into high-dose protocols (De Rosa *et al.,* 1996; Hahn *et al.,* 1996). High-dose protocols consisted of carboplatin, etoposide, and ifosfamide or double high-dose therapy with autologous stem cell rescue. In a review, Viens and Moraninchi (1995) reported a progression-free survival rate of 50% at 3 years posttransplantation in stage III/IV patients with chemosensitive disease. Unfortunately, patient numbers are small and the design of trials nonrandomized. However, the preliminary data are promising and should encourage larger randomized trials.

C. GERM CELL TUMORS

Germ cell malignancies comprise various seminomatous and nonseminomatous tumors, which most often affect young men. Although they account for less than 5% of all newly diagnosed malignancies per year, they are the second-most diagnosed tumors in young adult males. Even in advanced disease, 80% of the patients are cured with cisplatin-containing polychemotherapy (Einhorn, 1990). In relapse, still 30–40% of patients achieve long-term DFS or even cure with conventional salvage therapy consisting of cisplatin, ifosfamide, and etoposide (Einhorn *et al.,* 1992). Novel agents currently under investigation in the salvage setting are gemcitabine and the taxanes, taxol and taxotere (Bokemeyer *et al.,* 1994; Dunn *et al.,* 1997).

High-dose chemotherapy and SCT have been most often administered to relapsed, cisplatin-refractory patients or as consolidation following conventional salvage therapy. In refractory patients, the probability of long-term DFS was low (10–15%), whereas in patients responsive to salvage therapy, results of nonrandomized trials showed improved DFS rates of 51% following high-dose therapy (Motzer and Bosl, 1992; Siegert *et al.,* 1994). A few trials have incorporated high-dose chemotherapy as part of first-line treatment in subjects presenting with features suggesting poor prognosis, i.e., primary mediastinal disease, liver or bone metastases, β-human chorionic gonadotropin greater than 10,000 mIU/ml, and α-fetoprotein greater than 1000 ng/ml. One nonrandomized trial showed improved long-term DFS following high-dose chemotherapy

(Motzer *et al.,* 1993). Unfortunately, the other randomized trial, which did not include carboplatin, could not demonstrate superior results (Droz *et al.,* 1992). These initial studies show that the time is ripe for randomized prospective clinical trials to compare high-dose and conventional dose chemotherapy. Because of the low incidence of the disease, multicenter or even international studies will be required.

IV. PERIPHERAL BLOOD STEM CELL TRANSPLANTATION

A. MOBILIZATION STRATEGIES

Stem cell mobilization occurs after the administration of myelosuppressive chemotherapy, commonly cyclophosphamide in a dose of 4–7 g/m^2. During hematopoietic recovery there is up to a 50-fold increase in peripheral blood CFU-GM. The limitations of chemotherapy-induced mobilization consist of neutropenic septicemia, hemorrhage, and difficulty in predicting the time of recovery and harvest (Jagannath *et al.,* 1992; Kotasek *et al.,* 1992).

Several clinical trials demonstrated that the addition of hematopoietic growth factors results in an increased progenitor cell yield while myelotoxicity and its clinical sequelae are reduced (Gianni *et al.,* 1990; Haynes *et al.,* 1995). At present, granulocyte colony-stimulating factor (G-CSF) is the most commonly used growth factor. Schwartzberg described the additive mobilization effect of G-CSF in patients following high-dose cyclophosphamide with a four- to sixfold increase in the CD34$^+$ cell yield. Furthermore, the number of aphereses to achieve the threshold number of 20×10^4 CFU-GM/kg was significantly reduced from six to one (Schwartzberg, 1993). The recommended dose is 5 μg/kg/day subcutaneously. In many centers it is clinical practice to administer 300 μg per day. In our institution patients weighing up to 85 kg receive 300 μg whereas heavier patients receive 480 μg of G-CSF per day.

Granulocyte-macrophage colony-stimulating factor (GM-CSF) is similar in its mobilizing activity when given following chemotherapy but is less commonly used due to the more frequent adverse events such as fever, bone pain, and hypoxemia. Consequently, doses higher than the recommended 5 μg/kg are not indicated. Interleukin-3 (IL3) given alone does not effectively mobilize hematopoietic progenitor cells (Vose *et al.,* 1992). The combination with GM-CSF resulted in an increased number of CD34$^+$ cells, but no clear superiority over G-CSF could be demonstrated (Brugger *et al.,* 1992). PIXY 321, a GM-CSF/IL3 fusion protein, failed to show improved mobilization efficacy as compared with the parental agents (Winter *et al.,* 1995).

Hematopoietic growth factors can also be given alone. Sheridan *et al.* (1992) and Bensinger *et al.* (1993) reported the ability of G-CSF mobilized blood progenitor cells to reconstitute hematopoiesis. In both studies enhanced neutrophil recovery and platelet recovery as compared with autologous marrow were shown.

In allogeneic blood cell transplantation G-CSF has been used in almost all reported studies. The progenitor cell target is usually set at greater than 3×10^6 $CD34^+$ cells/kg. Generally, 10 μg/kg G-CSF is administered daily, and apheresis commenced at day 5. However, there seems to be a dose–yield relationship, and the level of $CD34^+$ cells is often sufficiently high for apheresis on day 4 (Hoglund *et al.,* 1996). In most of the patients one apheresis is sufficient to achieve the target number of $CD34^+$ cells.

Stem cell factor (SCF), the ligand for the c-kit receptor, has been shown in baboon experiments to act synergistically with G-CSF (Andrews *et al.,* 1991). Several clinical studies showed an increase in the $CD34^+$ cell yield when SCF was given together with G-CSF (Begley *et al.,* 1994; Basser *et al.,* 1995). The most significant finding was a 2.5- to 4-fold increase of peripheral $CD34^+$ cells in patients receiving both 25 μg/kg SCF and 10 μg/kg G-CSF as compared to patients receiving 10 μg/kg SCF with G-CSF or G-CSF alone (Glaspy *et al.,* 1995). The timing of apheresis is best guided by analysis of blood $CD34^+$ cell levels. Most centers use 20 $CD34^+$ cells/μl blood as the threshold for starting apheresis.

B. HEMATOPOIETIC RECONSTITUTION USING MOBILIZED HEMATOPOIETIC PROGENITORS

The majority of hematopoietic precursor cells mobilized into the peripheral blood are lineage committed with coexpression of CD33, CD45-RA, HLA-DR, and CD71 on $CD34^+$ cells (Haas *et al.,* 1994b). More primitive precursors, including pluripotent hematopoietic stem cells, have been defined by a lack of CD38 and HLA-DR and coexpression of Thy-1 (To *et al.,* 1994). The relative numbers of $CD34^+Thy\text{-}1^+lineage^-$ cells is 1.4-fold higher in mobilized peripheral blood than in bone marrow (Haas *et al.,* 1995).

Peripheral blood stem cells, which have been mobilized with growth factors, lead to an enhanced recovery of myeloid and megakaryocytic lineages, provided the number of infused $CD34^+$ cells is greater than a threshold of 2.0×10^6 cells/kg (To *et al.,* 1997). In comparison to autologous BMT, the duration of neutropenia as defined by an absolute neutrophil count of less than 0.5×10^9 cells/L, is about 10 days shorter, and unsupported platelet counts of greater than 20×10^9 cells/L are reached approximately 6 days earlier (Sheridan *et al.,* 1994). There seems to be a dose–response relationship for the platelet recovery as patients receiving greater than 5.0×10^6 $CD34^+$ cells had a significantly reduced duration of thrombocytopenia (Körbling *et al.,* 1995). The large number of T cells in unmodified PBSC grafts may result in an earlier reconstitution of T-cell-mediated immunity (Roberts *et al.,* 1993). Whether or not this translates into a decrease in infections remains an unanswered question.

Dercksen (1995) emphasized the role of L-selectin coexpression and recommended transplantation of greater than 2.1×10^6 $CD34^+$L-selectin$^+$ cells/kg to achieve rapid platelet recovery. Multiple studies have shown that addition of bone marrow to PBSC grafts does not shorten engraftment times (Ghalie *et al.*, 1994).

A major initial concern was the capacity of peripheral blood progenitor cells to maintain sustained hematopoiesis. PBSC transplant in the allogeneic setting resulted in rapid trilineage recovery and sustained hematopoiesis with donor genotype (Bensinger *et al.*, 1995; Schmitz *et al.*, 1995). This clearly indicates that circulating progenitors are capable of sustained hematopoiesis following transplantation.

In a randomized study comparing filgrastim-mobilized PBSCs and autologous bone marrow grafts, the median time to platelet recovery of greater than 20×10^9 cells/L maintained without platelet infusion was 16 versus 23 days, respectively (Schmitz *et al.*, 1996). Furthermore, earlier neutrophil recovery is a consistent finding in clinical studies using mobilized blood progenitor cells. More rapid neutrophil recovery translates into fewer febrile days and thus reduction in antibiotic treatment. Consequently, patients can be discharged earlier from the hospital. This has led to a more cost-efficient procedure (Henon *et al.*, 1992; To *et al.*, 1992; Peters *et al.*, 1993b). The conclusion might be drawn that autologous bone marrow transplantation is obsolete, a patient not responding to current mobilization protocols being the only exception.

V. MALIGNANT CONTAMINATION OF AUTOLOGOUS GRAFTS

Malignant contamination of the graft might be one of the reasons responsible for the higher relapse rate following autologous transplantation. Therefore, detection of malignant cells in either mobilized blood or bone marrow and subsequent removal from the graft constitute important issues. Novel, highly sensitive technologies such as multiparameter flow cytometry and rt-PCR are capable of detecting one malignant cell in up to 10^6 cells (Miller *et al.*, 1993; Reading *et al.*, 1993). In addition to these more sensitive tests, loss of chromosomal abnormalities has been used to define cytogenetic remission. Consequently, evidence of malignant contamination (or minimal residual disease, MRD) might be detected in patients being in complete remission as defined by clinical criteria and bone marrow morphology.

The source of tumor relapse may be related to contaminating tumor cells infused at the time of autologous SCT or failure of high-dose chemotherapy to eliminate residual disease. If the level of tumor cells contaminating autologous grafts is significant, then tumor cell purging techniques would be justified. To determine what are significant levels of tumor cell contamination of autologous grafts is difficult. Other than performing randomized clinical trials, which examine the efficacy of purging, gene-marking studies provide the only other direct method for establishing the source of relapse. At least in AML and neuroblastoma, gene-marking studies have provided evidence that contaminating tumor cells contribute to relapse (Brenner *et al.,* 1993; Ross *et al.,* 1993).

The development of sensitive techniques for detection of residual tumor cells has made a significant contribution to the field. In addition to determining the numbers of tumor cells in peripheral blood or bone marrow, some investigators have developed *in vitro* assays to measure the ability of contaminating tumor cells to proliferate and metastasize (Lapidot *et al.,* 1994; Campana and Pui, 1995).

A. DETECTION OF RESIDUAL DISEASE

In AML and ALL, immunophenotypic abnormalities can be detected by flow cytometry (Syrjala *et al.,* 1994). PCR-based techniques, which detect tumor-specific chromosomal aberrations, have an even higher sensitivity. The threshold of detection is approximately 1 malignant cell in 10^6–10^7 cells (Brisco *et al.,* 1994). In acute leukemia, residual malignant cells are detectable in bone marrow, mobilized PBSCs, or steady-state peripheral blood, although the number seems to be highest in bone marrow (Seriu *et al.,* 1995). This supports the findings of Gorin *et al.* (1991), who showed that purging of autografts resulted in a higher relapse-free survival in patients with AML. In ALL, the risk of relapse might be even higher, though the impact of purging on the clinical outcome has not been elucidated (Ramsay *et al.,* 1985).

MRD has also been detected in patients with malignant lymphoma and multiple myeloma. Both *in vitro* culture and molecular methods have been combined to determine the incidence of clonogenic tumor cells. Sharp *et al.* (1995) cultured lymphoma cells from bone marrow and mobilized peripheral harvests and showed that proliferating lymphoma cells harbored an immunoglobulin gene rearrangement identical to that of the original tumor. Negrin and Pesando (1994) showed that in 85% of patients with involved bone marrow, t(14;18) was detectable in the peripheral blood. However, Haas *et al.* (1994b) reported that B lymphoid precursor cells and mature $CD19^+$ cells were significantly less frequent in blood mobilized with chemotherapy and G-CSF compared with steady-state BM and PB (85- to 127-fold, respectively).

In multiple myeloma, the malignant cells are also characterized by a unique, monoclonal CDR-gene rearrangement. Using PCR, Dreyfus *et al.* (1995) ware able to detect a monoclonal population in 44% of 22 myeloma patients. Henry *et al.* (1996) and Witzig *et al.* (1995), using PCR, showed that myeloma cells were reduced in peripheral blood harvests as compared with autologous bone marrow. Bergsagel *et al.* (1995) demonstrated the presence of clonally related B lymphocytes in the peripheral blood of myeloma patients. These frequently detectable $CD19^+$ and $CD20^+$ cells exhibit a substantial resistance to adriamycin and bear an immunoglobulin gene rearrangement identical to that of the malignant plasma cells. Interestingly, cells with immunoglobulin gene rearrangements coexpress CD34 as detected by rt-PCR and surface antigen staining (Szczepek *et al.,* 1997). Obviously, these results, if confirmed by other investigators, will have impact on the efficacy of current CD34$^+$ selection procedures.

In epithelial tumors, i.e., breast, ovarian, and small-cell lung cancers, detection of contaminating tumor cells is most commonly based on immunocytochemistry (Schulze *et al.,* 1997). Cytokeratin antibodies seem to produce less false positive staining of mesenchymal cells than antibodies against epithelium-specific antigens (Schlimok *et al.,* 1987). One malignant cell per 10^6 cells can be detected by this technique (Osborne *et al.,* 1991). rt-PCR is thought to be even more sensitive, and *in situ* PCR might provide the best sensitivity and specificity by combining molecular detection with cell morphology.

Numerous publications have reported the detection of epithelial tumors such as breast cancer in bone marrow and peripheral blood (Datta *et al.,* 1994). Although there seems to be an increase in circulating tumor cells following mobilization, this almost always occurs in patients with preexisting disease in the marrow. In these cases, the amount of MRD was always higher in the bone marrow (Ross *et al.,* 1993; Brugger *et al.,* 1994; Ross *et al.,* 1995). Furthermore, Haas *et al.* (1997) showed that the risk of malignant contamination was much lower when the graft was collected following two cycles of chemotherapy compared with one cycle only. Thus *in vivo* purging by chemotherapy may also reduce the incidence of epithelial cell contamination of autologous SCT.

B. TUMOR-CELL PURGING

In AML, purging of the bone marrow with 4-hydroperoxycyclophosphamide (4HC) or mafosfamide seems to result in an improved long-term DFS. Gorin *et al.* (1991) showed that patients in CR1 transplanted with purged marrow had a significantly better leukemia-free survival as compared to the unpurged group, 63 versus 34% at 40 months posttransplant. Furthermore, Rizzoli demonstrated that adjustment of the mafosfamide dose according to the number of CFU-B is associated with a better clinical outcome (Gorin *et al.,* 1991). In this group, the probability of DFS 120 months posttransplant was 70% in the dose-adjusted mafosfamide arm and 50% in the standard-dose arm. These are very promising results, but a stratification into the three major risk groups would help to clarify whether patient prognosis may have contributed to the favorable outcome.

In contrast, an improved clinical outcome as measured by DFS has not been proven in ALL. In five studies using either immunomagnetic beads, monoclonal antibodies plus complement, or 4HC, the probability of DFS at 3 years posttransplant ranged between 17 and 32% (Ritz *et al.,* 1982; Rowley *et al.,* 1990; Gilmore *et al.,* 1991). Novel molecular purging strategies such as antisense nucleotides against p53 mRNA or the bcr-abl gene transcript particularly for

Ph^+ ALL are currently under clinical evaluation (Martiat *et al.,* 1993; de Fabritiis *et al.,* 1997).

Enrichment of $CD34^+$ cells by selection with immunomagnetic beads leads to a 3–5 log reduction of contaminating tumor cells (Schiller *et al.,* 1995). Therefore, $CD34^+$ cell selection represents a more widely applicable method for purging in CD34 antigen-negative malignant diseases (Shpall *et al.,* 1994). However, the detection of t(14;18) in the $CD34^+$ lymphopoietic progenitor cells of patients with follicular lymphoma indicates that $CD34^+$ selection alone might not entirely deplete the graft of tumor cells (Liu *et al.,* 1994). A multimodal approach combining positive selection with subsequent tumor depletion might achieve a graft free from malignant cells.

Clearly, stringent evaluation of the incidence of clonogenic tumor cells in autologous stem cell grafts is necessary. New technologies for detecting rare cell populations such as PCR have greatly assisted in detection of tumor cells at low frequency. In addition to identification of a malignant phenotype, the proliferative and metastatic potential of tumor cells requires evaluation using *in vitro* clonogenic assays and perhaps xenograft models of human engraftment. Ultimately, carefully controlled, randomized prospective clinical trials are required to examine the clinical efficacy of tumor cell purging methods.

VI. FUTURE DEVELOPMENTS

A. SINGLE APHERESIS

Collection of sufficient progenitor cells for the hematopoietic rescue of at least one high-dose chemotherapy cycle with a single apheresis is of clinical interest. Several investigators have shown the feasibility of such an approach (Hillyer *et al.,* 1991; Malachowski *et al.,* 1992). Recently, Murea *et al.* (1996) reported the results of large-scale apheresis, processing 20 L of peripheral blood in one collection during G-CSF-enhanced recovery following cytotoxic chemotherapy. An almost twofold increase in the number of $CD34^+$ cells was achieved and 74% of patients had collections greater than 2.5×10^6 $CD34^+$ cells/kg in one apheresis. This compared favorably with 10-L collections, where only 52% of patients reached this level. Interestingly, there was a trend toward higher numbers of $CD34^+$ cells in the second 10-L portion, suggesting a mobilization effect due to the procedure. The antigenic profile of the progenitor cells, i.e., expression of CD34, CD38, Thy-1, and HLA-DR, remained constant throughout the harvest. The use of central venous catheters designed for hemodialysis enabled higher flow rates (130–150 ml/min) and processing of the 20-L collection in approximately 2 h (Hahn *et al.,* 1995). No adverse events were reported. The authors conclude that these procedures should become standard and may be particularly helpful in difficult to mobilize patients.

B. *EX VIVO* EXPANSION

The discovery of hematopoietic growth factors has enabled *in vitro* techniques for the expansion of hematopoietic progenitor cells from autologous or allogeneic stem cell sources. Lineage differentiation is partially controllable by optimization of culture conditions, particularly the combinations of cytokines and, as yet, poorly defined adhesive interactions. From a clinical perspective, there would be many potential applications for these developing technologies. Expansion of the most primitive compartment, that is, cells that are responsible for long-term reconstitution of hematopoiesis following myeloablation, would represent the most significant advance. Morphologically, these cells are undifferentiated, $CD34^+$, Thy-1^+, $CD38^-$, and HLA-DR^-. They require multiple synergistic hematopoietic growth factors for proliferation and, because of their long-term persistence *in vivo,* are critical target cells in gene therapy (Piacibello *et al.,* 1997). Renewal of the stem cell phenotype *in vitro* is an important objective since retroviral vectors for gene delivery require cell division for stable integration to occur. Another objective would be to generate a graft with greater than 2.0×10^6 $CD34^+$ cells/kg from a small sample of bone marrow, peripheral, or cord blood. In addition to reducing the amount of harvested material required for transplantation, increased numbers of hematopoietic progenitors may allow multiple cycles of high-dose chemotherapy.

It appears that very primitive progenitors may be expanded using soluble cytokines alone, though much more experimental evidence is required to prove that stem cell progeny generated *in vitro* retain the ability to reconstitute hematopoiesis *in vivo.* The addition of Flt-3 ligand and thrombopoietin to cytokine cocktails may be crucial. This is best shown in a study where cord blood derived $CD34^+$ cells were stimulated by Flt-3 ligand and thrombopoietin in a stroma-free system. Over 20 weeks, long-term culture initiating cells (LTC-IC) and CFU-GM were expanded by 5–6 log (Piacibello *et al.,* 1997). Flt-3 ligand alone or in combination with other early-acting growth factors appears to increase the rate of cell cycle recruitment of quiescent $CD34^+CD38^-$ cells in adult marrow (Haylock *et al.,* 1997).

Lastly, little is known about the properties of cells that give rise to the rapid recovery in platelets and neutrophils following mobilized peripheral blood stem cell transplants. Expansion of these cell subsets could potentially hasten post-transplant recovery in situations where recovery is prolonged, i.e., cord blood transplantation. Production of granulocytes and their precursors from $CD34^+$ cells is possible using a combination of four growth factors (IL3, IL6, SCF, and G-CSF) (Haylock *et al.,* 1994). Three clinical studies showed the feasibility of large-scale expansion of hematopoietic progenitors for autologous transplantation in adults (Haylock *et al.,* 1994; Alcorn *et al.,* 1996; Williams *et al.,* 1996). However, significant reduction of the severity and duration of neutropenia was not shown, although the number of cells was less than 10^{11}, the number of progenitors which would be required to alter the rate of recovery as postulated by Haylock *et al.* (1994).

Adoptive cellular immunotherapy using natural killer (NK) cells or cytotoxic T lymphocytes to suppress MRD may have clinical potential. Following mobilization, the graft is enriched in NK cells and their incidence and activity increased throughout the entire collection period (Silva *et al.,* 1995; Verbik *et al.,* 1995). Their numbers can be expanded three- to fivefold by administration of IL2 (Silva *et al.,* 1995). Reinfusion of the NK cells 2 days following PBSC transplant did not adversely affect engraftment nor lead to additional toxicity (Lister *et al.,* 1993). Procedures for preparation of cell subsets that amplify GVT reactivity of autologous grafts are an important research objective.

C. GENE THERAPY

The aim of some gene therapy approaches is to achieve long-term expression of a therapeutic or marker gene by stable integration of the transgene into a target cell type that has the potential to persist *in vivo* in a quiescent or cycling state for years. Hematopoietic stem cells have these properties, and much research has focused on more efficient methods for gene transfer to this cell type. Although a high transduction rate has been reported *in vitro* using retroviral vectors, long-term expression posttransplant is rarely seen (Dunbar *et al.,* 1995). Human stromal cells might be an alternate target for gene therapy because they are secretory cells and can be manipulated *in vitro* more easily than hematopoietic stem cells (Simmons and Gronthos, 1996). At present, clinical indications for gene therapy comprise single-gene disorders such as severe combined immunodeficiency syndrome, chronic granulomatous disease, and Gaucher's disease (Kohn *et al.,* 1995; Malech *et al.,* 1995; Nimgaonkar *et al.,* 1995). Provided that safety issues are adequately addressed, introduction of the multidrug resistance (MDR) gene into normal CD34$^+$ cells represents a novel approach which would render the marrow resistant to chemotherapy-dose escalation (Hanania *et al.,* 1995).

D. NONMALIGNANT DISEASES

The indications of autotransplants in nonmalignant diseases are mostly, but not entirely, covered in the previous gene therapy section. Prince *et al.* (1995) reported that most of the mobilized CD34$^+$CD38$^-$ cells in patients with paroxysmal nocturnal hemoglobinuria (PNH) possess a normal surface phenotype (CD59$^+$DAF$^+$). This condition arises from somatic mutation of a bone marrow progenitor that disrupts glycosylinositol phospholipid (GPI) anchoring of cell surface proteins. GPI-anchored decay accelerating factor (DAF) and CD59 are absent in affected cells. The authors concluded that apheresis samples could serve as a source of unaffected stem cells for autologous marrow transplantation of PNH patients since the blood stem cell compartment, which is thought to reside within the $CD34^+CD38^-$ cell subset, is not affected by the mutation.

Severe autoimmune disorders such as systemic lupus erythematosis and rheumatoid arthritis are potential indications for myeloablative chemoradiotherapy (Marmont and van Bekkum, 1995; Marmont *et al.,* 1995). The complete eradication of T-cell-mediated immunity may encourage suppression of autoimmune T-cell clones during the recovery of T-cell lymphopoiesis following autologous SCT. T-cell depletion of mobilized PBSC grafts may be required since PBSCs and bone marrow transplants are rich in immunocompetent T lymphocytes. Additionally, TBI and anti-thymocyte globulin (ATG) should be integral parts of the high-dose conditioning regimen. The use of unmodified PBSCs and lack of TBI and ATG in all conditioning regimens may explain the disappointing outcome of five patients reported previously (Euler *et al.,* 1996).

VII. SUMMARY

In autologous SCT, hematopoietic progenitor cells harvested from mobilized blood have largely superseded bone marrow as the preferred source of stem cells. Due to the faster hematopoietic recovery, PBSC transplants allow for safer and more cost-effective high-dose therapy. Granulocyte colony-stimulating factor alone or in combination with cytotoxic chemotherapy are the most commonly used mobilization regimens. In adults, a single large-scale apheresis can yield greater than 2.0×10^6 $CD34^+$ cells/kg in most patients and should be sufficient to rescue at least one high-dose therapy cycle with a single collection procedure. The large number of hematopoietic progenitors in multiple blood stem cell harvests allows for two or three cycles of high-dose therapy with subsequent autologous rescue. Addition of novel hematopoietic growth factors such as stem cell factor, Flt-3 ligand, or thrombopoietin may provide higher progenitor cell numbers, especially in those who fail current mobilization protocols. The impact of various mobilization protocols on comobilization of malignant cells is still unknown.

Ex vivo technologies for $CD34^+$ cell selection, in addition to representing an elegant purging tool in $CD34^-$ malignancies, have provided a highly purified hematopoietic progenitor cell population for gene therapy. Developing technologies for *ex vivo* expansion of progenitor cell subsets may provide sources of cells for both short- and long-term hematopoietic reconstitution. In addition, culture techniques may provide methods for purging Ph^+ cells from autologous transplants in CML patients who have no suitable donor. Autologous transplantation is also likely to impact on the treatment of nonmalignant disease such as PNH and severe autoimmune disease. The goal of safe transplantation having been achieved, the next challenge will be to assess the utility of various *ex vivo* strategies for amplifying host immunological responses to malignancy.

REFERENCES

Alcorn, M. J., Holyoake, T. L., Richmond, L., *et al.* (1996). CD34-positive cells isolated from cryopreserved peripheral-blood progenitor cells can be expanded *ex vivo* and used for transplantation with little or no toxicity. *J. Clin. Oncol.* **14,** 1839–1847.

Anderson, J. E., Litzow, M. R., Appelbaum, F. R., *et al.* (1993). Allogeneic, syngeneic, and autologous marrow transplantation for Hodgkin's disease: The 21-year Seattle experience. *J. Clin. Oncol.* **11,** 2342–2350.

Andrews, G., Knitter, G. H., Bartelmex, S. H., *et al.* (1991). Recombinant human stem cell factor, a c-kit ligand, stimulates hematopoiesis in primates. *Blood* **78,** 1975–1980.

Atkins, C. D. (1993). A predictive model for aggressive non-Hodgkin's lymphoma. The International Non-Hodgkin's Lymphoma Prognostic Factors Project. *N. Engl. J. Med.* **329,** 987–994.

Attal, M., Harousseau, J. L., Stoppa, A. M., *et al.* (1996). A prospective randomized trial of autologous bone marrow transplantation and chemotherapy in multiple myeloma. *N. Engl. J. Med.* **335,** 91–97.

Barlogie, B. (1991). Toward a cure for multiple myeloma? *N. Engl. J. Med.* **325,** 1304–1306.

Baro, J., Richard, C., Sierra, J., *et al.* (1992). Autologous bone marrow transplantation in 22 adult patients with lymphoblastic lymphoma responsive to conventional dose chemotherapy. *Bone Marrow Transplant.* **10,** 33–38.

Barrett, A. J., Horowitz, M. M., Ash, R. C., *et al.* (1992). Bone marrow transplantation for Philadelphia chromosome-positive acute lymphoblastic leukemia. *Blood* **79,** 3067–3070.

Basser, R., Begley, C. G., Mansfield, R., *et al.* (1995). Mobilization of PBPC by priming with stem cell factor (SCF) before filgrastim compared to concurrent administration. *Blood* **86,** 687a.

Begley, C. G., Basser, R., Mansfield, R., *et al.* (1994). Randomized prospective study demonstrating a prolonged effect of SCF with G-CSF (filgrastim) on PBPC in untreated patients: Early results. *Blood* **84,** 25a.

Bensinger, W., Singer, J., Appelbaum, F., *et al.* (1993). Autologous transplantation with peripheral blood mononuclear cells collected after administration of recombinant granulocyte colony stimulating factor. *Blood* **81,** 3158–3163.

Bensinger, W. I., Weaver, C. H., Appelbaum, F. R., *et al.* (1995). Transplantation of allogeneic peripheral blood stem cells mobilized by recombinant human granulocyte colony-stimulating factor. *Blood* **85,** 1655–1658.

Bensinger, W. I., Buckner, C. D., Anasetti, C., *et al.* (1996). Allogeneic marrow transplantation for multiple myeloma: An analysis of risk factors on outcome. *Blood* **88,** 2787–2793.

Bergsagel, P. L., Masellis Smith, A., Smith, A., *et al.* (1995). In multiple myeloma clonotypic B lymphocytes are detectable among $CD19^+$ peripheral blood cells expressing CD38, CD56 and monotypic immunoglobulin light chain. *Blood* **85,** 436–447.

Bezwoda, W. R., Seymoure, L., and Dansey, R. D. (1995). High-dose chemotherapy with hematopoietic rescue as primary treatment for metastatic breast cancer: A randomized trial. *J. Clin. Oncol.* **13,** 2483–2489.

Bierman, P. J., Bagin, R. G., Jagannath, S., *et al.* (1993). High dose chemotherapy followed by autologous hematopoietic rescue in Hodgkin's disease: Long term follow-up in 128 patients. *Ann. Oncol.* **4,** 767–773.

Bloomfield, C. D. (1992). Prognostic factors for selecting curative therapy for adult acute myeloid leukemia. *Leukemia* **6,** 65–67.

Bloomfield, C. D., and de la Chapelle, A. (1987). Chromosome abnormalities in acute nonlymphocytic leukemia: Clinical and biologic significance. *Semin. Oncol.* **14,** 372–383.

Boccadoro, M., Marmont, F., Tribalto, M., *et al.* (1991). VMCP/VBAP alternating combination chemotherapy is not superior to melphalan and prednisone even in high-risk patients. *J. Clin. Oncol.* **9,** 444–448.

Bokemeyer, C., Schmcll, H. J., Natt, F., *et al.* (1994). Preliminary results of a phase I/II trial of paclitaxel in patients with relapsed or cisplatin-refractory testicular cancer. *J. Cancer Res. Clin. Oncol.* **120,** 754–757.

Bonadonna, G., Valagussa, P., and Santoro, A. (1986). Alternating non-cross-resistant combination chemotherapy or MOPP in stage IV Hodgkin's disease. *Ann. Intern. Med.* **104,** 739–746.

Bonadonna, G., Zambetti, M., and Valagussa, P. (1995). Sequential or alternating doxorubicin and CMF regimens in breast cancer with more than three positive nodes: Ten-year results. *JAMA* **273,** 542–547.

Boucheix, C., David, B., Sebban, C., *et al.* (1994). Immunophenotype of adult acute lymphoblastic leukemia, clinical parameters, and outcome: An analysis of a prospective trial including 562 tested patients (LALA87). French Group on Therapy for Adult Acute Lymphoblastic Leukemia. *Blood* **84,** 1603–1612.

Brenner, M. K., Rill, D. R., Moen, R. C., *et al.* (1993). Gene-marking to trace origin of relapse after autologous bone marrow transplantation. *Lancet* **341,** 85–86.

Brisco, M. J., Condon, J., Hughes, E., *et al.* (1994). Outcome prediction in childhood acute lymphoblastic leukaemia by molecular quantification of residual disease at the end of induction. *Lancet* **343,** 196–200.

Brugger, W., Bross, K., Frisch, J., *et al.* (1992). Mobilization of peripheral blood progenitor cells by sequential administration of interleukin-3 and granulocyte macrophage colony-stimulating factor following polychemotherapy with etoposide, ifosfamide and cisplatin. *Blood* **70,** 1193–1200.

Brugger, W., Bross, K. J., Glatt, M., *et al.* (1994). Mobilization of tumor cells and hematopoietic progenitor cells into peripheral blood of patients with solid tumors. *Blood* **83,** 636–640.

Campana, D., and Pui, C. H. (1995). Detection of minimal residual disease in acute leukemia: Methodologic advances and clinical significance. *Blood* **85,** 1416–1434.

Canellos, G. P., Anderson, J. R., Propert, K. J., *et al.* (1992). Chemotherapy of advanced Hodgkin's disease with MOPP, ABVD, or MOPP alternating with ABVD. *N. Engl. J. Med.* **327,** 1478–1484.

Carella, A. M., Podesta, M., Grassoni, F., *et al.* (1991). Selective overshoot of Ph-negative blood hemopoietic cells after an intensive idarubicin-containing regimen and their repopulating capacity after reinfusion. *Stem Cells* **11,** S67–S72.

Carlo-Stella, C., Mangoni, L., Piovani, G., *et al.* (1991). *In vitro* marrow purging in chronic myelogenous leukemia: Effect of mafosfamide and recombinant granulocyte-macrophage colony-stimulating factor. *Bone Marrow Transplant.* **8,** 265–273.

Clift, R. A., Buckner, C. D., Thomas, E. D., *et al.* (1987). The treatment of acute non-lymphoblast leukemia by allogeneic marrow transplantation. *Bone Marrow Transplant.* **2,** 243–258.

Coleman, C. N., Picozzi, V. J., Jr., Cox, R. S., *et al.* (1986). Treatment of lymphoblastic lymphoma in adults. *J. Clin. Oncol.* **4,** 1628–1637.

Cunningham, D., Paz-Ares, L., Milan, S., *et al.* (1994). High-dose melphalan and autologous bone marrow transplantation as consolidation in previously untreated myeloma. *J. Clin. Oncol.* **12,** 759–763.

Curtis, J. E., Messner, H. A., Hasselback, R., *et al.* (1984). Contributions of host- and disease-related attributes to the outcome of patients with acute myelogenous leukemia. *J. Clin. Oncol.* **2,** 253–259.

Datta, Y. H., Adams, P. T., Drobyski, W. R., *et al.* (1994). Sensitive detection of occult breast cancer by the reverse-transcriptase polymerase chain reaction. *J. Clin. Oncol.* **12,** 475–482.

de Fabritiis, P., Skorski, T., De Propris, M. S., *et al.* (1997). Effect of bcr-abl oligodeoxynucleotides on the clonogenic growth of chronic myelogenous leukaemia cells. *Leukemia* **11,** 811–819.

De Rosa, L., Montuoro, A., Amodeo, R., *et al.* (1996). Two successive high dose chemotherapy (HDC) with circulating progenitor cells (CPC) support in advanced ovarian cancer. *Bone Marrow Transplant.* **17,** S32.

Dercksen, M. W., Gerritsen, W. R., Rodenhuis, S., *et al.* (1995). Expression of adhesion molecules on $CD34^+$ cells: $CD34^+$ L-selectin$^+$ cells predict a rapid platelet recovery after peripheral blood stem cell transplantation. *Blood* **85,** 3313–3319.

Dreyfus, F., Ribrag, V., Leblond, V., *et al.* (1995). Detection of malignant B cells in peripheral blood stem cell collections after chemotherapy in patients with multiple myeloma. *Bone Marrow Transplant.* **15,** 707–711.

Droz, J., Pico, J., Biron, P., *et al.* (1992). No evidence of a benefit of early intensified chemotherapy (HDCT) with autologous bone marrow transplantation (ABMT) in first line treatment of poor risk non-seminomatous germ cell tumors. *Proc. Am. Soc. Clin. Oncol.* **11,** 197.

Dunbar, C. D., Cottler-Fox, M., O'Shaughnessy, J. A., *et al.* (1995). Retrovitally marked CD34-enriched peripheral blood and bone marrow cells contribute to long-term engraftment after autologous transplantation. *Blood* **85,** 3048–3057.

Dunn, T. A., Grunwald, V., Bokmeyer, C., *et al.* (1997). Pre-clinical activity of taxol in non-seminomatous germ cell tumor cell lines and nude mouse xenografts. *Invest. New Drugs* **15,** 91–98.

Einhorn, L. (1990). Treatment of testicular cancer: A new and improved model. *J. Clin. Oncol.* **8,** 1777–1781.

Einhorn, L., Weathers, T., Loehrer, P., *et al.* (1992). Second line chemotherapy with vinblastine, ifosfamide, and cisplatin after initial chemotherapy with cisplatin, VP-16 and bleomycin (PVP-16B) in disseminated germ cell tumors (GCT). *Proc. Am. Soc. Clin. Oncol.* **11,** 196.

Euler, H. H., Marmont, A. M., Bacigalupo, A., *et al.* (1996). Early recurrence or persistence of autoimmune diseases after unmanipulated autologous stem cell transplantation. *Blood* **88,** 3621–3625.

Fenaux, P., Lai, J. L., Miaux, O., *et al.* (1989). Burkitt cell acute leukemia (L3 ALL) in adults: A report of 18 cases. *Br. J. Haematol.* **71,** 371–376.

Ferrara, J. L. M., and Deeg, H. J. (1991). Graft versus host disease. *N. Engl. J. Med.* **324,** 667–674.

Frei, E., Antman, K., Teicher, B., *et al.* (1980). Dose: A critical factor in cancer chemotherapy. *Am. J. Med.* **69,** 585–593.

Freund, M., Heil, G., Arnold, R., *et al.* (1994). Treatment of relapsed or refractory adult acute lymphocytic leukemia. *Ann. Hematol.* **68,** S10.

Gaynor, J., Chapman, D., Little, C., *et al.* (1988). A cause-specific hazard rate analysis of prognostic factors among 199 adults with acute lymphoblastic leukemia: The Memorial Hospital experience since 1969. *J. Clin. Oncol.* **6,** 1014–1030.

Ghalie, R., Richman, C. M., Bender, J. G., *et al.* (1994). Sequential transplants using mobilized peripheral blood progenitor cells. *J. Clin. Apheresis* **9,** 176–182.

Gianni, A. M., Bregni, M., Siena, S., *et al.* (1990). Recombinant human granulocyte-macrophage colony-stimulating factor reduces hematologic toxicity and widens clinical applicability of high-dose cyclophosphamide treatment in breast cancer and non-Hodgkin's lymphoma. *J. Clin. Oncol.* **8,** 768–778.

Gianni, A. M., Bregni, M., Siena, S., *et al.* (1996). Is high dose better than standard-dose chemotherapy as initial treatment of poor-risk large cell lymphomas? A criticial analysis from available randomized trials. *Ann. Oncol.* **7,** S12.

Gianni, A. M., Siena, S., Bregni, M., *et al.* (1989). Granulocyte-macrophage colony-stimulating factor to harvest circulating haemopoietic stem cells for autotransplantation. *Lancet* **2,** 580–585.

Gianni, A. M., Siena, S., Bregni, M., *et al.* (1992). Growth-factor supported high-dose sequential (HDS) adjuvant chemotherapy in breast cancer with at least 10 positive nodes. *Proc. Am. Soc. Clin. Oncol.* **11,** 68.

Gilmore, M., Hamon, H., Prentice, H., *et al.* (1991). Failure of purged autologous bone marrow transplantion in high risk acute lymphoblastic leukemia in first complete remission. *Bone Marrow Transplant.* **8,** 19–26.

Glaspy, J., LeMaistre, C. F., Lill, M., *et al.* (1995). Dose-response of 7 day administration of recombinant methionyl human stem cell factor (SCF) in combination with filgrastim (G-CSF) for progenitor cell mobilization in patients with stage II–IV breast cancer. *Blood* **86,** 463a.

Goekbuget, N., and Hoelzer, D. (1995). Therapy of acute lymphoblastic leukemia in adults. *Onkologie* **18,** 292–300.

Goldman, J., McGlave, P., Szydlo, R., *et al.* (1992). Impact of disease duration and prior treatment on outcome of bone marrow transplantation for chronic myelogenous leukemia. *Exp. Hematol.* **6,** 473a.

Goldman, J. M., Th'ing, K. H., Park, D. S., *et al.* (1978). Collection, cryopreservation and subsequent viability of haemopoietic stem cells intended for treatment of chronic granulocytic leukaemia in transformation. *Br. J. Haematol.* **40,** 185–195.

Gorin, N. C., Labopin, M., Meloni, G., *et al.* (1991). Autologous bone marrow transplantation for acute myeloblastic leukemia in Europe: Further evidence of the role of marrow purging by mafosfamide. *Leukemia* **5,** 896–904.

Gratwohl, A., Hermans, J., and Baldomero, H. (1996). Hematopoietic precursor cell transplants in Europe: Activity in 1994. Report from the European Group for Blood and Marrow Transplantation (EBMT). *Bone Marrow Transplant.* **17,** 137–148.

Gratwohl, A., Hermans, J., Niederwieser, D., *et al.* (1991). Bone marrow transplantation for chronic myeloid leukemia: Long-term results. *Bone Marrow Transplant.* **12,** 509–516.

Grigg, A. P., Roberts, A. W., Raunow, H., *et al.* (1995). Optimizing dose and scheduling of filgrastim (granulocyte colony-stimulating factor) for mobilization and collection of peripheral blood progenitor cells in normal volunteers. *Blood* **86,** 4437–4445.

Haas, R., Möhle, R., Fruhauf, S., *et al.* (1994a). Patient characteristics associated with successful mobilizing and autografting of peripheral blood progenitor cells in malignant lymphoma. *Blood* **83,** 3787–3794.

Haas, R., Möhle, R., Murea, S., *et al.* (1994b). Characterization of peripheral blood progenitor cells mobilized by cytotoxic chemotherapy and recombinant human granulocyte colony-stimulating factor. *J. Hematother.* **3,** 323–330.

Haas, R., Möhle, R., Pforsich, M., *et al.* (1995). Blood derived autografts collected during granulocyte colony-stimulating factor-enhanced recovery are enriched with early Thy-1+ hematopoietic progenitor cells. *Blood* **85,** 949–955.

Haas, R., Moos, M., Karcher, A., *et al.* (1994c). Sequential high-dose therapy with peripheral-blood progenitor-cell support in low-grade non-Hodgkin's lymphoma. *J. Clin. Oncol.* **12,** 1685–1692.

Haas, R., Schmid, H., Hahn, U., *et al.* (1997). Tandem high-dose therapy with ifosfamide, epirubicin, carboplatin and peripheral blood stem cell support is an effective adjuvant treatment for high-risk primary breast cancer. *Eur. J. Cancer* **33,** 372–378.

Hagemeister, F. B., Tannir, N., McLaughlin, P., *et al.* (1987). MIME chemotherapy (methyl-GAG, ifosfamide, methotrexate, etoposide) as treatment for recurrent Hodgkin's disease. *J. Clin. Oncol.* **5,** 556–561.

Hahn, U., Goldschmidt, H., Salwender, H., *et al.* (1995). Large-bore central venous catheters for the collection of peripheral blood stem cells. *J. Clin. Apheresis* **10,** 12–16.

Hahn, U., Haas, R., Schmid, H., *et al.* (1996). Sequential high-dose chemotherapy with peripheral blood progenitor cell (PBPC) support in patients with advanced ovarian carcinoma. *Bone Marrow Transplant.* **17,** S33.

Haioun, C., Lepage, E., Gisselbrecht, C., *et al.* (1994). Comparison of autologous bone marrow transplantation with sequential chemotherapy for intermediate-grade and high-grade non-Hodgkin's lymphoma in first complete remission: A study of 464 patients. *J. Clin. Oncol.* **12,** 2543–2550.

Hanania, E. G., Kavanagh, J., Hortobagyi, G., *et al.* (1995). Recent advances in the application of gene therapy to human disease. *Am. J. Med.* **99,** 537–552.

Haylock, D. N., Horsfall, M. J., Dowse, T. L., *et al.* (1997). Increased recruitment of hemopoietic progenitor cells underlies the *ex vivo* expansion potential of FLT3 ligand. *Blood* **90,** 2260–2272.

Haylock, D. N., Makino, S., Dowse, T. L., *et al.* (1994). *Ex vivo* hematopoietic progenitor cell expansion. *Immunomethods* **5,** 217–225.

Haynes, A., Hunter, A., McQuaker, G., *et al.* (1995). Engraftment characteristics of peripheral blood stem cells mobilised with cyclophosphamide and the delayed addition of G-CSF. *Bone Marrow Transplant.* **16,** 359–363.

Henon, P. H. R., Liang, H., Beck-Wirth, G., *et al.* (1992). Comparison of hematopoietic and immune recovery after autologous bone marrow or blood stem cell transplants. *Bone Marrow Transplant.* **9,** 285–291.

Henry, J. M., Sykes, P. J., Brisco, M. H., *et al.* (1996). Comparison of myeloma cell contamination of bone marrow and peripheral blood stem cell harvests. *Br. J. Haematol.* **92,** 614–619.

Hillyer, C. D., Tiegerman, K. O., and Berkman, E. M. (1991). Increase in circulating colony-forming units-granulocyte-macrophage during large-volume leukapheresis: Evaluation of a new cell separator. *Transfusion* **31,** 327–332.

Hoelzer, D. (1994) Treatment of acute lymphoblastic leukemia *Semin. Hematol.* **31,** 1–15.

Hoelzer, D., Thiel, E., Loffler, H., *et al.* (1988). Prognostic factors in a multicenter study for treatment of acute lymphoblastic leukemia in adults. *Blood* **71,** 123–131.

Hoelzer, D., Thiel, E., Ludwig, W. D., *et al.* (1993). Follow-up of the first two successive German multicentre trials for adult ALL (01/81 and 02/84). *Leukemia* **7,** (Suppl. 2), S130–S134.

Hoglund, M., Smedmyr, B., Stimonsson, B., *et al.* (1996). Dose-dependent mobilisation of haematopoietic progenitor cells in healthy volunteers receiving glycosylated rHuG-CSF. *Bone Marrow Transplant.* **18,** 19–27.

Horowitz, M. M., Gale, R. P., Sondel, P. M., *et al.* (1990). Graft versus leukemia reaction after bone marrow transplantation. *Blood* **75,** 555–562.

Hryniuk, W. M., and Bush, H. (1984). The importance of dose intensity in chemotherapy of metastatic breast cancer. *J. Clin. Oncol.* **2,** 1281–1288.

Jagannath, S., Barlogie, B., Dicke, K. A., *et al.* (1990). Autologous bone marrow transplantation in multiple myeloma: Identification of prognostic factors. *Blood* **76,** 1860–1866.

Jagannath, S., Vesole, D. H., Glenn, L., *et al.* (1992). Low risk intensive therapy for multiple myeloma with combined autologous bone marrow and blood stem cell support. *Blood* **80,** 1666–1672.

Keating, M. J., Smith, T. L., Kantarjian, H., *et al.* (1988). Cytogenetic pattern in acute myelogenous leukemia: A major reproducible determinant of outcome. *Leukemia* **2,** 403–412.

Kessinger, A., Bierman, P. J., Vose, J. M., *et al.* (1991). High-dose cyclophosphamide, carmustine and etoposide followed by autologous peripheral stem cell transplantation for patients with relapsed Hodgkin's disease. *Blood* **77,** 2322–2325.

Koerbling, M., Hunstein, W., Fliedner, T. M., *et al.* (1989). Disease-free survival after autologous bone marrow transplantation in patients with acute myelogenous leukemia. *Blood* **74,** 1898–1904.

Kohn, D. B., Weinberg, K. I., Nolta, J. A., *et al.* (1995). Engraftment of gene-modified umbilical cord blood cells in neonates with adenosine deaminase deficiency. *Nat. Med.* **1,** 1017–1023.

Kolb, H. J., Mittermuller, J., Clemm, C., *et al.* (1990). Donor leukocyte transfusions for treatment of recurrent chronic myelogenous leukemia in marrow transplant patients. *Blood* **76,** 2462–2465.

Körbling, M., Burke, P., Braine, H., *et al.* (1978). Successful engraftment of blood derived normal hemopoietic stem cells in chronic myelogenous leukemia. *Exp. Hematol.* **9,** 684–690.

Körbling, M., Holle, R., Haas, R., *et al.* (1990). Autologous blood stem-cell transplantation in patients with advanced Hodgkin's disease and prior radiation to the pelvic site. *J. Clin. Oncol.* **8,** 978–985.

Körbling, M., Przepiorka, D., Huh, Y. O., *et al.* (1995). Allogeneic blood stem cell transplantation for refractory leukemia and lymphoma: Potential advantage of blood over marrow allografts. *Blood* **85,** 1659–1665.

Kotasek, D. D., Shepherd, K. M., Sage, R. E., *et al.* (1992). Factors affecting blood stem cell collections following high dose cyclophosphamide mobilization in lymphoma, myeloma and solid tumors. *Bone Marrow Transplant.* **9,** 11–17.

Lapidot, T., Sirard, C., Vormoor, J., *et al.* (1994). A cell initiating human acute myeloid leukaemia after transplantation into SCID mice. *Nature* **367,** 645–648.

Lister, J., Pincus, S. M., Rybka, W. B., *et al.* (1993). Adoptive immunotherapy immediately after peripheral blood stem cell transplantation does not adversely affect engraftment and results in amplified killer function post transplant. *Blood* **82,** 650a.

Liu, Y., Hernandez, A. M., and Shibata, D. (1994). BCL2 translocation frequency rises with age in humans. *Proc. Natl. Acad. Sci. USA* **91,** 8910–8914.

Lo Coco, F., Mandelli, F., Diverio, D., *et al.* (1990). Therapy-induced Ph1 suppression in chronic myeloid leukemia: Molecular and cytogenetic studies in patients treated with alpha-2a IFN, high-dose chemotherapy and autologous stem cell infusion. *Bone Marrow Transplant.* **6,** 253–258.

Magrath, I., Adde, M., Shad, A., *et al.* (1996). Adults and children with small non-cleaved-cell lymphoma have a similar excellent outcome when treated with the same chemotherapy regimen. *J. Clin. Oncol.* **12,** 925–934.

Malachowski, M. E., Comenzo, R. L., Hillyer, C. D., *et al.* (1992). Large-volume leukapheresis for peripheral blood stem cell collection in patients with hematologic malignancies. *Transfusion* **32,** 732–735.

Malech, H. L., Sekhsaria, S., Whiting, T., *et al.* (1995). Development of a phase I clinical trial of gene therapy for chronic granulomatous disease. *Blood* **85,** 295a.

Marmont, A. M., and Van Bekkum, D. W. (1995). Stem cell transplantation for severe autoimmune diseases: New proposals but still unanswered questions. *Bone Marrow Transplant.* **16,** 497–498.

Marmont, A., Tyndall, A., Gratwohl, A., *et al.* (1995). Haemopoietic precursor-cell transplants for autoimmune diseases. *Lancet* **345,** 978.

Martiat, P., Lewalle, P., Taj, A. S., *et al.* (1993). Retrovirally transduced antisense sequences stably suppress P210 Bcr-abl expression and inhibit the proliferation of BCR/ABL-containing cell lines. *Blood* **81,** 502–509.

Maurer, J., Janssen, J. W. G., Thiel, E., *et al.* (1991). Detection of chimeric BCR-ABL genes in acute lymphoblastic leukemia by the polymerase chain reaction. *Lancet* **337,** 1055–1058.

McGlave, P. B. (1991). Therapy for chronic myelogenous leukemia with bone marrow transplantation. *Hematol. Oncol. Clin. North Am.* **4,** 235–258.

McGlave, P. B., Arthur, D., Miller, W. J., *et al.* (1990). Autologous transplantation for CML using marrow treated *ex vivo* with recombinant human interferon gamma. *Bone Marrow Transplant.* **6,** 115–120.

Miller, A. (1991). Incidence and demographics: Radiation risk. *In* "Breast Diseases" (J. R. Harris, Ed.), pp. 229–232. Lippincott, Philadelphia.

Miller, W. H., Jr., Levine, K., DeBlasio, A., *et al.* (1993). Detection of minimal residual disease in acute promyelocytic leukemia by reverse transcription polymerase chain reaction assay for the PML/RAR-alpha fusion mRNA. *Blood* **82,** 1689–1694.

Motzer, R., and Bosl, G. (1992). High dose chemotherapy for resistant germ cell tumors: Recent advances and future directions. *J. Natl. Cancer Inst.* **84,** 1703–1709.

Motzer, R., Mazumdar, M., Gulati, S., *et al.* (1993). Phase II trial of high dose carboplatin and etoposide with autologous bone marrow transplantation in first line therapy for patients with poor-risk germ cell tumors. *J. Natl. Cancer Inst.* **85,** 1828–1835.

Murea, S., Goldschmidt, H., Hahn, U., *et al.* (1996). Successful collection and transplantation of peripheral blood stem cells in cancer patients using large-volume leukaphereses. *J. Clin. Apheresis* **11,** 185–194.

Negrin, R. S., and Pesando, J. (1994). Detection of tumor cells in purged bone marrow and peripheral-blood mononuclear cells by polymerase chain reaction amplification of bcl-2 translocation. *J. Clin. Oncol.* **12,** 1021–1027.

Nimgaonkar, M., Mierski, J., Beeler, M., *et al.* (1995). Cytokine mobilization of peripheral blood stem cells in patients with Gaucher disease with a view to gene therapy. *Exp. Hematol.* **23,** 1633–1641.

Osborne, M. P., Wong, G. Y., Asina, S., *et al.* (1991). Sensitivity of immunocytochemical detection of breast cancer cells in human bone marrow. *Cancer Res.* **51,** 2706–2709.

Peters, W. P., Rosner, G., Ross, M., *et al.* (1993a). Comparative effects of granulocyte-macrophage colony-stimulating factor (GM-CSF) and granulocyte colony-stimulating factor (G-CSF) on priming peripheral blood progenitor cells for use with autologous bone marrow after high-dose chemotherapy. *Blood* **81,** 1709–1719.

Peters, W. P., Ross, M., Vredenburgh, J. J., *et al.* (1993b). High-dose chemotherapy with autologous bone marrow support after standard-dose adjuvant therapy for high-risk breast cancer. *J. Clin. Oncol.* **11,** 1132–1143.

Peters, W. P., Shpall, E. J., Jones, R. B., *et al.* (1988). High-dose combination alkylating agents with bone marrow support as initial treatment for metastatic breast cancer. *J. Clin. Oncol.* **6,** 1368–1376.

Pfreundschuh, M. G., Rueffer, U., Lathan, B., *et al.* (1994). Dexa-BEAM in patients with Hodgkin's disease refractory to multidrug chemotherapy regimes: A trial of the German Hodgkin's Disease Study Group. *J. Clin. Oncol.* **12,** 580–586.

Philip, T., Armitage, J. O., Spitzer, G., *et al.* (1987). High-dose therapy and autologous bone marrow transplantation after failure of conventional chemotherapy in adults with intermediate-grade or high-grade non-Hodgkin's lymphoma. *N. Engl. J. Med.* **316,** 1493–1498.

Philip, T., Guglielmi, C., Hagenbeek, A., *et al.* (1995). Autologous bone marrow transplantation as compared with salvage chemotherapy in relapses of chemotherapy-sensitive non-Hodgkin's lymphoma. *N. Engl. J. Med.* **333,** 1540–1545.

Piacibello, W., Sanavio, F., Garetto, L., *et al.* (1997). Extensive amplication and self-renewal of human primitive hematopoietic stem cells from cord blood. *Blood* **89,** 2644–2653.

Pilarski, L. M., and Belch, A. J. (1994). Circulating monoclonal B cells expressing P glycoprotein may be a reservoir of multidrug-resistant disease in multiple myeloma. *Blood* **83,** 724–736.

Porter, D. L., Roth, M. S., McGarigle, C., *et al.* (1994). Induction of graft-versus-host disease as immunotherapy for relapsed chronic myeloid leukemia. *N. Engl. J. Med.* **330,** 100–106.

Prince, G. M., Nguyen, M., Lazarus, H. M., *et al.* (1995). Peripheral blood harvest of unaffected CD34+CD38- hematopoietic precursors in paroxysmal nocturnal hemoglobinuria. *Blood* **86,** 3381–3386.

Ramsay, N., LeBien, T., Nesbit, M., *et al.* (1985). Autologous bone marrow transplantation for patients with ALL in second or subsequent remission: Results of bone marrow treated with monoclonal antibodies with BA-1, BA-2 and BA-3 plus complement. *Blood* **66,** 508–513.

Reading, C. L., Estey, E. H., Huh, Y. O., *et al.* (1993). Expression of unusual immunophenotype combinations in acute myelogenous leukemia. *Blood* **81,** 3083–3090.

Reece, D. E., Barnett, M. J., Connors, J. M., *et al.* (1991). Intensive chemotherapy with cyclophosphamide, carmustine, and etoposide followed by autologous bone marrow transplantation for relapsed Hodgkin's disease. *J. Clin. Oncol.* **9,** 1871–1879.

Rieder, H., Ludwig, W. D., Gassmann, W., *et al.* (1993). Chromosomal abnormalities in adult acute lymphoblastic leukemia: Results of the German ALL/AUL study group. *Recent Results Cancer Res.* **131,** 133–148.

Ritz, J., Sallan, S. E., Bast, R. C., *et al.* (1982). Autologous bone marrow transplantation in CALLA-positive acute lymphoblastic leukemia after *in vitro* treatment with J5 monoclonal antibody and complement. *Lancet* **2,** 60–63.

Roberts, M. M., To, L. B., Gillis, D., *et al.* (1993). Immune reconstitution following peripheral blood stem cell transplantation, autologous bone marrow transplantation and allogeneic bone marrow transplantation. *Bone Marrow Transplant.* **12,** 469–475.

Ross, A. A., Cooper, B. W., Lazarus, H. M., *et al.* (1993). Detection and viability of tumor cells in peripheral blood stem cell collections from breast cancer patients using immunocytochemical and clonogenic assay techniques. *Blood* **82,** 2605–2610.

Ross, A. A., Miller, G. W., Moss, T. J., *et al.* (1995). Immunocytochemical detection of tumor cells in bone marrow and peripheral blood stem cell collections from patients with ovarian cancer. *Bone Marrow Transplant.* **15,** 929–933.

Rowley, S. D., Jones, R. J., Miller, C. B., *et al.* (1990). Acute lymphoblastic leukemia (ALL): A phase I study of autologous bone marrow transplantation with combination drug purging. *Proc. Am. Soc. Clin. Oncol.* **9,** 202.

Russell, J. A., Brown, C., Brown, T., *et al.* (1996). Allogeneic blood cell transplants for hematological malignancy: Preliminary comparison of outcomes with bone marrow transplantation. *Bone Marrow Transplant.* **17,** 703–708.

Schiffer, C. A., Lee, E. J., Tomiyasu, T., *et al.* (1989). Prognostic impact of cytogenetic abnormalities in patients with *de novo* acute nonlymphocytic leukemia. *Blood* **73,** 263–270.

Schiller, G., Vescio, R., Freytes, C., *et al.* (1995). Transplantation of $CD34^+$ peripheral blood progenitor cells after high-dose chemotherapy for patients with advanced multiple myeloma. *Blood* **86,** 390–397.

Schlimok, G., Funke, I., Holzmann, B., *et al.* (1987). Micrometastatic cancer cells in bone marrow: *In vitro* detection with anti-cytokeratin and *in vivo* labeling with anti-17-1A monoclonal antibodies. *Proc. Natl. Acad. Sci. USA* **84,** 8672–8676.

Schmitz, N., Dreger, P., Suttorp, M., *et al.* (1995). Primary transplantation of allogeneic peripheral blood progenitor cells mobilized by filgrastim (granulocyte colony-stimulating factor). *Blood* **85,** 1666–1672.

Schmitz, N., Glass, B., Dreger, P., *et al.* (1993). High-dose chemotherapy and hematopoietic stem cell rescue in patients with relapsed Hodgkin's disease. *Ann. Hematol.* **66,** 251–256.

Schmitz, N., Linch, D. C., Dreger, P., *et al.* (1996). Randomised trial of filgrastim-mobilised peripheral blood progenitor cell transplantation versus autologous bone-marrow transplantation in lymphoma patients. *Lancet* **347,** 353–357.

Schulze, R., Schulze, M., Wischnik, A., *et al.* (1997). Tumor cell contamination of peripheral blood stem cell transplants and bone marrow in high-risk breast cancer patients. *Bone Marrow Transplant.* **19,** 1223–1228.

Schwartzberg, L. S. (1993). Peripheral blood stem cell mobilization in the out-patient setting. *In* "Peripheral Blood Stem Cell Autografts" (E. W. Wunder and P. R. Henon, Eds.), pp. 177–184. Springer-Verlag, Heidelberg.

Seriu, T., Yokota, S., Nakao, M., *et al.* (1995). Prospective monitoring of minimal residual disease during the course of chemotherapy in patients with acute lymphoblastic leukemia, and detection of contaminating tumor cells in peripheral blood stem cells for autotransplantation. *Leukemia* **9,** 615–623.

Sharp, J. G., Mann, S. L., Murphy, B., *et al.* (1995). Culture methods for the detection of minimal tumor contamination of hematopoietic harvests: A review. *J. Hematother.* **4,** 141–148.

Sheridan, W. P., Begley, C. G., Juttner, C. A., *et al.* (1992). Effect of peripheral-blood progenitor cells mobilised by filgrastim (G-CSF) on platelet recovery after high-dose chemotherapy. *Lancet* **339,** 640–644.

Sheridan, W. P., Begley, C. G., To, L. B., *et al.* (1994). Phase II study of autologous filgrastim (G-CSF)-mobilized peripheral blood progenitor cells to restore hemopoiesis after high-dose chemotherapy for lymphoid malignancies. *Bone Marrow Transplant.* **14,** 105–111.

Shpall, E. J., Jones, R. B., Bearman, S. I., et al. (1994). Transplantation of enriched CD34-positive autologous marrow into breast cancer patients following high-dose chemotherapy: Influence of CD34-positive peripheral-blood progenitors and growth factors on engraftment. *J. Clin. Oncol.* **12,** 28–36.

Siegert, W., Beyer, J., Strohscheer, I., *et al.* (1994). High-dose treatment with carboplatin, etoposide, and ifosfamide followed by autologous stem-cell transplantation in relapsed or refractory germ cell cancer: A phase I/II study. *J. Clin. Oncol.* **12,** 1223–1231.

Silva, M. R. G., Parreira, A., and Ascensão, J. L. (1995). Natural killer cell numbers and activity in mobilized peripheral blood stem cell grafts: Conditions for *in vitro* expansion. *Exp. Hematol.* **23,** 1676–1681.

Simmons, P. J., and Gronthos, S. (1996). The biology and application of human bone marrow stromal cell precursors. *J. Hematother.* **5,** 15–23.

Socinski, M. A., Cannistra, S. A., Elias, A., *et al.* (1988). Granulocyte-macrophage colony stimulating factor expands the circulating haemopoietic progenitor cell compartment in man. *Lancet* **1,** 1194–1198.

Soussain, C., Patte, C., Ostronoff, M., *et al.* (1995). Small noncleaved cell lymphoma and leukemia in adults: A retrospective study of 65 adults treated with the LMB pediatric protocols. *Blood* **85,** 664–674.

Stiff, P. J., Murgo, A. J., Wittes, R. F., *et al.* (1983). Quantification of the peripheral blood colony forming unit-culture rise following chemotherapy. Could leukocytaphereses replace bone marrow for autologous transplantation? *Transfusion* **23,** 500–503.

Stone, R. M., and Mayer, R. J. (1993). Treatment of newly diagnosed adults with *de novo* acute myeloid leukemia. *Hematol. Oncol. Clin. North Am.* **4,** 47–64.

Storb, R., Deeg, H. J., Anasetti, C., *et al.* (1989). Pathophysiology of marrow graft failure; a review of the Seattle data: bone marrow transplantation. *In* "UCLA Symposium on Bone Marrow Transplantation: Current Controversies" (R. P. Gale and R. Champlin, Eds.), pp. 19–30. A. R. Liss, New York.

Syrjala, M., Anttila, V. J., Ruutu, T., *et al.* (1994). Flow cytometric detection of residual disease in acute leukemia by assaying blasts co-expressing myeloid and lymphatic antigens. *Leukemia* **8,** 1564–1570.

Szczepek, A. J., Bergsagel, P. L., Axelsson, L., *et al.* (1997). $CD34^+$ cells in the blood of patients with multiple myeloma express CD19 and IgH mRNA and have patient-specific IgH VDJ rearrangements. *Blood* **89,** 1824–1833.

Thomas, E., Storb, R., Clift, R. A., *et al.* (1975). Bone-marrow transplantation. *N. Engl. J. Med.* **292,** 832–843.

To, L. B., Haylock, D. N., Dowse, T., *et al.* (1994). A comparative study of the phenotype and proliferative capacity of peripheral blood (PB) $CD34^+$ cells mobilized by four different protocols and those of steady-phase PB and bone marrow $CD34^+$ cells. *Blood* **84,** 2930–2939.

To, L. B., Roberts, M. M., Haylock, D. N., *et al.* (1992). Comparison of hematological recovery times and supportive care requirements of autologous recovery phase peripheral blood stem cell transplants, autologous bone marrow transplants and allogeneic bone marrow transplants. *Bone Marrow Transplant.* **9,** 277–284.

To, L. B., Roberts, M. M., and Rawling, C. M. (1997). Establishment of a clinical threshold cell dose: Correlation between CFU-GM and duration of aplasia. *In* "Hematopoietic Stem Cells: The Mulhouse Manual" (T. A. Bock, W. Brugger, and W. Kanz, Eds.), pp. 15–20. AlphaMed Press, Dayton, OH.

To, L. B., Sheppard, K. M., Haylock, D. N., *et al.* (1989). Single high doses of cyclophosphamide enable the collection of high numbers of hemopoietic stem cells from the peripheral blood. *Exp. Hematol.* **18,** 442–447.

Tricot, G., Vesole, D. H., Jagannath, S., *et al.* (1996). Graft-versus-myeloma effect: Proof of principle. *Blood* **87,** 1196–1198.

Troussard, X., Leblond, V., Kuentz, M., *et al.* (1990). Allogeneic bone marrow transplantation in adults with Burkitt's lymphoma or acute lymphoblastic leukemia in first complete remission. *J. Clin. Oncol.* **8,** 809–812.

Turhan, A. G., Humphries, R. K., Eaves, C. J., *et al.* (1990). Detection of breakpoint cluster region-negative and nonclonal hematopoiesis *in vitro* and *in vivo* after transplantation of cells selected in cultures of chronic myeloid leukemia marrow. *Blood* **76,** 2404–2410.

Udomsakdi, C., Eaves, C. J., Swolin, B., *et al.* (1992). Rapid decline of chronic myeloid leukemic cells in long-term culture due to a defect at the leukemic stem cell level. *Proc. Natl. Acad. Sci. USA* **89,** 6192–6196.

Van Hoff, D. D., Clark, G. M., Weiss, G. R., *et al.* (1986). Use of *in vitro* dose response effects to select anti-neoplastics for high-dose or regional administration regimens. *J. Clin. Oncol.* **4,** 1827–1834.

Verbik, D. J., Jackson, J. D., Pirruccello, S. J., *et al.* (1995). Functional and phenotypic characterization of human peripheral blood stem cell harvests. *Blood* **85,** 1964–1970.

Viens, P., and Maraninchi, D. (1995). High-dose chemotherapy and autologous marrow transplantation for common epithelial ovarian carcinoma. *In* "High-Dose Cancer Therapy" (J. O. Armitage and K. H. Antman, Eds.), pp. 847–854. Williams & Wilkins, Baltimore.

Vose, J. M., Kessinger, A., Bierman, P. J., *et al.* (1992). The use of rhIL-3 for mobilization of peripheral blood stem cells in previously treated patients with lymphoid malignancies. *Int. J. Cell Cloning* **10,** S62–S64.

Weiden, P. L., Sullivan, K. M., Flournoy, N., *et al.* (1981). Antileukemic effect of chronic GVHD. Contribution to improved survival after allogeneic marrow transplantation. *N. Engl. J. Med.* **304,** 1529–1533.

Williams, S. F., Lee, W. J., Bender, J. G., *et al.* (1996). Selection and expansion of peripheral blood $CD34^+$ cells in autologous stem cell transplantation for breast cancer. *Blood* **87,** 1687–1691.

Wingard, J. R., Mellis, E. D., Sostrin, M. B., *et al.* (1988). Interstitial pneumonitis after allogeneic bone marrow transplantation: Nine years of experience at a single institution. *Medicine* **67,** 175–186.

Winter, J. N., Lazarus, H. M., Rademaker, A. A. F., *et al.* (1995). Comparison of PIXY321 and GM-CSF for mobilization of peripheral blood progenitor cells (PBPC) in advanced breast cancer. *Blood* **86,** S2301.

Witzig, T. E., Gertz, M. A., Pineda, A. A., *et al.* (1995). Detection of monoclonal plasma cells in the peripheral blood stem cell harvests of patients with multiple myeloma. *Br. J. Haematol.* **89,** 640–642.

Zittoun, R., Mandelli, F., Willemze, R., *et al.* (1995). Autologous or allogeneic bone marrow transplantation compared with intensive chemotherapy in acute myelogenous leukemia. *N. Engl. J. Med.* **332,** 217–223.

7

Cord Blood Stem Cell Transplantation

Marcus R. Vowels

Haematology/Oncology
Sydney Children's Hospital
Randwick, Sydney, New South Wales 2031, Australia

I. Introduction

Allogeneic bone marrow transplantation (BMT) is a curative treatment for patients with leukemia, some cancers, aplastic anemia, and genetic diseases. The most suitable donor is a sibling that is matched for all six human leukocyte antigens (HLA, 6/6), the next being a 5/6 or 6/6 HLA-matched relative. However, only 30% of patients have such a family donor (Vowels *et al.*, 1992).

II. UNRELATED BONE MARROW DONOR TRANSPLANTS

For those 70% of patients with no family donor, registries of unrelated bone marrow donors have been established worldwide and now contain over 4 million donors. Only 25% of patients searching unrelated donor registries actually receive a transplant because either no donor is identified or during the average 4 months taken to identify a donor, the patient becomes unsuitable for BMT. Thus, this leaves approximately 50% of patients in need of an alternate and more rapidly accessible source of stem cells.

Patients who receive a matched-unrelated donor transplant have a greater than 50% risk of acute graft-versus-host disease (GVHD, grades II–IV) when unprocessed bone marrow is used and a 35% risk when T-cell-depleted bone marrow is used, resulting in significant morbidity and mortality (Kernan *et al.,* 1993; Balduzzi *et al.,* 1995; Casper *et al.,* 1995; Cornish *et al.,* 1995; Davies *et al.,* 1995). Chronic GVHD occurs in 40% of patients receiving either unprocessed or T-cell-depleted marrow.

Over 60% of adult marrow donors are infected with CMV, which is transmitted in the marrow transplant, causing significant problems posttransplant. Whether T-deplete or T-replete marrow is used, 1 year event free survival ranges from 40 to 60% for leukemia and from 20 to 40% for aplastic anemia (Kernan *et al.,* 1993; Balduzzi *et al.,* 1995; Casper *et al.,* 1995; Cornish *et al.,* 1995; Davies *et al.,* 1995).

Ethnic minorities are not adequately represented in registries in Caucasian countries, despite efforts to develop this. Thus, ethnic minorities have a reduced chance of finding a suitable donor in a registry based in a country of primarily Caucasian background. This chapter will examine the role of cord blood as an alternate source of hematopoietic stem cells for transplantation.

III. CORD BLOOD

Cord blood was identified in the early 1980s as having a high content of marrow stem cells (Broxmeyer *et al.,* 1989). A significant volume of cord blood can be collected from the placenta via needling of the umbilical cord veins and a small additional amount from placental veins. This procedure is performed after the baby is born, either while the placenta is *in situ* in the womb or after the placenta is delivered.

On average, 112 ml can be collected, containing 14×10^8 nucleated cells ($(1-68) \times 10^8$), 1.7×10^5 CFU-GM ($(0-25) \times 10^5$), and 1.8×10^5 BFU-E ($(0-14) \times 10^5$) (Thierry *et al.,* 1992). These cells can be frozen whole or concentrated into a smaller volume without significant loss of stem cells (Charbord *et al.,* 1992; Rubinstein *et al.,* 1995) and maintained at −180°C for 10 years or more.

In vitro data indicate both qualitative and quantitative differences exist between cord blood and bone marrow lymphocytes (Harris *et al.,* 1992; Risdon *et al.,* 1994; Keever *et al.,* 1995). These differences suggest that GVHD might be less frequent and less severe compared with that seen with the use of bone marrow. Thus, cord blood is attractive as a source of stem cells for transplant.

IV. RELATED CORD BLOOD TRANSPLANTS

Concern that there may not be enough hematopoietic stem cells in a donation of cord blood was dispelled after the successful engraftment of a child with Fanconi anemia in Paris in 1988 (Gluckman *et al.,* 1989). This patient survives over 8 years posttransplant.

Forty four related matched and mismatched cord blood transplants—reported to the International Cord Blood Transplant Registry—and 78 transplants from European Centers have been performed for malignant and nonmalignant diseases in patients ranging in age from 0.2 to 20 years (there is some duplication of patients in the two reports) (Wagner *et al.,* 1995; Gluckman *et al.,* 1997). Engraftment occurred in 27/31 (87%) and 61/71 (86%) of patients receiving HLA-identical transplants, respectively. GVHD grades II–IV occurred in 1/28 (4%) and 5/60 (8%) of patients for HLA-identical transplants and 2/6 (33%) and 9/18 (50%) of patients for mismatched transplants, respectively. The most striking observation is that GVHD occurred in only 4–8% of matched transplants, a figure that is in contradistinction to the 20% seen in children receiving matched-sibling bone marrow transplants. As well, GVHD grades II–IV occurred in less than 50% of patients receiving mismatched transplants. Taken together, the clinical and *in vitro* data support the contention that GVHD will be less frequent and less severe with cord blood compared to bone marrow transplants.

This type of transplant, however, may have a relatively limited use because of the higher rate of engraftment failure compared to bone marrow. The reduced potential for GVHD may also be associated with less graft-versus-leukemia reactivity compared to allogeneic bone marrow transplants, though this hypothesis has not been clinically tested. Equally importantly, HLA-matched related cord blood does not add to the available sources of marrow stem cells, since the donor's marrow would have been used anyway prior to the availability of cord blood. Related cord blood may, however, supplement the bone marrow dose and decrease the amount of bone marrow needing to be collected, a fact that is particularly relevant since the donor is small. Its use requires that a mother is pregnant at the time that a stem cell transplant is needed or that cord blood is collected at each delivery in a family where a child may need a transplant at some later time.

V. UNRELATED CORD BLOOD TRANSPLANTS

The potential to use unrelated cord blood was made possible by the establishment of an unrelated cord blood bank at the New York Blood Center in February 1993 (Rubinstein *et al.,* 1993). The use of unrelated cord blood is based on the following premises:

1. Despite there being large numbers of unrelated adult donors in bone marrow donor registries, only a proportion of patients benefit, and a new source of stem cells is needed.
2. There will be a lower frequency and severity of GVHD than with a compatible-unrelated bone marrow donor.
3. Because of this lower risk of GVHD, cord blood that is more mismatched than unrelated bone marrow can be used with reasonable safety, thus making unrelated cord blood a versatile product.

The first unrelated cord blood transplant was performed at the Duke University Medical Center in September 1993 using cord blood from the New York Cord Blood Bank (NYCBB) (Kurtzberg *et al.,* 1996). Over subsequent years, the rate of accrual of cord blood transplants using cord blood from the NYCBB has escalated. As of March 1997, over 370 cord blood transplants had been performed worldwide using cord blood from the NYCBB (355 transplants) and 3 European banks (18 transplants) (Haut *et al.,* 1996; Kurtzberg *et al.,* 1996; Migliaccio *et al.,* 1996; Rubinstein *et al.,* 1996; Slone *et al.,* 1996; Smith *et al.,* 1996; Wagner *et al.,* 1996; Gluckman *et al.,* 1997).

The single largest experience has been that of the Duke University Medical Center, North Carolina, where 100 transplants have been performed (Kurtzberg *et al.,* 1996; Wagner *et al.,* 1997). The mean age of patients was 7 years (0.3–58 years) and the mean weight 17 kg (5–82 kg). Recipients included patients up to the age of 34 years and with body weight greater than 79 kg. Cord blood was HLA-mismatched for one to three antigens in 94% of patients. GVHD prophylaxis was with cyclosporin and steroids, and G-CSF was given posttransplant. The majority of patients received conditioning with total body irradiation and melphalan or busulphan and melphalan, depending on age and underlying diagnosis.

Engraftment occurred in 90% of patients; neutrophils reached 0.5×10^9 per liter in a mean of 22 days (14–37), and platelet counts were 50×10^9 per liter in a mean of 82 days. A new thawing method appears to have shortened the time to engraftment (Rubinstein *et al.,* 1995; Kurtzberg *et al.,* 1996). Fifty percent of patients developed acute GVHD, grades II–IV, a low incidence, considering that over 90% of transplants were from mismatched donations. One-year disease-free survival was 50%. It is of interest that the mean times to identify a cord blood unit and to proceed to transplant were less than 18 days and less than 115 days, respectively, periods comparable to or shorter than unrelated bone marrow donor searches.

Risk factors are starting to emerge. Cell dose and CFU-GM counts correlate with potential for engraftment (Kurtzberg *et al.,* 1996; Migliaccio *et al.,* 1996; Gluckman *et al.,* 1997; Wagner *et al.,* 1997), with cell dose below 3.5×10^7 associated with a lower engraftment rate. Increasing frequency of GVHD correlates with increasing degree of HLA mismatching (Table 7.1) (Kurtzberg *et al.,* 1996; Rubinstein *et al.,* 1996). Increased survival is related to recipient weight (<40 kg), age (<12 years), cell dose ($>3.5 \times 10^7$/kg), and degree of mismatching (<2 HLA) (Kurtzberg *et al.,* 1996; Rubinstein *et al.,* 1996).

This clinical experience highlights that it is possible to engraft unrelated patients of varying weights and ages with unrelated cord blood and that, despite mismatching, engraftment, GVHD, and survival appear to be similar to that seen with fully matched, unrelated bone marrow donors, supporting the role of unrelated cord blood in the treatment of malignant and nonmalignant diseases in children. The current primary role of unrelated cord blood transplantation is where there is no family or unrelated matched donor or where the delay in transplantation associated with search of unrelated marrow donor registries will worsen the patient's prognosis. Unrelated cord blood may indeed become a substitute for unrelated bone marrow.

What lies ahead is to define the cell dose, degree of mismatching, and weight of patients that result in acceptable engraftment and survival. These data will assist in determining which is the safest unit to use. A case comparison is planned within the International Bone Marrow Donor Registry (IBMTR) and the International Cord Blood Transplant Registry (ICBTR) to define these parameters and to define the equivalent degree of mismatching that is associated with a similar outcome to a matched unrelated bone marrow donor.

An equally pressing question is whether cord blood can be used in adult recipients with the same outcome as in children. Of 18 adults transplanted using cord blood from the New York bank, ten out of eleven patients that were evaluated engrafted, but only three-survive (Kurtzberg *et al.,* 1996; Rubinstein *et al.,* 1996). Nine patients weighing greater than 40 kg are reported and five survive (Laughlin *et al.,* 1996). The rate of engraftment failure appears to be similar to that of children.

TABLE 7.1 Risk of GVHD Correlated with Degree of Matching

Matching	GVHD frequency (grades II–IV) (%)
6/6	18
5/6	50
4/6	56
(3–4)/6	73

It is not clear whether disease status and clinical well-being at the time of transplant, greater risk of GVHD in adults compared to children, or the rate of engraftment is relevant in interpreting these early data in larger recipients. However, these studies raise concern that cell dose and mismatching may be more important issues in this age group. Thus, the role of cord blood transplant in adults is still to be defined.

Ex vivo expansion of cord blood is being investigated both in the laboratory (Broxmeyer *et al.,* 1992; Hows *et al.,* 1992) and in planned clinical studies. These studies will help define the safety of cord blood stem cell expansion techniques and whether *in vitro* expansion will shorten the period of engraftment and increase engraftment rates, both in small and in large patients.

VI. UNRELATED CORD BLOOD BANKING

The first unrelated cord blood bank was established at the New York Blood Center in February 1993 (Rubinstein *et al.,* 1993, 1995). The NYCBB accrued 5000 donations by December 1995, 6000 donations by August 1996, and 7000 donations by early March 1997. Banks have since been or are being established in the USA (>7000 collections), France (>500), Belgium (>600), Holland (570), Germany (>2000), Italy (>2000), Israel (100), Australia (>1000), Spain (750), Great Britain (650), Poland, Japan, Southeast Asia, and New Zealand.

At the time cord blood is collected, the hospital has historically had the responsibility for disposing of the placenta. However, to bank unrelated cord blood, it is a necessity to obtain informed consent both for collection and for use of cord blood. Anonymity of the donor is essential to protect the identity of the child and thereby prevent an approach for a bone marrow donation from the donor (the baby) at some later time (McCullough *et al.,* 1994).

An assessment of the quality and safety of the product is required. Thus, it is necessary to obtain a medical history for infectious and genetic diseases, which may be transmitted in the cord blood, to obtain follow-up information on the baby, and to test the mother and cord blood for infectious diseases. The cord blood can be screened by examination of the blood count and film and viability assays such as CFU-GM and $CD34^+$ cell number. HLA typing of A, B, and DR antigens is performed and allows rapid identification of suitable donations for confirmatory typing.

Suitable methods for reduction of volume and for thawing of cord blood with minimal loss of stem cells have been developed (Charbord *et al.,* 1992; Rubinstein *et al.,* 1995). Volume reduction of cord blood prior to freezing is an important issue to resolve as it allows 2–3 times as many cord bloods to be stored in each storage tank, with consequent savings in cost and space.

Unrelated bone marrow donor registries in Caucasian countries have an ethnic mix that does not always reflect the ethnic mix in the community; however, this has not been a significant problem in the NYCBB. Although there are religious, cultural, and language barriers to obtaining informed consent for banking, these do not appear to impede banking of cord blood from ethnic minority groups.

VII. ADVANTAGES AND DISADVANTAGES OF CORD BLOOD

Cord blood can be collected with no risk to the neonate or mother. There is a potential for reduced frequency and severity of GVHD, and therefore cord blood may be used across tissue-typing barriers hitherto not possible with adult bone marrow. The restriction of having to find a fully matched donor is relaxed, so that compared to unrelated bone marrow, banked cord blood can be utilized more readily. Cryopreserved stem cells in a cord blood bank are immediately available "off the shelf," without the delay inherent in finding a volunteer unrelated bone marrow donor.

However, unlike bone marrow, the volume and cell content of cord blood is limited by the placental yield. As well, genetic disease may be unknowingly transmitted. Fortunately, there are few genetic diseases which will result in clinical manifestations through hematopoietic engraftment. These are immune deficiency and storage diseases, frequency less than 1 in 100,000, and hemoglobinophathies, 1 in 3500 in the relevant ethnic groups. Precautions can be implemented to further decrease this risk. These include a blood count and film on the cord blood donation, a medical history from the mother, and follow-up of the baby. There is the potential for contamination of the cord blood collection with maternal blood cells, specifically lymphocytes, and consequent risk of GVHD. Low contamination with maternal lymphocytes has been reported (Socie *et al.,* 1994; Hall *et al.,* 1995; Scaradavou *et al.,* 1996), though no adverse events have been ascribed to this finding.

With increasing international interest and activity in cord blood transplantation, the number and antigenic diversity of donations to be banked need to be addressed. Factors that impact on this question are the quality of the product stored (cell count and CFU-GM), cost-effectiveness of the activity, and safety of use (risk of GVHD).

There is a direct correlation between risk of GVHD, decreasing chance of survival, and increasing number of HLA mismatches (Table 7.1). Thus, a strategy with the banking of cord blood is to aim to provide a cord blood unit which is matched at a level that will provide an acceptable outcome. It would be reasonable to plan to bank that number of cord blood units which would provide a majority of patients units that are matched for 5/6 or 6/6 HLA.

The experience of the NYCBB and predictions from the American National Marrow Donor Program (NMDP) are useful in identifying the potential number of cord blood units needing to be stored for a particular racial group to provide 5/6 or 6/6 HLA matched units. The NYCBB could provide serological matches (5/6 or 6/6 HLA) for 58% of 2800 patients searching the bank when the bank contained 6000 donations. Data based on the NMDP indicate the number of 5/6 or 6/6 HLA-matched Caucasian donors that will be identified as a bank size increases (Table 7.2) (Beatty *et al.,* 1995).

The rate of accrual of new haplotypes increases as the size of the bank increases, but beyond a certain number of collections, the rate of accrual of new haplotypes falls off rapidly. Seven percent of new haplotypes are added for the first 1000 units stored, 32% for the next 19,000 stored, and 14% for the next 25,000. However, once a bank size reaches 30,000, only 6% of new haplotypes are added for a further 15,000 donations. Thus, banking 20,000 units of a particular ethnic group should provide 5/6 or 6/6 HLA matches for over 80% of patients searching the bank and covers 40% of haplotype combinations.

The size and utilization of the NYCBB have increased since 1995. One hundred cord blood transplants had been performed worldwide by December 1995 (5000 donations). By early March 1997, 355 transplants had been performed from a bank size of 7000 donations. Thus, the crude usage rate has increased from 1:50 in 1995 to 1:20 in 1997.

VIII. CONCLUSIONS AND FUTURE PROGRESS

Related cord blood transplants are associated with a low risk of GVHD. However, they also have a lower rate of engraftment and in patients with leukemia, the low risk of GVHD may also be associated with reduced graft-versus-leukemia benefit. Related cord blood does not add to the sources of hematopoietic stem cells but does offer the potential to avoid collecting bone marrow from a small infant with its attendant potential risks.

The use of unrelated cord blood has outstripped that of related cord blood. Unrelated cord blood banks are now widely established and offer a new source of stem cells. GVHD appears to occur at a frequency that is similar to or less than unrelated bone marrow with similar engraftment rates. However, more data are required to confirm these preliminary observations.

The role of cord blood in adults is yet to be defined. Although engraftment failure rates are similar to those in children, engraftment appears delayed and

TABLE 7.2 Relationship between Increasing Bank Size and Number of Donations Identified

Bank size	Percentage of donors identified
1,000	25
5,000	48
10,000	70
20,000	82

complications greater. *In vitro* expansion of cord blood stem cells may assist engraftment by reducing the rate of graft failure and shortening engraftment times, leading to a potential role in larger patients.

As the number of cord blood units stored increases, the potential to provide 5/6 or 6/6 HLA-matched donations will also increase. The availability and use of more closely matched units may ultimately decrease transplant-related morbidity and mortality.

REFERENCES

Balduzzi, A., Gooley, T., Anasetti, C., *et al.* (1995). Unrelated donor marrow transplantation in children. *Blood* **86,** 3247–3256.

Beatty, P. G., Mori, M., and Milford, E. (1995). Impact of racial genetic polymorphism on the probability of finding an HLA-matched donor. *Transplantation* **60,** 778–783.

Broxmeyer, H. E., Douglas, G. W., Hangoc, G., *et al.* (1989). Human umbilical cord blood as a source of transplantable hematopoietic stem/progenitor cells. *Proc. Natl. Acad. Sci. USA* **86,** 3828–3832.

Broxmeyer, H. E., Hangoc, G., Cooper, S. H., *et al.* (1992). Growth characteristics and expansion of human umbilical cord blood and estimation of its potential for transplantation in adults. *Proc. Natl. Acad. Sci. USA* **89,** 4109–4123.

Casper, J., Camitta, B., Truitt, R., *et al.* (1995). Unrelated bone marrow donor transplants in children with leukemia or myelodysplasia. *Blood* **85,** 2354–2363.

Charbord, P., Newton, I., Schaal, J. P., *et al.* (1992). The separation of human cord blood by density gradient does not induce a major loss of progenitor cells. *Bone Marrow Transplant.* **9,** 109–110.

Cornish, J., Potter, M. N., Steward, D., *et al.* (1995). Unrelated donor bone marrow transplant for acute lymphoblastic leukemia in childhood. *Blood* **86**(Suppl 1), 382a.

Davies, S. M., Shu, X. O., Blazer, B. R., *et al.* (1995). Unrelated donor bone marrow transplantation: Influence of HLA A and B incompatibility on outcome. *Blood* **86,** 1636–1642.

Gluckman, E., Broxmeyer, H. E., Auerbach, A. D., *et al.* (1989). Hematopoietic reconstitution in a patient with Fanconi's anemia by means of umbilical cord blood from a HLA-identical sibling. *N. Engl. J. Med.* **321,** 1174–1178.

Gluckman, E., Rocha, V., Boyer-Chammard, A., *et al.* (1997). Outcome of cord-blood transplantation from related and unrelated donors. *N. Engl. J. Med.* **337,** 373–381.

Hall, J. M., Lingenfelter, P., Adams, S. L., *et al.* (1995). Detection of maternal cells in human umbilical cord blood using fluorescence *in situ* hybridization. *Blood* **86,** 2829–2832.

Harris, D. T., Schumacher, M. J., Locascio, J., *et al.* (1992). Phenotypic and functional immaturity of human umbilical cord blood T lymphocytes. *Proc. Natl. Acad. Sci. USA* **89,** 10006–10010.

Haut, P., Atlas, M., Rubinstein, P., Morgan, E., Kletzel, M. (1996). Mismatched cord blood stem cell transplant. *Blood* **88**(Suppl 1), 265a.

Hows, J. M., Bradley, B. A., Marsh, J. C. W., *et al.* (1992). Growth of human umbilical-cord blood in longterm haemopoietic cultures. *Lancet* **340,** 73–76.

Keever, C. A., Abu-Hajir, M., Graf, W., *et al.* (1995). Characterization of the alloreactivity and anti-leukemia reactivity of cord blood mononuclear cells. *Bone Marrow Transplant.* **15,** 407–419.

Kernan, N. A., Bartsch, G., Ash, R. C., *et al.* (1993). Analysis of 462 transplantations from unrelated donors facilitated by the National Marrow Donor Program. *N. Eng. J. Med.* **328,** 593–602.

Kurtzberg, J., Laughlin, M., Graham, M. L., *et al.* (1996). Placental blood as a source of hematopoietic stem cells for transplantation into unrelated recipients. *N. Engl. J. Med.* **335,** 157–166.

Laughlin, M. J., Smith, C. A., Martin, P., *et al.* (1996). Hematopoietic engraftment using placental cord blood (PCB) unrelated donor transplantation in recipients $>$40 kg weight. *Blood* **88**(Suppl 1), 266a.

McCullough, J., Clay, M. E., Fautsch, S., *et al.* (1994). Proposed policies and procedures for the establishment of a cord blood bank. *Blood Cells* **20,** 609–626.

Migliaccio, A. R., Adamson, J. W., Rubinstein, P., *et al.* (1996). Placental/cord blood stem cell transplantation: Correlation between the progenitor cell dose and time to myeloid engraftment in 55 unrelated transplants. *Blood* **88,** 286a.

Risdon, G., Gaddy, J., Stehman, F. B., *et al.* (1994). Proliferative and cytotoxic responses of human cord blood T lymphocytes following allogenic stimulation. *Cell Immunol.* **154,** 14–24.

Rubinstein, P., Carrier, C., Adamson, J., *et al.* (1996). New York Blood Center's program for unrelated placental/umbilical cord blood (PCB) transplantation: 243 transplants in the first 3 years. *Blood* **88,** 142a.

Rubinstein, P., Dobrila, L., Rosenfeld, R. E., *et al.* (1995). Processing and cryopreservation of placental/umbilical cord blood for bone marrow reconstitution. *Proc. Natl. Acad. Sci. USA* **92,** 10119–10122.

Rubinstein, P., Rosenfeld, R. E., Adamson, J. W., *et al.* (1993). Stored placental blood for unrelated bone marrow reconstitution. *Blood* **81,** 1679–1690.

Scaradavou, A., Carrier, C., Mollen, N., *et al.* (1996). Detection of maternal DNA in placental/umbilical cord blood by locus-specific amplification of the noninherited maternal HLA gene. *Blood* **88,** 1494–1500.

Slone, V., Abu-Gosh, A., Goldman, S., *et al.* (1996). Delayed platelet, but comparable myeloid engraftment following unrelated cord blood transplantation (UCBT): Decreased megakaryocytic lineage ($CD34^+$/$CD41^+$) stem cells in cord blood. *Blood* **88**(Suppl 1), 114a.

Smith, F. O., Robertson, K. A., Lucas, K. G., *et al.* (1996). Umbilical cord blood transplantation from HLA-mismatched unrelated donors: The Indiana University experience. *Blood* **88**(Suppl 1), 266a.

Socie, G., Gluckman, E., Carosella, E., *et al.* (1994). Search for maternal cells in human umbilical cord blood by polymerase chain reaction amplification of two mini satellite sequences. *Blood* **83,** 340–344.

Thierry, D., Traineau, R., Adam, M., *et al.* (1992). Study on the hematopoietic stem cells from umbilical cord blood. *Bone Marrow Transplant.* **9,** S101–S105.

Vowels, M. R., Honeyman, M., Ziegler, J., *et al.* (1992). Searches for matched and closely matched related and unrelated marrow donors undertaken in a paediatric unit. *J. Paediatr. Child Health* **28,** 379–382.

Wagner, J. E., Defor, T., Rubenstein, P., *et al.* (1997). Transplantation of unrelated donor umbilical cord blood (UCB): Outcomes and analysis of risk factors. *Blood* **90**(Suppl 1), 398a.

Wagner, J. E., Kernan, N. A., Steinbuch, M., *et al.* (1995). Allogenic sibling umbilical-cord blood transplantation in children with malignant and non-malignant disease. *Lancet* **346,** 214–219.

Wagner, J. E., Rosenthal, J., Sweetman, R., *et al.* (1996). Successful transplantation of HLA-matched and HLA-mismatched umbilical cord blood from unrelated donors: Analysis of engraftment and acute graft-versus-host disease. *Blood* **88,** 795–802.

8

Reconstitution of Immunity by Adoptive Immunotherapy with T Cells

Stanley R. Riddell,
Edus Houston Warren,
Deborah Lewinsohn, Cassian Yee,
and Philip D. Greenberg

Fred Hutchinson Cancer Research Center
Seattle, Washington 98109

I. INTRODUCTION

The term cellular adoptive immunotherapy has been used to describe the transfer of effector cells of the immune system to augment or restore immune responses for the treatment of malignant or infectious diseases. The rationale for developing this approach as a therapy for human diseases is based upon the demonstration that progression of virus infections and virus-induced malignancies is associated with deficiencies in virus-specific immune responses and upon the identification of antigens expressed by tumors that can be recognized by the host immune system (Reusser *et al.,* 1991; Li *et al.,* 1994; Van Pel *et al.,* 1995; Henderson and Finn, 1996; Lucas *et al.,* 1996). Detailed experimentation in animal models has supported the principle that immune effector cells can be used therapeutically for virus infections and malignancy, but several impediments have hindered clinical applications, and adoptive immunotherapy has only recently begun to achieve therapeutic success in patients. The impediments to earlier successful intervention in humans were related to (a) the difficulty in isolating effector cells from the host or a suitable donor with reactivity for the relevant tumor or pathogen, (b) the expansion of effector cells *in vitro* to numbers sufficient to mediate therapeutic effects following adoptive transfer, and (c) the requirements that the reinfusion of effector cells be accomplished without toxicity to the host and the transferred cells persist *in vivo* for a sufficient duration to eradicate the tumor or control the pathogen. These impediments have in part been resolved, and strategies have been developed to proceed with clinical investigation, providing renewed optimism that the promise exhibited in animal models of cellular immunotherapy can be realized in humans.

There are distinct types of effector cells with unique functional properties that have been shown to have therapeutic potential in animal models of adoptive immunotherapy, including lymphokine-activated killer (LAK) cells, major histocompatibility complex (MHC) restricted, antigen-specific cytotoxic and helper T cells, and activated monocytes (Bukowski *et al.,* 1985; Mule *et al.,* Ada and Jones, 1986; Shinomiya *et al.,* 1989). MHC-restricted T cells that recognize processed tumor-associated antigens or virally encoded antigens have proven to be the most potent and effective effector cells in adoptive immunotherapy of cancer and viral infections in murine models. This review will focus on the development of cellular immunotherapy with T cells for human diseases.

An important milestone for the development of cellular immunotherapy in humans was the use of this approach to prevent or treat infections caused by EBV (Epstein–Barr virus) and CMV (cytomegalovirus) in immunocompromised bone marrow transplant (BMT) patients (Riddell *et al.,* 1992b; Walter *et al.,* 1995; Heslop *et al.,* 1996). In this setting it was possible to isolate from the donor, T cells specific for target antigens encoded by these viruses and to develop effective *in vitro* culture techniques to expand the T cells for administration to the BMT recipient. These studies provided the first evidence that T-cell immunity could be transferred in humans by adoptive immunotherapy and established principles that could be applied to the therapy of malignant diseases. A second milestone was the identification in tumor-bearing patients of autologous T cells that recognized antigens expressed by the patient's tumor (Table 8.1) (Van Pel *et al.*, 1995; Henderson and Finn, 1996). The discovery of genes encoding tumor antigens in malignant melanoma proceeded rapidly due to the efforts of Boon *et al.*, who transfected antigen-negative target cells with a cDNA library derived from a tumor expressing the antigen combined with plasmids encoding the class I HLA restricting allele and subsequently screened transfected target cells with tumor-specific T-cell clones to identify cDNAs encoding tumor antigens (van der Bruggen *et al.*, 1991; Traversari *et al.*, 1992). Remarkably, the majority of antigens expressed in melanoma cells have been shown to be normal self-proteins that are relatively tumor specific such as MAGE-1 (van der Bruggen *et al.*, 1991) and MAGE-3 (Gaugler *et al.*, 1994) or associated with melanocyte differentiation such as tyrosinase (Brichard *et al.*, 1993),MART1 (Kawakami *et al.*, 1994a), gp100 (Kawakami *et al.*, 1994b), and gp75 (Wang *et al.*, 1996), although epitopes derived from mutated proteins such as CDK4 (Wolfel *et al.*, 1995) and β-catenin (Robbins *et al.*, 1996) have also been identified. Antigens expressed by other malignancies have also been identified and include the unique immunoglobulin idiotype expressed in each B-cell lymphoma (Wilson *et al.*, 1990; Kwak *et al.*, 1992), fusion products of the bcr/abl and PML/RARα translocations in chronic myelogenous leukemia (CML) and acute promyelocytic leukemia (APL), respectively (Chen *et al.*, 1992; Gambacorti-Passerini *et al.*, 1993), mutated proteins such as Ras and p53 (Noguchi *et al.*, 1994; Peace *et al.*, 1994), and proteins encoded by viruses involved in malignant transformation, including EBV and human papilloma virus (HPV) (Murray *et al.*, 1992; Sing *et al.*, 1997).

The use of allogeneic BMT for hematopoietic malignancies presents a unique opportunity to utilize cellular immunotherapy as an adjunctive antitumor therapy to the transplant conditioning regimen, since HLA-identical allogeneic donors and recipients differ in expression of minor histocompatibility (H) antigens that may be expressed on the recipient's tumor. The infusion of polyclonal donor

TABLE 8.1 Classes of Human Tumor Antigens

Mutated proteins expressed as part of the malignant transformation, e.g., bcr/abl in CML, PML/RARα in APL, and CDK and β-catenin in melanoma
Normal self-proteins either preferentially expressed or overexpressed, e.g., MAGE-1, tyrosinase, gp 100, and gp75 in melanoma and HER-2/neu in breast and ovarian cancer
Unique proteins related to cell of origin, e.g., immunoglobulin idiotype in B-cell lymphoma and multiple myeloma
Viral proteins in virus-associated maligancies, e.g., Epstein–Barr virus, LMP-1 and LMP-2 in Hodgkin's disease, and human papilloma virus proteins in cervical carcinoma

peripheral blood mononuclear cells (PBMC) induces a complete hematologic and molecular remission in a majority of patients with relapse of CML and a significant fraction of patients with relapse of other hematopoietic malignancies after allogeneic BMT. Minor H antigens are known to be the targets of both graft-versus-host (GVH) and graft-versus-leukemia (GVL) responses, but several studies demonstrate that some minor H antigens exhibit restricted expression to recipient hematopoietic cells, suggesting that GVH and GVL responses may be separable (Goulmy, 1996; Warren *et al.,* 1998). Thus, potential targets for T-cell therapy of cancer have been identified, and efforts to develop cellular immunotherapy using *ex vivo* expanded T cells or vaccination strategies to induce or augment host, tumor-reactive, T-cell responses are now in progress.

II. BIOLOGICAL AND MOLECULAR BASIS OF T-CELL RECOGNITION OF VIRUSES AND TUMORS

The molecular basis for recognition by host T cells of virus-infected cells and cells undergoing malignant transformation has been the subject of intense investigation for several decades. The study of T cell responses to viruses and nominal antigens has proven to be particularly informative for elucidating the pathways of antigen presentation, the role of costimulatory molecules in T cell activation, and the dichotomy in T cell phenotype and function based upon the recognition of antigens presented by class I or class II MHC molecules, respectively (Mosmann *et al.,* 1986; Yewdell and Bennink, 1990; Mondino *et al.,* 1996).

A. EFFECTOR FUNCTIONS OF $\alpha\beta$ T CELLS

There are two major subsets of mature T cells in the peripheral blood, which express CD3 and the $\alpha\beta$ T-cell receptor (TcR). These subsets are distinguished by the surface expression of either CD4, defining the T helper (Th) subset, or CD8, defining the cytotoxic (CTL) subset. A third subset of T cells which is present in low frequency in the peripheral blood, but in higher frequency in some tissue sites, expresses CD3 and the $\gamma\delta$ TcR but lacks both CD4 and CD8 (Doherty *et al.,* 1992). $\gamma\delta$ T cells have been shown to recognize cells infected with HSV (Sciammas *et al.,* 1994) and to lyse tumor cells from colonic carcinoma (Maeurer *et al.,* 1996). However, strategies for isolating and propagating $\gamma\delta$ T cells reactive with viral or tumor antigens, and the adoptive transfer of $\gamma\delta$ T cells, have been less extensively studied in animal models. Thus, the potential therapeutic role of this subset of T cells will not be considered further in this review.

The recognition structure for the $\alpha\beta$ TcR of the $CD3^+CD4^+CD8^-$ Th subset consists of a peptide fragment of a protein antigen presented in association with class II MHC on an antigen-presenting cell (APC) (Fig. 8.1). Class II MHC

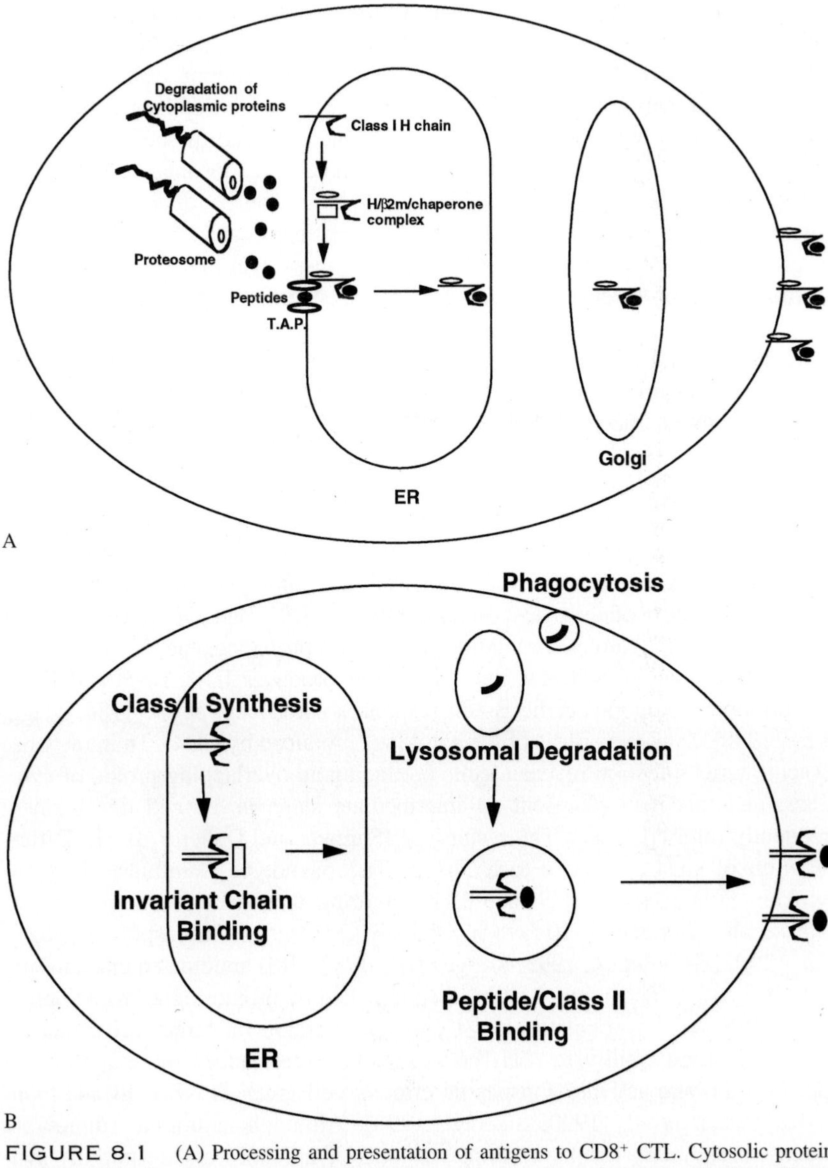

FIGURE 8.1 (A) Processing and presentation of antigens to CD8$^+$ CTL. Cytosolic proteins are degraded by the proteosome to peptide fragments that enter the endoplasmic reticulum (ER) via a specific peptide transporter (TAP). In the ER the peptides associate with class I MHC molecules and are transported via the Golgi stacks to the cell surface for recognition by CD8$^+$ CTL. (B) Processing and presentation of antigens to CD4$^+$ Th cells by APC. Proteins are endocytosed by phagocytic cells and degraded in lysosomes to peptides. In general, class II MHC molecules do not bind peptides in the ER because of the presence of the invariant chain, which occupies the peptide-binding groove. However, as class II transits the Golgi, it enters a distal compartment containing peptides and the invariant chain is removed, allowing peptide binding. The class II MHC/peptide complex then transits to the cell surface for recognition by CD4$^+$ Th.

molecules are not expressed on many normal cells that are targets for virus infection or malignant transformation, although expression may be up-regulated by cytokines such as γIFN (γ-interferon) and TNF (tumor necrosis factor). Typically, antigens presented by class II MHC for recognition by CD4$^+$ T cells are endocytosed by professional APC, which include dendritic cells, monocyte/macrophages, and B cells, and then degraded in acidic endosomal compartments to peptides that associate with the class II MHC αβ heterodimer and are transported to the cell surface (Morrison *et al.,* 1986; Neefjes and Ploegh, 1992). The absence of class II MHC on most host cells suggests that the effector functions of virus or tumor-specific CD4$^+$ Th are not dependent on direct interaction with the tumor or virally infected cell but rather are mediated after recognition of APC at the local site. Indeed, in murine models, CD4$^+$ Th cells specific for antigens expressed by virus-induced malignancies can be therapeutically effective even though the tumor cells lack detectable expression of class II MHC (Greenberg, 1991).

The major function attributed to CD4$^+$ T cells is the production of cytokines that serve to orchestrate a local and systemic immune response. Studies in mice and humans have identified two major subsets of CD4$^+$ Th cells, distinguished by the cytokines they produce following TcR activation (Mosmann *et al.,* 1986). The Th1 subset produces IL-2 (interleukin-2), γIFN, and GM-CSF (granulocyte–macrophage colony-stimulating factor) and promotes delayed-type hypersensitivity responses, whereas the Th2 subset produces IL-4, IL-5, and IL-10 and promotes antigen-specific B-cell responses (Mosmann *et al.,* 1986; Maggi *et al.,* 1988). *In vivo* analysis of the cytokines produced by CD4$^+$ Th in response to acute virus infection revealed cells producing an overlapping profile of cytokines, and this may represent an intermediate stage in Th cell development frequently referred to as a Th0 phenotype (Sarawar and Doherty, 1994). Differentiation of CD4$^+$ Th cells to a Th1 or Th2 phenotype is influenced by the cytokine milieu, with γIFN and IL-12 favoring the development of Th1 responses and IL-4 and IL-10 favoring the development of Th2 responses (Maggi *et al.,* 1992; Seder *et al.,* 1993; Rogge *et al.,* 1997). It is anticipated that antiviral and antitumor effects might best be accomplished by promoting a Th1 response, since γIFN will up-regulate class I MHC expression on target cells, thereby increasing susceptibility to CTL lysis, and IL-2 will induce the expansion of specific CTL and activate nonspecific effector cells such as NK cells and monocytes (Biron *et al.,* 1990; Staeheli, 1990). Although additional studies are needed, preliminary results in animal models of influenza and respiratory syncytial virus infection support this hypothesis. In mice infected with these viruses, the adoptive transfer of Th2 CD4$^+$ T cells was either ineffective in or enhanced morbidity, whereas the transfer of Th1 cells was effective in resolving infection (Alwan *et al.,* 1994; Graham *et al.,* 1994).

CD3$^+$CD8$^+$ T cells comprise the second major class of peripheral αβ T cells and recognize via the TcR peptide fragments of antigens presented by class I

MHC molecules on the surface of target cells (Yewdell and Bennink, 1992) (Fig. 8.1). The peptides associated with class I MHC are derived from both self-proteins and foreign cytosolic proteins which have been degraded to peptides by the proteosome complex (Yewdell and Bennink, 1992). The peptides are then transported into the endoplasmic reticulum (ER) by the heterodimeric transmembrane peptide transporter complex (TAP1/TAP2), which is physically associated with class I MHC molecules on the lumenal surface of the ER (Spies and DeMars, 1991). Peptides with an amino acid sequence containing an appropriate binding motif may then bind to the cleft in the class I MHC heavy chain, leading to a stable heavy chain/β_2-microglobulin/peptide complex that is transported to the cell surface (Yewdell and Bennink, 1992).

CD8$^+$ CTL recognition leads to activation and release of cytolytic granules containing perforins, serine proteases, and calcium-binding proteins in the local vicinity of the target cell (Doherty, 1993a). Perforin damages the target cell membrane and facilitates the entry of the granzymes into the target cell nucleus, which induces apoptosis (Doherty, 1993a; Shi *et al.,* 1997). Target cells expressing Fas may also be induced to undergo apoptosis after engagement of the Fas receptor with Fas ligand expressed by activated CTL. CD8$^+$ CTL produce multiple cytokines after activation. Although mature differentiated CTL do not produce IL-2 after antigen stimulation, γIFN, GM-CSF, and TNF are produced and serve to promote direct antiviral effects by activation and recruitment of additional effector cells to the local site (Doherty, 1993b). Consistent with these functional abilities, CD8$^+$ CTL have proven to be highly effective in animal models for promoting tumor regression, resolving acute virus infections and maintaining protective immunity to persistent or latent viruses.

B. ROLE OF COSTIMULATION IN THE INDUCTION OF T-CELL IMMUNITY

The initiation of a T-cell-dependent immune response requires engagement of the T-cell antigen receptor by peptide bound to a MHC molecule and the engagement of costimulatory receptors. The best characterized costimulatory receptor on T cells is the CD28 molecule, which binds to B7-1 and B7-2 ligands on APC (Linsley *et al.,* 1992; Linsley and Ledbetter, 1993). The expression of these costimulatory molecules is limited to professional APC including monocyte/macrophages, dendritic cells, and activated B cells, illustrating the importance of these cells in initiating the host response. The delivery of costimulation to T cells appears to be a highly regulated event. CTLA4 is a receptor related to CD28 and is expressed on T cells early after activation and also binds B7-1 and B7-2. However, rather than delivering a positive signal, CTLA4 appears to deliver a negative signal and inhibit T-cell proliferation (Thompson and Allison, 1997).

III. IMMUNOLOGIC MEMORY MEDIATED BY $\alpha\beta$ T CELLS

An important characteristic of the host T-cell response to virus infection is the establishment of immunologic memory, which is characterized by the maintenance of an increased frequency of T cells responsive to antigens expressed by the pathogen and results in protection or attenuation of infection following repeat exposure to the pathogen. The majority of effector T cells elicited to antigenic stimulation ultimately die, but a sufficient pool of reactive T cells is able to persist and provide memory (Ahmed, 1992; Akbar *et al.*, 1993). In animal models of malignancy, immunization with γ-irradiated tumor cells will elicit tumor-specific T-cell responses that persist and protect the host from subsequent tumor challenge (Greenberg, 1991). The factors that govern the transition from the effector to memory phase are not well understood, but one objective of cellular immunotherapy for persistent human virus infections or tumors would be to provide the host with sufficient numbers of persisting T cells to control episodes of virus reactivation and eliminate minimal residual tumor deposits.

Considerable controversy has existed over the mechanisms by which T-cell memory is maintained, and several possibilities have experimental support. These include (a) the presence of persistent reservoirs of antigen in the host that restimulate primed T cells and/or activate naïve T cells, (b) the intermittent stimulation of primed T cells by cross-reacting environmental antigens or self-peptides, a consequence of the lower threshold for activation of primed T cells as compared to naïve T cells, and (c) the presence of differentiated long-lived memory T cells that can remain in a resting state until they reencounter antigen (Gray and Matzinger, 1991; Ahmed 1992; Matzinger, 1994; Tough and Sprent, 1994). Early studies analyzing the maintenance of $CD4^+$ T-cell memory to KLH (keyhole limpet hemocyanin) and of $CD8^+$ T-cell memory to the male HY antigen or to LCMV (lymphocytic choriomeningitis) demonstrated that $CD4^+$ Th responses persisted for less than 6 weeks and $CD8^+$ CTL responses persisted for less than 16 weeks in the absence of priming antigen in the host (Gray and Matzinger; 1991; Oehen *et al.*, 1992). However, more recent studies using congenic mouse strains or β_2-microglobulin $-/-$ mice as recipients of adoptively transferred purified $CD8^+$ CTL specific for LCMV, Sendai virus, or influenza virus have concluded that $CD8^+$ CTL memory can be maintained long-term in the absence of priming antigen (Hou *et al.*, 1994; Lau *et al.*, 1994). These studies have not elucidated if the transferred memory CTL persist in a resting state or require intermittent restimulation by cross-reacting environmental antigens, self-peptides, or cytokines (Tough *et al.*, 1996). These will be important issues to fully resolve, since a comprehensive understanding of the mechanisms by which T-cell memory is induced and maintained should provide insights into how to manipulate antigen-specific T cells isolated for use in adoptive immunotherapy and to improve the probability that long-term protective immunity will be restored following adoptive transfer.

IV. MECHANISMS OF EVASION OF T-CELL RESPONSES BY VIRUSES AND TUMORS

Viruses have evolved mechanisms to evade $\alpha\beta$ T cells, to prolong episodes of acute infection and facilitate the establishment of a persistent infection in the host (Table 8.2). Understanding the mechanisms utilized by individual viruses to evade T-cell recognition is relevant for the development of cellular immunotherapy since T cells of a particular phenotype or antigen specificity may need to be selected for effective therapy. These mechanisms may also be instructive for studies of immunotherapy for cancer, since tumors may utilize similar principles to evade host immunity.

A. VIRUS EVASION MECHANISMS

1. Latency

Several viruses such as EBV, CMV, VZV (Varicella-zoster virus), HSV (herpes simplex virus), and HIV (human immunodeficiency virus) are successful in establishing a lifelong persistent infection in the host, at least in part due to their ability to enter a state of latency in which host cells contain the viral genome but are not expressing viral proteins (Stevens, 1989; Middleton *et al.*, 1991; Chun *et al.*, 1995). The absence of viral antigens presented by class I or class II MHC in such latently infected cells precludes their recognition by $\alpha\beta$ T cells, even at the peak of the host-immune response. Reactivation of herpes viruses from latency should not pose a life-threatening problem to the immunocompetent host, since the T-cell memory response will contain infection and prevent dissemination, but sufficient replication may occur to permit transmission to a new host. Reactivation in immunosuppressed hosts can result in the rapid dissemination of virus to vital organs and is responsible for the severe

TABLE 8.2 Immune Evasion Mechanisms of Viruses

Latency (HIV, herpes viruses)
Interference with antigen processing and presentation (HSV, CMV, adenovirus)
Mutations in peptide epitopes (EBV, HIV, hepatitis B virus)
Secreted viral proteins that suppress the host-immune response (pox viruses, vaccinia)

CMV, VZV, and HSV infections in allogeneic bone marrow transplant recipients (Meyers, 1990).

EBV appears to have a form of latency in which a single viral protein, EBNA-1, is expressed in the infected B cells (Klein, 1989). These cells can persist in immunocompetent hosts because EBNA-1 is not a target antigen for $CD8^+$ CTL due to the presence in the protein of a unique amino acid sequence consisting of repeats of glycine and alanine (Levitskaya *et al.,* 1995). This sequence precludes the processing and presentation of the EBNA-1 protein via the class I antigen processing pathway, and $CD8^+$ CTL to EBNA-1 are absent in EBV-seropositive individuals.

2. Interference with Antigen Processing and Presentation

Given the ability of $CD8^+$ CTL to directly destroy virally infected cells, it is not surprising that viruses have evolved strategies to interfere with the display of class I MHC molecules bearing viral peptides at the cell surface. Viruses such as adenovirus and HSV reduce the transcription and/or translation of host proteins including class I MHC and other cellular proteins involved in antigen processing (Schrier *et al.,* 1983; Paabo *et al.,* 1989; Read *et al.,* 1993). However, viral proteins produced after infection also interfere selectively with events involved in the generation and transport of preformed class I MHC/peptide complexes. The adenovirus E3 gp 19 protein was the first viral protein described to inhibit class I antigen presentation and does so by inserting itself into the ER membrane and binding to the class I heavy chain, thereby retaining the class I molecules in the ER (Fig. 8.1A) (Korner and Burgert, 1994). HSV also expresses a protein that acts downstream of transcription and translation of class I MHC. HSV ICP47 is produced at immediate early (IE) times and binds to TAP to interfere with the transport of peptides into the ER (York *et al.,* 1994; Hill *et al.,* 1995).

Cytomegalovirus encodes multiple proteins that interfere with class I antigen processing in a coordinate fashion. US3 is expressed at the IE phase of infection and functions to bind and retain class I MHC molecules in the ER (Jones *et al.,* 1996). US2 and US11 are expressed at early (E) times postinfection and result in the reverse translocation of the class I heavy chain from the ER into the cytosol, where these molecules are degraded by the proteosome (Wiertz *et al.,* 1996a, 1996b). Finally, US6, which is expressed during the late (L) phase of infection, functions on the lumenal side of the ER to block the transport of peptides into the ER (Ahn *et al.,* 1997). The cumulative effect of these CMV evasion strategies is a profound absence of $CD8^+$ CTL specific for viral proteins synthesized after the IE stage of virus replication. The overwhelming majority of $CD8^+$ CTL in immunocompetent hosts with protective immunity are directed against structural virion proteins such as pp65 and pp150 that enter the cell with the virion and are processed for presentation prior to the virus-induced blockade in antigen processing (Riddell *et al.,* 1991; Wills *et al.,* 1996). It is anticipated that many viruses will utilize similar strategies to evade recognition by $CD8^+$

CTL, and it may prove to be important for successful cellular immunotherapy with virus-specific T cells, to target those antigens presented before the virus has subverted the antigen processing machinery.

3. Mutations in Peptide Epitopes

The binding of peptides to MHC molecules and the interaction between the T-cell receptor and the MHC/peptide complex are highly specific. Thus, alterations in peptide sequence as a result of mutations in the viral genome may change the binding affinity of the peptide for the MHC molecule, resulting in less efficient or absent presentation, or if the peptide retains MHC binding, sequence changes in TcR contact residues may preclude TcR recognition or deliver aberrant or antagonist signals to the T cell (Klenerman *et al.,* 1994; Koup, 1994; Goulder *et al.,* 1997). Virus mutational escape of T-cell recognition was first shown to occur in situations where intense selective pressure against a single epitope was created by infecting mice transgenic for a TcR specific for a single LCMV epitope, with a large inoculum of LCMV. A variant LCMV emerged that was no longer recognized by the host-immune system because of a mutation in the epitope targeted by the transgenic TcR (Pircher *et al.,* 1990).

Mutational escape has also been documented in natural human virus infections. The identification of a mutant EBV strain with sequence variation in an immunodominant epitope presented by HLA A11 in a geographic area where HLA A11 is a prevalent allele suggests the mutant virus achieved a biologic advantage in the population (de-Campos-Lima *et al.,* 1993).

The high error rate of the HIV-1 reverse transcriptase results in the rapid diversification of the infecting virus inoculum. Progressive infection is characterized by the emergence of virus variants with mutations in epitopes recognized by $CD8^+$ CTL, rendering these CTL ineffective in control of virus replication. Some mutant peptides still bind to the class I MHC but deliver an antagonist signal to the T cell, which inhibits its ability to lyse target cells expressing the wild-type epitope (Phillips *et al.,* 1991; Klenerman *et al.,* 1994; Koup, 1994; Goulder *et al.,* 1997). This "peptide antagonism" has also been observed in natural infection with hepatitis B virus (HBV) and may permit the continued replication in the host of both mutant and wild-type viruses (Bertoletti *et al.,* 1994; Klenerman *et al.,* 1994). The progression of infection with HIV and HBV in immunocompetent hosts and the frequent identification of isolates with mutations in T-cell epitopes highlight the importance of virus variation and identify an obstacle for effective cellular immunotherapy of such infections. A rational strategy to overcome mutational escape would be to isolate T cells specific for multiple epitopes for simultaneous administration, to reduce the probability the virus can continue to replicate by virtue of the generation of escape mutants.

4. Secreted Viral Proteins That Suppress Host-Immune Responses

In large part, the binding of cytokines to specific receptors regulates the activation, differentiation, and effector functions of immune cells. Many viruses have evolved strategies to interfere with these signaling processes that are essential for optimal function and regulation of a host-immune response. Pox viruses are particularly adept at such interference, perhaps because of the critical role of cytokine-induced inflammatory responses in controlling infection with these viruses (Ramsay *et al.,* 1993). Thus, Shope fibroma virus produces a soluble TNF binding protein that inhibits TNF binding to cell surface TNF receptors (Upton *et al.,* 1991). Vaccinia virus produces an IL-1 binding protein that inhibits the activity of IL-1 (Alcam'i and Smith, 1992). Myxoma virus encodes a homolog to the γIFN receptor (Upton *et al.,* 1992).

Perhaps of greater relevance to adoptive therapy in humans is the finding that EBV produces a homolog of IL-10 (Hsu *et al.,* 1990). The viral IL-10 promotes differentiation of $CD4^+$ Th cells toward a Th2 phenotype and has recently been shown to interfere with TAP activity (de-Waal-Malefyt *et al.,* 1991; Zeidler *et al.,* 1997). Thus, the production of viral IL-10 at the site of infection may interfere with the efficacy of $CD8^+$ CTL by decreasing the magnitude of the response due to a limited production of IL-2, a Th1 cytokine, and by interfering with the presentation of viral epitopes.

B. EVASION OF T-CELL IMMUNITY BY TUMORS

The identification of potentially immunogenic proteins in human tumors has aroused considerable interest in understanding the mechanisms by which tumor cells escape immune surveillance and persist in immunocompetent hosts. As with virus infections, the identification of various evasion strategies will provide insights into how to best utilize cellular adoptive immunotherapy to achieve therapeutic effects. Additional insights are likely to be derived from an analysis of tumors, which are resistant to adoptive immunotherapy or emerge after initially successful therapy with T cells that are specific for a tumor-associated antigen.

1. Down-regulation of Tumor Cell Class I MHC Expression

Down-regulation of class I MHC expression is observed in many fresh tumor specimens of epithelial origin, suggesting the emergence of tumor variants that might have escaped $CD8^+$ CTL recognition. Benign tumors derived from epithelial tissues typically express normal levels of class I MHC, whereas the invasive or metastatic counterparts have lost or reduced levels of class I MHC expression. Analysis of tumors of hematopoietic origin indicates that loss of class I MHC molecules occurs less frequently. Class I MHC is expressed on follicular lymphomas of B-cell origin (Schultze *et al.,* 1997), and studies in our laboratory indicate that the overwhelming majority of acute myeloid leukemias express high levels of class I MHC (Warren and Riddell, unpublished data). Of interest, a graft-versus-tumor effect after allogeneic bone marrow transplantation, presumed to be mediated by T cells specific for minor histocompatibility antigens presented as peptides by class I MHC, is pronounced for both AML and lymphoma (Horowitz *et al.*, 1990; Jones *et al.*, 1991).

The decline in class I expression on malignant cells may result from several mechanisms, including point mutations (Zeff *et al.*, 1990), genomic loss (Browning *et al.*, 1993), alterations in methylation and chromatin structure (Boucraut *et al.*, 1993), altered binding of transcription factors to class I regulatory sequences mediated by oncogenic molecules (Bernards 1991), and decreased levels of β_2-microglobulin (Wang *et al.*, 1993; Bicknell *et al.*, 1994). Defects in the expression of the TAP genes have also been described in small-cell lung cancer (Restifo *et al.*, 1993) and in cervical carcinomas (Cromme *et al.*, 1994a, 1994b).

The frequency of alterations in class I MHC expression described on human tumors suggests a role in evading host immune surveillance similar to the strategies utilized by persistent human viruses and may pose a serious obstacle for effective cellular immunotherapy with $CD8^+$ CTL. However, class I loss is often allele specific, presumably to preserve some resistance of the tumor cells to recognition and elimination by NK cells (Hoglund *et al.*, 1997). Thus, the identification and isolation of T cells reactive with epitopes of tumor-associated antigens presented by the expressed MHC alleles and/or the use of cytokines that may up-regulate class I expression may be useful.

2. Selection of Tumor Antigen Loss Variants

The outgrowth of antigen loss variants has been observed in animal tumor models and in patients receiving immunomodulatory therapy (Lehmann *et al.*, 1995; Dudley and Roopenian, 1996). Thus, it is desirable whenever possible to target antigens that are expressed uniformly in the tumor cell population. The expression of several melanoma antigens has been examined by *in situ* immunophenotyping of histologic sections from melanoma biopsies. Tyrosinase and MART-1 are expressed uniformly whereas gp100 and gp75 exhibit heterogeneous expression. These results suggest that it may be beneficial to direct T-cell therapy against multiple antigens to minimize the possibility that antigen loss variants will escape elimination.

For tumors such as chronic myeloid leukemia with chromosome translocations, which encode fusion proteins that are essential for transformation it would be attractive to direct T-cell therapy against the novel fusion epitopes. However, despite considerable efforts by several groups, it has proven to be difficult to isolate such T cells from most individuals.

3. Suppressor T Cells and Soluble Suppressive Factors

In animal models of tumor-specific T-cell therapy, the efficacy of $CD8^+$ CTL may be inhibited by the presence of suppressor T cells, and it is often necessary for tumor eradication to reduce this suppressor activity by preadministering cyclophosphamide (Greenberg, 1991). It is not known if similar suppressor T cells are elicited in patients with spontaneously developing tumors, but the administration of chemotherapy to cytoreduce the tumor prior to cellular immunotherapy may be necessary anyway and would limit the potential inhibitory capacity of suppressor cells.

Immunosuppressive cytokines such as TGFβ (transforming growth factor β) or IL-10 have been shown to be produced by some human tumors and may suppress T-cell proliferation and function at the local tumor site. In an animal model, an immunogenic tumor transfected to express TGFβ escaped immune surveillance by failing to induce tumor-specific T-cell responses (Torre-Amione *et al.,* 1990). Such a mechanism may be most inhibitory to the induction of tumor-specific T-cell responses and could potentially be overcome by the infusion of a large number of tumor-specific T cells generated and expanded *ex vivo* and by reducing the tumor burden by administering chemotherapy prior to immunotherapy.

4. Lack of Costimulation by Tumor Cells

The induction of tumor-specific T-cell responses requires the activation of the TcR by recognition of MHC/peptide complexes present on the tumor cell and the engagement of the CD28 molecule by B7-1 or B7-2 to provide costimulation. Thus, even if malignant cells express tumor-associated antigens, they may fail to function as APC because they lack expression of the requisite costimulatory molecules essential to induce a T-cell response. The absence of costimulatory molecules has been demonstrated as a mechanism whereby pre-B-cell ALL and follicular lymphomas fail to induce T-cell responses, and absence of costimulatory molecules may be a general property of human tumors that facilitates their escape from immune surveillance (Schultze *et al.,* 1995; Cardoso *et al.,* 1996). Strategies to restore costimulation by transfecting tumor cells with B7 or by providing bystander cells that express B7 have been successful for activating and expanding tumor-specific T cells *in vitro* (Schultze *et al.,* 1995; Cardoso *et al.,* 1996, 1997). Animal model studies indicate that transduction of tumor cells to express B7 or GM-CSF, which will recruit professional APC-expressing costimulatory molecules to the site, is effective in inducing tumor-specific T-cell immunity *in vivo* (Dranoff *et al.,* 1993; Townsend and Allison, 1993). The lack of costimulation by tumor cells is a potential obstacle for the induction of T-cell responses *in vivo* and their isolation *in vitro.* However, the absence of costimulation may be less of a hindrance to the efficacy of adoptively transferred T cells, since a very strong and sustained response could potentially be achieved by repeated infusions.

5. Expression of Fas Ligand by Tumor Cells

The interaction of Fas ligand (FasL) on the surface of activated T cells with Fas on target cells provides one mechanism for T-cell-induced cytotoxicity (Kondo *et al.,* 1997). However, T cells also express Fas and expression of FasL on selected tissues such as the testes may provide a mechanism for immune privilege by mediating the active destruction of T cells recognizing antigens at these sites (Nagata, 1996). Recently, the expression of FasL has been described on several types of human tumors, including melanoma (Hahne *et al.,* 1996), hepatocellular carcinoma (Strand *et al.,* 1996), and myeloma (Villunger *et al.,* 1997). These FasL-positive tumor cells mediate the destruction of Fas-expressing T lymphocytes. Therefore, tumor cells may dampen the efficacy of immune attack by killing—via FasL–Fas interactions—T cells that react with tumor-derived antigenic determinants. Such a mechanism may contribute to the relatively weak tumor-specific T-cell responses that develop in tumor-bearing hosts and in patients with large tumor burdens and could compromise the survival and efficacy of adoptively transferred T cells. Transferred $CD8^+$ CTL may still lyse the FasL-expressing tumor cells by perforin and granzyme mediated apoptosis, but it may be necessary to infuse large numbers of T cells, reduce the tumor burden with chemotherapy, or provide repeated infusions of T cells to overcome the loss of a proportion of transferred T cells by tumor FasL-mediated killing.

V. METHODS OF SELECTION OF ANTIGEN-SPECIFIC T-CELL CLONES

A. ISOLATION AND EXPANSION OF VIRUS-SPECIFIC $CD4^+$ AND $CD8^+$ T CELLS FOR ADOPTIVE IMMUNOTHERAPY

The adoptive transfer of virus-specific T cells to restore or augment protective host responses has been investigated for CMV and EBV infection in allogeneic BMT recipients using allogeneic T cells isolated *in vitro* from the peripheral blood of the immunocompetent donor and for HIV infection using autologous T cells isolated from the peripheral blood of the patient. The principal methodology for isolating T cells specific for individual viruses relies on the presentation of viral antigen by an autologous APC to activate virus-specific T cells, followed by culture in IL-2 to expand reactive T cells. Most laboratories have utilized virus-infected cells as APCs, although in the setting of adoptive immunotherapy for HIV, APCs pulsed with peptides derived from viral proteins previously defined to be targets of CTL responses have also been used as *in vitro* stimulators (Lieberman *et al.,* 1997). The strategies for isolating $CD8^+$ and $CD4^+$ T-cell clones for use in adoptive immunotherapy of CMV infection will be discussed as an example of the principles involved. Studies of adoptive immunotherapy for EBV infection have used polyclonal T-cell lines, and methods for generating EBV-reactive T cells are discussed by Heslop *et al.* (1996).

1. Isolation of $CD8^+$ CMV-Specific CTL Clones

CMV infection after allogeneic BMT progresses to life-threatening disease in patients with a deficiency of virus-specific T-cell responses due to the delay in recovery of mature functional T cells (Reusser *et al.,* 1991; Li *et al.,* 1994). To hasten the recovery of CMV-specific T-cell immunity and to potentially provide protection from CMV disease, CMV-specific T cells have been isolated from the bone marrow donor, expanded *in vitro,* and adoptively transferred to the recipient early after BMT. If polyclonal populations of T cells were used in therapy, there would be a risk of inducing graft-versus-host disease (GVHD) due to the presence of contaminating T cells specific for recipient minor H antigens. Thus, our laboratory has focused on the development of culture methods that permit the isolation and expansion of individual virus-specific T-cell clones for use in adoptive immunotherapy to permit a rigorous evaluation of safety and to more precisely identify the effector cells required for therapeutic efficacy.

To isolate $CD8^+$ CMV-specific CTL, a cell that could be permissively infected with CMV was required as an APC. Dermal fibroblasts were an attractive cell for this function because they support a lytic CMV infection, express class I but not class II MHC molecules, and could be established in culture from a skin biopsy specimen obtained from the bone marrow donor. After generation of a fibroblast line, an aliquot was infected with CMV and used for stimulation of PBMC obtained from the donor. The cultures were restimulated 7 days later and supplemented with low-dose IL-2 (2-5 U/ml) to expand CMV-reactive CTL. Typically, strong class I restricted CMV-specific cytolytic activity was demonstrated after two stimulations (Fig. 8.2), and at this time $CD8^+$ T cells were cloned by plating at limiting dilution. Several factors were identified to improve cloning efficiency of human $CD8^+$ CTL, including the use of round-bottom microwells, the addition of both γ-irradiated autologous PBMC and LCL (lymphoblastoid cell lines) as feeder cells, the use of anti-CD3 monoclonal antibody (mAb) in place of CMV-infected fibroblasts to provide T-cell receptor stimulation, and the use of IL-2 at concentrations of 25–50 U/ml to support T-cell expansion (Walter *et al.,* 1995). Fourteen days after plating, wells positive for growth were easily identified by scanning with light microscopy, and the T cells from these wells were transferred to larger wells for expansion. To avoid the potential carryover of infectious CMV, subsequent expansion of the T-cell clones was accomplished using anti-CD3 mAb as the stimulus for activation via the TcR instead of CMV-infected fibroblasts (Riddell and Greenberg 1990). Culture conditions containing the optimal number of feeder cells were defined and provided a 500 to 1000-fold expansion in T-cell number over a 12- to 14-day stimulation cycle. Thus, individual T-cell clones were readily expanded to greater than 10^9 cells for use in adoptive transfer. The procedure for isolating and expanding $CD8^+$ CMV-specific T-cell clones is depicted schematically in Fig. 8.3.

2. Isolation of $CD4^+$ CMV-Specific Th Clones

To isolate $CD4^+$ CMV-specific T cells, γ-irradiated autologous PBMC containing class II MHC positive monocytes and dendritic cells were pulsed with a

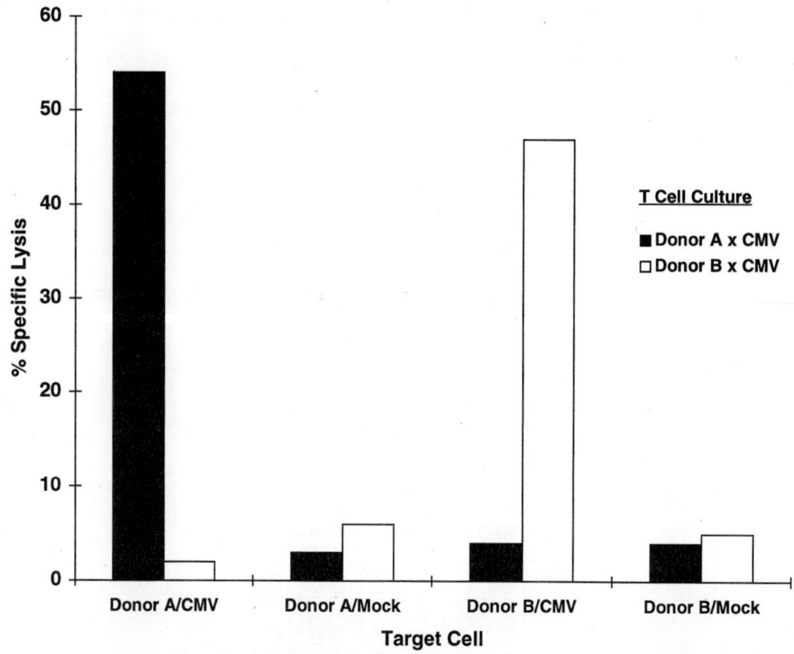

FIGURE 8.2 Generation of class I MHC-restricted CTL from normal CMV-seropositive donors. Peripheral blood lymphocytes obtained from two class I MHC mismatched donors, A and B, were stimulated with autologous CMV-infected fibroblasts and assayed for lytic activity against autologous and mismatched CMV and uninfected fibroblasts at an effector to target ratio of 10:1.

CMV antigen extract prepared from infected fibroblasts and used to stimulate T cells at a responder to stimulator ratio of 2:1. The cultures were restimulated after 7 days with fresh γ-irradiated PBMC pulsed with CMV antigen, and low-dose IL-2 (2 U/ml) was added during the second stimulation cycle to promote proliferation of activated T cells. Seven days after restimulation, T cells from the cultures were plated in limiting dilution cultures in round-bottom plates with γ-irradiated autologous PBMC, LCL, CMV antigen, and IL-2 (25–50 U/ml). Wells positive for growth were identified after 14 days, and the T-cell clones were restimulated with anti-CD3 mAb using the same culture conditions described for expansion of $CD8^+$ CTL.

The peak incidence of CMV disease after allogeneic BMT is 6–8 weeks, and disease rarely occurs in the first 5 weeks after transplant. Since the time from culture initiation to having sufficient clonally derived, CMV-specific T cells (greater than 5×10^9) was approximately 8 weeks, the cultures were initiated from the donor 3–4 weeks before the BMT so that cells would be available prior to the period of highest risk of CMV disease.

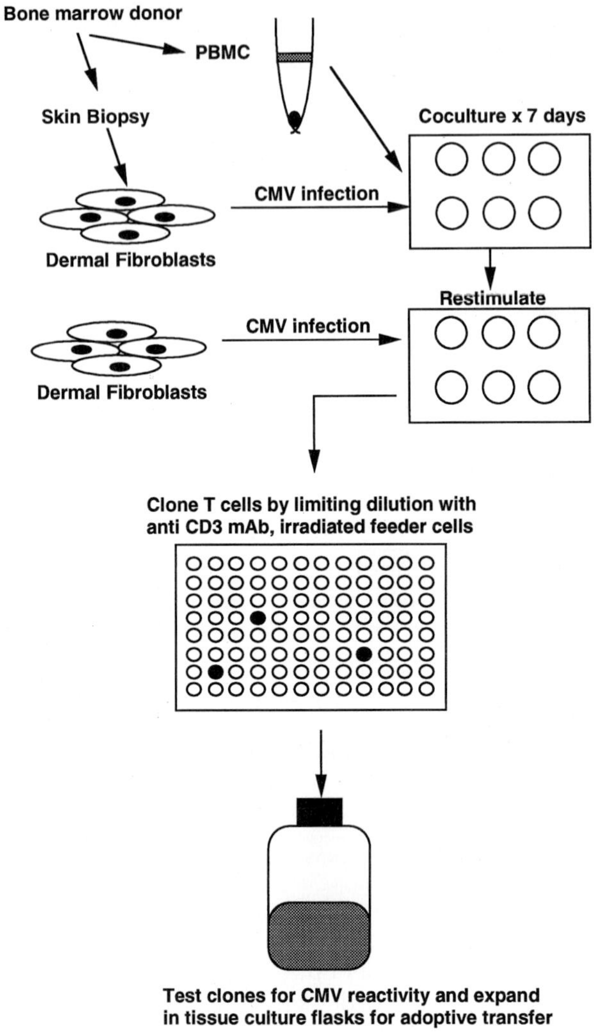

FIGURE 8.3 Strategy for isolation of $CD8^+$ CMV-specific CTL clones from CMV-seropositive bone marrow donors.

B. ISOLATION AND EXPANSION OF TUMOR-REACTIVE T CELLS FOR USE IN ADOPTIVE IMMUNOTHERAPY

A major issue to be considered in generating tumor-reactive T cells for use in adoptive immunotherapy is the selection of potential target antigens expressed by the tumor. There are now several human tumors for which antigens recognized by T cells have been identified (Table 8.1), and many laboratories are developing strategies to generate tumor-reactive T cells from patients. Our laboratory has focused on developing T-cell therapy in two settings: malignant melanoma, using T cells specific for defined melanoma antigens such as tyrosinase and gp 100, and relapsed acute leukemia after allogeneic BMT, using allogeneic T cells specific for minor histocompatibility antigens expressed by recipient hematopoietic cells including the leukemic blasts (Warren *et al.,* 1998). The initial studies are directed at investigating the role of $CD8^+$ cytolytic T cells, although in both settings $CD4^+$ Th have been identified, and will be evaluated in subsequent studies.

1. Isolation of Melanoma-Reactive $CD8^+$ T Cells

T cells that recognize melanoma cells were initially isolated from the blood obtained from melanoma patients by stimulation with tumor cells *in vitro* or from tumor biopsies by culturing tumor-infiltrating lymphocytes with high doses of IL-2 (Aebersold *et al.,* 1991; van der Bruggen *et al.,* 1991). These strategies were frequently unsuccessful, but the occasional isolation of $CD8^+$ CTL was important in facilitating the identification of the tumor-associated antigens (Van Pel *et al.,* 1995). Because it was difficult to reproducibly obtain tumor cell lines for use as stimulators from tumor biopsy samples, several alternative approaches have been developed to selectively present immunogenic tumor antigens or antigenic peptides to T cells and improve the identification of tumor-reactive T cells.

A widely used strategy to generate T cells specific for defined melanoma antigens is to pulse dendritic cells (DCs), obtained from the patient's blood after culture *in vitro* with GM-CSF and IL-4, with previously defined immunogenic peptides and use these DCs to stimulate autologous peripheral blood T cells (Celis *et al.,* 1994; Rivoltini *et al.,* 1995). Algorithms, based on previously identified anchor residues for peptides binding to individual class I alleles, have been used to identify peptide sequences within tumor-associated proteins that will bind to class I MHC with high affinity. This strategy has been combined with peptide-stimulation protocols to isolate T cells recognizing subdominant epitopes that were not recognized by CTL isolated by stimulation with tumor cells alone (Tsai *et al.,* 1997). Such an approach offers the potential to obtain T cells reactive with multiple epitopes of a tumor antigen, reducing the possibility that tumor variants lacking a single immunodominant epitope would escape T-cell therapy. A problem with this approach is the propensity to isolate low-avidity T cells that recognize peptide pulsed-target cells but not the tumor cells that presumably express lower numbers of MHC/peptide complexes than is achieved by pulsing DC with the peptide (Alexander-Miller *et al.,* 1996). The use of cytokines such as IL-10 to modulate the outgrowth of T cells, or low doses of peptide to pulse DC, has resulted in the more frequent isolation of T cells with tumor reactivity (Tsai *et al.,* 1997).

The second major strategy under investigation is to express the gene encoding the tumor antigen in an appropriate autologous APC. Many methods for expressing the antigen have been utilized, including transduction with retroviral vectors, transfection with DNA or RNA, and infection with recombinant viral vectors such as vaccinia or adenovirus (Yang *et al.,* 1995; Boczkowski *et al.,* 1996; Yee *et al.,* 1996). The advantage of expressing the tumor antigen in an APC is that the approach should be useful for patients of all HLA types and will select for high-avidity T cells that are more frequently tumor reactive. Our laboratory has used a vaccinia recombinant virus encoding tyrosinase to infect autologous peripheral blood monocytes and successfully isolate tyrosinase-specific CD8$^+$ and CD4$^+$ T-cell clones that react with melanoma cells (Yee *et al.,* 1996). However, this method is not successful in all patients and tyrosinase-reactive T cells must be isolated by cloning to separate them from the T cells responding to vaccinia antigens. Thus, additional refinements in the methodology are needed to improve the feasibility of isolating T cells for adoptive immunotherapy of melanoma. Studies are now in progress in which melanoma patients are being immunized with peptides or recombinant vectors encoding known melanoma antigens in an effort to augment or elicit a T-cell response, and, if successful, this may improve the feasibility of isolating such T cells for *in vitro* expansion and reinfusion to further augment the tumor-reactive T-cell response.

2. Isolation and Characterization of CD8$^+$ CTL Specific for Minor Histocompatibility (H) Antigens

The identification of minor H antigens that are expressed in hematopoietic cells, including leukemic cells but not in fibroblasts and other tissue types, has suggested that such tissue-restricted antigens could potentially serve as targets for T-cell immunotherapy to enhance graft-versus-leukemia activity without inducing graft-versus-host disease (de Bueger *et al.,* 1992; van der Harst *et al.,* 1994; Dolstra *et al.,* 1997). To explore the feasibility of this strategy, several laboratories have attempted to isolate donor T cells specific for recipient minor H antigens, but such T cells were found to be present in low frequency in the blood of unprimed donors and were difficult to isolate directly from donor PBMC samples. However, the frequency of allogeneic minor H antigen-specific T cells in PBMC increases after BMT due to *in vivo* priming by recipient antigens. Further amplification of donor-derived T-cells that are specific for recipient minor H antigens is accomplished by stimulation with γ-irradiated PBMC obtained and stored from the recipient pretransplant (Goulmy *et al.,* 1983; de Bueger *et al.,* 1992). Several cycles of stimulation are often required, and T-cell cloning must be performed to distinguish those clones recognizing minor H antigens whose expression is restricted to recipient hematopoietic cells.

A recent study of 10 BMT donor–recipient pairs identified 12 minor H antigens that exhibited expression restricted to hematopoietic cells, and these antigens were also expressed on leukemic blasts (Warren *et al.,* 1998). Dermal fibroblasts and keratinocytes have been used as representative nonhematopoietic target cells for *in vitro* cytotoxicity assays to analyze tissue-restricted expression of the minor H antigens recognized by these CTL clones. However, defining the expression of minor H antigens in tissues with this method is limited by the difficulties in obtaining and culturing samples from all tissue sites.

Thus, a critical issue for the development of adoptive immunotherapy to selectively augment GVL activity is to demonstrate that the expression of genes encoding candidate minor H antigen targets is truly limited to hematopoietic cells. This will require identification of the genes encoding minor H antigens so that molecular techniques can be used to screen tissue libraries for their expression. Biochemical methods have been used to isolate and sequence the minor H. antigen peptides bound to MHC molecules (den Haan *et al.,* 1995; Wang *et al.,* 1995; Meadows *et al.,* 1997), but the short peptide sequence obtained does not ensure that the gene encoding the antigen will be identified in available databases. A second approach to identifying genes encoding CTL-defined antigens being implemented in our laboratory is based on the cDNA expression cloning strategy that has been successfully applied to the identification of melanoma antigens (Van Pel *et al.,* 1995). With this strategy cDNA expression libraries are prepared from antigen-positive cells, and pools containing approximately 100 cDNAs are cotransfected with a plasmid encoding the class I-restricting allele into antigen-negative COS cells. CTL clones are then cocultured with the transfectants and the supernatants screened for cytokine production indicating that the transfectants now express the minor H antigen recognized by the CTL. Subdividing positive pools into smaller pools will allow identification of the cDNA encoding the antigen. The advantage of this approach is that the gene encoding the antigen is identified directly and the reagents are immediately available to screen tissues for expression.

VI. GENETIC MODIFICATION OF T CELLS TO IMPROVE THE SAFETY AND/OR EFFICACY OF ADOPTIVE IMMUNOTHERAPY

The development of methods to introduce genes into T cells using retrovirus-mediated gene transfer allows the introduction of a marker gene as a means of analyzing *in vivo* persistence of transferred T cells and the introduction of genes that confer novel functions and may improve the safety and/or efficacy of cellular immunotherapy. As discussed earlier, the infusion of polyclonal donor T cells to bone marrow transplant patients with relapsed leukemia was effective in some patients but caused GVHD in a majority of patients. The introduction of a gene encoding herpes virus thymidine kinase (TK) into polyclonal T cells to be used in therapy renders the cells susceptible to ablation by administration of ganciclovir or acyclovir (Heyman *et al.,* 1989). This strategy was recently used in clinical trials to ablate transferred polyclonal T cells in patients developing severe GVHD following donor lymphocyte infusions (Bonini *et al.,* 1997). A similar strategy was used by our laboratory in initial studies of adoptive immunotherapy with autologous $CD8^+$ HIV-specific CTL clones for HIV infection because of the potential for immunopathology. However, the transferred TK-modified CTL were eliminated after 5–8 weeks by a host-immune response to the foreign transgene product (Riddell *et al.,* 1996). These results suggest that in immunocompetent hosts the TK gene may be useful to establish the safety of cellular immunotherapy in settings where toxicity is possible but will limit the long-term persistence of infused T cells because of the immunogenicity of TK.

A potential limitation of adoptive immunotherapy with $CD8^+$ CTL alone is the failure of this subset to produce IL-2, which may be necessary for *in vivo* proliferation and persistence (Riddell and Greenberg, 1995; Walter *et al.,* 1995; Riddell *et al.,* 1996). $CD8^+$ CTL can potentially be modified by gene transfer to provide an autocrine growth signal, and such modified CTL could provide persistent immunity even in $CD4^+$-deficient hosts. One strategy to accomplish this involves modifying the IL-2 receptor to make it responsive to cytokines such as GM-CSF that are normally produced by $CD8^+$ CTL after antigen stimulation. The IL-2 signal is delivered by heterodimerization of the cytoplasmic regions of the IL-2 β and γ chains after binding of IL-2 to the trimeric $\alpha\beta\gamma$ receptor complex. Thus, chimeric IL-2 receptors were designed to contain the transmembrane and cytoplasmic domains of the IL-2 receptor γ and β chains and the extracellular domains of the GM-CSF receptor α and β chains, respectively. T cells expressing these chimeric cytokine receptors after transfection were induced to proliferate upon exposure to GM-CSF (Nelson *et al.,* 1994). Thus, the production of GM-CSF by such modified $CD8^+$ CTL as a result of antigen stimulation could potentially provide an autocrine IL-2 signal to the CTL and improve *in vivo* survival. It is anticipated that T cells with such enhanced functions will be clinically evaluated in the future and may provide recipients with more effective responses than those achieved with unmodified T-cell therapy.

VII. CLINICAL STUDIES OF CELLULAR ADOPTIVE IMMUNOTHERAPY FOR VIRUSES AND VIRUS-INDUCED MALIGNANCIES

A. T-CELL THERAPY OF CYTOMEGALOVIRUS INFECTION

CMV is a ubiquitous herpes virus that causes an often-unrecognized primary infection in 50–70% of the population. However, in patients with iatrogenic or acquired immunodeficiency, reactivation of CMV in CMV-seropositive hosts or the acquisition of primary CMV infection often leads to progressive infection and visceral disease.

1. CMV-Specific T-Cell Immunity and CMV Disease in Immunosuppressed Transplant Recipients

The hypothesis that progressive CMV infection in BMT recipients was related to a quantitative deficiency of virus-specific T-cell responses has been examined in several studies. All published reports have identified a correlation between the presence of MHC-restricted T-cell responses to CMV antigens and protection from the subsequent occurrence of CMV disease (Quinnan *et al.*, 1984; Reusser *et al.*, 1991; Li *et al.*, 1994). Studies in our laboratory and in animal models of CMV infection have suggested that the CD8$^+$ CTL subset is a crucial component of this protective host T-cell response (Reddehase *et al.*, 1985; Reusser *et al.*, 1991; Li *et al.*, 1994).

2. Specificity of CD8$^+$ CMV-Reactive T Cells

Definition of the specificity of CTL responses in individuals with protective immunity to CMV is essential for the development of effective adoptive immunotherapy to permit the selection of T cells that reflect the immunodominant responses elicited and maintained in immunocompetent hosts. CMV expresses its genes in a temporal sequence, with discrete phases of gene expression termed the immediate early (IE), early (E), and late (L) phases. The timed addition of inhibitors to block viral protein or RNA synthesis was employed to determine if CTL preferentially recognized proteins produced at IE, E, or L stages of the replicative cycle. Surprisingly, even if the production of newly synthesized viral proteins was completely blocked in target cells with an RNA synthesis inhibitor, CD8$^+$ CMV-specific CTL lines and the majority of CTL clones isolated from normal CMV-seropositive individuals still efficiently lysed these targets (Riddell *et al.*, 1991). Thus, rather than proteins synthesized at IE, E, or L times, virion proteins introduced into the cytoplasm of the target cell following viral entry were the immunodominant-target antigens of the host CTL response. Studies to assess the specificity of the CD8$^+$ CMV-reactive CTL that recovered after allogeneic BMT and conferred protection from subsequent CMV disease showed these CTL were also specific for epitopes derived from structural virion proteins (Li *et al.*, 1994). The contribution of individual virion proteins as antigens for CTL has been assessed by pulsing peptide fragments of purified proteins onto target cells or infecting target cells using recombinant vaccinia viruses encoding a single CMV gene. The matrix protein pp65 has been identified as the target of the immunodominant-host response most frequently, although major responses to a second matrix protein, pp150, are also observed in some individuals (McLaughlin-Taylor *et al.*, 1994; Wills *et al.*, 1996). CMV-infected target cells are rapidly (less than 1 h) sensitized for lysis by CD8$^+$ CTL specific for either pp65 or pp150, and the infected cell remains a target throughout the entire replicative cycle (Fig. 8.4). Thus, such CTL should be effective in limiting virus dissemination by promptly eliminating newly infected cells. CTL specific for pp65 and/

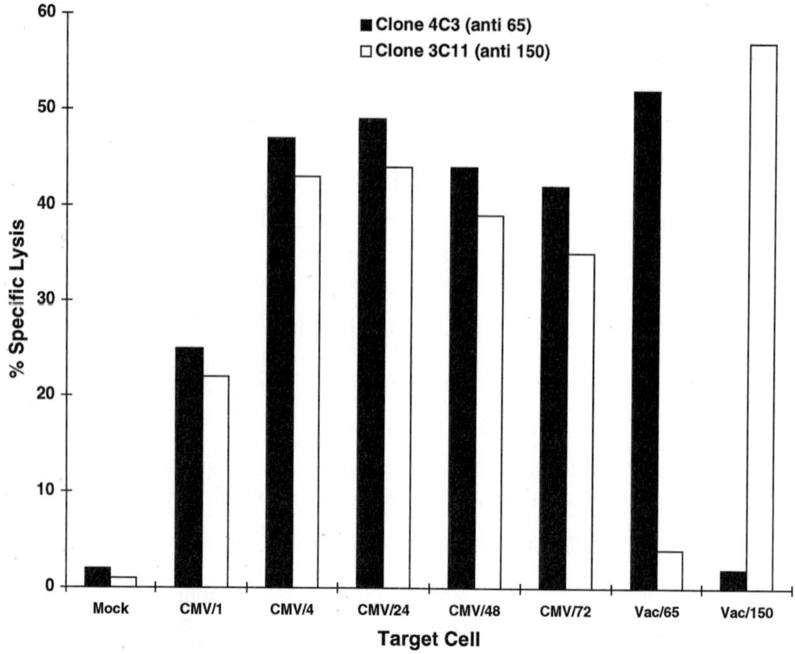

FIGURE 8.4 CD8+ CTL clones specific for CMV pp65 or pp150 lyse CMV-infected targets at all stages of the virus replicative cycle. T-cell clones specific for CMV pp65 (clone 4C3) and for CMV pp150 (clone 3C11) were assayed against autologous target cells infected with CMV for 1–72 h and against autologous target cells infected with vaccinia/CMV 65 recombinant virus or vaccinia/CMV150 recombinant virus. The effector to target ratio is 10:1.

or pp150 are maintained in high frequency for life in normal CMV-seropositive individuals, suggesting that virus reactivation occurs intermittently but presumably remains subclinical because of rapid control by the host-immune response.

3. Adoptive Transfer of CD8+ CMV-Specific T-Cell Clones to BMT Recipients

The first study to evaluate adoptive immunotherapy with virus-specific T cells in humans was conducted in allogeneic BMT recipients with the objectives of defining the safety of infusing escalating doses of CD8+ CMV-specific CTL clones isolated from the bone marrow donor and determining the ability of transferred CTL to persist *in vivo* and restore functional CMV-specific CTL responses. CD8+ CTL alone were administered in the initial study to permit a clear definition of the safety and potential efficacy of this subset of T cells. The T cells were infused as prophylaxis for CMV infection, and patients did not receive prophylactic or preemptive ganciclovir unless they had evidence of persistent CMV reactivation. CD8+ CMV-specific CTL clones were selected for reactivity against structural virion proteins, since this was representative of the immunodominant-CTL response in immunocompetent individuals, and these antigens were found to be processed and presented by infected cells prior to the virus-induced down-regulation in class I MHC expression (Riddell *et al.,* 1991). Fourteen patients were enrolled in the study, and each received a weekly infusion of CTL for 4 weeks at cell doses of 3.3×10^7, 1×10^8, 3.3×10^8, and 1×10^9 per m^2 of body surface area, respectively, beginning 28–42 days after BMT. No serious acute or late toxicity was attributed to the CTL infusions. Minor toxicity was seen in two patients; one patient developed a transient fever 2 h after the fourth infusion and one patient experienced chills during three of the four infusions.

To determine if transferred CTL were effective in restoring measurable immune responses and persisted long-term *in vivo,* blood samples were analyzed for CMV-specific CTL reactivity at multiple time points before, during, and after therapy. Eleven of the fourteen patients had absent CTL responses immediately prior to the first infusion, but after the 4-week treatment period, all exhibited CTL responses equivalent to those in the donor (Fig. 8.5). In a subset of three patients, the T-cell receptor Vβ gene rearrangements expressed in CTL clones administered to the patient were used as a unique molecular marker to conclusively identify the contribution of infused CTL to the responses observed in the recipients and to determine the duration the transferred CTL persisted *in vivo.* This analysis demonstrated that the infused CTL persisted for at least 12 weeks after administration (Walter *et al.,* 1995).

CTL responses equivalent to those in the immunocompetent marrow donor were achieved in all patients immediately after the fourth infusion and remained detectable for at least 12 weeks after therapy, but the magnitude of the response declined in the subset of patients who developed grade II GVHD as a consequence of the BMT and required prolonged treatment with cyclosporin and prednisone. This finding may reflect the adverse effects of intense immunosuppressive drug therapy on the *in vivo* survival of transferred CTL. However, these same patients failed to recover endogenous $CD4^+$ CMV-specific Th responses, and experimental models of persistent virus infection in CD4-deficient mice have shown that adoptively transferred virus-specific $CD8^+$ CTL fail to persist long-term (Matloubian *et al.,* 1994). Thus, a second study, described shortly, is investigating the administration of both CMV-specific $CD8^+$ CTL and $CD4^+$ Th clones to determine if long-term persistence of the CD8 subset can be improved by coadministering $CD4^+$ Th cells.

4. Virologic Monitoring of BMT Patients Receiving Adoptive Immunotherapy with $CD8^+$ CMV-Specific CTL

All patients in this initial study of adoptive immunotherapy as prophylaxis for CMV infection were followed for virus reactivation by cultures of the blood, urine, and throat to provide insight into the potential efficacy of therapy. Transient excretion of CMV was detected in throat or urine cultures in three patients

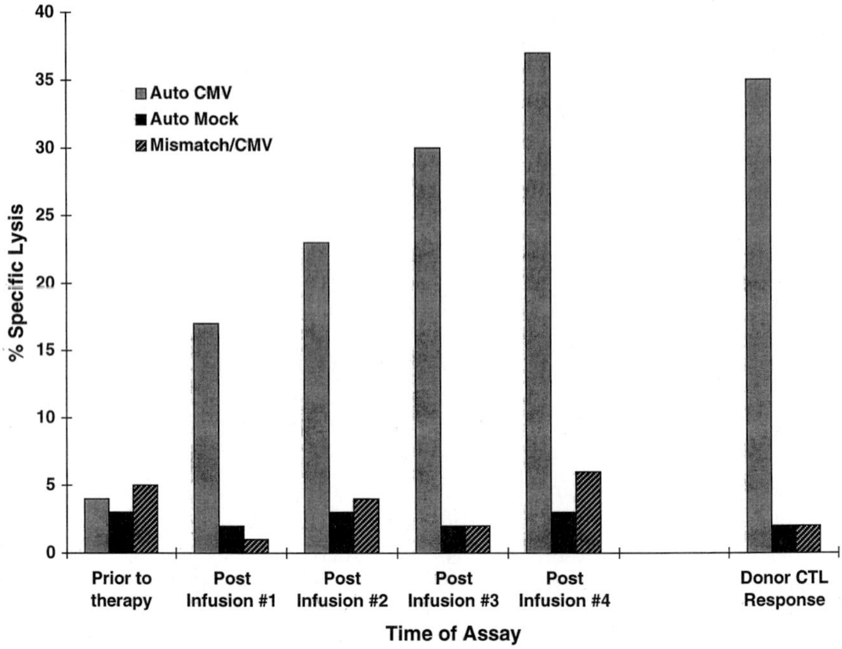

FIGURE 8.5 Cell dose related reconstitution of CD8$^+$ CMV-specific CTL responses after adoptive transfer. Peripheral blood lymphocytes were obtained from patients prior to and 2 days after each infusion of CMV-specific CD8$^+$ CTL clones, and CMV-specific CTL generation was assessed. Target cells include autologous CMV-infected, autologous mock-infected, and class I MHC-mismatched, CMV-infected fibroblasts. Data are shown for an effector to target ratio of 10:1.

but no patient had evidence of CMV viremia, and none of the fourteen patients developed CMV disease (Walter *et al.,* 1995). Thus, this initial study demonstrated that the transfer of CD8$^+$ CMV-specific T-cell clones was safe and effective in restoring normal levels of CTL in allogeneic BMT recipients and provided sufficiently encouraging evidence of antiviral activity to proceed with additional studies.

5. Adoptive Immunotherapy of CMV with CD8$^+$ CTL and CD4$^+$ Th Clones

The administration of both CD8$^+$ CTL and CD4$^+$ Th clones to immunodeficient BMT recipients may restore both arms of the host T-cell-immune response and improve the persistence of the transferred CD8$^+$ CTL subset. A study evaluating the prevention of CMV disease in allogeneic BMT recipients by adoptive immunotherapy with both CMV-specific CD8$^+$ CTL and CD4$^+$ Th clones is now in progress and will enroll 30 allogeneic BMT recipients who are CMV$^+$ and receive bone marrow or peripheral blood stem cells from a MHC-matched, CMV$^+$-related donor. Patients will not receive prophylactic ganciclovir

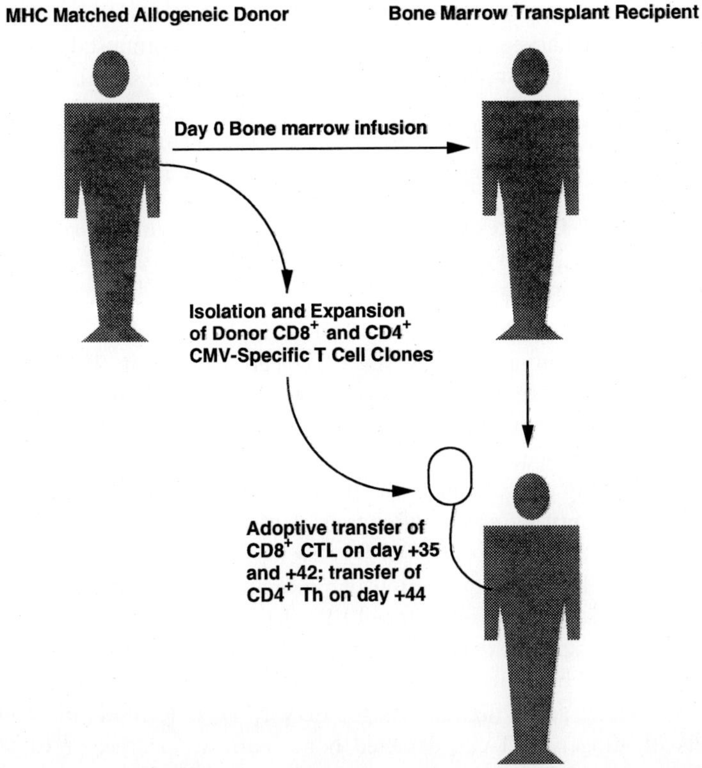

FIGURE 8.6 Strategy for reconstituting CD8$^+$ and CD4$^+$ CMV-specific T-cell immunity after allogeneic bone marrow transplant.

and will receive two infusions of CD8$^+$ CTL at a cell dose of $1 \times 10^9/m^2$ beginning 28–35 days after transplant, followed by a single infusion of CD4$^+$ Th cells at a cell dose of $1 \times 10^9/m^2$ administered 2 days after the second infusion of CD8$^+$ CTL (Fig. 8.6). The patients will be monitored for CMV-specific CTL and Th responses after therapy, and those with incomplete reconstitution will receive a third CD8$^+$ CTL infusion ($2 \times 10^9/m^2$) and a second CD4$^+$ Th infusion (2×10^9) 2 days apart.

B. T-CELL THERAPY OF EBV-INDUCED LYMPHOPROLIFERATIONS AFTER BONE MARROW TRANSPLANT

EBV infects approximately 90% of the population and establishes a persistent latent infection in B cells and oropharyngeal epithelium. EBV infection of immunocompetent hosts is occasionally associated with the later occurrence of malignancies, including Burkitt's lymphoma, a proportion of cases of Hodgkin's disease, and nasopharyngeal carcinoma. In immunocompromised hosts, EBV causes immunoblastic lymphomas (O'Reilly *et al.,* 1997). All EBV-associated malignancies have been shown to express some EBV proteins; thus immunotherapy with T cells specific for EBV has been contemplated as a potential therapy for both immunocompetent and immunodeficient hosts.

1. EBV-Specific T-Cell Responses and EBV-Induced Lymphoproliferation

A strong EBV-specific CD8$^+$cytotoxic T-cell response is elicited in immunocompetent hosts following primary infection with EBV and correlates with the resolution of the clinical manifestation of EBV. Similar to studies of the endogenous reconstitution of CMV-specific T-cell response in BMT recipients, Lucas *et al.* (1996) have analyzed the temporal recovery of EBV-specific CTL responses following unmodified or T-cell-depleted BMT and determined that deficiencies of EBV-specific CD8$^+$ CTL were observed at 3 months after transplant in the majority of individuals and recovery of these responses was delayed in most patients until 6 months after transplant. Most cases of posttransplant EBV-lymphoproliferative disease (LPD) developed in the first 4 months after transplant, coincident with the period of profound deficiency of EBV-specific CTL, and those patients who developed EBV LPD were found to have weak or undetectable EBV-specific cytolytic activity (Lucas *et al.,* 1996). Patients receiving T-cell-depleted bone marrow to prevent GVHD or T-cell-specific antibodies to treat GVHD are at the highest risk for EBV LPD (O'Reilly *et al.,* 1997). In recipients of allogeneic T-cell-depleted bone marrow, the risk of EBV LPD varies from 11 to 26%, depending on the method of T-cell depletion and the type of posttransplant immunosuppression. Recipients of unmodified bone marrow rarely ($<$1%) develop EBV LPD unless they require intense posttransplant immunosuppression with anti-T-cell monoclonal antibodies or antithymocyte globulin for GVHD. These observations suggest a critical role for T cells in preventing the outgrowth of EBV-infected B cells.

Insights into the ability of different T-cell populations to control EBV LPD have been derived from studies in SCID mice inoculated intravenously or subcutaneously with EBV-transformed B cells (LCL) to promote the development of disseminated or localized EBV lymphomas. The infusion of CD8$^+$ EBV-specific T-cell lines or clones completely prevented or delayed the outgrowth of EBV lymphoma, although CD4$^+$ EBV-specific Th cells also mediated partial protection against these class II MHC$^+$ tumors (Rowe *et al.,* 1991; Lacerda *et al.,* 1996; O'Reilly *et al.,* 1997).

2. Specificity of CD8$^+$ EBV-Reactive CTL

CD8$^+$ EBV-specific CTL responses can readily be elicited *in vitro* from EBV-seropositive donors using γ-irradiated autologous EBV-transformed LCL as stimulator cells. EBV LCL express only six EBV nuclear antigens, termed EBNA-1, EBNA-2, EBNA-3A, EBNA-3B, EBNA-3C, and EBNA-LP, and two membrane antigens, LMP-1 and LMP-2. These eight EBV antigens are also expressed in the B cells of EBV-associated lymphoproliferations in immunosuppressed hosts. Vaccinia recombinant viruses encoding these individual EBV proteins have been constructed and used to assess the specificity of EBV-reactive CTL elicited after stimulation of peripheral blood lymphocytes with EBV LCL. In the majority of individuals studied, one or more of EBNA-3A, -3B, and -3C antigens were immunodominant, with lesser responses against EBNA-2, EBNA-LP, LMP-1, and LMP-2 (Murray *et al.,* 1992).

3. Adoptive Transfer of T Cells to Immunodeficient Hosts with EBV-Induced Lymphoproliferation

The pathogenesis of EBV-lymphoproliferative syndromes in severely immunosuppressed BMT recipients appears to reflect the loss of T-cell immune responses required to maintain EBV as a latent infection. The EBV-infected B cells proliferating in these individuals express high levels of class I MHC and EBV proteins known to be targets of $CD8^+$ CTL. Thus, adoptive immunotherapy with EBV-reactive T cells has the potential to reconstitute the host with the necessary effector cells to eliminate EBV-transformed cells and control EBV infection. This strategy received support from a study by Papadopoulous and colleagues (1994) in which five patients who developed EBV-lymphoproliferative syndromes after receiving a T-cell-depleted allogeneic BMT received infusions of unirradiated mononuclear cells obtained from the peripheral blood of their EBV^+ bone marrow donors. The dose of donor leukocytes administered was relatively low (1×10^6 $CD3^+$ T cells/kg of patient body weight) due to the concern that these unselected donor T cells might cause GVHD. Therapy was associated with complete regression of the EBV lymphoproliferation in all five patients within 14–30 days. Three of the five patients became long-term survivors, although two of the three developed chronic GVHD. Fourteen additional patients have been treated utilizing cell doses as low as 1×10^5/kg and 89% of the patients have had complete eradication of their EBV LPD. The incidence of acute and chronic GVHD was 16 and 42%, respectively [O'Reilly *et al.,* 1977]. The patients experienced dramatic increases in circulating EBV-specific CTL, consistent with rapid expansion of this subset after antigen stimulation *in vivo.*

The risk of GVHD associated with T-cell therapy for EBV LPD should be reduced if donor T cells are selected for reactivity with EBV by *in vitro* stimulation with EBV^+ cells or if donor EBV-specific T-cell clones were used in therapy. Recently, the administration of EBV-specific T-cell lines with cytolytic activity for EBV LCL has been used for the treatment and prevention of EBV LPD after T-cell-depleted BMT. In these studies, the cell lines were transduced with a retroviral vector encoding neomycin phosphotransferase to permit analysis of the persistence and migration of T cells after adoptive transfer. The enrichment of polyclonal T cells by *in vitro* culture was effective in eliminating GVHD as a toxicity of therapy, and the infused gene-marked CTL induced regressions of established EBV LPD at cell doses of $1 \times 10^8/m^2$ and prevented the development of EBV LPD in susceptible individuals (Heslop *et al.*, 1996). Moreover, the marked CTL response persisted for over 18 months after infusion (Heslop *et al.*, 1996).

C. ADOPTIVE TRANSFER OF CD8+ HIV-SPECIFIC CTL CLONES TO HIV-INFECTED INDIVIDUALS

1. Antiviral Activity of CD8+ HIV-Specific CTL in HIV-Infected Individuals

Studies of host-immune responses to HIV and their contribution to the control of virus replication suggest that the vigorous CD8+ HIV-specific CTL responses that develop in infected individuals exert significant antiviral activity. Following primary infection, resolution of viremia and recovery of CD4+ T-cell counts to near-normal levels correlate with the appearance of CD8+ HIV-specific CTL and occurs prior to the detection of HIV-specific antibodies with neutralizing activity (Borrow *et al.*, 1994; Koup *et al.*, 1994). Even oligoclonal CTL responses to a single epitope may induce a dramatic decline in virus replication, although the selection of variants with escape mutations in the epitope may herald the recurrence of high-level viremia (Borrow *et al.*, 1997; Price *et al.*, 1997). Individuals who have a prolonged (greater than 10 years) asymptomatic phase of infection characteristically maintain strong CTL responses to HIV, and the residual virus in the patient remains susceptible to these CTL, suggesting immunologic control of virus replication may be a factor in nonprogression (Klein *et al.*, 1995; Harrer *et al.*, 1996; Greenough *et al.*, 1997). By contrast, individuals with progressive HIV infection frequently demonstrate the accumulation of HIV variants that escape recognition by the host CTL response, and loss of detectable CD8+ CTL is associated with progression to AIDS (Carmichael *et al.*, 1993; Koup, 1994; Goulder *et al.*, 1997). These data suggest that CD8+ CTL, which are capable of eliminating productively infected cells by direct cytolysis and of inhibiting viral spread by secretion of chemokines, contribute to the down-regulation of HIV viremia observed in most individuals in the transition from the acute to the chronic phase of infection, but virus replication is rarely, if ever, completely curtailed. It is apparent from studies examining the dynamics of virus replication and CD4+ T-cell turnover that during the chronic phase HIV replication continues at a considerable level, and lymphoid organs serve as a major site of both sequestration of virus particles in the follicular dendritic network and ongoing virus replication in resident CD4+ T cells and macrophages (Embretson *et al.*, 1993; Pantaleo *et al.*, 1997).

2. Adoptive Transfer of Autologous HIV-Specific CTL Clones

Based on the observations that CD8+ HIV-specific CTL mediate considerable but incomplete antiviral activity, we have pursued adoptive therapy with CD8+ HIV-specific CTL clones as a strategy to augment this partially effective host response and to study mechanisms of virus evasion. To address patient safety as well as the persistence and migration of transferred T cells, the initial study was performed with autologous Gag-specific CTL clones which were genetically modified to express the hygromycin phosphotransferase (Hy) and herpes simplex virus thymidine kinase (TK) genes (Riddell *et al.,* 1992a). Transduction of CTL with HyTK accomplished two objectives. First, the CTL contained a genetic marker for tracking migration and persistence *in vivo* and second, they were rendered sensitive to ganciclovir and could be eliminated if they caused toxicity. The infusion of HyTK-modified, Gag-specific $CD8^+$ CTL to six HIV^+ patients was associated with only minor side effects, but persistence of transferred CTL was limited by the induction of host-CTL responses to Hy and TK (Riddell *et al.,* 1996). This identified an obstacle for gene therapy and necessitated the study to be redesigned to fully evaluate safety, persistence and migration, and antiviral activity of transferred HIV-specific CTL. Thus, in an ongoing study, unmodified autologous $CD8^+$ Gag-specific CTL clones are being infused in escalating doses to establish safety. The patients then receive two infusions of $CD8^+$ CTL modified with the LN retrovirus to evaluate the migration of CTL to lymph nodes. Preliminary results of this study demonstrate that (a) host-CTL responses in PBL were dramatically augmented for 1 week after adoptive transfer but then declined, (b) infused CTL accumulated in lymph node sites adjacent to HIV-infected cells, and (c) CTL infusions were associated with a temporary reduction in circulating $CD4^+$ HIV-infected cells. Strategies to improve the persistence of unmodified $CD8^+$ HIV-specific CTL including the use of low-dose IL-2 administration or $CD4^+$ HIV-specific Th cells genetically modified to resist HIV infection will be evaluated in future studies.

VIII. POTENTIAL FOR ADOPTIVE IMMUNOTHERAPY WITH TUMOR-REACTIVE T CELLS

A. IMMUNOTHERAPY OF ADVANCED MELANOMA WITH TUMOR-INFILTRATING LYMPHOCYTES

The first efforts to exploit the therapeutic potential of adoptive immunotherapy for cancer were undertaken by Rosenberg and colleagues in patients with advanced malignancies (Rosenberg *et al.,* 1988). Until the recent developments in culture techniques described earlier, it was only possible in rare circumstances to isolate MHC-restricted, tumor-specific T cells from the peripheral blood of tumor-bearing patients. Thus, studies were done to determine if such T cells might be enriched in the lymphocyte population infiltrating the tumor. For most tumors, these efforts did not identify tumor-specific T cells, with the notable exception of melanoma, where T cells with MHC-restricted tumor specificity were cultured from tumor biopsies in approximately 30% of patients (Topalian *et al.,* 1989). A culture method utilizing high concentrations of IL-2 was described for the large-scale expansion of lymphocytes derived from biopsies of melanoma, sarcoma, and adenocarcinoma, but the majority of these cultures did not demonstrate specific lytic activity for the autologous tumor (Topalian *et al.*, 1987).

The initial study of the adoptive transfer of tumor-infiltrating lymphocytes (TIL) enrolled 20 patients with advanced melanoma and administered a single dose of cyclophosphamide to reduce the tumor burden and suppressor T cells, followed by the infusion of TIL and systemic IL-2. A 55% response rate was observed, and responses occurred in patients who had previously failed IL-2 alone (Rosenberg *et al.*, 1988). Subsequent studies confirmed a similar response rate and correlated the responses to the presence in the TIL preparation of lytic activity for the autologous tumor (Aebersold *et al.*, 1991). The introduction of a marker gene into a subpopulation of TIL has shown that adoptively transferred TIL can localize to the sites of tumor and persist *in vivo* for at least 2 months (Rosenberg *et al.*, 1990). These results were encouraging and have resulted in continued development of approaches for specific T-cell therapy of melanoma.

The identification of several defined antigens recognized by melanoma-specific CTL has suggested it may be possible to utilize APC expressing recombinant tumor antigens, rather than tumor cells, potentially for *in vivo* immunization or *in vitro* stimulation to improve the probability of obtaining tumor-reactive CTL from the blood for *ex vivo* expansion and use in adoptive therapy (van der Bruggen *et al.*, 1991; Traversari *et al.*, 1992; Brichard *et al.*, 1993; Kawakami *et al.*, 1994c). Culture conditions to expand tumor-reactive T cells that rely on lower doses of IL-2, may result in T cells that have an increased capacity for retaining specific lytic activity and remaining viable after adoptive transfer. These strategies are now being investigated by several laboratories, and clinical studies of T-cell therapy for melanoma using $CD8^+$ CTL clones specific for defined antigens expressed by the tumor are in progress.

B. IMMUNOTHERAPY FOR HEMATOLOGIC MALIGNANCIES AFTER ALLOGENEIC BONE MARROW TRANSPLANT WITH T CELLS SPECIFIC FOR MINOR H ANTIGENS

The clear demonstration of a graft-versus-tumor effect associated with allogeneic BMT and the ability to isolate minor H antigen-specific T-cell clones that recognize antigens selectively expressed in hematopoietic cells including leukemic blasts from BMT recipients suggests that the adoptive transfer of such T cells may be effective for treating or preventing relapse. Preliminary studies in NOD/SCID mice engrafted with human AML demonstrate that such minor H antigen-specific CTL exhibit significant antileukemic activity. However, for clinical application, the introduction of a suicide gene may be essential for the initial infusions in case unexpected GVH responses are induced. This strategy is currently being developed in our laboratory for the treatment of relapsed AML and ALL after BMT.

IX. CONCLUSIONS

Results of animal model studies have established the principles and identified the therapeutic potential of T-cell adoptive immunotherapy for infectious and malignant diseases. Studies of the immunobiology of CMV and EBV infection have identified the crucial role of host $\alpha\beta$ T-cell responses for protective immunity and led to the first studies of adoptive T-cell transfer to restore deficient T-cell immunity in immunocompromised patients. These initial studies have defined the safety and efficacy of this approach for augmenting virus-specific T-cell immunity and have implications for the extension of this therapy to virus-associated malignancy and malignant diseases. The natural-host T-cell response to HIV is only partially effective in controlling virus replication, and amplifying this response by the transfer of HIV-specific T-cell clones may also be beneficial. However, it is probable that the obstacles that limit the efficacy of the natural-host response to HIV would similarly interfere with the efficacy of transferred T cells. Thus, it may be necessary to genetically modify the transferred T cells to confer functions that promote antiviral activity, efficient migration to sites of infection, or viral resistance. Efforts to evaluate immunotherapy of HIV infection with gene-modified T cells have already been initiated and should provide insights into the potential of this approach.

Considerable challenges remain in translating the successes achieved in adoptive immunotherapy for virus infections to the treatment of malignancy. However, progress has been made in developing this approach for malignant melanoma and for acute and chronic leukemia, and the results of studies now in progress will assist in defining future directions for adoptive cellular therapy of malignancy.

REFERENCES

Ada, G. L., and Jones, P. D. (1986). The immune response to influenza infection. *Curr. Top. Microbiol. Immunol.* **128,** 1–54.

Aebersold, P., Hyatt, C., Johnson, S., *et al.* (1991). Lysis of autologous melanoma cells by tumor-infiltrating lymphocytes: Association with clinical response. *J. Natl. Cancer Inst.* **83,** 932–937.

Ahmed, R. (1992). Immunological memory against viruses. *Semin. Immunol.* **4,** 105–109.

Ahn, K., Gruhler, A., Galocha, B., *et al.* (1997). The ER-luminal domain of the HCMV glycoprotein US6 inhibits peptide translocation by TAP. *Immunity* **6,** 613–621.

Akbar, A. N., Salmon, M., Savill, J., *et al.* (1993). A possible role for bcl-2 in regulating T-cell memory: A "balancing act" between cell death and survival. *Immunol. Today* **14,** 526–532.

Alcam'i, A., and Smith, G. L. (1992). A soluble receptor for interleukin-1 beta encoded by vaccinia virus: A novel mechanism of virus modulation of the host response to infection. *Cell* **71,** 153–167.

Alexander-Miller, M. A., Leggatt, G. R., and Berzofsky, J. A. (1996). Selective expansion of high- or low-avidity cytotoxic T lymphocytes and efficacy for adoptive immunotherapy. *Proc. Natl. Acad. Sci. USA* **93,** 4102–4107.

Alwan, W. H., Kozlowska, W. J., and Openshaw, P. J. (1994). Distinct types of lung disease caused by functional subsets of antiviral T cells. *J. Exp. Med.* **179,** 81–89.

Bernards, R. (1991). N-myc disrupts protein kinase C-mediated signal transduction in neuroblastoma. *EMBO J.* **10,** 1119–1125.

Bertoletti, A., Sette, A., Chisari, F. V., *et al.* (1994). Natural variants of cytotoxic epitopes are T-cell receptor antagonists for antiviral cytotoxic T cells. *Nature* **369,** 407–410.

Bicknell, D. C., Rowan, A., and Bodmer, W. F. (1994). Beta 2-microglobulin gene mutations: A study of established colorectal cell lines and fresh tumors. *Proc. Natl. Acad. Sci. USA* **91,** 4751–4756.

Biron, C. A., Young, H. A., and Kasaian, M. T. (1990). Interleukin 2-induced proliferation of murine natural killer cells *in vivo. J. Exp. Med.* **171,** 173–188.

Boczkowski, D., Nair, S. K., Snyder, D., *et al.* (1996). Dendritic cells pulsed with RNA are potent antigen-presenting cells *in vitro* and *in vivo. J. Exp. Med.* **184,** 465–472.

Bonini, C., Ferrari, G., Verzeletti, S., *et al.* (1997). HSV-TK gene transfer into donor lymphocytes for control of allogeneic graft-versus-leukemia. *Science* **276,** 1719–1724.

Borrow, P., Lewicki, H., Hahn, B. H., *et al.* (1994). Virus-specific $CD8^+$ cytotoxic T-lymphocyte activity associated with control of viremia in primary human immunodeficiency virus type 1 infection. *J. Virol.* **68,** 6103–6110.

Borrow, P., Lewicki, H., Wei, X., *et al.* (1997). Antiviral pressure exerted by HIV-1-specific cytotoxic T lymphocytes (CTLs) during primary infection demonstrated by rapid selection of CTL escape virus. *Nat. Med.* **3,** 205–211.

Boucraut, J., Guillaudeux, T., Alizadeh, M., *et al.* (1993). HLA-E is the only class I gene that escapes CpG methylation and is transcriptionally active in the trophoblast-derived human cell line JAR. *Immunogenetics* **38,** 117–130.

Brichard, V., Van Pel, A., Wolfel, T., *et al.* (1993). The tyrosinase gene codes for an antigen recognized by autologous cytolytic T lymphocytes on HLA-A2 melanomas. *J. Exp. Med.* **178,** 489–495.

Browning, M. J., Krausa, P., Rowan, A., *et al.* (1993). Tissue typing the HLA-A locus from genomic DNA by sequence-specific PCR: Comparison of HLA genotype and surface expression on colorectal tumor cell lines. *Proc. Natl. Acad. Sci. USA* **90,** 2842–2845.

Bukowski, J. F., Warner, J. F., Dennert, G., *et al.* (1985). Adoptive transfer studies demonstrating the antiviral effect of natural killer cells *in vivo. J. Exp. Med.* **161,** 40–52.

Cardoso, A. A., Schultze, J. L., Boussiotis, V. A., *et al.* (1996). Pre-B acute lymphoblastic leukemia cells may induce T-cell anergy to alloantigen. *Blood* **88,** 41–48.

Cardoso, A. A., Seamon, M. J., Afonso, H. M., *et al.* (1997). *Ex vivo* generation of human anti-pre-B leukemia-specific autologous cytolytic T cells. *Blood* **90,** 549–561.

Carmichael, A., Jin, X., Sissons, P., *et al.* (1993). Quantitative analysis of the human immunodeficiency virus type 1 (HIV-1)-specific cytotoxic T lymphocyte (CTL) response at different stages of HIV-1 infection: Differential CTL responses to HIV-1 and Epstein–Barr virus in late disease. *J. Exp. Med.* **177,** 249–256.

Celis, E., Tsai, V., Crimi, C., *et al.* (1994). Induction of anti-tumor cytotoxic T lymphocytes in normal humans using primary cultures and synthetic peptide epitopes. *Proc. Natl. Acad. Sci. USA* **91,** 2105–2109.

Chen, W., Peace, D. J., Rovira, D. K., *et al.* (1992). T-cell immunity to the joining region of p210 BCR-ABL protein. *Proc. Natl. Acad. Sci. USA* **89,** 1468–1472.

Chun, T. W., Finzi, D., Margolick, J., *et al.* (1995). *In vivo* fate of HIV-1-infected T cells: Quantitative analysis of the transition to stable latency. *Nat. Med.* **1,** 1284–1290.

Cromme, F. V., Airey, J., Heemels, M. T., *et al.* (1994a). Loss of transporter protein, encoded by the TAP-1 gene, is highly correlated with loss of HLA expression in cervical carcinomas. *J. Exp. Med.* **179,** 335–340.

Cromme, F. V., van Bommel, P. F., Walboomers, J. M., *et al.* (1994b). Differences in MHC and TAP-1 expression in cervical cancer lymph node metastases as compared with the primary tumours. *Br. J. Cancer* **69,** 1176–1181.

de Bueger, M., Bakker, A., Van Rood, J. J., *et al.* (1992). Tissue distribution of human minor histocompatibility antigens. Ubiquitous versus restricted tissue distribution indicates heterogeneity among human cytotoxic T lymphocyte-defined non-MHC antigens. *J. Immunol.* **149,** 1788–1794.

de-Campos-Lima, P. O., Gavioli, R., Zhang, Q. J., *et al.* (1993). HLA-A11 epitope loss isolates of Epstein–Barr virus from a highly A11^{+} population. *Science* **260,** 98–100.

den Haan, J. M., Sherman, N. E., Blokland, E., *et al.* (1995). Identification of a graft versus host disease-associated human minor histocompatibility antigen. *Science* **268,** 1476–1480.

de-Waal-Malefyt, R., Haanen, J., Spits, H., *et al.* (1991). Interleukin 10 (IL-10) and viral IL-10 strongly reduce antigen-specific human T cell proliferation by diminishing the antigen-presenting capacity of monocytes via downregulation of class II major histocompatibility complex expression. *J. Exp. Med.* **174,** 915–924.

Doherty, P. C. (1993a). Cell-mediated cytotoxicity. *Cell* **75,** 607–612.

Doherty, P. C. (1993b). Inflammation in virus infections. *Semin. Virol.* **4,** 117–122.

Doherty, P. C., Allan, W., Eichelberger, M., *et al.* (1992). Roles of alpha beta and gamma delta T cell subsets in viral immunity. *Annu. Rev. Immunol.* **10,** 123–151.

Dolstra, H., Fredrix, H., Preijers, F., *et al.* (1997). Recognition of a B cell leukemia-associated minor histocompatibility antigen by CTL. *J. Immunol.* **158,** 560–565.

Dranoff, G., Jaffee, E., Lazenby, A., *et al.* (1993). Vaccination with irradiated tumor cells engineered to secrete murine granulocyte-macrophage colony-stimulating factor stimulates potent, specific, and long-lasting anti-tumor immunity. *Proc. Natl. Acad. Sci. USA* **90,** 3539–3543.

Dudley, M. E., and Roopenian, D. C. (1996). Loss of a unique tumor antigen by cytotoxic T lymphocyte immunoselection from a 3-methylcholanthrene-induced mouse sarcoma reveals secondary unique and shared antigens. *J. Exp. Med.* **184,** 441–447.

Embretson, J., Zupancic, M., Ribas, J. L., *et al.* (1993). Massive covert infection of helper T lymphocytes and macrophages by HIV during the incubation period of AIDS. *Nature* **362,** 359–362.

Gambacorti-Passerini, C., Grignani, F., Arienti, F., *et al.* (1993). Human CD4 lymphocytes specifically recognize a peptide representing the fusion region of the hybrid protein pml/RAR alpha present in acute promyelocytic leukemia cells. *Blood* **81,** 1369–1375.

Gaugler, B., Van den Eynde, B., van der Bruggen, P., *et al.* (1994). Human gene MAGE-3 codes for an antigen recognized on a melanoma by autologous cytolytic T lymphocytes. *J. Exp. Med.* **179,** 921–930.

Goulder, P. J., Phillips, R. E., Colbert, R. A., *et al.* (1997). Late escape from an immunodominant cytotoxic T-lymphocyte response associated with progression to AIDS. *Nat. Med.* **3,** 212–217.

Goulmy, E. (1996). Human minor histocompatibility antigens. *Curr. Opin. Immunol.* **8,** 75–81.

Goulmy, E., Gratama, J. W., Blokland, E., *et al.* (1983). A minor transplantation antigen detected by MHC-restricted cytotoxic T lymphocytes during graft-versus-host disease. *Nature* **302,** 159–161.

Graham, M. B., Braciale, V. L., and Braciale, T. J. (1994). Influenza virus-specific CD4^{+} T helper type 2 lymphocytes do not promote recovery from experimental virus infection. *J. Exp. Med.* **180,** 1273–1282.

Gray, D., and Matzinger, P. (1991). T cell memory is short-lived in the absence of antigen. *J. Exp. Med.* **174,** 969–974.

Greenberg, P. D. (1991). Adoptive T cell therapy of tumors: Mechanisms operative in the recognition and elimination of tumor cells. *Adv. Immunol.* **49,** 281–355.

Greenough, T. C., Brettler, D. B., Somasundaran, M., *et al.* (1997). Human immunodeficiency virus type 1-specific cytotoxic T lymphocytes (CTL), virus load, and CD4 T cell loss: Evidence supporting a protective role for CTL *in vivo. J. Infect. Dis.* **176,** 118–125.

Hahne, M., Rimoldi, D., Schroter, M., *et al.* (1996). Melanoma cell expression of Fas(Apo-1/CD95) ligand: Implications for tumor immune escape. *Science* **274,** 1363–1366.

Harrer, T., Harrer, E., Kalams, S. A., *et al.* (1996). Cytotoxic T lymphocytes in asymptomatic long-term nonprogressing HIV-1 infection. Breadth and specificity of the response and relation to *in vivo* viral quasispecies in a person with prolonged infection and low viral load. *J. Immunol.* **156,** 2616–2623.

Henderson, R. A., and Finn, O. J. (1996). Human tumor antigens are ready to fly. *Adv. Immunol.* **62,** 217–256.

Heslop, H. E., Ng, C. Y., Li, C., *et al.* (1996). Long-term restoration of immunity against Epstein–Barr virus infection by adoptive transfer of gene-modified virus-specific T lymphocytes. *Nat. Med.* **2,** 551–555.

Heyman, R. A., Borrelli, E., Lesley, J., *et al.* (1989). Thymidine kinase obliteration: Creation of transgenic mice with controlled immune deficiency. *Proc. Natl. Acad. Sci. USA* **86,** 2698–2702.

Hill, A., Jugovic, P., York, I., *et al.* (1995). Herpes simplex virus turns off the TAP to evade host immunity. *Nature* **375,** 411–415.

Hoglund, P., Sundback, J., Olsson-Alheim, M. Y., *et al.* (1997). Host MHC class I gene control of NK-cell specificity in the mouse. *Immunol. Rev.* **155,** 11–28.

Horowitz, M. M., Gale, R. P., Sondel, P. M., *et al.* (1990). Graft-versus-leukemia reactions after bone marrow transplantation. *Blood* **75,** 555–562.

Hou, S., Hyland, L., Ryan, K. W., *et al.* (1994). Virus-specific $CD8^+$ T-cell memory determined by clonal burst size. *Nature* **369,** 652–654.

Hsu, D. H., de-Waal-Malefyt, R., Fiorentino, D. F., *et al.* (1990). Expression of interleukin-10 activity by Epstein–Barr virus protein BCRF1. *Science* **250,** 830–832.

Jones, R. J., Ambinder, R. F., Piantadosi, S., *et al.* (1991). Evidence of a graft-versus-lymphoma effect associated with allogeneic bone marrow transplantation. *Blood* **77,** 649–653.

Jones, T. R., Wiertz, E. J., Sun, L., *et al.* (1996). Human cytomegalovirus US3 impairs transport and maturation of major histocompatibility complex class I heavy chains. *Proc. Natl. Acad. Sci. USA* **93,** 11327–11333.

Kawakami, Y., Eliyahu, S., Delgado, C. H., *et al.* (1994a). Cloning of the gene coding for a shared human melanoma antigen recognized by autologous T cells infiltrating into tumor. *Proc. Natl. Acad. Sci. USA* **91,** 3515–3519.

Kawakami, Y., Eliyahu, S., Delgado, C. H., *et al.* (1994b). Identification of a human melanoma antigen recognized by tumor-infiltrating lymphocytes associated with *in vivo* tumor rejection. *Proc. Natl. Acad. Sci. USA* **91,** 6458–6462.

Kawakami, Y., Eliyahu, S., Sakaguchi, K., *et al.* (1994c). Identification of the immunodominant peptides of the MART-1 human melanoma antigen recognized by the majority of HLA-A2-restricted tumor infiltrating lymphocytes. *J. Exp. Med.* **180,** 347–352.

Klein, G. (1989). Viral latency and transformation: The strategy of Epstein–Barr virus. *Cell* **58,** 5–8.

Klein, M. R., van Baalen, C. A., Holwerda, A. M., *et al.* (1995). Kinetics of Gag-specific cytotoxic T lymphocyte responses during the clinical of HIV-1 infection: A longitudinal analysis of rapid progressors and long-term asymptomatics. *J. Exp. Med.* **181,** 1365–1372.

Klenerman, P., Rowland-Jones, S., McAdam, S., *et al.* (1994). Cytotoxic T-cell activity antagonized by naturally occurring HIV-1 Gag variants. *Nature* **369,** 403–407.

Kondo, T., Suda, T., Fukuyama, H., *et al.* (1997). Essential roles of the Fas ligand in the development of hepatitis. *Nat. Med.* **3,** 409–413.

Korner, H., and Burgert, H. G. (1994). Down-regulation of HLA antigens by the adenovirus type 2 E3/19K protein in a T-lymphoma cell line. *J. Virol.* **68,** 1442–1448.

Koup, R. A. (1994). Virus escape from CTL recognition. *J. Exp. Med.* **180,** 779–782.

Koup, R. A., Safrit, J. T., Cao, Y., *et al.* (1994). Temporal association of cellular immune responses with the initial control of viremia in primary human immunodeficiency virus type 1 syndrome. *J. Virol.* **68,** 4650–4655.

Kwak, L. W., Campbell, M. J., Czerwinski, D. K., *et al.* (1992). Induction of immune responses in patients with B-cell lymphoma against the surface-immunoglobulin idiotype expressed by their tumors. *N. Engl. J. Med.* **327,** 1209–1215.

Lacerda, J. F., Ladanyi, M., Louie, D. C., *et al.* (1996). Human Epstein–Barr virus (EBV)-specific cytotoxic T lymphocytes home preferentially to and induce selective regressions of autologous EBV-induced B cell lymphoproliferations in xenografted C.B-17 scid/scid mice. *J. Exp. Med.* **183,** 1215–1228.

Lau, L. L., Jamieson, B. D., Somasundaram, T., *et al.* (1994). Cytotoxic T-cell memory without antigen. *Nature* **369,** 648–652.

Lehmann, F., Marchand, M., Hainaut, P., *et al.* (1995). Differences in the antigens recognized by cytolytic T cells on two successive metastases of a melanoma patient are consistent with immune selection. *Eur. J. Immunol.* **25,** 340–347.

Levitskaya, J., Coram, M., Levitsky, V., *et al.* (1995). Inhibition of antigen processing by the internal repeat region of the Epstein–Barr virus nuclear antigen-1. *Nature* **375,** 685–688.

Li, C. R., Greenberg, P. D., Gilbert, M. J., *et al.* (1994). Recovery of HLA-restricted cytomegalovirus (CMV)-specific T-cell responses after allogeneic bone marrow transplant: Correlation with CMV disease and effect of ganciclovir prophylaxis. *Blood* **83,** 1971–1979.

Lieberman, J., Skolnik, P. R., Parkerson, G., *et al.* (1997). Safety of autologous, *ex vivo*-expanded human immunodeficiency virus (HIV)-specific cytotoxic T-lymphocyte infusion in HIV-infected patients. *Blood* **90,** 2196–2206.

Linsley, P. S., Greene, J. L., Tan, P., *et al.* (1992). Coexpression and functional cooperation of CTLA-4 and CD28 on activated T lymphocytes. *J. Exp. Med.* **176,** 1595–1604.

Linsley, P. S., and Ledbetter, J. A. (1993). The role of the CD28 receptor during T cell responses to antigen. *Annu. Rev. Immunol.* **11,** 191–212.

Lucas, K. G., Small, T. N., Heller, G., *et al.* (1996). The development of cellular immunity to Epstein–Barr virus after allogeneic bone marrow transplantation. *Blood* **87,** 2594–2603.

Maeurer, M. J., Martin, D., Walter, W., *et al.* (1996). Human intestinal Vdelta1^{+} lymphocytes recognize tumor cells of epithelial origin. *J. Exp. Med.* **183,** 1681–1696.

Maggi, E., Del-Prete, G., Macchia, D., *et al.* (1988). Profiles of lymphokine activities and helper function for IgE in human T cell clones. *Eur. J. Immunol.* **18,** 1045–1050.

Maggi, E., Parronchi, P., Manetti, R., *et al.* (1992). Reciprocal regulatory effects of IFN-gamma and IL-4 on the *in vitro* development of human Th1 and Th2 clones. *J. Immunol.* **148,** 2142–2147.

Matloubian, M., Concepcion, R. J., and Ahmed, R. (1994). CD4^{+} T cells are required to sustain CD8^{+} cytotoxic T-cell responses during chronic viral infection. *J. Virol.* **68,** 8056–8063.

Matzinger, P. (1994). Immunology. Memories are made of this? *Nature* **369,** 605–606.

McLaughlin-Taylor, E., Pande, H., Forman, S., *et al.* (1994). Identification of the major late human cytomegalovirus matrix protein pp65 as a target antigen for CD8^{+} virus-specific cytotoxic T lymphocytes. *J. Med. Virol.* **43,** 103–110.

Meadows, L., Wang, W., den Haan, J. M., *et al.* (1997). The HLA-A*0201-restricted H-Y antigen contains a posttranslationally modified cysteine that significantly affects T cell recognition. *Immunity* **6,** 273–281.

Meyers, J. D. (1990). Infections in marrow transplant recipients. *In* "Principles and Practices of Infectious Diseases" (G. L. Mandell, R. G. Douglas, and J. E. Bennett, Eds.), pp. 2291–2294. Churchill, New York.

Middleton, T., Gahn, T. A., Martin, J. M., *et al.* (1991). Immortalizing genes of Epstein–Barr virus. *Adv. Virus Res.* **40,** 19–55.

Mondino, A., Khoruts, A., and Jenkins, M. K. (1996). The anatomy of T-cell activation and tolerance. *Proc. Natl. Acad. Sci. USA* **93,** 2245–2252.

Morrison, L. A., Lukacher, A. E., Braciale, V. L., *et al.* (1986). Differences in antigen presentation to MHC class I- and class II-restricted influenza virus-specific cytolytic T lymphocyte clones. *J. Exp. Med.* **163,** 903–921.

Mosmann, T. R., Cherwinski, H., Bond, M. W., *et al.* (1986). Two types of murine helper T cell clone. I. Definition according to profiles of lymphokine activities and secreted proteins. *J. Immunol.* **136,** 2348–2357.

Mule, J. J., Shu, S., and Rosenberg, S. A. (1985). The anti-tumor efficacy of lymphokine-activated killer cells and recombinant interleukin 2 *in vivo. J. Immunol.* **135,** 646–652.

Murray, R. J., Kurilla, M. G., Brooks, J. M., *et al.* (1992). Identification of target antigens for the human cytotoxic T cell response to Epstein–Barr virus (EBV): Implications for the immune control of EBV-positive malignancies. *J. Exp. Med.* **176,** 157–168.

Nagata, S. (1996). Fas ligand and immune evasion. *Nat. Med.* **2,** 1306–1307.

Neefjes, J. J., and Ploegh, H. L. (1992). Intracellular transport of MHC class II molecules. *Immunol. Today* **13,** 179–184.

Nelson, B. H., Lord, J. D., and Greenberg, P. D. (1994). Cytoplasmic domains of the interleukin-2 receptor beta and gamma chains mediate the signal for T-cell proliferation. *Nature* **369,** 333–336.

Noguchi, Y., Chen, Y. T., and Old, L. J. (1994). A mouse mutant p53 product recognized by $CD4^+$ and $CD8^+$ T cells. *Proc. Natl. Acad. Sci. USA* **91,** 3171–3175.

Oehen, S., Waldner, H., Kundig, T. M., *et al.* (1992). Antivirally protective cytotoxic T cell memory to lymphocytic choriomeningitis virus is governed by persisting antigen. *J. Exp. Med.* **176,** 1273–1281.

O'Reilly, R. J., Small, T. N., Papadopoulos, E., *et al.* (1997). Biology and adoptive cell therapy of Epstein–Barr virus-associated lymphoproliferative disorders in recipients of marrow allografts. *Immunol. Rev.* **157,** 195–216.

Paabo, S., Severinsson, L., Andersson, M., *et al.* (1989). Adenovirus proteins and MHC expression. *Adv. Cancer Res.* **52,** 151–163.

Pantaleo, G., Demarest, J. F., Schacker, T., *et al.* (1997). The qualitative nature of the primary immune response to HIV infection is a prognosticator of disease progression independent of the initial level of plasma viremia. *Proc. Natl. Acad. Sci. USA* **94,** 254–258.

Papadopoulos, E. B., Ladanyi, M., Emanuel, D., *et al.* (1994). Infusions of donor leukocytes to treat Epstein–Barr virus-associated lymphoproliferative disorders after allogeneic bone marrow transplantation. *N. Engl. J. Med.* **330,** 1185–1191.

Peace, D. J., Smith, J. W., Chen, W., *et al.* (1994). Lysis of ras oncogene-transformed cells by specific cytotoxic T lymphocytes elicited by primary *in vitro* immunization with mutated ras peptide. *J. Exp. Med.* **179,** 473–479.

Phillips, R. E., Rowland-Jones, S., Nixon, D. F., *et al.* (1991). Human immunodeficiency virus genetic variation that can escape cytotoxic T cell recognition. *Nature* **354,** 453–459.

Pircher, H., Moskophidis, D., Rohrer, U., *et al.* (1990). Viral escape by selection of cytotoxic T cell-resistant virus variants *in vivo. Nature* **346,** 629–633.

Price, D. A., Goulder, P. J., Klenerman, P., *et al.* (1997). Positive selection of HIV-1 cytotoxic T lymphocyte escape variants during primary infection. *Proc. Natl. Acad. Sci. USA* **94,** 1890–1895.

Quinnan, G. V., Masur, H., Rook, A. H., *et al.* (1984). Herpesvirus infections in the acquired immune deficiency syndrome. *JAMA* **252,** 72–77.

Ramsay, A. J., Ruby, J., and Ramshaw, I. A. (1993). A case for cytokines as effector molecules in the resolution of virus infection. *Immunol. Today* **14,** 155–157.

Read, G. S., Karr, B. M., and Knight, K. (1993). Isolation of a herpes simplex virus type 1 mutant with a deletion in the virion host shutoff gene and identification of multiple forms of the vhs (UL41) polypeptide. *J. Virol.* **67,** 7149–7160.

Reddehase, M. J., Weiland, F., Munch, K., *et al.* (1985). Interstitial murine cytomegalovirus pneumonia after irradiation: Characterization of cells that limit viral replication during established infection of the lungs. *J. Virol.* **55,** 264–273.

Restifo, N. P., Esquivel, F., Kawakami, Y., *et al.* (1993). Identification of human cancers deficient in antigen processing. *J. Exp. Med.* **177,** 265–272.

Reusser, P., Riddell, S. R., Meyers, J. D., *et al.* (1991). Cytotoxic T-lymphocyte response to cytomegalovirus after human allogeneic bone marrow transplantation: Pattern of recovery and correlation with cytomegalovirus infection and disease. *Blood* **78,** 1373–1380.

Riddell, S. R., and Greenberg, P. D. (1990). The use of anti-CD3 and anti-CD28 monoclonal antibodies to clone and expand human antigen-specific T cells. *J. Immunol. Methods* **128,** 189–201.

Riddell, S. R., and Greenberg, P. D. (1995). Principles for adoptive T cell therapy of human viral diseases. *Annu. Rev. Immunol.* **13,** 545–586.

Riddell, S. R., Elliott, M., Lewinsohn, D. A., *et al.* (1996). T-cell mediated rejection of gene-modified HIV-specific cytotoxic T lymphocytes in HIV-infected patients. *Nat. Med.* **2,** 216–223.

Riddell, S. R., Greenberg, P. D., Overell, R. W., *et al.* (1992a). Phase I study of cellular adoptive immunotherapy using genetically modified $CD8^+$ HIV-specific T cells for HIV seropositive patients undergoing allogeneic bone marrow transplant. *Hum. Gene. Ther.* **3,** 319–338.

Riddell, S. R., Rabin, M., Geballe, A. P., *et al.* (1991). Class I MHC-restricted cytotoxic T lymphocyte recognition of cells infected with human cytomegalovirus does not require endogenous viral gene expression. *J. Immunol.* **146,** 2795–2804.

Riddell, S. R., Watanabe, K. S., Goodrich, J. M., *et al.* (1992b). Restoration of viral immunity in immunodeficient humans by the adoptive transfer of T cell clones. *Science* **257,** 238–241.

Rivoltini, L., Kawakami, Y., Sakaguchi, K., *et al.* (1995). Induction of tumor-reactive CTL from peripheral blood and tumor-infiltrating lymphocytes of melanoma patients by *in vitro* stimulation with an immunodominant peptide of the human melanoma antigen MART-1. *J. Immunol.* **154,** 2257–2265.

Robbins, P. F., El-Gamil, M., Li, Y. F., *et al.* (1996). A mutated beta-catenin gene encodes a melanoma-specific antigen recognized by tumor infiltrating lymphocytes. *J. Exp. Med.* **183,** 1185–1192.

Rogge, L., Barberis-Maino, L., Biffi, M., *et al.* (1997). Selective expression of an interleukin-12 receptor component by human T helper 1 cells. *J. Exp. Med.* **185,** 825–831.

Rosenberg, S. A., Aebersold, P., Cornetta, K., *et al.* (1990). Gene transfer into humans: Immunotherapy of patients with advanced melanoma, using tumor-infiltrating lymphocytes modified by retroviral gene transduction. *N. Engl. J. Med.* **323,** 570–578.

Rosenberg, S. A., Packard, B. S., Aebersold, P. M., *et al.* (1988). Use of tumor-infiltrating lymphocytes and interleukin-2 in the immunotherapy of patients with metastatic melanoma. A preliminary report. *N. Engl. J. Med.* **319,** 1676–1680.

Rowe, M., Young, L. S., Crocker, J., *et al.* (1991). Epstein–Barr virus (EBV)-associated lymphoproliferative disease in the SCID mouse model: Implications for the pathogenesis of EBV-positive lymphomas in man. *J. Exp. Med.* **173,** 147–158.

Sarawar, S. R., and Doherty, P. C. (1994). Concurrent production of interleukin-2, interleukin-10, and gamma interferon in the regional lymph nodes of mice with influenza pneumonia. *J. Virol.* **68,** 3112–3119.

Schrier, P. I., Bernards, R., Vaessen, R. T., *et al.* (1983). Expression of class I major histocompatibility antigens switched off by highly oncogenic adenovirus 12 in transformed fat cells. *Nature* **305,** 771–775.

Schultze, J. L., Cardoso, A. A., Freeman, G. J., *et al.* (1995). Follicular lymphomas can be induced to present alloantigen efficiently: A conceptual model to improve their tumor immunogenicity. *Proc. Natl. Acad. Sci. USA* **92,** 8200–8204.

Schultze, J. L., Seamon, M. J., Michalak, S., *et al.* (1997). Autologous tumor infiltrating T cells cytotoxic for follicular lymphoma cells can be expanded *in vitro*. *Blood* **89,** 3806–3816.

Sciammas, R., Johnson, R. M., Sperling, A. I., *et al.* (1994). Unique antigen recognition by a herpesvirus-specific TCR-gamma delta cell. *J. Immunol.* **152,** 5392–5397.

Seder, R. A., Gazzinelli, R., Sher, A., *et al.* (1993). Interleukin 12 acts directly on $CD4^+$ T cells to enhance priming for interferon gamma production and diminishes interleukin 4 inhibition of such priming. *Proc. Natl. Acad. Sci. USA* **90,** 10188–10192.

Shi, L., Mai, S., Israels, S., *et al.* (1997). Granzyme B (GraB) autonomously crosses the cell membrane and perforin initiates apoptosis and GraB nuclear localization. *J. Exp. Med.* **185,** 855–866.

Shinomiya, H., Shinomiya, M., Stevenson, G. W., *et al.* (1989). Activated killer monocytes: Preclinical model systems. *Immunol. Ser.* **48,** 101–126.

Sing, A. P., Ambinder, R. F., Hong, D. J., *et al.* (1997). Isolation of Epstein–Barr virus (EBV)-specific cytotoxic T lymphocytes that lyse Reed–Sternberg cells: Implications for immune-mediated therapy of EBV^+ Hodgkin's disease. *Blood* **89,** 1978–1986.

Spies, T., and DeMars, R. (1991). Restored expression of major histocompatibility class I molecules by gene transfer of a putative peptide transporter. *Nature* **351,** 323–324.

Staeheli, P. (1990). Interferon-induced proteins and the antiviral state. *Adv. Virus Res.* **38,** 147–200.

Stevens, J. G. (1989). Human herpesviruses: A consideration of the latent state. *Microbiol. Rev.* **53,** 318–332.

Strand, S., Hofmann, W. J., Hug, H., *et al.* (1996). Lymphocyte apoptosis induced by CD95 (APO-1/Fas) ligand-expressing tumor cells—A mechanism of immune evasion? *Nat. Med.* **2,** 1361–1366.

Thompson, C. B., and Allison, J. P. (1997). The emerging role of CTLA-4 as an immune attenuator. *Immunity* **7,** 445–450.

Topalian, S. L., Muul, L. M., Solomon, D., *et al.* (1987). Expansion of human tumor infiltrating lymphocytes for use in immunotherapy trials. *J. Immunol. Methods* **102,** 127–141.

Topalian, S. L., Solomon, D., and Rosenberg, S. A. (1989). Tumor-specific cytolysis by lymphocytes infiltrating human melanomas. *J. Immunol.* **142,** 3714–3725.

Torre-Amione, G., Beauchamp, R. D., Koeppen, H., *et al.* (1990). A highly immunogenic tumor transfected with a murine transforming growth factor type beta 1 cDNA escapes immune surveillance. *Proc. Natl. Acad. Sci. USA* **87,** 1486–1490.

Tough, D. F., and Sprent, J. (1994). Turnover of naive- and memory-phenotype T cells. *J. Exp. Med.* **179,** 1127–1135.

Tough, D. F., Borrow, P., and Sprent, J. (1996). Induction of bystander T cell proliferation by viruses and type I interferon *in vivo*. *Science* **272,** 1947–1950.

Townsend, S. E., and Allison, J. P. (1993). Tumor rejection after direct costimulation of $CD8^+$ T cells by B7-transfected melanoma cells. *Science* **259,** 368–370.

Traversari, C., van der Bruggen, P., Luescher, I. F., *et al.* (1992). A nonapeptide encoded by human gene MAGE-1 is recognized on HLA-A1 by cytolytic T lymphocytes directed against tumor antigen MZ2-E. *J. Exp. Med.* **176,** 1453–1457.

Tsai, V., Southwood, S., Sidney, J., *et al.* (1997). Identification of subdominant CTL epitopes of the GP 100 melanoma-associated tumor antigen by primary *in vitro* immunization with peptide-pulsed dendritic cells. *J. Immunol.* **158,** 1796–1802.

Upton, C., Macen, J. L., Schreiber, M., *et al.* (1991). Myxoma virus expresses a secreted protein with homology to the tumor necrosis factor receptor gene family that contributes to viral virulence. *Virology* **184,** 370–382.

Upton, C., Mossman, K., and McFadden, G. (1992). Encoding of a homolog of the IFN-gamma receptor by myxoma virus. *Science* **258,** 1369–1372.

van der Bruggen, P., Traversari, C., Chomez, P., *et al.* (1991). A gene encoding an antigen recognized by cytolytic T lymphocytes on a human melanoma. *Science* **254,** 1643–1647.

van der Harst, D., Goulmy, E., Falkenburg, J. H., *et al.* (1994). Recognition of minor histocompatibility antigens on lymphocytic and myeloid leukemic cells by cytotoxic T-cell clones. *Blood* **83,** 1060–1066.

Van Pel, A., van der Bruggen, P., Coulie, P. G., *et al.* (1995). Genes coding for tumor antigens recognized by cytolytic T lymphocytes. *Immunol. Rev.* **145,** 229–250.

Villunger, A., Egle, A., Marschitz, I., *et al.* (1997). Constitutive expression of Fas (Apo-1/CD95) ligand on multiple myeloma cells: A potential mechanism of tumor-induced suppression of immune surveillance. *Blood* **90,** 12–20.

Walter, E. A., Greenberg, P. D., Gilbert, M. J., *et al.* (1995). Reconstitution of cellular immunity against cytomegalovirus in recipients of allogeneic bone marrow by transfer of T-cell clones from the donor. *N. Engl. J. Med.* **333,** 1038–1044.

Wang, R. F., Parkhurst, M. R., Kawakami, Y., *et al.* (1996). Utilization of an alternative open reading frame of a normal gene in generating a novel human cancer antigen. *J. Exp. Med.* **183,** 1131–1140.

Wang, W., Meadows, L. R., den Haan, J. M., *et al.* (1995). Human H-Y: A male-specific histocompatibility antigen derived from the SMCY protein. *Science* **269,** 1588–1590.

Wang, Z., Cao, Y., Albino, A. P., *et al.* (1993). Lack of HLA class I antigen expression by melanoma cells SK-MEL-33 caused by a reading frameshift in beta 2-microglobulin messenger RNA. *J. Clin. Invest.* **91,** 684–692.

Warren, E. H., Greenberg, P. D., and Riddell, S. R. (1998). Cytotoxic T lymphocyte-defined human minor histocompatibility antigens with a restricted tissue distribution. *Blood* **91,** 2197–2207.

Wiertz, E. J., Jones, T. R., Sun, L., *et al.* (1996a). The human cytomegalovirus US11 gene product dislocates MHC class I heavy chains from the endoplasmic reticulum to the cytosol. *Cell* **84,** 769–779.

Wiertz, E. J., Tortorella, D., Bogyo, M., *et al.* (1996b). Sec61-mediated transfer of a membrane protein from the endoplasmic reticulum to the proteasome for destruction. *Nature* **384,** 432–438.

Wills, M. R., Carmichael, A. J., Mynard, K., *et al.* (1996). The human cytotoxic T-lymphocyte (CTL) response to cytomegalovirus is dominated by structural protein pp65: Frequency, specificity, and T-cell receptor usage of pp65-specific CTL. *J. Virol.* **70,** 7569–7579.

Wilson, A., George, A. J., King, C. A., *et al.* (1990). Recognition of a B cell lymphoma by anti-idiotypic T cells. *J. Immunol.* **145,** 3937–3943.

Wolfel, T., Hauer, M., Schneider, J., *et al.* (1995). A p16INK4a-insensitive CDK4 mutant targeted by cytolytic T lymphocytes in a human melanoma. *Science* **269,** 1281–1284.

Yang, N. S., Sun, W. H., Yannelli, J. R., *et al.* (1995). Gene gun and other non-viral approaches for cancer gene therapy. *Nat. Med.* **1,** 481–483.

Yee, C., Gilbert, M. J., Riddell, S. R., *et al.* (1996). Isolation of tyrosinase-specific $CD8^+$ and $CD4^+$ T cell clones from the peripheral blood of melanoma patients following *in vitro* stimulation with recombinant vaccinia virus. *J. Immunol.* **157,** 4079–4086.

Yewdell, J. W., and Bennink, J. R. (1990). The binary logic of antigen processing and presentation to T cells. *Cell* **62,** 203–206.

Yewdell, J. W., and Bennink, J. R. (1992). Cell biology of antigen processing and presentation to major histocompatibility complex class I molecule-restricted T lymphocytes. *Adv. Immunol.* **52,** 1–123.

York, I. A., Roop, C., Andrews, D. W., *et al.* (1994). A cytosolic herpes simplex virus protein inhibits antigen presentation to $CD8^+$ T lymphocytes. *Cell* **77,** 525–535.

Zeff, R. A., Nakagawa, M., Mashimo, H., *et al.* (1990). Failure of cell surface expression of a class I major histocompatibility antigen caused by somatic point mutation. *Transplantation* **49,** 803–808.

Zeidler, R., Eissner, G., Meissner, P., *et al.* (1997). Downregulation of TAP1 in B lymphocytes by cellular and Epstein–Barr virus-encoded interleukin-10. *Blood* **90,** 2390–2397.

9

Exogenous Gene Transfer into Lymphoid and Hematopoietic Progenitor Cells

Lun Quan Sun, Michelle Miller, and Geoff Symonds

Johnson and Johnson Research
G.P.O. Box 3331, Sydney, New South Wales 2001, Australia

I. Introduction

An understanding of hematopoiesis is central to the development of strategies for gene transfer to hematopoietic and lymphoid cells. Hematopoiesis is the tightly regulated process of cell proliferation and maturation that results in the production of terminally differentiated cell types (T and B lymphocytes, monocytes, granulocytes, eosinophils, basophils, megakaryocytes, and erythrocytes) which are derived from hematopoietic stem cells (HSCs) (Scholfield, 1983;

Daniel and Dexter, 1989; Moore and Shapiro, 1994). Hematopoietic stem cells are described as pluripotent since they have the potential to give rise to progeny cell differentiation along any of the cell-lineage pathways summarized in Fig. 9.1. The cell types, which express an introduced transgene, will include the transduced parental cell type as well as the consequent more differentiated progeny. Theoretically, stable integration in the HSC has the potential for gene expression in all hematopoietic lineages.

Proliferation and differentiation of primitive hematopoietic cells are driven by hematopoietic growth factors (HGFs), some of which have pleomorphic actions, acting at more than one stage of lineage development as well as stimulating production of more than one cell type. Production of maximal numbers of cells from primitive hematopoietic precursors *in vitro* requires synergistic combinations of growth factors, particularly those HGFs that act at the initial stages of lineage development (Metcalf, 1994).

In addition to the property of pluripotency, it has been shown that, *in vivo,* only a minor proportion of HSCs are cycling, and relatively small numbers of cycling stem cells contribute to the process of steady-state hematopoiesis (Jordan and Lemischka, 1990; Keller and Snodgrass, 1990; Dunbar *et al.,* 1995). Thus, the process of hematopoiesis is sustained by the maintenance of a quiescent HSC pool, which undergo divisions relatively infrequently unless stimulated by HGFs. *In vitro* stimulation of HSCs with multiple HGFs is associated with cellular differentiation but may also result in production of cells with, at least, *in vitro* experimental properties of HSCs (i.e., renewal of long-term culture initiating cells (LTC-IC)).

These biological properties make HSCs a suitable target cell type for gene therapy. The use of retroviral vectors for gene transfer to HSCs has presented a

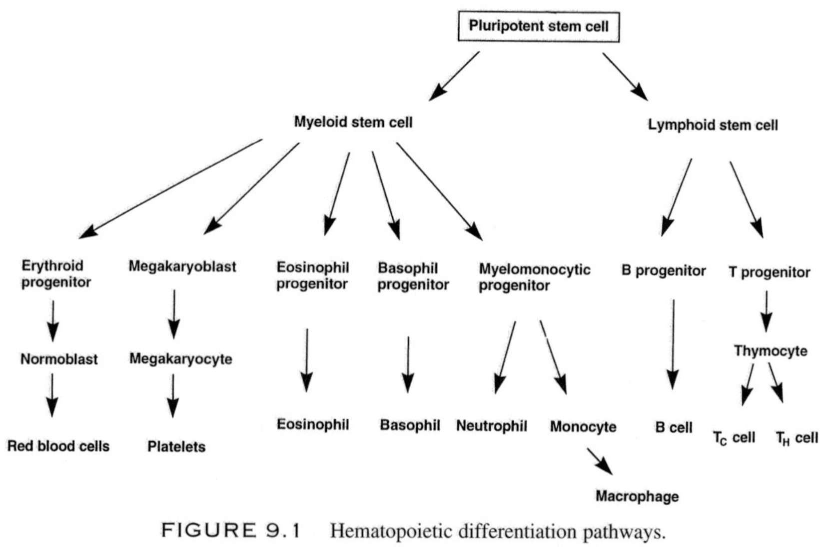

FIGURE 9.1 Hematopoietic differentiation pathways.

TABLE 9.1 Lymphocytes and Hematopoietic Stem Cells (HSC) as Targets for Gene Therapy

Biological property	Lymphocytes	HSC
Long-term survival *in vivo*	Long-term survival as "memory" T or B lymphocytes	Reside in bone marrow in the quiescent state
Phenotype	Differentiated: T helper ($CD3^+CD4^+$) or cytotoxic T cell ($CD3^+CD8^+$) subsets	Undifferentiated pluripotent cell with potential for multilineage engraftment
Entry into cell cycle (for retroviral vectors)	IL-2 and T-cell receptor cross-linking antibodies	Multiple HGFs; may lose HSC phenotype on HGF stimulation

significant challenge, since entry into the cell cycle is required for efficient integration, and as discussed herein, optimal conditions for this process have not been fully determined.

Retroviral-gene transfer to lymphoid cells has in some ways been less problematic. T cells in peripheral blood are stimulated to divide by interleukin-2 (IL-2) and cross-linking T-cell receptor antibodies. Because lymphocytes are more easily manipulated *in vitro* than HSCs, many groups have initially focused on gene transfer to these cells. The biological properties that make lymphocytes and HSCs suitable targets for gene therapy are summarized in Table 9.1.

Gene transfer technologies not only are applicable to single-gene disorders but also provide a potentially powerful and versatile approach for manipulation of cellular biochemical pathways to achieve a variety of therapeutic goals. These include gene marking (Brenner *et al.,* 1993a), modification of sensitivity to chemotherapeutic agents, inhibition of viral replication, and enhancement of T-cell effector function (Table 9.2). Retroviral vectors are currently the most

TABLE 9.2 Gene Therapy Strategies Using Hematopoietic or Lymphoid Cells

Single-gene disorders
ADA deficiency, Gaucher's disease, Hurler's syndrome, Fanconi's anemia, chronic granulomatous disease
Transplantation and cancer
Gene marking, suicide genes, chemotherapy-resistant genes, tumor purging by proteins that block the expression of oncogenes, wild-type tumor suppressor proteins (p53), toxins controlled by a tumor-specific promoter
Immunotherapy
Genetic vaccines, cytokine enhancement of lymphocyte effector function, chimeric growth factor receptors
Acquired immune deficiency syndrome
Catalytic ribozymes, antisense RNA, dominant repressors of HIV genes, RNA decoys

widespread method for transfer of therapeutic genes into somatic cells (Miller, 1992).

II. VECTOR DESIGN

A number of retroviral vectors have been developed for gene transfer into human hematopoietic cells and are generally based on the Moloney murine leukemia virus (MoMLV) backbone. The goals of vector design have been to (i) minimize the probability of recombination events leading to production of replication-competent retrovirus (RCR), (ii) achieve high transduction efficiency, and (iii) allow long-term gene expression *in vivo.* A wide variety of vector constructs, derived from primary vector designs such as N2 and LNL6, have been described. These vectors include promoters for gene transcription such as retroviral long terminal repeats (LTRs) (Hock *et al.,* 1989), internal promoters (Overell *et al.,* 1988), in some cases with the gene of interest expressed in reverse orientation (Dzierzak *et al.,* 1988), self-inactivating (SIN) LTR (Yee *et al.,* 1987), pol III and pol I promoters (Sullenger *et al.,* 1990; Palmer *et al.,* 1993), with the expression cassette inverted in the LTR (double-copy vectors) (Hantzopoulos *et al.,* 1989), and the use of modified LTRs (Hawley *et al.,* 1989). However, it has gradually become apparent that the more complicated vector designs are difficult to produce in high titer and have a propensity for retroviral vector genome rearrangement. This has led back to the use of simpler, single-gene vector designs. In our laboratory, a simple design of retroviral vector based on LNL6 has been consistently found to be very efficient in the transduction of human peripheral blood lymphocytes in terms of vector titers, transduction efficiency as measured by G418 resistance, duration of transgene expression, and stability of transgene transcripts.

III. PACKAGING CELL LINES

The retroviral vector RNA genome lacks critical structural genes required for infectivity and replication. Packaging cell lines have been engineered to express structural proteins (*gag/pol* and *env*) that are required for formation of an infectious particle. Such expression occurs from an endogenous expression construct(s) within the packaging cell line. There is the possibility of homologous recombination leading to production of replication-competent retrovirus (RCR) or "helper virus," especially if there are regions of homology between helper and vector genomes. To reduce the possibility of recombination, the PA317 packaging cell line was designed with deletions not only of the ψ packaging signal but also of two other regions in the 5′ and 3′ LTR (Miller and Buttimore, 1986). In addition to modifications of the packaging cell line, a modified version of the N2 vector, LNL6 (Bender *et al.,* 1987), was designed. The start codon of the *gag* gene was replaced by a stop codon, and some of the MoMLV sequences of the LTR were replaced by sequences of the Moloney murine sarcoma virus (MSV) to decrease homology between the vector and helper genomes. This packaging cell line (PA317) and retroviral vector (LNL6) combination has been used most commonly in clinical studies to date. Other amphotropic packaging cell lines such as ψ-CRIP and AM12 have deletions of the packaging signal as well as expression of the *gag/pol* and *env* genes on two separate plasmids to decrease the possibility of recombination (Danos and Mulligan, 1988; Markowitz *et al.,* 1988).

In addition to RCR and viral titer, another key issue in the design of packaging cell lines is cell tropism conferred by the affinity of the viral envelope protein for the target cell receptor. The recently developed packaging cell line PG13 (Miller *et al.,* 1991) can pseudotype the MoMLV vector with the gibbon ape leukemia virus (GALV) envelope protein. These pseudotyped retroviral vectors appear able to infect larger numbers of human hematopoietic cells compared to MoMLV-based amphotropic retroviruses, since the cognate receptor for GALV, Glvr-1, appears to be expressed at higher levels on hematopoietic cells (von Kalle *et al.,* 1994).

Although significant effort is being devoted to development of procedures for purification and transduction of human hematopoietic stem cells (see later), peripheral blood lymphocytes (PBLs) are presently considered one of the safest and most efficient cellular delivery vehicles for human gene therapy.

IV. GENE TRANSFER INTO LYMPHOID CELLS

Lymphoid cells have been used as targets for gene transfer for hereditary diseases such as defects in purine metabolism and leukocyte adhesion or acquired diseases such as acquired immunodeficiency syndrome (AIDS).

A. RETROVIRAL VECTOR-MEDIATED GENE TRANSFER INTO PERIPHERAL BLOOD LYMPHOCYTES

To date, many studies have examined the factors that influence retroviral transduction efficiency using immortalized human T-cell lines, and the data from these studies are not always applicable to the transduction of PBLs. In the case of primary blood lymphocytes, a viral titer greater than 10^6 cfu/ml is required to obtain 2–10% transduction efficiency. In some cases, G418 selection of cells expressing the neomycin resistance gene (*neo*) may be necessary to increase the relative transduction efficiency, though care must be taken to optimize the dose and minimize the toxic effects of G418 (Sun *et al.,* 1995).

The scheme for gene transfer into PBLs for clinical applications is depicted in Fig. 9.2. Patient PBLs are collected by apheresis. After separation on a Ficoll/Hypaque gradient, $CD8^+$ T cells are depleted from buffy coat by binding to antibody-coated plates and then stimulated with a CD3 cross-linking antibody (OKT3). Following stimulation, PBLs are transduced with retroviral vectors in the presence of IL-2.

The culture systems employed for transduction of cells in clinical trials include tissue culture flasks, gas-permeable bags, and hollow-fiber bioreactors. Tissue culture flasks or gas-permeable bags (Woffendin *et al.,* 1996) (e.g., LifeCell™, Baxter, Deerfield, IL) are the simplest method for culturing cells, though larger-volume cultures can be labor-intensive. Furthermore, there is a potential risk of contamination when media is exchanged or cells are harvested. The CELLMAX™ system (Cellco, Inc., Germantown, MD) is a high-density culture system. Media is continuously exchanged via an internal hollow-fiber array (Packard, 1990). Cartridges with a moderate pore size have been developed for growth of OKT3-stimulated PBL and have a capacity to produce up to 5×10^9 lymphoid cells per cartridge. The advantages of this system over conventional methods are (1) minimal operator intervention, with a lower chance of culture contamination, and (2) a high-density culture system which requires less

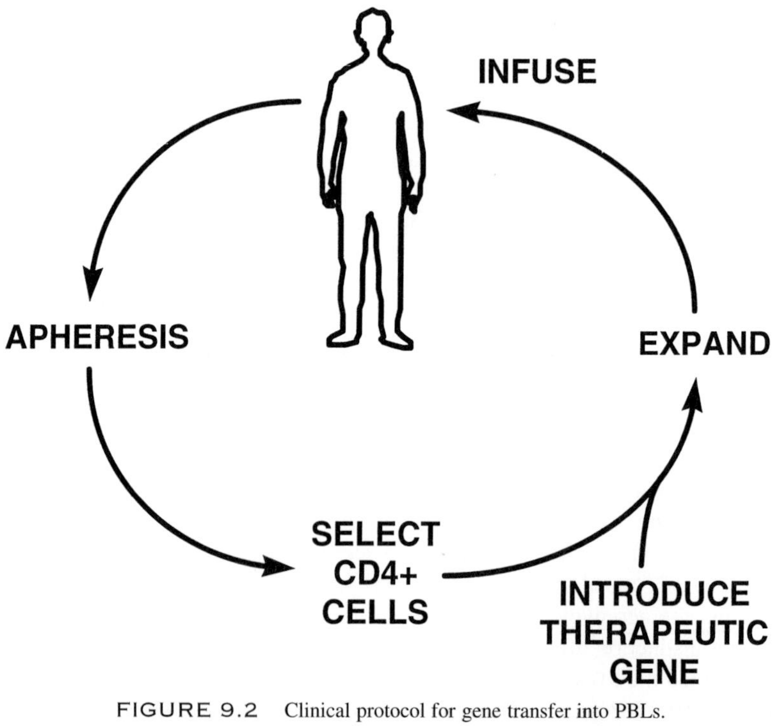

FIGURE 9.2 Clinical protocol for gene transfer into PBLs.

retroviral supernatant than standard culture systems. We have demonstrated the utility of this system in clinical applications (Knop *et al.,* 1998).

B. T-CELL GENE THERAPY

The concepts for a genetic type of therapy were developed almost 20 years ago and have recently been transformed (albeit with limited success) into clinical reality. The first gene therapy trial for the treatment of adenosine deaminase (ADA) deficiency in 1990 was followed rapidly by a growing number of gene therapy trials for other genetic disorders, AIDS, and cancer.

1. Adenosine Deaminase Deficiency

This rare disease is characterized by an inherited severe combined immunodeficiency syndrome (SCID) resulting from the absence of T and B cells, probably due to the toxic accumulation of deoxadenosine, a substrate for ADA. In 1990, two children were treated by retroviral-mediated gene transfer of the ADA gene into autologous PBL, demonstrating the safety and potential efficacy of this gene therapy approach (Blaese *et al.,* 1995).

2. Acquired Immunodeficiency Syndrome

Despite the advent of combination drug therapy for human immunodeficiency virus (HIV-1), the AIDS epidemic shows no sign of abating (Barre-Sinoussi *et al.,* 1983; Gallo *et al.,* 1984). The HIV-1 retrovirus exhibits tropism and cytotoxicity for $CD4^+$ T lymphocytes, the helper T-cell subset that is responsible for amplifying and sustaining B-cell and cytotoxic T-cell responses. A deficiency of the helper T subset results in susceptibility to disseminated viral, fungal, and protozoan infections as well as certain tumors. Following HIV-1 infection of host cells, the virus integrates into the genome of the infected cells, with the provirus becoming an inherited feature of these cells. Therefore AIDS can be viewed as a genetic disease and, as such, is an obvious candidate for gene therapy.

At present, there are four genetic approaches that target HIV-1 replication. These are as follows: expression of ribozymes that catalytically cleave and inactivate HIV-1 RNA molecules (Sarver *et al.,* 1990; Weerasinghe *et al.,* 1991; Lo *et al.,* 1992; Ojwang *et al.,* 1992; Sun *et al.,* 1994, 1995; Wang *et al.,* 1998), antisense RNAs that inhibit reverse transcription, processing, and translation of HIV-1 RNA molecules (Chatterjee *et al.,* 1992; Crisell *et al.,* 1993), mutant HIV-1 structural or regulatory genes with dominant repressor activity (Woffendin *et al.,* 1994), and RNA decoys to inhibit HIV-1 transcription, processing, and packaging (Lisziewicz *et al.,* 1993).

In our laboratory, we are developing the use of ribozymes to target HIV-1 genes in infected PBL and peripheral $CD34^+$ hematopoietic stem cells. Ribozymes are RNA molecules that can be engineered to enzymatically cleave and inactivate specific genes. We have shown that ribozymes that target either the HIV *tat* gene or the ψ packaging site of HIV-1 (Sun *et al.,* 1994, 1995; Wang *et al.,* 1998) protect T-lymphocyte SupT1 cells from subsequent HIV-1 infection (assayed by syncytia and p24 ELISA antigen assay). In a second cell line system, CEM T4 cells were retrovirally transduced (LNL6) with either an anti-*tat* ribozyme (Rz2) and the neomycin resistance gene (*neo*) marker or *neo* alone. This ribozyme construct elicited an antiviral effect, reducing p24 antigen levels by 70–80% compared to vector controls (*neo* alone) (Wang *et al.,* 1998).

These results were recapitulated using non-HIV-1-infected PBL (Sun *et al.,* 1995). Both total and CD4$^+$-enriched PBL were transduced with the LNL6 vector containing the *neo* marker gene (mock) or the vector containing both the *neo* gene and Rz2 ribozyme (RRz2). Transduced cells were selected by G418 resistance and subjected to HIV-1 challenge. The RRz2 retrovital construct was shown to inhibit HIV-1 infection by 70–90% in PBL (Sun *et al.,* 1995).

Recent findings suggest that even a modest reduction in viral load will have a beneficial effect on the immune system of AIDS patients (Ho *et al.,* 1995; Wei *et al.,* 1995). Therefore, it is postulated that infusion of transduced, HIV-1-resistant, CD4$^+$ PBLs may provide enough functional CD4$^+$ T cells to result in a clinical benefit. We are now conducting clinical trials which will address safety issues, efficacy, and the potential emergence of escape mutants in HIV-1-infected patients.

3. Cancer

A variety of genetic approaches for the treatment of cancer are now possible with the advent of retroviral gene transfer techniques. Potential strategies include the introduction of genes that encode cytokines to enhance immune cell effector function, tumor antigens to enhance immunogenicity of the tumor (antitumor vaccine), and suicide genes which provide a method for elimination of cytotoxic lymphocytes should they have significant toxicity such as graft-versus-host disease. Cancer cells may be transduced with proteins that block the expression of oncogenes, wild-type tumor suppressor proteins such as p53, and toxins that are controlled by tumor-specific promoters, though this therapy would require safe and efficient methods for delivery of the transgene to tumor cells *in vivo* (Beaudry *et al.,* 1996).

Genetically engineered T lymphocytes have been a major focus for the delivery of therapeutic agents, since tumor-specific T lymphocytes may have the capacity to home to tumor sites. One such approach is to transfer cytokine genes to T lymphocytes and to enhance the immune response at the site of the tumor, thereby avoiding the potential toxicity of systemically administered cytokines (Hwu *et al.,* 1993; Cai *et al.,* 1995).

The incorporation of the herpes simplex virus thymidine kinase (HSV-TK) "suicide gene" into the vector provides a method for elimination of transduced cells. If side effects develop, then transduced cells can be eliminated with ganciclovir. In allogeneic bone marrow transplantation, donor lymphocytes play a central therapeutic role in both graft-versus-leukemia reactivity and immune reconstitution, though these cells may also cause graft-versus-host disease (GVHD). Eight patients who relapsed or developed Epstein–Barr virus-induced lymphoma after T-cell-depleted bone marrow transplant were treated with donor lymphocytes transduced with HSV-TK, and it was shown that when GVHD developed in three patients, disease activity could be effectively controlled with ganciclovir (Bonini *et al.,* 1997).

V. GENE TRANSFER INTO HEMATOPOIETIC PROGENITOR CELLS

Although it is possible to derive therapeutic benefit from *ex vivo* transduction and infusion of differentiated hematopoietic cells such as lymphocytes, long-term maintenance of gene expression using this approach will probably require repeated infusions of transduced populations of cells. As already described, gene transfer of ADA into T lymphoid cells of ADA-deficient patients has demonstrated the relative success of this approach.

The pluripotential nature of HSCs makes them ideal targets for introducing genes into multiple hematopoietic lineages, and a wider range of diseases have the potential to be treated, including malignancies, genetic disorders, and AIDS (Kohn, 1995). Currently, retroviral vectors are most commonly used for delivery of exogenous genes into hematopoietic cells (Hodgson, 1995). However, in contrast to lymphoid cells, retroviral transduction of HSCs presents a number of technical problems: (1) HSCs are rare cells types (less than 1:100 of bone marrow mononuclear cells) and require high-performance cell separation devices for their purification; (2) they reside in a quiescent state and require growth factor stimulation for retroviral transduction; and (3) considerable effort will be required to define *ex vivo* systems which stimulate HSC renewal divisions, though there are recent indications that this may be possible using specific growth factor combinations.

A. CHARACTERIZATION OF HSC POPULATIONS

Under normal conditions, HSCs reside in the bone marrow and are rare in steady-state peripheral blood (less than 0.02% of peripheral blood mononuclear cells). Myelosuppressive chemotherapy or exogenous recombinant growth factors such as granulocyte colony-stimulating factor (G-CSF) or granulocyte–macrophage colony-stimulating factor (GM-CSF) stimulate the release of HSCs into the peripheral circulation (Dunbar *et al.,* 1995; To *et al.,* 1997) and are readily harvested by apheresis. A third source of HSCs is umbilical cord blood (Broxmeyer *et al.,* 1989). Although cord blood has obvious applications in pediatric transplants, its application in adults has been limited by the relatively small volume of collections.

Simplistically, HSCs are often represented as the antecedent cell type that presides over the hematopoietic hierarchy (Fig. 9.1), and though this has been a useful concept, it is probably only possible to define HSCs in functional terms (Uchida *et al.,* 1993). Characterization of human HSCs is difficult due to their low frequency in bone marrow as well as the absence of simple *in vitro* assays that reflect their potential for hematopoietic reconstitution. Much of what is known about HSCs has been derived from mouse studies, where well-established transplantation models have provided a means for testing the identity of HSCs as defined by other less time-consuming surrogate assays (Visser and van Bekkum, 1990; Dolnikov *et al.,* 1998).

Human HSCs are characterized by expression of the cell surface glycoprotein CD34 (Civin *et al.,* 1990), though there are recent reports that HSCs may also reside outside of the $CD34^+$ cell compartment (Goodell *et al.,* 1997). CD34 expression is not limited to primitive HSCs but is also expressed on more differentiated progenitors which are already restricted to lineage development. In combination with other surface markers (CD38, CD33, HLA-DR, Thy-1, and lineage markers) or dye efflux, more primitive precursors have been defined. Studies in SCID-mice models of human engraftment have shown that lineage negative, $CD34^+Thy\text{-}1^{lo}$ cells can give rise to sustained long-term reconstitution of the human hematopoietic system (Baum *et al.,* 1992). Other methods for identifying primitive hematopoietic progenitor subpopulations with the potential for long-term reconstitution include dye efflux studies using rhodamine 123 (Rh) and Hoechst 33342 (Ho). Studies with murine bone marrow have shown that the most primitive cells are Rh/Ho^{dull} (1/13,000 of unseparated bone marrow cells), whereas Ho^{dull} Rh^{medium} and Ho^{dull} Rh^{bright} fractions represent more differentiated hematopoietic progenitor cells (Bradford *et al.,* 1997).

The mechanism for Hoechst 33342 staining is complex and depends on both the activity of the multiple-drug resistance pump and the state of DNA condensation (metachromatic red shift on binding to condensed DNA). Bone marrow stained with Hoechst 33342 and analyzed simultaneously for red and blue emission reveals a rare subpopulation of primitive cells which are able to reconstitute long-term, lethally irradiated recipient mice (Goodell *et al.,* 1996). Both quiescent (G0/G1) and cycling (S/G2/M) subfractions were shown to reconstitute hematopoiesis (Goodell *et al.,* 1996). Although HSCs are said to be quiescent, a small fraction do appear to cycle, albeit at a very low rate (average turnover of 30 days) (Spangrude and Johnson, 1990; Bradford *et al.,* 1997).

A number of surrogate functional assays for HSCs have been developed and may correlate with the potential for multilineage engraftment (Spangrude *et al.,* 1988). The Dexter culture system uses a stromal "feeder" layer to support hematopoietic cells in long-term culture (at least 3 months). This *in vitro* model of hematopoiesis has provided a basis for development of the long-term culture initiating cell (LTC-IC) assay (Sutherland *et al.,* 1989; Lemieux *et al.,* 1995). *In vitro* assays to directly measure lymphoid precursor cells require the use of thymic stromal layers derived from fetal tissue (Freedman *et al.,* 1996) or fetal thymus organ cultures (Yeoman *et al.,* 1993).

In vivo competitive repopulation assays for human HSCs using human–animal chimeras have provided a model of human engraftment, which can be manipulated to determine the hematopoietic potential of candidate HSC populations. Xenograft mouse models may employ a grafted-human tissue fragment such as bone to provide a microenvironment for support of hematopoietic cells (Lapidot *et al.,* 1993; Kaneshima *et al.,* 1994). Human–sheep chimeras have been created by transplantation of human HSCs into fetuses *in utero* (Zanjani *et al.,* 1994). These competitive xenograft assays probably provide the best measure of long-term repopulating potential short of human transplantation and will be important experimental models for the preclinical evaluation and optimization of gene transfer methods.

B. ISOLATION OF HSC POPULATIONS

Stem cells constitute a small proportion of human bone marrow, mobilized peripheral blood, or cord blood, so it is often desirable to enrich for the $CD34^+$ population (1–3% of mononuclear cells) as used in the majority of HSC gene therapy protocols. Enrichment is required to purge tumor cells from autologous bone marrow or mobilized peripheral blood transplants for various malignancies, as well as for practical reasons, to reduce the number of target cells per volume of retroviral supernatant. Methods for stem cell enrichment include density gradient centrifugation, counterflow centrifugal elutriation, and adhesion to lectins such as wheat germ agglutinin (Visser and van Bekkum, 1990). Devices that have been developed for use in clinical protocols utilize monoclonal antibodies against CD34 and are based on magnetic separation or streptavidin–biotin affinity labeling (Lansdorp and Thomas, 1991; Lord and Marsh, 1993).

C. DISEASE TARGETS FOR HSC GENE THERAPY

Single-gene genetic disorders are ideal candidates for stem cell gene therapy. To date, genetic diseases which have been targets for HSC gene therapy protocols include severe combined immunodeficiency (SCID), Gaucher's disease, Hurler's syndrome, Fanconi's anemia, and chronic granulomatous disease (Marcel and Grausz, 1996). The ADA stem cell trials have now been running for several years, and although there have been no documented "cures," the therapy has been shown to have some degree of efficacy. Three children diagnosed *in utero* with SCID were infused after birth with their own cord blood progenitors transduced with the ADA gene (Kohn *et al.,* 1995). All children showed evidence of engraftment of transduced progenitors, and genetically marked cells were found in their peripheral blood up to 3 years posttransplant. Although the children have all been maintained with PEG-ADA supplements, gradual reductions in the level of PEG-ADA infusions have resulted in increased proportions of ADA-positive lymphocytes in their peripheral blood. This remarkable finding is evidence that the genetically marked cells have a selective advantage *in vivo,* with the capacity to proliferate under selective pressure.

The majority of HSC gene therapy stem cell trials in malignant diseases have involved genetic marking of HSCs with the *neo* marker gene. These trials were aimed at determining the source of tumor relapse following transplantation of autologous marrow that had been genetically marked. Although marked cells have been found in a proportion of relapsing patients in a number of trials (Brenner, 1996), a low level of transduction may result in a significant incidence of false-negative results. Importantly, these gene-marking trials have yielded data which show the long-term persistence of the transduced gene in a small proportion of HSCs (Brenner *et al.,* 1993b; Dunbar *et al.,* 1995).

A number of gene therapy approaches have been developed for treatment for HIV-1 (Sarver *et al.,* 1990; Weerasinghe *et al.,* 1991; Chatterjee *et al.,* 1992; Lo *et al.,* 1992; Ojwang *et al.,* 1992; Crisell *et al.,* 1993; Lisziewicz *et al.,* 1993; Sun *et al.,* 1994; Woffendin *et al.,* 1994; Sun *et al.,* 1995; Wang *et al.,* 1998). In most trials, including our own, PBLs have been the target for gene transfer (Knop *et al.,* 1998). An alternate approach is to transduce $CD34^+$ progenitors. Expression of the transgene in the HSC pool could protect cells other than T cells from HIV infection and may also result in more persistent expression of the transferred gene.

D. GENE TRANSFER INTO HSCS

Gene therapy protocols targeting human HSCs have primarily used recombinant retroviruses as delivery vehicles. HSC-based therapies require long-term expression of the transgene, limiting the choice of vector to those that allow stable integration into the host genome. The success of gene transfer in the murine system has led to studies in humans. Although *in vitro* assays such as LTC-IC have demonstrated relatively high transduction levels, the percentage of transduced cells engrafting in the bone marrow as well as the numbers of marked cells in the peripheral blood of the patients remains disappointingly low.

One of the difficulties in translating results from the murine system to humans has been developing similar protocols for clinical implementation. The high levels of gene transfer routinely reported for murine progenitors were often achieved by cocultivation of target cells with retroviral producer cells. Although transduction by cocultivation has been shown to be superior to cell-free viral supernatant (Havenga *et al.,* 1997), there is the risk of contamination of cellular infusions with producer cells, which is currently not acceptable for clinical application (Gordon, 1988). This problem can be partially overcome by multiple rounds of infection with viral supernatant, which is now standard in most clinical gene therapy protocols. Altering the retroviral tropism by pseudotyping with an *env* gene derived from GALV or the G protein of vesicular stomatitis virus

(VSV) has also been successful in increasing gene transfer into human HSCs (von Kalle *et al.,* 1994; Yang *et al.,* 1995).

Most clinical protocols include overnight preculture in media supplemented with the growth factors interleukin-3 and -6 and stem cell factor (SCF) to induce cycling, followed by addition of retroviral supernatant daily for 2–4 days (Fig. 9.3). A significant concern is that this growth factor combination will act primarily to differentiate cells, not to promote self-renewal divisions of HSCs. Recent studies have shown that FLT-3/FLK-2 ligand, SCF, and thrombopoietin–megakaryocyte growth and development factor (MGDF) are capable of expanding progenitor cell numbers while maintaining the reconstituting ability of these cells (Rasko *et al.,* 1997; Yonemura *et al.,* 1997; Amado *et al.,* 1998).

As discussed earlier, it has recently become apparent that stem cells cycle relatively infrequently. Therefore some investigators have adopted a long-term culture protocol (Dube *et al.,* 1995). These cultures were run for 21 days, and retrovirus was added on days 0, 8, and 15. Up to 100% of cells were positive for the provirus at day 21, with successful gene transfer into 70–80% of clonogenic precursors. Canine HSCs transduced in this way were able to successfully engraft in nonmyeloablated dogs (Bienzle *et al.,* 1994). Gene-marking trials in myeloma patients using this long-term protocol have resulted in higher

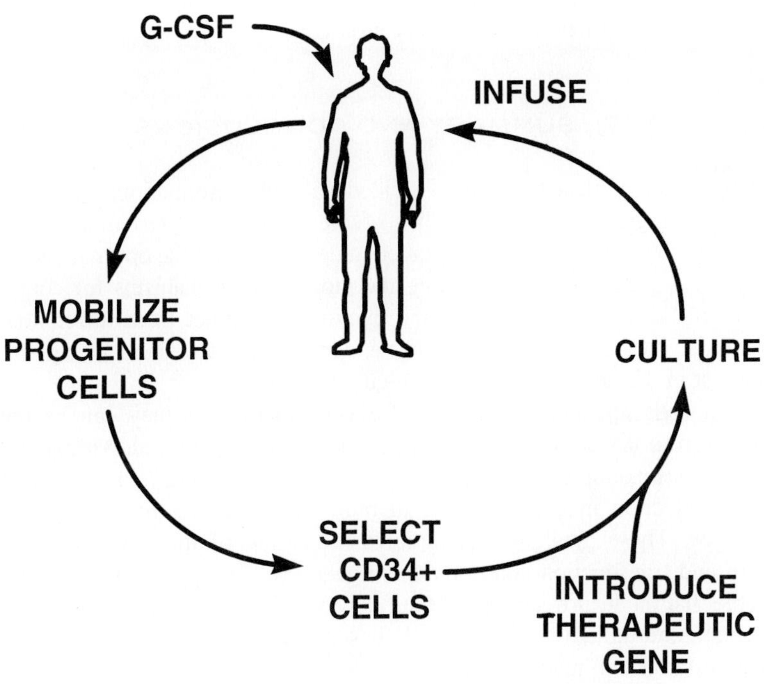

FIGURE 9.3 Protocol for gene transfer into CD34$^+$ cells.

levels of genetically marked cells in the peripheral blood than short-term protocols (Stewart *et al.,* 1996).

In addition to low levels of gene transfer *in vitro* contributing to poor gene expression *in vivo,* it is also possible that transduced cells may be immunogenic. For example, transferred HSV-TK-modified cytotoxic lymphocytes were eliminated after 5–8 weeks by a host-immune response to the foreign transgene product (Riddell *et al.,* 1996). Therefore, even though retroviral vectors are relatively nonimmunogenic, it may be possible that cells expressing the transgene will be cleared bv the host's immune system.

As previously stated, one of the major obstacles in developing successful gene therapy protocols has been the absence of models that accurately predict the *in vivo* reconstituting potential of transduced HSCs. Although most groups, including our own, assess transduction efficiency by assaying proviral integration and expression in clonogenic CFC (colony-forming cells) and LTC-IC, these do not appear to correlate well with repopulation in the clinical setting.

The development of vectors with reporter genes that can be detected by flow cytometry is an encouraging development. These transgene-encoded markers include murine CD2, CD4, heat-stable antigen (HSA) and its human homolog CD24, human nerve growth factor receptor (NGFR), and green fluorescent protein (GFP) from the jellyfish *Acquoria victoria* (Chalfie *et al.,* 1993; Bauer and Hickstein, 1997). The use of these vectors in combination with flow cytometric analysis of the transduced stem cell population may facilitate the development of more efficient gene therapy protocols.

VI. SUMMARY AND CONCLUSIONS

Preliminary clinical trials have yielded valuable information, and after initially encouraging results, a number of areas for future research have been identified. These include a more detailed understanding of the optimal conditions required for gene transfer, including the molecular mechanisms for control of hematopoietic stem cell renewal and cycling as distinct from differentiation/maturation. New reporter genes and xenograft models of human engraftment will assist in the development of these culture systems.

Blood cells are attractive targets for gene therapy as they can be readily removed from the body, manipulated *ex vivo,* and reintroduced. Although there are numerous examples of genetic transformation of such cells, including retroviral transduction in ongoing gene therapy clinical trials, there remain many challenges. These challenges include identification of the optimal cell type, growth and transduction conditions, and gene transfer vectors. It is not inconceivable that medical treatment of the future will have cell therapy as part of its armament, and though it is not clear how soon that will be, the pace of advance indicates that it will indeed come.

ACKNOWLEDGMENTS

Work in the authors' laboratories has been supported, in part, by Gene Shears Pty Ltd.

REFERENCES

Amado, R. G., Symonds, G., Jamieson, B. D., *et al.* (1998). Effects of megakaryocyte growth and development factor on survival and retroviral transduction of T lymphoid progenitor cells. *Hum. Gene Ther.* **9,** 173–183.

Barre-Sinoussi, F., Chermann, J. C., Rey, F., *et al.* (1983). Isolation of a T-lymphotropic retrovirus from a patient at risk for acquired immune deficiency syndrome (AIDS). *Science* **220,** 868–871.

Bauer, T. R., and Hickstein, D. D. (1997). Transduction of human hematopoietic cells and cell lines using a retroviral vector containing a modified murine CD4 reporter gene. *Hum. Gene Ther.* **8,** 243–252.

Baum, C. M., Weissman, I. L., Tsukamoto, A. S., *et al.* (1992). Isolation of a candidate human hematopoietic stem-cell population. *Proc. Natl. Acad. Sci. USA* **89,** 2804–2808.

Beaudry, G. A., Bertelsen, A. H., and Sherman, M. I. (1996). Therapeutic targeting of the p53 tumor suppressor gene. *Curr. Opin. Biotechnol.* **7,** 592–600.

Bender, M. A., Palmer, T. D., Gelinas, R. E., *et al.* (1987). Evidence that the packaging signal of Moloney murine leukemia virus extends into the gag region. *J. Virol.* **61,** 1639–1646.

Bienzle, D., Abrams-Ogg, A. C. G., Kruth, S. A., *et al.* (1994). Gene transfer into hematopoietic stem cells: Long-term maintenance of *in vitro* activated progenitors without marrow ablation. *Proc. Natl. Acad. Sci. USA* **91,** 350–354.

Blaese, R. M., Culver, K. W., Miller, A. D., *et al.* (1995). T lymphocyte-directed gene transfer for ADA-SCID: Initial trial results after 4 years. *Science* **270,** 475–480.

Bonini, C., Ferrari, G., Verzeletti, S., *et al.* (1997). HSV-TK gene transfer into donor lymphocytes for control of allogeneic graft-versus-leukemia. *Science* **276,** 1719–1724.

Bradford, G. B., Williams, B., Rossi, R., *et al.* (1997). Quiescence, cycling, and turnover in the primitive hematopoietic stem cell compartment. *Exp. Hematol.* **25,** 445–453.

Brenner, M. K. (1996). Gene marking. *Hum. Gene Ther.* **7,** 1927–1936.

Brenner, M. K., Rill, D. R., Holladay, M. S., *et al.* (1993a). Gene marking to determine whether autologous marrow infusion restores long-term haemopoiesis in cancer patients. *Lancet* **342,** 3048–3057.

Brenner, M. K., Rill, D. R., Moen, R. C., *et al.* (1993b). Gene-marking to trace origin of relapse after autologous bone marrow transplantation. *Lancet* **341,** 85–86.

Broxmeyer, H. E., Douglas, G. W., Hangoc, G., *et al.* (1989). Human umbilical cord blood as a potential source of transplantable hemopoietic stem/progenitor cells. *Proc. Natl. Acad. Sci. USA* **86,** 3828–3832.

Cai, Q., Rubin, J. T., and Lotze, M. T. (1995). Genetically marking human cells—Results of the first clinical gene transfer studies. *Cancer Gene Ther.* **2,** 125–136.

Chalfie, M., Tu, Y., Eukirchen, G., *et al.* (1993). Green fluorescent protein as a marker for gene expression. *Science* **263,** 802–805.

Chatterjee, S., Johnson, P. R., and Wong, K. K., Jr. (1992). Dual-target inhibition of HIV-1 *in vitro* by means of an adeno-associated virus antisense vector. *Science* **258,** 1485–1488.

Civin, C. I., Trischmann, T. M., Fackler, M. J., *et al.* (1990). Report on the CD34 cluster workshop. *In* "Leukocyte Typing IV" (W. Knapp, Ed.), Oxford Univ. Press, Oxford.

Crisell, P., Thompson, S., and James, W. (1993). Inhibition of HIV-1 replication by ribozymes that show poor activity *in vitro. Nucleic Acids Res.* **21,** 5251–5255.

Daniel, C. P., and Dexter, T. M. (1989). The role of growth factors in haemopoietic development: Clinical and biological implications. *Cancer Metastasis Rev.* **8,** 253–262.

Danos, O., and Mulligan, R. C. (1988). Safe and efficient generation of recombinant retroviruses with amphotropic and ecotropic host ranges. *Proc. Natl. Acad. Sci. USA* **85,** 6460–6464.

Dolnikov, A., Shounan, Y. Millington, M., *et al.* (1998). V-*myc* in a simple, single gene retroviral vector causes rapid induction of leukemia and concomitant apoptosis following bone marrow transplantation. *Leukemia* **12,** 542–553.

Dube, I., Stewart, K., Lutzko, C., *et al.* (1995). Gene transfer into marrow-derived hematopoietic stem cells. *Gene Ther* **2,** 579–580.

Dunbar, C. E., Cottler-Fox, M., and Joyce, A. (1995). Retrovirally marked CD34-enriched peripheral blood and bone marrow cells contribute to long-term engraftment after autologous transplantation. *Blood* **85,** 3048–3057.

Dzierzak, E. A., Papayannopoulou, T., and Mulligan, R. C. (1988). Lineage-specific expression of a human β-globin gene in murine bone marrow transplant recipients reconstituted with retrovirus-transduced stem cells. *Nature* **331**, 35–41.

Freedman, A. R., Zhu, H., Levine, J. D., *et al.* (1996). Generation of human T lymphocytes from bone marrow $CD34^+$, cells *in vitro*. *Nat. Med.* **2,** 46–51.

Gallo, R. C., Salahuddin, S. Z., Popovic, M., *et al.* (1984). Frequent detection and isolation of cytopathic retroviruses (HTLV-III) from patients with AIDS and at risk for AIDS. *Science* **224,** 500–503.

Goodell, M. A., Brose, K., Paradis, G., *et al.* (1996). Isolation and functional properties of murine hematopoietic stem cells that are replicating *in vivo*. *J. Exp. Med.* **183,** 1797–1806.

Goodell, M. A., Rosenzweig, M., Kim, H., *et al.* (1997). Dye efflux studies suggest that hematopoietic stem cells expressing low or undetectable levels of CD34 antigen exist in multiple species. *Nat. Med.* **3,** 1337–1345.

Gordon, M. Y. (1988). Adhesive properties of haemopoietic stem cells. *Br. J. Haematol.* **68,** 149–151.

Hantzopoulos, P. A., Sullenger, B. A., Ungers, G., *et al.* (1989). Improved gene expression upon transfer of the adenosine deaminase minigene outside the transcriptional unit of a retroviral vector. *Proc. Natl. Acad. Sci. USA* **86,** 3519–3523.

Havenga, M., Hoogerbrugge, P., Valerio, D., *et al.* (1997). Retroviral stem cell gene therapy. *Stem Cells* **15,** 162–179.

Hawley, R. G., Sabourin, L. A., and Hawley, T. S. (1989). An improved retroviral vector for gene transfer into undifferentiated cells. *Nucleic Acids Res.* **17,** 4001–4001.

Ho, D. D., Neumann, A. U., Perelson, A. S., *et al.* (1995). Rapid turnover of plasma virions and CD4 lymphocytes in HIV-1 infection. *Nature* **373,** 123–126.

Hock, R. A., Miller, A.D., and Osborne, W. R. A. (1989). Expression of human adenosine deaminase from various strong promoters after gene transfer into human haematopoietic cell lines. *Blood* **74,** 876–881.

Hodgson, C. (1995). The vector void in gene therapy. *Bio/Technology* **13,** 222–225.

Hwu, P., Yannelli, J., Kriegler, M., *et al.* (1993). Functional and molecular characterization of tumor-infiltrating lymphocytes transduced with tumor necrosis factor-alpha cDNA for the gene therapy of cancer in humans. *J. Immunol.* **150,** 4104–4115.

Jordan, C. T., and Lemischka, I. R. (1990). Clonal and systemic analysis of long-term hematopoiesis in the mouse. *Genes Dev.* **4,** 220–232.

Kaneshima, H., Namikawa, R., and McCune, J. M. (1994). Human hematolymphoid cells in SCID mice. *Curr. Opin. Immunol.* **6,** 327–333.

Keller, G., and Snodgrass, R. (1990). Life span of multipotential hematopoietic stem cells *in vivo*. *J. Exp. Med.* **171,** 1407–1418.

Knop, A. E., Arndt, A., Raponi, M., *et al.* (1998). Artificial capillary culture: Expansion and retroviral transduction of $CD4^+$ T-lymphocytes for clinical application. *Hum. Gene Ther.* (in press).

Kohn, D. B. (1995). The current status of gene therapy using hematopoietic stem cells. *Curr. Opin. Pediatr.* **7,** 56–63.

Kohn, D. B., Weinberg, K. I., and Nolta, J. A. (1995). Engraftment of gene-modified umbilical cord blood cells in neonates with adenosine deaminase deficiency. *Nat. Med.* **1,** 1017–1023.

Lansdorp, P. M., and Thomas, T. E. (1991). Selection of human hematopoietic stem cells. *In* "Bone Marrow Processing and Purging: A Practical Guide" (A. P. Gee, Ed.), pp. 351–362. CRC Press, Boca Raton, FL.

Lapidot, T., Pflumio, F., and Dick, J. E. (1993). Modeling human hematopoiesis in immunodeficient mice. *Lab. Anim. Sci.* **43,** 147–150.

Lemieux, M. E., Rebel, V. I., Lansdorp, P.M., *et al.* (1995). Characterization and purification of a primitive hematopoietic cell type in adult mouse marrow capable of lymphomyeloid differentiation in long-term marrow "switch" cultures. *Blood* **86,** 1339–1347.

Lisziewicz, J., Sun, D., Smythe, J., *et al.* (1993). Inhibition of human immunodeficiency virus type 1 replication by regulated expression of a polymeric Tat activation response RNA decoy as a strategy for gene therapy in AIDS. *Proc. Natl. Acad. Sci. USA* **90,** 8000–8004.

Lo, K. M. S., Biasolo, M. A., Dehni, G., *et al.* (1992). Inhibition of replication of HIV-1 by retroviral vectors expressing tat-antisense and anti-tat ribozyme RNA. *Virology* **190,** 176–183.

Lord, B. I., and Marsh, J. C. W. (1993). Enrichment of progenitor cell populations from murine and human haemopoietic tissue. *In* "Haemopoiesis" (N. G. Testa and G. Molineux, Eds.), pp. 21–42 IRL Press, Oxford.

Marcel, T., and Grausz, J. D. (.1996). The TMC Worldwide Gene Therapy Enrollment Report (June 1996). *Hum. Gene Ther.* **7,** 2025–2046.

Markowitz, D., Goff, S., and Bank, A. (1988). Construction and use of a safe and efficient amphotropic packaging cell line. *Virology* **167,** 400–406.

Metcalf, D. (1994). "Clonal Culture of Hemopoietic Cells: Techniques and Application" Elsevier, Amsterdam/New York.

Miller, A.D. (1992). Human gene therapy comes of age. *Nature* **357,** 455–460.

Miller, A. D., and Buttimore, C. (1986). Redesign of retrovirus packaging cell lines to avoid recombination leading to helper virus production. *Mol. Cell. Biol.* **6,** 2895–2902.

Miller, A. D., Garcia, J. V., von Suhr, N., *et al.* (1991). Construction and properties of retrovirus packaging cells based on gibbon ape leukemia virus. *J. Virol.* **65,** 2220–2224.

Moore, M. A. S., and Shapiro, F. (1994). Regulation and function of hematopoietic stem cells. *Curr. Opin. Hematol.* **1,** 180–186.

Ojwang, J. O., Hampel, A., Looney, D. J., *et al.* (1992). Inhibition of human immunodeficiency virus type 1 expression by a hairpin ribozyme. *Proc. Natl. Acad. Sci. USA* **89,** 10802–10806.

Overell, R. W., Wesser, K. E., and Cosman, D. (1988). Stably transmitted triple-promoter retroviral vectors and their use in transformation of primary mammalian cells. *Mol. Cell. Biol.* **8,** 1803–1808.

Packard, B. S. (1990). Mitogenic stimulation of human tumor-filtrating lymphocytes by secreted factors from human tumor cell lines. *Proc. Natl. Acad. Sci. USA* **87,** 4058–4062.

Palmer, T. D., Miller, A. D., Reeder, R. H., *et al.* (1993). Efficient expression of a protein coding gene under the control of an RNA polymerase I promoter. *Nucleic Acids Res.* **21,** 3451–3457.

Rasko, J. E., O'Flaherty, E., and Begley, C. G. (1997). Mpl ligand (MGDF) alone and in combination with stem cell factor (SCF) promotes proliferation and survival of human megakaryocyte, erythroid and granulocyte/macrophage progenitors. *Stem Cells* **15,** 33–42.

Riddell, S. R., Elliott, M., Lewinsohn, D. A., *et al.* (1996). T-cell mediated rejection of gene-modified HIV-specific cytotoxic T lymphocytes in HIV-infected patients. *Nat. Med.* **2,** 216–223.

Sarver, N., Cantin, E. M., Chang, P. S., *et al.* (1990). Ribozymes as potential anti-HIV-1 therapeutic agents. *Science* **247,** 1222–1225.

Scholfield, R. (1983). The stem cell system. *Biomed. Pharmacother.* **37,** 375.

Spangrude, G. J., and Johnson, G. R. (1990). Resting and activated subsets of mouse multipotent hematopoietic stem cells. *Proc. Natl. Acad. Sci. USA* **87,** 7433–7437.

Spangrude, G. J., Heimfeld, S., and Weissman, I. L. (1988). Purification and characterization of mouse hematopoietic stem cells. *Science* **241,** 58–62.

Stewart, A. K., Nanji, S., Krygsman, P., *et al.* (1996). Engraftment of gene marked long term marrow culture (LTMC) cells in myeloma patients. *Blood* **88,** 1068a.

Sullenger, B. A., Lee, T. C., Smith, C. A., *et al.* (1990). Expression of chimeric tRNA-driven antisense transcripts renders NIH 3T3 cells highly resistant to Moloney murine leukemia virus replication. *Mol. Cell. Biol.* **10,** 6512–6523.

Sun, L. Q., Pyati, J., Wang, L., *et al.* (1995). Resistance to HIV-1 infection conferred by transduction of human peripheral blood lymphocytes with ribozyme, antisense or polyTAR constructs. *Proc. Natl. Acad. Sci. USA* **92,** 7272–7276.

Sun, L. Q., Warrilow, D., Wang, L., *et al.* (1994). Ribozyme-mediated suppression of Moloney murine leukemia virus and human immunodeficiency virus type 1 replication in permissive cell lines. *Proc. Natl. Acad. Sci. USA* **91,** 9715–9719.

Sutherland, H. J., Eaves, C. J., Eaves, A. C., *et al.* (1989). Characterization and partial purification of human marrow cells capable of initiating long-term hemopoiesis *in vitro. Blood* **74,** 1563–1570.

To, L. B., Haylock, D. N., Simmons, P. J., *et al.* (1997). The biology and clinical uses of blood stem cells. *Blood* **89,** 2233–2258.

Uchida, N., Fleming, W. H., Alpern, E. J., *et al.* (1993). Heterogeneity of hematopoietic stem cells. *Curr. Opin. Immunol.* **5,** 177–184.

Visser, J. W. M., and van Bekkum, D. W. (1990). Purification of pluripotent hemopoietic stem cells: Past and present. *Exp. Hematol.* **18,** 248–256.

von Kalle, C., Kiem, H. P., Goehle, S., *et al.* (1994). Increased gene transfer into haematopoietic cells by extended *in vitro* exposure to a pseudotyped retroviral vector. *Blood* **84,** 2890–2897.

Wang, L., Witherington, C., King, A., *et al.* (1998). Preclinical characterization of an anti-tat ribozyme for therapeutic application. *Hum, Gene Ther.* **9,** 1283–1291.

Weerasinghe, M., Liem, S. E., Asad, S., *et al.* (1991). Resistance to human immunodeficiency virus type 1 (HIV-1) infection in human $CD4^+$ lymphocyte-derived cell lines conferred by using retroviral vectors expressing an HIV-1 RNA-specific ribozyme. *J. Virol.* **65,** 5531–5534.

Wei, X., Ghosh, S. K., Taylor, M. E., *et al.* (1995). Viral dynamics in human immunodeficiency virus type 1 infection. *Nature* **373,** 117–122.

Woffendin, C., Ranga, U., Yang, Z. Y., *et al.* (1996). Expression of a protective gene prolongs survival of T cells in human immunodeficiency virus-infected patients. *Proc. Natl. Acad. Sci. USA* **93,** 2889–2894.

Woffendin, C., Yang, Z. Y., Ranga, U., *et al.* (1994). Nonviral and viral delivery of a human immunodeficiency virus protective gene into human T cells. *Proc. Natl. Acad. Sci. USA* **91,** 11581–11585.

Yang, Y., Vanin, E. F., Whitt, M. A., *et al.* (1995). Inducible, high-level production of infectious murine leukemia retroviral vector particles pseudotyped with vesicular stomatitis virus G envelope protein. *Hum. Gene Ther.* **6,** 1203–1213.

Yee, J. K., Moores, J. C., Jolly, D. J., *et al.* (1987). Gene expression from transcriptionally disabled retroviral vectors. *Proc. Natl. Acad. Sci. USA* **84,** 5179–5201.

Yeoman, H., Gress, R. E., Bare, C. V., *et al.* (1993). Human bone marrow and umbilical cord blood cells generate $CD4^+$ and $CD8^+$ single-positive T cells in murine fetal thymus organ culture. *Proc. Natl. Acad. Sci. USA* **90,** 10778–10782.

Yonemura, Y., Ku, H., Lyman, S. D., *et al.* (1997). *In vitro* expansion of hematopoietic progenitors and maintenance of stem cells: Comparison between FLT3/FLK-2 ligand and KIT ligand. *Blood* **89,** 1915–1921.

Zanjani, E. D., Silva, M. R. G., and Flake, A. W. (1994). Retention and multilineage expression of human hematopoietic stem cells in human–sheep chimeras. *Blood Cells* **20,** 331–340.

10

PROCESS DEVELOPMENT FOR *EX VIVO* CELL THERAPY

CHARLES DOWDING, TOM LEEMHUIS, AUDREY JAKUBOWSKI, AND CHRISTOPHER READING

SyStemix Inc. Palo Alto, California 94304

I. INTRODUCTION

SyStemix has developed a Good Manufacturing Process (GMP) cell-processing scheme for isolating, transducing, and cryopreserving highly purified hematopoietic stem cells (HSCs). The HSC purification process uses multiparameter high-speed, fluorescence-activated cell sorting to purify $CD34^{+}Thy\text{-}1^{+}$ cells and is designed to accommodate incoming tissues of variable cell composition from local or distant sites for processing at SyStemix's facilities in a reproducible, controlled manner. Cellular products are being manufactured by this process in several clinical development studies in the United States and France, which are intended to lead to product marketing. In this chapter, we discuss the utility of highly purified autologous and allogeneic stem cell products in filling unmet clinical needs in the treatment of hematopoietic malignancy, autoimmune disorders, selected inherited genetic abnormalities, and human immunodeficiency virus (HIV) disease. We detail the current regulatory environment, quality assurance issues, and the development process that led to our current clinical trials and briefly discuss future directions in GMP development.

II. CLINICAL UTILITY OF TRANSPLANTATION OF PURIFIED HSCs

A. AUTOLOGOUS TRANSPLANTATION

The cell surface phenotype of human HSCs has been defined as $CD34^+Thy\text{-}1^+$ (Baum *et al.,* 1992). Pure populations of these cells have been shown to result in multilineage engraftment in the severe combined immunodeficiency disease-human (SCID-Hu) thymus/liver and bone mouse models and fetal sheep xenografts. The use of highly purified autologous HSCs might prove to be clinically useful for reconstituting hematopoiesis following high-dose (myeloablative) anticancer chemotherapy/radiotherapy in patients with hematopoietic diseases such as multiple myeloma and non-Hodgkin's lymphoma (NHL), since few, if any, tumor cells would be reinfused with the HSCs. Purified HSCs for hematopoietic support might also prove useful for nonhematopoietic malignancies where tumor cells are known to be present in the blood or bone marrow (breast cancer) or when a high degree of T-lymphocyte elimination from the supporting stem cell fraction is required (autoimmune disorders). Purified HSCs are ideal targets for gene therapy, since, under appropriate conditions, they are pluripotent and self-renewing.

I. B-Cell Malignancies

SyStemix is participating in clinical trials that test the ability of $CD34^+Thy\text{-}1^+$ cells to provide hematopoietic support after myeloablative therapy for cancer. Although there are diseases or stages of disease where $CD34^+$ cell selection alone might result in a tumor-free autograft, the purification of $CD34^+Thy\text{-}1^+$ HSCs from a "parental" $CD34^+$ cell population invariably results in greater passive elimination ("negative" selection) of undesirable cells. There are certain cancers that probably mandate using highly purified HSCs. For example, there is considerable evidence that the lesion underlying the malignant plasma cell proliferation observed in multiple myeloma occurs in an immature B-lymphocyte precursor that expresses both CD34 and CD19/20 (Belch *et al.,* 1994; Bergsagel *et al.,* 1994; Takishita *et al.,* 1994; Gazitt *et al.,* 1995; Szczepek *et al.,* 1997). Patient-specific oligonucleotide probes for the complementarity-determining region III (CDRIII) have been used to demonstrate elimination of residual tumor cells below detectable levels in the $CD34^+Thy\text{-}1^+$ cell fraction, whereas tumor cells can remain detectable in the $CD34^{+}Thy\text{-}1^{-}$ cell fraction (Gazitt *et al.,* 1995). Thus, autologous transplantation with highly purified $CD34^{+}Thy\text{-}1^{+}$ cells would result in the reinfusion of significantly fewer tumor-propagating cells compared with transplantation of $CD34^{+}$ cell selected autografts. As previously mentioned, greater "passive" elimination of plasma cells ($CD34^{-}$) from the graft is achieved during the purification of $CD34^{+}Thy\text{-}1^{+}$ cells compared with $CD34^{+}$ cells. The current phase I clinical trial in myeloma addresses HSC dose requirement for rapid neutrophil and platelet reconstitution (Archimbaud *et al.,* 1997). Whether the use of purified HSCs will affect long-term survival of multiple myeloma patients remains to be determined. However, it seems reasonable to believe that clinical benefit might result from a combination of (1) substantial elimination of tumor cells from the graft, (2) more effective pretransplant antitumor therapy, and (3) the use of "immunotherapy" directed at eliminating minimal residual disease soon after HSC reinfusion (e.g., anti-CD 19/20/56 monoclonal antibodies or adoptive transfer of tumor-specific lymphocytes).

The early stages of B-lymphocyte differentiation, such as IgD_{H} rearrangement, occur before surface expression of CD19 in cells that are expressing CD34 and CD10 (Bertrand *et al.,* 1997; Davi *et al.,* 1997). Consequently, molecular lesions associated with immunoglobulin-gene rearrangement might be expected to occur in cells with the immature B-lymphocyte phenotype. There is evidence that tumor-propagating cells in certain types of NHL and B-cell chronic lymphocytic leukemia express the CD34 molecule (Macintyre *et al.,* 1995; Gahn *et al.,* 1997). It is not known whether these tumor-propagating cells segregate into the $CD34^{+}Thy\text{-}1^{-}$ (presumably $CD19^{+}$) population but this seems likely. Since purification of the $CD34^{+}Thy\text{-}1^{+}$ cell fraction from normal bone marrow (BM) results in the elimination of 5–6 logs of $CD19^{+}$ B lymphocytes, the use of $CD34^{+}Thy\text{-}1^{+}$ cells in support of high-dose chemotherapy would result in the reinfusion of substantially fewer tumor cells than other strategies using unfractionated autologous cells or the $CD34^{+}$ cell fraction, regardless of whether tumor-propagating cells have a $CD34^{+}19^{+}$ phenotype. Clinical trials with purified HSCs are ongoing and clinical benefit remains to be determined.

2. Other Malignancies

Trials designed to evaluate the clinical utility of purified autologous HSCs for hematopoietic support for high-dose chemotherapy in metastatic breast cancer are also in progress (Negrin *et al.,* 1997). Tumor depletion results are encouraging: with immunofluorescence detection of anticytokeratin antibodies as a method for detecting residual disease, a 5- to 6- log depletion of tumor cells is achieved during HSC purification. However, there is evidence that in a small proportion of breast cancer patients tumor cells expressing cytokeratin express CD34 and possibly Thy-1 (Kohler *et al.,* 1994; Wessman *et al.,* 1995), mandating prescreening of the tumor phenotype to avoid inappropriate patient recruitment. As with B-lymphocyte malignancies, there is probably merit in using immunotherapy to target residual disease concomitantly with antitumor high-dose chemotherapy radiotherapy and purified HSC support.

3. Severe Autoimmune Disease

There are anecdotal reports of remittance of underlying autoimmune disease following autologous transplantation of nonpurified HSC grafts following cytotoxic treatment for cancer (Cooley *et al.,* 1997; Marmont, 1997). The use of purified HSCs in humans with severe autoimmune disease might have clinical utility in a myeloablative setting because of the 5- to 6- log T-lymphocyte and B-lymphocyte depletion associated with HSC purification. Typically, a purified HSC autograft containing approximately 1×10^6 HSCs/kg would contain a T-lymphocyte dose of 1×10^3/kg. Theoretically, a highly purified HSC autograft should be superior to a CD34$^+$ cell selected autograft for treating autoimmune disorders due to the fewer T lymphocytes being reinfused. To test this hypothesis, clinical trials are planned in severe autoimmune diseases.

Overall, the long-term clinical benefits associated with purified HSC autotransplants await results from ongoing and future clinical trials. However, transplantation of autologous HSCs in a myeloablative setting results in platelet, neutrophil, and erythrocyte engraftment rates comparable to those reported for CD34$^+$ cell selected cells, provided a threshold dose of $(0.7-1.0) \times 10^6$ HSCs/kg is reinfused. There is a risk of delayed engraftment if insufficient HSCs are collected from the patient. Therefore, systems need to be in place to exclude patients who mobilize poorly from this procedure. There is also the possibility that use of purified HSCs will reduce autologous anticancer immunity; thus, monitoring residual disease and survival following transplantation is important. On the other hand, the use of lymphocyte and natural killer (NK) cell free HSCs for hematopoietic support facilitates clinical development of tumor-specific cell therapies given concomitantly with stem cells.

B. ALLOGENEIC TRANSPLANTATION

Certain leukemias and myelodysplastic syndromes can be cured by allogeneic transplantation following myeloablative conditioning. The use of purified HSCs, essentially free of alloreactive T and B lymphocytes, permits clinical studies using human leukocyte-associated antigen (HLA)-disparate family members, HLA-matched or -unmatched unrelated individuals, and HLA-matched sibling donors. The potential benefit of expanding the allogeneic donor pool is offset by the increased risk of graft rejection through the lack of graft-facilitating cells and a reduction in graft-versus-tumor (GVT) activity. Therefore, purified HSC allografts will likely only be clinically useful when combined with lymphocyte addback strategies that provide sufficient engraftment-facilitating and/or GVT cell populations. SyStemix is participating in clinical trials using allogeneic transplantation of purified HSCs and purified-lymphocyte populations for the treatment of patients with acute leukemia or myelodysplastic syndrome. Transduction of the herpes simplex thymidine kinase gene into these lymphocyte populations to render the cells sensitive to gancyclovir is planned so that potential undesirable side effects can be controlled. It is hoped that these studies will lead to a greater understanding of the cell types and interactions involved in hematologic and immune reconstitution, graft-versus-host disease (GVHD), and GVT activity.

The use of allogeneic HSCs for *in utero* transplantation offers a good opportunity to cure certain genetic metabolic disorders while the fetus is still tolerant to HLA disparity. Genetic correction results from the engraftment of normal stem cells that either compete with, suppress, or eradicate ineffective host cells. The low frequency of alloreactive donor T lymphocytes in a purified HSC allograft, compared with a $CD34^+$ cell selected allograft, should contribute to clinical utility.

Immune rejection of solid organ grafts remains a considerable problem that might be prevented by transplanting a purified HSC allograft obtained from the organ donor prior to transplanting the target organ. Animal models indicate that engraftment of purified HSCs prior to solid organ transplant can induce host tolerance and prevent graft rejection (Gandi *et al.,* 1996).

C. GENE-MODIFIED HSCS IN AUTOLOGOUS TRANSPLANTATION

The use of HSCs as targets for somatic gene therapy offers the possibility of a lifetime cure due to the self-renewal potential of HSCs. Preclinical studies at SyStemix have demonstrated that transduction of the transdominant *rev* mutant gene REVM10 into purified HSCs results in inhibition of HIV replication in (1) chronically infected T lymphocytes (Escaich *et al.,* 1995) and (2) T lymphocytes differentiating from HSCs transplanted into SCID-Hu thymus/liver mice (Bonyhadi *et al.,* 1997). SyStemix is participating in a clinical trial using REVM10-transduced HSCs for the treatment of HIV disease and should prove whether clinically relevant genetic modification of HSCs is possible using current technologies. The likelihood of *ex vivo* transduced HSCs engrafting in a normocellular or HIV-related hypocellular bone marrow is unknown, but the trial will evaluate limited myelosuppressive conditioning of the HIV^+ recipient. Once stable, long-term genetic modification of HSCs is possible, diseases resulting from enzyme deficiencies, such as factor VIII, adenosine deaminase, and glucocerebrosidase, might be cured through transduction of the appropriate normal genes into HSCs.

III. REGULATION OF *EX VIVO* CELL PROCESSING

The past decade has witnessed a great surge in the development of new cellular therapy and cell-based gene therapy products, posing a challenge for regulators who are compelled simultaneously to support scientific innovation in the development of new therapeutic modalities and to protect public health and safety. Furthermore, some precedents for harm from *ex vivo* manipulations, donor incompatibilities, and sample cross-contamination have added to the hurdles that must be addressed by both product developers and regulators.

A. U.S. REQUIREMENTS

1. Biological Investigational New Drug

Many of the early cell therapy strategies were conceived in academic research laboratories and were incorporated into complex clinical trials with very little oversight by the Food and Drug Administration (FDA). Traditionally, study approval was granted by the FDA via investigator-sponsored investigational new drug (IND) applications, and studies were governed primarily by institutional internal review boards (IRB). The advent of these new therapies challenged regulators to assess the existing options for regulating these products. The FDA published a draft "Points to Consider" document in August 1991 outlining the regulatory framework for cell and gene therapy and aligned this class of product with other biotechnology products such as monoclonal antibodies. According to this outline, the existing IND, product license application (PLA), and establishment license application (ELA) framework, already in place for biotechnology and pharmaceutical products, would be implemented for somatic cell therapy products, with the FDA Center for Biologics Evaluation and Research (CBER) assuming the role of review. This approach followed the traditional route to commercialization, which required an IND to initiate clinical trials and licensure of the manufacturing process and the manufacturing establishment prior to marketing. Substantial emphasis was placed on control of the final cellular product by the requirement for lot-to-lot release based on approved specifications for purity, identity, potency, and sterility (microbial and endotoxin).

Although the foregoing policy has been in effect since 1991, a 1993 Federal Register notice detailed the extent of FDA statutory authority over products of this type and reinforced the requirement for conformity to existing drug and device regulations, including current GMP (cGMP). The notice clarified requirements for biological products as well as manufacturing processes and devices used in producing these products. During this time frame, some manufacturers argued successfully to FDA that their stem cell isolation technology was a medical device and should be regulated under the device regulations. Because of the perceived complexity of the SyStemix stem cell isolation process, FDA and SyStemix agreed to manage this cell isolation system as a biological product under the IND regulations. In practice, the technical reviews for products of this type are performed by relevant agency scientists, frequently involving specialists in both biologics and devices, making the distinction somewhat moot.

As a result of pressures from the transplant community as well as political pressures to downsize government, FDA (1997) proposed a radical shift in the regulatory paradigm, with the publication on February 28, 1997, of a draft proposal for the regulation of cellular- and tissue-based products, which outlined distinct areas in which no FDA submissions are required and where FDA oversight is limited. With this proposal FDA does not relinquish review and approval of any new device or process used in the "manufacture" of these products but does remove the need for sites to be licensed to perform such manipulations, provided they use FDA-approved processes or devices. The proposal emphasizes FDA's interest in protecting public health and worker safety and acknowledges a need to focus resources to address other emerging technologies such as *ex vivo* cell expansion and gene modification. Details of the current and proposed regulations can be obtained from the foregoing FDA reference, by writing to the Agency, or via the Internet at http://www.FDA.gov.

2. Current Good Manufacturing Practices (cGMP)

Under the current U.S. regulatory framework, all investigational products, whether drugs, biological products, or medical devices, must be manufactured under cGMP to provide assurance of the quality of the product. The regulations defining what constitutes cGMP provide general guidelines to a manufacturer but demand a level of interpretation when applied to new and emerging technologies. It is clearly the manufacturer's responsibility to ensure conformance to cGMP through appropriate control of facilities, personnel, equipment, labels, and records, using precedent established in the traditional pharmaceutical setting. This may, in the case of many new technologies, offer limited guidance. Many companies turn to external consultants for guidance, among who are many former agency inspectors and reviewers, who can provide valuable insights.

3. Foundation for the Accreditation of Hematopoietic Cell Therapy (FAHCT)

The 1993 codification of a very broad FDA authority to regulate all aspects of somatic cell therapy was in keeping with the role assumed by the FDA under Commissioner David T. Kessler. In response to a generic drug scandal in which several drug manufacturers were found criminally liable for fraud, mislabeling, and misbranding, FDA reacted by assuming an expanded role, adding large numbers of investigative staff and performing extensive surveillance of the manufacturing sites. Not surprisingly, FDA's announcement of its control over virtually every aspect of somatic cell therapy drew a response from a number of constituents. Blood banks, large academic medical centers, and transplant clinics argued forcefully that this level of regulatory control was unnecessary, was unwarranted, and would stifle the innovation needed to see this technology evolve to a clinically useful standard. They emphasized that good clinical practices were largely "standard practice," with IRB and protocol review committees in some form in place at virtually every institution. Additionally, tissue banking centers felt that additional controls were not warranted, having become well advanced in the development of internal quality assurance practices and standards based on the blood bank model that evolved several years earlier.

Transplant centers followed this lead and created their own organization (FAHCT, homepage http://www.uchsc.edu//ctrsinst/fahct) charged with developing quality assurance standards as well as an infrastructure to accredit and thereby self-regulate the practice of stem cell transplantation. Their plan included active dialogue with the FDA to establish independent credibility with the agency and to validate their interest in the principle and acceptability of a peer-reviewed process. FAHCT gained the FDA's agreement, in principle, for having its own accrediting body which would perform systems audits on member clinics and accredit those sites that met the standards defined by the organization. These include both organizational standards to ensure appropriate medical control of patient care and laboratory standards for collecting, processing, and storing patient tissues. These standards are noticeably similar to the current quality assurance/good manufacturing practices defined for blood and tissue banks by FDA in the late 1980s. They reflect the wisdom of FAHCT's leadership in choosing to accept the GMP style of accountability and standards and in focusing on the value of maintaining FAHCT enforcement control by using a periodic site inspection procedure.

B. EUROPEAN REQUIREMENTS

While the U.S. regulatory environment has experienced major shifts over the past decade, the European environment was shaped by the device manufacturers' efforts to classify cell isolation products and their associated reagent kits as medical devices. Their success is documented by the fact that the first cell isolation device was registered across Europe (as a CE-marked device) several years prior to obtaining FDA approval. Two additional cell isolation devices have obtained the CE mark in Europe, although regulatory approval in the United States is still pending. Interestingly, there has been fragmented national activity across Europe in the evolution of national regulation of clinical trials and viral safety reviews. The French, largely through the Agence du Graffe, have taken keen interest in the regulation of cellular products as a new product class, with standards yet to be defined. This proposal remains to be addressed in a French legal framework and the status of its impact on the European framework is unclear.

IV. QUALITY ASSURANCE

A. QUALITY CONTROL

It is essential that a system be in place to ensure that all cell processing is handled according to GMP aimed at preventing contamination and preserving integrity and function. Most often, final product quality control (QC) testing is limited by the amount of product available; therefore, a comprehensive quality support system needs to be in place to supplement final product testing. For example, environmental monitoring programs and process validation experiments are crucial if there are insufficient cell numbers to adequately test for sterility on a lot-to-lot basis. To effectively argue that products are sterile, a company must implement a quality assurance program that starts with proper facility and process design and includes equipment validation, raw material testing, documentation control, vendor qualification, personnel training, and internal auditing. Proving that each lot manufactured meets preestablished quality standards consumes a significant percentage of the effort and cost required to manufacture the cellular therapy product. Quality standards will likely resemble those imposed on traditional pharmaceutical products regardless of whether the FDA or individual professional societies hold ultimate responsibility for regulating these products. Ultimately, each individual company is responsible for assuring that its product is both safe and effective for its intended use. Because external guidelines are continuing to evolve, an opportunity still exists for each company to influence the extent and scope of the regulations that will affect them.

An effective and knowledgeable QC testing group is an essential part of a comprehensive quality assurance program. The QC group is generally responsible for incoming raw material testing, environmental monitoring, and final product testing. Raw material specifications are established by the development and manufacturing groups to ensure that each incoming lot of material is acceptable for use. The degree to which a particular raw material is in contact with the final product underlies the extent of testing required. Depending on the company's past history with a particular supplier, raw material testing can involve extensively testing each incoming lot to ensure that the material meets the company's specification. Over time, it may be possible to qualify vendors on the basis of their ability to consistently provide product that meets specification. In such cases, vendor audits and periodic raw material testing can be substituted.

Cellular products must be handled in controlled environments that must be routinely monitored for possible contamination. This testing is expensive but provides an ongoing assessment of both worker and product safety. The development of closed systems for manufacturing will reduce considerably the need and extent of environmental monitoring. Currently, all open-container manipulations occur in a Class 100 biosafety cabinet that is disinfected between each use. Monitoring the air within the biosafety cabinet for microbial contamination during processing can provide assurance of process control. Environmental monitoring, analyzed for trends, is particularly useful in monitoring process or facility aberrations.

Final product testing focuses on testing for cell number, sterility, viability, purity, and biopotency. Since clinical trials with cellular products depend on these assays for clinical decisions, methods used must be appropriately validated.

The sterility and viability assays are most important in early-phase clinical trials, which emphasize patient safety as their endpoint.

B. DOCUMENTATION

The most salient feature of a company's document control procedures should be that all manufacturing and quality control personnel operate from the same set of current effective procedures. All raw material specifications, test methods, standard operating procedures, and manufacturing batch records should be controlled in a quality assurance database that is capable of tracking the current status of each document. A copy of all relevant procedures should be kept in the manufacturing or testing areas. All effective procedures should require interdepartmental sign off and justification for any changes. A history file should be kept on each document so that it is possible to retrospectively determine which procedures were in effect at the time a particular lot of material was manufactured.

C. FACILITIES

Proper facility design contributes greatly to ensuring final product quality. Ideally, the manufacturing area is designed for safely handling potentially infectious agents to protect the products from cross-contamination as well as to protect the worker and surrounding environment from the spread of infectious agents. The air-handling units within the facility should provide single-pass (100% outside) HEPA-filtered air at a slightly negative pressure relative to adjacent corridors. Segregation of activities into separate processing suites, with separate air-handling systems, gowning rooms, material flow patterns, cleaning stations, and waste flow patterns, will minimize cross-contamination. For example, segregating viral vectors used for gene therapy from other cellular purification processes will ensure that one product does not cross-contaminate another. With regard to biosafety, whenever possible, a sterile-disposable material should be used for processing. If reuseable product contact equipment is required, labor-intensive and expensive validation of all cleaning and sterilization procedures is necessary to rule out the likelihood of cross-contamination.

V. PROCESS DEVELOPMENT

A. PROCESS PROOF-OF-CONCEPT LEADING TO PROCESS QUALIFICATION

The primary technical challenge facing those trying to develop devices and procedures for the purification of specific cell types is to be able to manufacture a highly purified and safe product. This mandates a product that is sterile and of sufficient dose or yield to perform the function required of it. Prior to developing a manufacturing process, consideration should be given to several variables: source and composition of starting cell population, its size and shelf life, availability of or ability to manufacture GMP reagents and devices, and the shelf life and potential toxicity of the manufactured product. The resources available, licensing and patent issues surrounding ancillary reagents and devices, and the clinical timetable driving the development process need to be integrated into the overall development plan. Initial process proof-of-concept studies need to evaluate target cell purity, yield, and postprocess biological function within the framework of a potentially feasible process. Once feasibility seems likely, reagents and documents supporting a GMP need to be developed. Quality assurance requirements (for example, QC testing of raw materials and in-process testing) are integrated at this time. Finally, the process reproducibility/systems qualification is evaluated using tissue that is equivalent to the anticipated clinical raw material.

B. SOURCE OF HSCS

Clinically useful numbers of HSCs can theoretically be harvested from a variety of tissues, including bone marrow (BM), cord blood (CB), fetal liver, and mobilized peripheral blood (MPB). Based on the tissue origin, either multiple processes are required or the same basic process is modified at the beginning to normalize the starting population prior to selection. The widespread use of BM as a source of HSCs has largely been supplanted by the use of granulocyte colony-stimulating factor (G-CSF) MPB in autologous and, more recently, in allogeneic transplantation. This is due to the greater numbers of HSCs obtained from MPB, which lead to more rapid engraftment and shorter periods of hospitalization. Fetal liver is rarely used, and the HSC number in cord blood is so small that any purification-associated loss in HSC recovery could significantly compromise its utility.

Any HSC isolation process development effort begins with understanding the specifics of the collection procedures at the clinical sites. Typically for MPB, apheresis is used to harvest HSCs. Differential centrifugation separates the formed elements from plasma and returns the plasma, erythrocytes, and a significant proportion of the platelets and mature granulocytes to the donor. After 4–6 days of G-CSF mobilization (Seong *et al.*, 1997), the apheresis cell product typically contains from 2×10^{10} to 5×10^{10} cells and the HSC frequency ranges from 0.1 to 1.5%. Depending on the degree of HSC mobilization and specific use for HSCs, from one to several aphereses are required to achieve the desired HSC dose. Alternatively, a combination of cytokines and chemotherapeutic agents (for example, G-CSF and cytoxan) is useful for HSC mobilization in patients who have received multiple courses of chemotherapy/radiotherapy prior to mobilization. Each apheresis product after cytokine plus cytoxan mobilization typically contains between 1×10^9 and 10×10^9 cells, with the HSC frequency ranging from 1 to 10%. This method of collecting HSCs is associated with pronounced myelosuppression. Daily flow cytometric monitoring of the HSC frequency in the blood is required to determine when to begin apheresis, making the procedure useful only in cases where patients are recalcitrant to mobilization by G-CSF or where the chemotherapeutic agent provides antitumor benefit. Alternatively, collection of HSCs from BM requires general anesthesia of the donor. The BM harvest consists of aspirating multiple 2- to 20-mL volumes from the iliac crest and sternal bones. A target total nucleated cell number of 1×10^8 to 3×10^8 cells/kg of recipient body weight is typically achieved after collecting a BM volume of approximately 1 L during a single 2- to 3-h procedure.

In either case, significant effort is required by the sponsoring institution to be sure that the collection procedure is being performed optimally and according to preestablished quality standards. Procedures must be written which accommodate individual clinics' needs, while assuring uniform performance. Unique reagents and supplies have to be available at the sites, the technologists require training, data report forms need to be generated, and the sponsoring institution should conduct occasional audits of the sites' laboratory facilities.

Centralized cell-processing facilities might be able to justify expensive facilities, equipment, and dedicated staff if high volume throughput is achieved or if clinical practice or the regulatory environment demands a high degree of manufacturing process control (Dooley *et al.,* 1993). Unlike local cell-processing laboratories, centralized cell-processing facilities must contend with long-distance shipment of the raw material. Significant development effort was required to optimize transport medium to maintain optimal viability of the HSCs during shipment to SyStemix facilities. We recommend that all MPB or BM be diluted to a final concentration of less than 5×10^7 cells per mL with a tissue culture medium, human albumin and heparin. Once diluted, the cells can be shipped at 20–24°C in an insulated box in which HSC viability is maintained for 24–36 h. Extensive experimentation was required to optimize the shipping conditions, and an external vendor was contracted to validate that the insulated shipping container could maintain a steady temperature of 20–24°C for up to 36 h regardless of the outside temperature. Since airline pilots have the right to refuse carrying the cell products, detailed package labeling that complies with Federal shipping regulations and airline practice is critical to preventing delay in shipment.

C. SYSTEMIX'S ORIGINAL HSC MANUFACTURING PROCESS

Our clinical program began with a manufacturing process that has since been substantially improved to increase the number and quality of the HSC infusion product. Our objective was to develop a process to manufacture a HSC infusion product from MPB derived from multiple myeloma patients treated with GM-

CSF and cytoxan (Reading *et al.,* 1994). Counterflow centrifugal elutriation and phenylalanine methyl ester (PME) lysis were developed as the presort processes to eliminate erythrocytes and mature myeloid cells prior to using a high-speed, fluorescence-activated clinical cell sorter to purify $CD34^+Thy\text{-}1^+Lin^-$ (lineage negative) cells. A cell fraction depleted of erythrocytes and mature granulocytes was collected in a large Sanderson elutriation chamber in a Beckman J6-MI centrifuge. This mononuclear cell fraction was incubated for 45 min at room temperature with an equal volume of 5 mM PME dissolved in tissue culture medium. This granulocyte- and monocyte-lysed cell fraction was centrifuged through isotonic 20% Percoll, separating the cellular debris from the lysed myeloid cells within the supernatant.

Purification of $CD34^+Thy\text{-}1^+Lin^-$ HSCs from the erythrocyte- and myeloid-depleted cell fraction was achieved by labeling the cells with four monoclonal antibodies (mAbs) reacting against CD14, CD15, CD34, and Thy-1 (CDw90) antigens and using SyStemix's proprietary Clinical Cell Sorter (CCS) to select for $CD34^+CD14^-CD15^-Thy\text{-}1^+$ cells. All mAbs were directly conjugated to fluorophores, with the exception of anti-Thy-1, which was conjugated with biotin so that detection of Thy-1 binding could be enhanced through a streptavidin, phycoerythrin (PE)–biotin amplification sandwich. The CCS was set up to exclude particles $<6\ \mu m$ in diameter, exclude Lin^+ cells, and select for $CD34^+Thy\text{-}1^+$ cells. Cells were maintained at 2–8°C in DPBS (Dulbecco's phosphate-buffered saline) sheath fluid during sorting. The purified HSC product was tested for purity, viability, biopotency, and sterility and then cryopreserved in a 2-mL volume of tissue culture medium containing dimethyl sulfoxide (DMSO), Hetastarch, and human serum albumin (HSA), in polypropylene vials using a controlled rate freezer. To minimize the possibility of microbial cross-contamination during storage in liquid nitrogen, the vials were heat-sealed within CryoFlex tubing. The HSC infusion product was shipped to the clinical sites in a nitrogen dry shipper and, upon arrival, placed in a storage rack in liquid nitrogen until infusion. When the patient is ready to receive the HSC product, the vials are removed from liquid nitrogen and thawed rapidly at 37°C. The cells are diluted in a proprietary tissue culture medium, drawn into a syringe, and reinfused via a central venous catheter. Trypan Blue dye exclusion is used with a small sample to determine the postthaw viable cell dose and overall cell viability.

D. SYSTEMIX' CURRENT HSC MANUFACTURING PROCESSES

SyStemix is currently using what is essentially a single manufacturing process to purify three different HSC products (Fig. 10.1).

1. Autologous or allogeneic HSC derived from MPB: Autologous or allogeneic HSC products derived from MPB are under clinical development as support

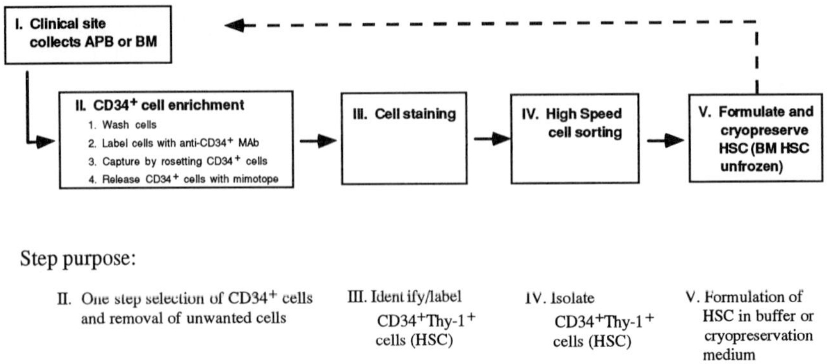

FIGURE 10.1. Overview of SyStemix's HSC isolation process.

for high-dose chemotherapy/radiotherapy in clinical trials in multiple myeloma, NHL, and breast cancer (autologous) and acute leukemia and myelodysplastic syndrome (allogeneic). All HSC products are cryopreserved.

2. *Gene-modified autologous HSC derived from MPB:* An autologous gene-modified HSC product, manufactured by transducing the HSC with a mutant *rev* gene to inhibit retroviral replication prior to cryopreservation, is under clinical development for patients with HIV. The manufacturing process was originally that presented in Fig. 10.1, with isolation and transduction of HSCs followed by cryopreservation. The current process isolates and transduces a CD34$^+$ cell fraction, thereby increasing HSC yield through omission of the high-speed sort step; the presence of CD34$^+$Thy$^-$ and CD34$^-$ cells during the gene transduction process does not appear to reduce the transduction efficiency or yield of HSCs.

3. *Fresh (noncryopreserved) allogeneic HSC derived from BM:* A study aimed at correcting severe immunodeficiency and chronic granulomatous disease by transplanting allogeneic BM-derived HSCs *in utero* is under clinical development. The process for isolating HSCs from BM is identical with that for MPB except a buffy coat is prepared as the isolation process initial step. Because of potential DMSO toxicity to the fetus, the purified HSCs are held under refrigeration in buffer for up to 24 h prior to transplantation rather than being cryopreserved.

1. Shipping Cells

As reported in the feasibility study section, the MPB cells are shipped to SyStemix for processing. Special attention is given to shipping of human cells derived from HIV$^+$ donors. The International Air Transportation Authority (IATA) regulations-concerning the passenger aircraft shipment of unfrozen HIV-infected products require leakproof, "double" containment of cells with less than 500 mL of tissue per container (class 6.2, subsection 4.4). To comply with this regulation, the diluted MPB product is split equally between two 500-mL bags, which are then placed into separate rigid plastic screw-capped containers within a SafTpak container (Edmonton, Canada). In addition, sufficient material to absorb a possible leak is a required inclusion in the packaging. A biohazard-approved nitrogen dry shipper is used to return the transduced HSC product from HIV^+ donors to the clinical sites for reinfusion. The 15-mL reinfusion product again meets the IATA requirement for passenger aircraft shipment of tissue from HIV^+ subjects.

2. Preparation of Cells for Fluorescence-Activated Cell Sorting

The predictable mobilization kinetics experienced with G-CSF administration underlies the choice of this method for harvesting HSCs for the current oncology and HIV clinical trials. The elutriation/chemical lysis process has now been replaced with a $CD34^+$ cell selection system (Hohaus *et al.,* 1997) because, compared with mobilization by GM-CSF and cytoxan, the typical G-CSF-mobilized blood product is composed of relatively few (approximately 35%) PME-sensitive myeloid cells among a relatively large number ($(2-5) \times 10^{10}$) of nucleated cells. The original procedure would have had only a modest cell-debulking effect, making sorting virtually impossible.

The new presort process consists of four steps (Hohaus *et al.,* 1997) (Step II., Fig. 10.1): First, the MPB cells are labeled with an anti-CD34 mAb in excess human immunoglobulin to prevent nonspecific binding. These sensitized cells are thoroughly washed. Second, MPB cells are incubated with paramagnetic Dynal beads coated with sheep anti-mouse IgG, whereby $CD34^+$ cells form rosettes with Dynal beads. Third, the $CD34^+$ cells rosetted with Dynal beads are drawn to a magnet and $CD34^-$ cells drain from the system. Fourth, $CD34^+$ cells are released from the Dynal beads using a peptide mimotope that competes with the mAb binding site of the CD34 molecule. The selection process uses a DPBS washing buffer containing HSA and sodium citrate (6.8 mM) and requires approximately 5 h. Processing experience at SyStemix using G-CSF mobilization in oncology patients to date ($n = 42$) has resulted in a $CD34^+$ cell fraction with a median purity of 73.4%, representing an average 245-fold enrichment of the $CD34^+$ cells and a median $CD34^+$ cell recovery of 53.0%. Higher $CD34^+$ cell purity (88%) and recovery (78%) are achieved from processing G-CSF-mobilized tissues derived from normal donors ($n = 50$), probably due to the higher $CD34^+$ cell frequency in the normal donor MPB tissues. Cells are prepared for sorting as reported in the feasibility study section except that the anti-CD14 and anti-CD15 mAbs are omitted from the labeling cocktail; it was found that <2% of $CD34^+Thy^+$ derived from MPB express CD14 and/or CD15, rendering these reagents superfluous to the HSC isolation process.

3. Clinical High-Speed, Fluorescence-Activated Cell Sorting

The current process does not differ from that described in the feasibility study section except the relatively high $CD34^+Thy^+$ frequency in the post-$CD34^+$ cell fraction (ranging from 25 to 70%) has permitted a lowering of the sorter coincidence setting from 2.0 to 1.0. This has resulted in an increased HSC yield.

4. Cryopreservation

The HSC products for the oncology patients are cryopreserved as described in the feasibility study section. The gene-modified HSCs for use in HIV$^+$ subjects are frozen in cryopreservation bags using the same cryoprotectant medium and freezing program as used for freezing in vials. After the addition of refrigerated cryoprotectant to the transduced HSCs, the cells are infused into a chilled bag. The bag is then heat sealed within a secondary bag before freezing. The secondary bag serves four functions: (1) to minimize the risk of exposure of any handler of the HSC product to potential pathogens, (2) to minimize the risk of cross-contamination of pathogens during storage in liquid nitrogen, (3) to prevent the loss of cells during thawing in the event of a leak in the Cryocyte-50 bag, and (4) to permit a simple thaw–hang process at the clinical site.

5. Transplantation

The process of thawing and infusion of the cryopreserved HSC products has not changed from that used for the original feasibility study, which is to rapidly thaw, dilute, and infuse. Cryopreservation of the gene-transduced HSCs in bags greatly simplifies the thawing and reinfusion; the cryopreservation bag containing cells is thawed in a water bath while sealed within the secondary bag. Once thawed, the secondary bag is cut open and the cryopreservation bag is removed. Before reinfusion, cells are diluted with two volumes of saline and an aliquot is removed for assessment of viable cell dose using the Trypan Blue exclusion test.

E. FUTURE PROCESS IMPROVEMENTS

Because of the ever present but small risk of contamination of the HSC product with adventitious pathogens, we are working to eliminate the use of animal proteins from any part of our manufacturing process. Studies are currently underway in an attempt to replace these reagents with autologous plasma, serum, or recombinant protein.

The ability to develop a closed-system manufacturing process will have a major impact on product and worker safety. Since it will streamline and make safer the thaw–infusion process, our experience in using the cryopreservation bag for storing the gene-modified HSC product has led us to consider the integration of this container into the process for storing the HSC products. Automated and functionally closed-cell washing machines and automated CD34$^+$ cell selection devices are also being evaluated.

REFERENCES

Archimbaud, E., Philip, I., Michallet, M., *et al.* (1997). Selected autologous peripheral blood $CD34^+Thy\text{-}1^+Lin^-$ hematopoietic stem cell (HSC) transplantation in multiple myeloma: A European study. *Blood* **90,** 394a.

Baum, C., Weissman, I., Tsukamoto, A., *et al.* (1992). Isolation of a candidate human hematopoietic stem-cell population. *Proc. Natl. Acad. Sci. USA* **89,** 2804–2808.

Belch, A., Bergsagel, P., Szczepek, A., *et al.* (1994). $CD34^+$ B cells in the blood of patients with multiple myeloma express clonotypic IgH sequences. *Blood* **84,** 385a.

Bergsagel, P., Belch, A., and Pilarski, L. (1994). The blood B cells and bone marrow plasma cells in a patient with multiple myeloma include cells with the same N-ras mutation. *Blood* **84,** 524a.

Bertrand, I. F., Billips, L., Burrows, P., *et al.* (1997). Ig DH gene segment transcription and rearrangement before surface expression of the pan-B cell marker CD19 in normal human bone marrow. *Blood* **90,** 736–744.

Bonyhadi, M., Moss, K., Voytovich, A., *et al.* (1997). RevM10-expressing T cells derived *in vivo* from transduced human hematopoietic stem-progenitor cells inhibit human immunodeficiency virus replication. *J. Virol.* **6,** 4707–4716.

Cooley, H., Snowden, J., Grigg, A., *et al.* (1997). Outcome of rheumatoid arthritis and psoriasis following autologous stem cell transplantation for hematologic malignancy. *Arthritis Rheum.* **9,** 1712–1715.

Davi, F., Faili, A., Gritti, C., *et al.* (1997). Early onset of immunoglobulin heavy chain gene rearrangements in normal human bone marrow $CD34^+$ cells. *Blood* **90,** 4014–4021.

Dooley, D., Peetoom, F., Mead, J., *et al.* (1993). Autologous transplant programs in community hospitals: Role of centralized stem cell processing laboratories. *J. Hematother.* **2,** 87–92.

Escaich, S., Kalfoglou, C., Plavec, I., *et al.* (1995). RevM10-mediated inhibition of HIV-1 replication in chronically infected T cells. *Hum. Gene Ther.* **6,** 625–634.

Gahn, B., Schafer, C., Neef, J., *et al.* (1997). Detection of trisomy 12 and Rb-deletion in $CD34^+$ cells in patients with B cell chronic lymphocytic leukemia. *Blood* **89,** 4275–4281.

Gandi, K., Morrison, S., Reitz, B., *et al.* (1996). Mice reconstituted with purified allogeneic hematopoietic stem cells are tolerant of solid organ grafts. *Blood* **88,** 594a.

Gazitt, Y., Reading, C., Hoffman, R., *et al.* (1995). Purified $CD34^+Lin^-Thy^+$ stem cells do not contain clonal myeloma cells. *Blood* **86,** 381–389.

Hohaus, S., Pforsich, M., Murea, S., *et al.* (1997). Immunomagnetic selection of $CD34^+$ peripheral blood stem cells for autografting in patients with breast cancer. *Br. J. Haematol.* **97,** 881–888.

Kohler, G., Brugger, W., Schaeffer, H., *et al.* (1994). Expression of CD34 stem cell antigen on tumor cells. *Blood* **84,** 497a.

Macintyre, E., Belanger, C., Debert, C., *et al.* (1995). Detection of clonal $CD34^+19^+$ progenitors in bone marrow of BCL2-IgH-positive follicular lymphoma patients. *Blood* **86,** 4691–4698.

Marmont, A. (1997). Stem cell transplantation for severe autoimmune disorders, with special reference to rheumatic disease. *J. Rheumatol.* **48,** S13–S18.

Negrin, R., Tierney, K., Stockerl-Goldstein, K., *et al.* (1997). Rapid hematopoietic engraftment following transplantation of purified $CD34^+Thyl^+$ stem cells in patients with metastatic breast cancer. *Blood* **90,** 593a.

Reading, C., Sasaki, D., Leemhuis, T., *et al.* (1994). Clinical scale purification of $CD34^+Thy^+Lin^-$ stem cells from mobilized peripheral blood by high speed fluorescence-activated cell sorting for use as an autograft for multiple myeloma patients. *Blood* **84,** 399a.

Seong, C., Durett, A., Mirza, N., *et al.* (1997). Mobilization kinetics of $CD34^+$Thy-1dim progenitor cells during recombinant human granulocyte-colony-stimulating factor administration in normal donors. *Transfusion* **37,** 406–410.

Szczepek, A., Bergsagel, P., Axelsson, L., *et al.* (1997). $CD34^+$ cells in the blood of patients with multiple myeloma express CD19 and IgH mRNA and have patient-specific IgH VDJ gene rearrangements. *Blood* **89,** 1824–1833.

Takishita, M., Kosaka, M., Goto, T., *et al.* (1994). Cellular origin and extent of clonal involvement in multiple myeloma: Genetic and phenotypic studies. *Br. J. Haematol.* **87,** 735–742.

Wessman, M., Schnell, J., Van den Berg, D., *et al.* (1995). Fluorescent detection of metastatic breast cancer cells in bone marrow and mobilized peripheral blood by immunofluorescence and by high speed flow cytometry. *Blood* **86,** 233a.

11

Cell Separation

Robert E. Nordon and
Klaus Schindhelm

Graduate School of Biomedical Engineering
University of New South Wales
Sydney, New South Wales 2052, Australia

I. INTRODUCTION

The aim of cell separation is to either positively select a target cell type (enrichment) or remove an unwanted cell type (depletion). For example, the role of cell enrichment may be to debulk the cell collection and so reduce the quantity of reagents used in downstream processes such as gene transfer, cryopreservation, and cell expansion. The removal of potentially pathogenic cell subsets (e.g., tumor cells from autologous stem cell transplants) is achieved either by positive selection of normal cells or by direct depletion of the unwanted phenotype. This review of cell separation methods will include an analysis of their physical basis and performance. Device performance is dictated by clinical requirements, particularly where cell dose determines the success or failure of therapy. Therefore this chapter also reviews the application of cell separation processes to clinical problems.

Cells can be separated by physical properties such as sedimentation rate or, more selectively, from the specificity of monoclonal antibody binding to cell surface receptors as defined by CD (cluster of differentiation) nomenclature. The selectivity of separation methods relates to the physical mode of separation (Fig. 11.1), whether it be by centrifugal elutriation (cell density and size) or fluorescent-activated cell sorting (light scatter and surface, cytoplasmic, or nuclear molecular markers). In addition to selectivity considerations, device separation performance is determined by the rate of cellular throughput, cell enrichment factor, yield, and purity.

II. CELL SEPARATION METHODS

A preliminary discussion of performance parameters is justified as it will facilitate quantitative comparisons of cell separation technologies and assist in identifying those physicochemical factors that influence separation efficiency. A classic "black box" representation of a cell separation device can be used to define standard performance parameters (Fig. 11.2). It is assumed that separation is a batch process with cells denoted as either (marker) positive or negative. During the depletion phase positive cells are captured while negative cells pass freely through the device. Positive cells may then be recovered by an enrichment step. The following parameters are commonly used to report cell separation device performance:

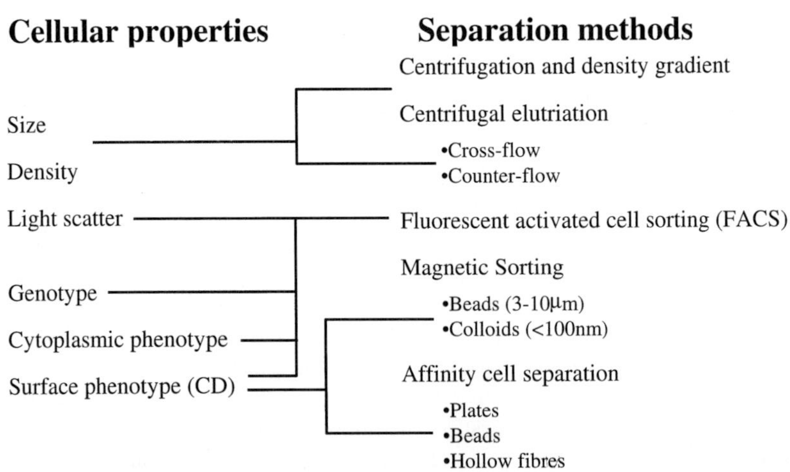

FIGURE 11.1 Cell separation methods.

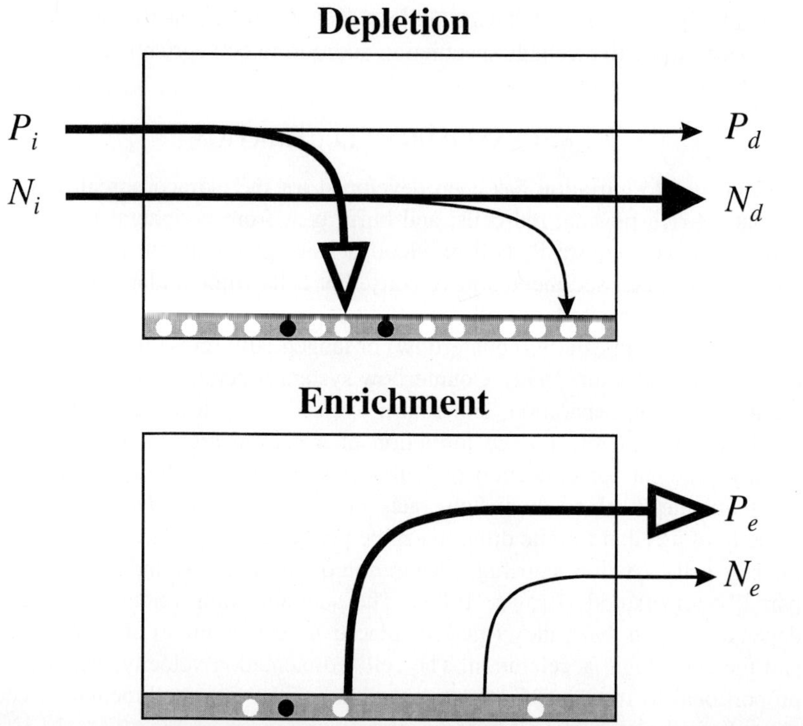

FIGURE 11.2 "Black box" representation of cell separation techniques. See text for definition of symbols.

$$\text{Depletion} = \frac{p[\text{positive cell recovery}]}{p[\text{negative cell recovery}]} = \frac{P_d/P_i}{N_d/N_i}$$

$$\text{Enrichment} = \frac{p[\text{positive cell capture and release}]}{p[\text{negative cell capture and release}]} = \frac{P_e/P_i}{N_e/N_i}$$

$$\text{Yield} = p[\text{positive cell capture and release}] = P_e/P_i,$$

where $p[\ \ ]$ is the probability, P and N refer to positive and negative cell numbers, and the subscripts i, e, and d refer to input, enriched, and depleted fractions, respectively. Purity is often quoted without reference to the preenrichment (input) incidence of positive cells:

$$\text{Purity} = \frac{P_e}{P_i + N_i} = \frac{\text{Enrichment}}{\text{Enrichment} + (N_i/P_i)}.$$

It is clear from this expression that purity will depend on the input incidence of positive cells in addition to device enrichment performance. Therefore a more complete specification of device performance requires that the preenrichment incidence of positive cells is specified in addition to cell enrichment purity.

A. CENTRIFUGAL ELUTRIATION

Centrifugal elutriation has been developed for the extracorporeal harvesting of platelet-rich plasma, red cells, and buffy coat from peripheral blood. These systems do not require Percoll or Ficoll density gradients but rely on relative differences in cell sedimentation velocity. For cells with similar densities, these separations are related to cell size (Sharpe, 1988). The centrifugal force is directed either oppositely (counterflow) or tangentially (cross-flow) to the direction of flow (Wagner, 1992). Counterflow systems operate semicontinuously by sequential draw, separation, and return cycles. Cross-flow techniques operate continuously and rely on the formation of a packed red cell bed to create a density gradient for collection of leukocytes. Continuous-flow centrifugal devices operate at blood-draw flow rates up to 100 ml/min. The sedimentation velocity of a cell at infinite dilution can be predicted from Stokes' law; however, the physical basis for centrifugal separation of whole blood and marrow is only partially understood (Brown, 1989). The sedimentation velocity of a cell is dependent on its buoyancy (mass displaced by cell volume) and viscous drag and the centrifugal acceleration. The cell sedimentation velocity, V_c, is directly proportional to the centrifugal acceleration, g, where the sedimentation coefficient, S_c, is the proportionality constant.

$$V_c = -gS_c.$$

The sedimentation coefficient depends on the cells' relative density and size (buoyancy) and the viscosity of the surrounding fluid. An estimate of this parameter using Stokes' Law assumes that cells are perfect spheres, have a Reynolds number less than one, and are surrounded by a Newtonian fluid (Brown, 1989).

$$S_c = \frac{2(\rho_c - \rho)a^2}{9\mu},$$

where ρ_c is the cell density, ρ is the fluid density, a is the cell radius, and μ is the fluid viscosity. In the case where a cell has a density less than that of surrounding fluid ($S_c < 0$), the cell is displaced by the denser surrounding fluid and migrates in the opposite direction to the applied centrifugal force.

Table 11.1 shows sedimentation rates for erythrocytes, granulocytes, lymphocytes, and platelets when at low concentration (Brown, 1989). Granulocytes and lymphocytes have sedimentation coefficients greater than red blood cells. Formation of a buffy coat layer containing leukocytes over the packed red cell mass after centrifugation may seem paradoxical; however, the packed red cell bed has an average density approaching that of a red cell (1.1 g/cm^3), greater than the

TABLE 11.1 Sedimentation Properties of Cells at Infinite Dilution at 37°C in Citrate Anticoagulated Plasma

	Erythrocytes	Platelets	Lymphocytes	Granulocytes
Frequency (μl^{-1})	5,000,000	230,000	2500	4500
Radius (10^{-4} cm)	2.7–2.9	1–1.2	4.5–9	5.0–7.5
Volume fraction (%)	40–50	0.1–0.2	0.1–0.8	0.2–0.8
Density (g/cm^3)	1.1	1.05	1.06	1.08
Sedimentation coefficient (10^{-9} s)	90–100	4–7	130–530	230–540

density of granulocytes or lymphocytes (1.06–1.08 g/cm^3). Therefore leukocytes are displaced by the surrounding red cell mass and emerge at the red cell–plasma interface to form the buffy coat layer. Cross-flow centrifugal elutriation uses the density gradient created by the packed red cell mass to separate platelets, mononuclear cells, and granulocytes from whole blood. Counterflow devices have a tapering cross-section along the flow path, which creates a fluid velocity gradient along the flow chamber. Thus, a cell's sedimentation velocity and device residence time depend on its sedimentation coefficient.

Platelets are collected relatively free of leukocyte contamination, with collection efficiencies around 50% (Rock *et al.,* 1983). Higher collection efficiencies can be achieved, but the platelet product begins to show significant leukocyte contamination (Rock *et al.,* 1985). Leukocyte contamination is normally around 0.1% (Anderson *et al.,* 1991), which may be enough to cause alloimmunization. Lymphocytes can be collected with near 100% collection efficiencies using ordinary anticoagulants (AuBuchon *et al.,* 1986) whereas granulocyte collections have only been achieved with strong rouleauxing (red cell clumping) agents, and even then tend to be limited to efficiencies of about 50% (Hester *et al.,* 1983).

Hematopoietic progenitors from pooled bone marrow are collected efficiently with negligible loss of mononuclear cells, CD34$^+$ cells, or CD33$^+$ cells with significant depletion of harvest volume, mature myeloid cells, platelets, and red blood cells. The hematocrit of mononuclear cell concentrates ranges from 5 to 45% and depends on the elutriation protocol and device (Davis *et al.,* 1993; Pierelli *et al.,* 1993; Hester *et al.,* 1995). This level of red cell depletion may be inadequate in many cases of allogeneic marrow transplantation, where the presence of ABO-incompatible red blood cells may lead to hemolytic transfusion reactions (Gremba *et al.,* 1994). The harvesting of peripheral blood stem cells is emerging as a central application for centrifugal elutriation. The main advantages of harvesting peripheral blood stem cells rather than bone marrow include the avoidance of an operative harvest procedure and the shorter duration of neutropenia and thrombocytopenia following high-dose chemotherapy. Apheresis is the preferred harvest procedure if the yield of stem cells from patients' marrow is limited because of marrow fibrosis, prior pelvic radiotherapy, or hard bones (Klumpp, 1992; Korbling *et al.,* 1994). Although apheresis procedures in healthy patients entail minimal risks, repetitive leukapheresis in patients with active malignancy is associated with a spectrum of complications. A retrospective study of 75 consecutive patients with hematological malignancy or breast cancer undergoing peripheral blood stem cell harvesting had a significant incidence of complications (Goldberg *et al.,* 1995). The main complications were catheter occlusion requiring thrombolytic therapy (50%) and catheter infection (16%). Transfusion support following apheresis was required for anemia (31%) and transient thrombocytopenia (15%), especially if stem cells had been mobilized using chemotherapy. Insertion of central venous catheters for adequate venous access was associated with a low incidence of pneumothorax or superior vena cava perforation (2%).

Centrifugal elutriation will continue to play a central role as a large-scale harvesting procedure for *ex vivo* cell therapies. Even though the level of selectivity is limited, this high-throughput, extracorporeal technology can provide the cell types required for current cell therapies such as adoptive cellular immunotherapy and stem cell transplantation. Downstream processing of leukocyte products will entail further cell selection using monoclonal antibody-based techniques followed by other procedures such as cryopreservation or manipulation of cells in culture.

B. FLUORESCENT-ACTIVATED CELL SORTING

Fluorescent-activated cell sorting (FACS) is without question one of the most selective cell separation methods available (Fig. 11.1). Cells are labeled with fluorescent marker molecules, individually analyzed (up to 30,000 cells/s), and electronically sorted on the basis of light scatter properties (forward and side scatter) and fluorescence (up to five colors). Multiparameter data are displayed as a series of uni- and bivariate frequency histograms. Cell subsets are defined by selecting gating regions on histograms and combining them using Boolean logic. These regions serve to define gates for fluorescent-activated, electrostatic droplet sorting (Shapiro, 1988). Flow cytometry has enabled extensive development of the CD system for classification of cell phenotype, which has led to correlating specialized cellular function with cell surface phenotype.

Even though FACS has the highest selectivity, sensitivity, and precision of all separation methods, cellular throughput is limited by the rate of electrostatic droplet sorting. Purity and yield are related to the number of cells within a droplet or sort package.

Poisson statistics have been used to predict cell purity and yield for a given cell event rate assuming that system "dead time" (i.e., time taken for analysis and sort decision) is negligible (Nordon, 1994). If the sorting electronics is set to sort any droplet containing positive events (100% yield, Fig. 11.3), then

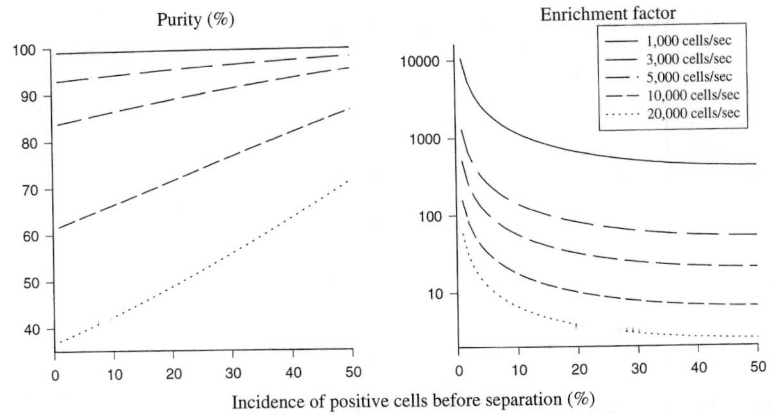

FIGURE 11.3 Sorting for 100% yield. Influence of cell event rate on FACS purity and enrichment (Poisson statistical model). The frequency of droplet formation is 30 kHz, with sort decisions made at a frequency of 10 kHz. Purity is directly related to the initial incidence of positive cells. Purity and enrichment factor decline at higher cell event rates.

purity is inversely related to the cell event rate. Another sort strategy is to collect only positive cells (100% purity, Fig. 11.4), with yields decreasing at higher cell event rates. Therefore the main parameters that influence cell purity, yield, and throughput are droplet formation rate, cell event rate, and sorting logic. Most conventional sorters have a droplet generation rate of 30,000 Hz, each sort packet consisting of three droplets. Yield and purity will be adequate provided that the cell event rate is less than 4000 cells/s. Most commonly, FACS is used as a non clinical research instrument; however, a clinical high-speed FACS has been developed by SyStemix, Inc., for isolation of transplantable human hematopoietic stem cells (van den Engh and Stokdijk, 1989). A full description of SyStemix's facility and quality control procedures is given in Chapter 10. Their design is based on the concepts originally developed by van den Engh and Stokdijk (1989), incorporating parallel processing to reduce system dead time (5 μs), high sheath pressures (40 psi), and high drop formation frequencies (60 kHz). The system operates at rates that are about five times faster than those of conventional commercially available sorters (20,000 events/s). Even though the sort rate is much faster, a preenrichment step is still required to process human stem cell transplants (10^{10} cells).

C. IMMUNOAFFINITY CELL SEPARATION

Immunoaffinity cell separation is based on an immunoaffinity substrate created by immobilization of specific ligands such as monoclonal antibody, lectins, or other binding domains to a material with little intrinsic cell affinity. Cells selectively attach to the substrate via bonds between cell-surface molecules and

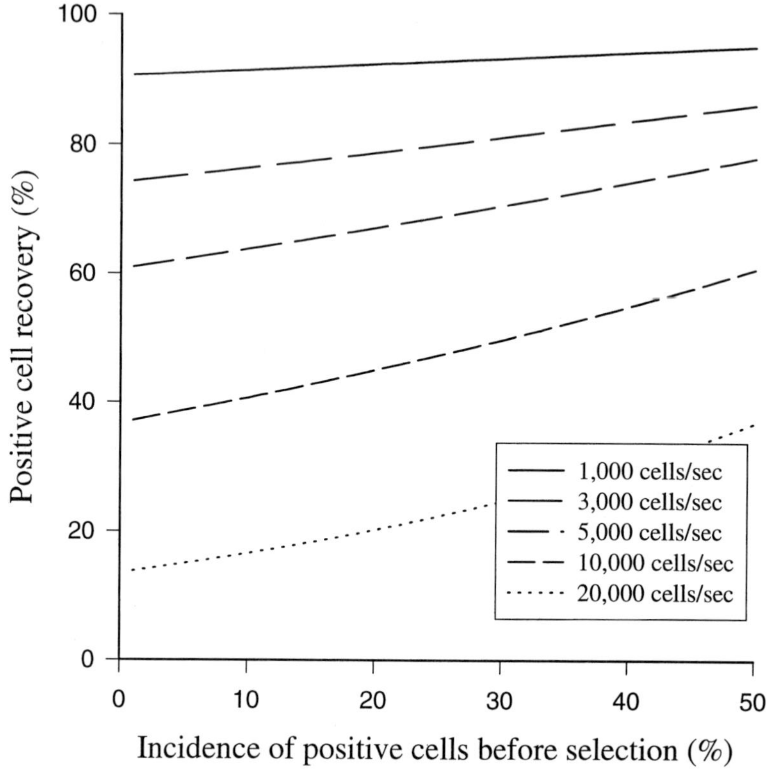

FIGURE 11.4 Sorting for 100% purity. Influence of cell event rate on positive cell recovery. Sort decisions are aborted if there are also negative cells in the sort packet. Yield decreases with increasing cell event rate.

ligand bound to the solid-phase support. Captured cells are detached using shear stress or chemical agents that disrupt physically or chemically the cell–substrate bond.

1. Physicochemical Factors That Influence Separation Performance

The efficiency of cell separation using immunoaffinity techniques is related to the physical processes of cell transport, binding, and release. Transport processes for cells include convection, sedimentation (Brown, 1989), inertial migration (Cox and Brenner, 1968), and shear-induced diffusion (Davis and Leighton, 1987). Unless acted upon by an external force, cells will have a trajectory that follows fluid streamlines. External forces that can influence the trajectory include gravity, centrifugal acceleration, or magnetic force in the case where cells are magnetically labeled. In addition to these body forces, close approach to other cells or the affinity substrate will influence the cell trajectory. The rate of diffusion of particles is inversely related to their radii (Stokes–Einstein equation) (Bird *et al.,* 1960a), and in contrast to solute separations, where mass transfer is assisted by diffusion, cells greater than 5 μm in diameter have negligible Brownian motion. Therefore, efficient cell recovery is reliant on the generation of fluid shear at surfaces where cells are deposited or attached.

Once cells interact with the immunoadsorbent, the relative levels of specific and nonspecific affinity will determine selectivity. Substrates, which have a low level of nonspecific cell adsorption, include bacteriological-grade polystyrene (Ellingsen *et al.,* 1990), polyacrylamide (Braun *et al.,* 1982), cross-linked polysaccharides (Sepharose) (Schrempf-Decker *et al.,* 1980), and cellulose (Nordon *et al.,* 1996). The constituents of the cell separation buffer will also influence the level of nonspecific adsorption. Albumin reduces nonspecific binding by poorly understood mechanisms. Random orientation of immobilized monoclonal antibody may lead to attachment of cells that express Fc receptors. Binding to Fc receptors can be blocked by incubation of cells with immunoglobulin. Calcium-dependent protein–protein interactions are blocked by the use of Ca^{2+}-free buffers or Ca^{2+}-chelating agents.

Chemical agents cleaving specific bonds without deleterious effects on cell viability are summarized in Table 11.2. These dissociation mechanisms include reduction of disulfide bridges (Grimsley *et al.,* 1993), enzymatic proteolysis (Civin, 1992; Sutherland *et al.,* 1992), and competitive binding using peptides (Tseng-Law *et al.,* 1995) or polyclonal antibody (Rasmussen *et al.,* 1992). The use of two consecutive selection processes such as immunoaffinity adsorption followed by targeted dissociation of antigen positive cells will significantly increase purity. The likelihood of cell attachment depends on the rate of bond formation and the cell detachment force. A number of models which attempt to relate ligand-association kinetic parameters to adhesion strength and fluid mechanics have been developed (Hammer and Lauffenburger, 1987; Cozens-Roberts *et al.,* 1990; Saterbak *et al.,* 1993). These theories are based on the assumptions initially proposed by Bell (1978). Application of bond force, f, results in an increase in the reverse rate constant, k_r:

TABLE 11.2 Dissolution of Antigen–Antibody Bonds

Agent	Mechanism	Reference
Cleland's reagent	Reduction of disulfide bridge	Grimsley *et al.*, 1993
Pasteurella haemolytica glycoprotease	Proteolysis of O-glycosylated-rich domains	Sutherland *et al.*, 1992
Chymopapain	Proteolysis of cysteine-rich domains	Civin, 1992
Peptide release	Competitive binding to idiotype	Tseng-Law *et al.*, 1995
Polyclonal anti-mouse Fab	Competitive binding to Fab	Rasmussen *et al.*, 1992

$$k_r = k_r^0 \exp\left[\frac{\lambda f}{k_b T}\right],$$

where k_r^0 is the reverse rate constant for unstressed bonds, λ is the range of interaction of the bond, k_b is the Boltzmann constant, and T is the temperature. At equilibrium, ligand–receptor bonds at the attachment interface balance hydrodynamic forces, which exert torque and drag on the cell. This theory has been difficult to prove directly, though all experimental systems show a strong dependence of cell attachment on fluid shear stress.

The shear stress at the immunoadsorbent surface is a critical control parameter and is directly proportional to the velocity gradient (shear rate) and the viscosity of the fluid (Bird *et al.*, 1960b).

$$\text{Shear stress} = \tau = -\mu \frac{dv}{dy},$$

where y is distance perpendicular to the surface, v is the fluid velocity parallel to the surface, and μ is the fluid viscosity. The forces generated by fluid flow over an attached cell are calculated approximatively by assuming that the cell is a hard sphere in contact with a plane surface in a linear shear flow field (O'Neill, 1968):

$$\text{Drag} = 1.7 \times 6\pi a^2 \tau$$

$$\text{Torque} = 0.94 \times 8\pi a^3 \tau.$$

The shear force which is transmitted to the sphere by fluid viscosity is directly related to the wall shear stress, τ, and the radius of the sphere, a ($Re < 1$). If cells are attached by relatively few receptor–ligand interactions, then the level of wall shear stress will determine the probability of cell detachment. Cozens-Roberts and Lauffenburger (1990) studied the detachment of marker-coated latex beads from ligand-coated glass surfaces and found that the median shear stress of detachment was directly proportional to marker and ligand surface density. In the context of cell separation, the shear stress to recover depleted and enriched cell fractions determines cell enrichment yield and purity (Nordon *et al.*, 1994). Washing and positive cell recovery steps require controllable and reproducible levels of shear stress.

An even distribution of flow throughout the adsorbent matrix is required to utilize fully immunoadsorbent surface area and to prevent sequestration of cells in regions of flow stasis. The Ceprate SC™ system (CellPro Inc., Seattle, WA) uses a fluidized bed reactor design to control cell attachment and recovery from beads. Gentle magnetic stirring of the suspension of beads and cells increases the probability of cell–bead interaction while ensuring that surface shear stress is low so that once bound, cells will remain attached to beads. Higher flow and stirring rates are used to generate greater shear stress to detach positive cells. For a hollow-fiber system, surface fluid shear stress is directly proportional to the device flow rate, and headers are designed to distribute flow uniformly to the hollow-fiber array (Nordon and Schindlielm, 1997).

These general design principles are summarized in Table 11.3. The specific application and performance of flat-plate, bead, and hollow-fiber immunoadsorption systems follow.

2. Flask Panning

The "panning" method for cell selection, first described by Wysocki, is an example of a flat-plate system (Wysocki and Sato, 1978). Ligand is physically adsorbed onto polystyrene Petri dishes (bacteriological grade) and following cell adsorption, the negative cell fraction is collected by gentle washes. Positive cells are recovered manually by vigorous pipetting of the surface. A more recent derivative of the cell panning technique is the use of polystyrene culture flasks which have been activated chemically for covalent attachment of monoclonal antibody (Morecki *et al.,* 1990). These flasks are commercially available (CELLector™, Applied Immune Sciences, Menlo Park, CA). The main utility of the CELLector™ system is to combine a cell selection process with *in vitro* culture. Bound cells are detached by striking the side of the flask. If cells are not detached by this procedure, culture for several hours may result in antibody capping and detachment of bound cells.

Holyoake *et al.* (1994) assessed the performance of Applied Immune Sciences flasks for the isolation of human $CD34^+$ hematopoietic progenitors. $CD34^+$ cells were isolated from the mononuclear cell fraction of bone marrow samples using a two-stage separation process. Soybean agglutinin flasks depleted T and B cells, fibroblasts, and stromal cells as well as some myeloid cells. This preenrichment step was followed by positive selection with flasks coated with anti-CD34 monoclonal antibody. The input incidence of $CD34^+$ cells was 1–5%, with the purity of the $CD34^+$ cell enriched fraction being around 80% (100-

TABLE 11.3 Factors Influencing Performance of Affinity Cell Separations

Device geometry	Uniform and controllable shear stress High surface area to volume ratio Uniform bulk flow Minimize flow stasis
Immunoadsorbent	Low nonspecific affinity High specific affinity
Mobile phase	Agents which block nonspecific cellular interactions Selective cleaving reagents
Depletion	Low shear stress High immunoadsorbent affinity
Enrichment	Optimal levels of shear stress for high purity and yield Selective cleaving reagents

fold enrichment). They also found that these flasks preferentially bound cells with high levels of CD34.

The capacity of flask panning is related to surface area ((0.5–1) × 10^6 cells/cm^2). About one hundred 150-cm^2 flasks would be required to process a bone marrow or peripheral stem cell graft (10^{10} mononuclear cells). Compared to other systems, flask panning has a low surface area to volume ratio.

3. Column Chromatography

Most bead or chromatographic based techniques rely on diffusive mechanisms for transport of molecules to and from adsorptive surfaces. In contrast to molecularly sized particles, Brownian diffusion for cells is insignificant, and cell release is dependent on the generation of fluid shear stress over the bead surface. In contrast to solute separations where molecules diffuse from affinity beads, the performance of packed columns for cell separation has been poor since cells are trapped in regions of flow stasis (Wigzell and Andersson, 1969; Wofsy and Kimura, 1971; Braun *et al.,* 1982; Kataoka *et al.,* 1988). Separation is more efficient if beads are kept in suspension and washing steps are assisted by gentle agitation (Berenson *et al.,* 1988; Maruyama *et al.,* 1989). Methods for recovery of bound cells include the use of fluid shear, competitive elution (Wofsy and Kimura, 1971; Rasmussen *et al.,* 1992; Tseng-Law *et al.,* 1995), chemical lysis (Kiefer, 1975; Grimsley *et al.,* 1993), or enzymes (Civin, 1992; Sutherland *et al.,* 1992).

Dreger *et al.* (1995) examined the performance of the Ceprate LC34™ system (CellPro) for selection of $CD34^+$ cells from G-CSF-mobilized peripheral blood progenitor cells. Ceprate™ affinity columns are based on biotin–avidin-mediated immunoadsorption of target cells (Dreger *et al.,* 1995). Adsorbed cells are detached by fluid shear stress generated by agitation of the bead suspension. The input incidence of $CD34^+$ cells was 0.65 ± 0.1% (mean ± standard error, $n = 6$), with a $CD34^+$ cell enrichment yield of 26 ± 1% and purity of 62 ± 3% (400-fold enrichment) (Dreger *et al.,* 1995). The surface area to volume ratio for bead systems is high, and scale-up for clinical separations has been feasible. CellPro produces a large-scale system for clinical $CD34^+$ cell selection (Ceprate SC™). Its performance will be compared with that of other large-scale systems at the conclusion of this section.

4. Hollow-Fiber Affinity Cell Separation

A hollow-fiber module consists of a parallel array of regenerated cellulose (Cuprophan) hollow fibers (internal diameter 200 μm) "potted" in polyurethane resin within a polycarbonate cylindrical shell. Conical-shaped inlet and outlet headers direct flow uniformly along the lumen of hollow fibers. Cuprophan hollow-fiber modules used clinically for hemodialysis typically have an internal surface area of 1 m^2. When these modules are applied to cell separation, surface fluid shear stress at the attachment interface can be controlled directly using device flow rate (Nordon *et al.,* 1994).

Cuprophan minidialyzers (internal surface area 200 cm^2, AKZO, Wuppertal, Germany) were modified by coating the inside of hollow fibers with monoclonal antibody to CD34 antigen (Nordon and Schindhelm, 1997). $CD34^+$ cells were enriched from mononuclear cell concentrates harvested by apheresis from patients undergoing stem cell mobilization using G-CSF and chemotherapy. After cell immunoadsorption, unbound cells were recovered at 25 dyn/cm^2. The remaining $CD34^+$ cells were recovered at 5 dyn/cm^2 after enzymatic lysis of the CD34 molecule with chymopapain (Baxter Immunotherapy Division, Munich, Germany). The preselection incidence of $CD34^+$ cells was $1.6 \pm 0.6\%$ ($\pm$SEM, $n = 7$). Enrichment purity was $94 \pm 3\%$ (1200-fold enrichment), and $61 \pm 9\%$ of input $CD34^+$ cells were recovered in enriched fractions (Nordon and Schindhelm, 1997).

The number of cells that can be separated by this system is related to the surface area of the hollow-fiber modules (cells per square centimeter). A device with a surface area of 2 m^2 would be sufficient for clinical applications (10^{10}-cell capacity). Although the surface area to volume ratio is less than that in the bead adsorption system, hollow-fiber modules in a configuration similar to that used for hemodialysis (1–2 m^2) may provide sufficient surface area in a compact configuration for clinical applications.

D. MAGNETIC SORTING

Cells become magnetically susceptible by labeling with an affinity substrate that contains paramagnetic or ferromagnetic material. Colloidal (20–100 nm) or monodispersed polymers (1–100 μm) are conjugated with ligands that bind to cell surface molecules. For depletion magnetically labeled cells are captured by a magnetic field, and labeled cells are recovered by removal of the magnetic field (enrichment). Separation performance is related to device geometry, magnetic label specificity, magnetic susceptibility of labeled cells, and the level of cell trapping in regions of flow stasis (voids).

1. Magnetic Labeling Materials

Magnetic cell separation technology has been made possible by the development of superparamagnetic labeling particles that contain magnetite (Fe_3O_4). This ferromagnetic crystal has greater magnetic susceptibility than paramagnetic materials and will not retain a permanent magnetic dipole if the size of crystals is smaller than a magnetic domain (less than 30 nm) (Bean and Livingston, 1959). Ferrofluids (Elmore, 1938) are colloidal solutions containing magnetite crystals, which are combined with various other materials. The particles range in size from 5 to 60 nm, which is sufficiently small for diffusion to prevent deposition or flocculation (Liberty and Feeley, 1991).

Superparamagnetic colloids have been prepared by forming magnetite in the presence of proteins (Owen and Sykes, 1984; Yau *et al.*, 1990), antibodies (Owen *et al.*, 1989), or dextrans (Molday and Mackenzie, 1982). The formation of these colloids involves mixing aqueous solutions of the material with ferrous and ferric chlorides (2:1) and gradual alkalization of the solution using sodium hydroxide. Ligands have been coupled to dextran colloids using chemical or immunoaffinity-based methods (Thomas *et al.,* 1989).

Ugelstad and co-workers have described the preparation of monosized magnetic particles which are formed by the process of activated swelling (Ugelstad *et al.,* 1983, 1992, 1993). Very small seed particles (0.5–1 μm) contain an activating compound that greatly increases the absorption of monomer during the swelling process. Particles may swell by as much as 1000 times their original volume to form highly uniform spherical beads. Iron oxides are precipitated within the pores of cross-linked beads, which are then coated with a polymeric material to seal pores and provide chemical groups for binding of proteins. Dynal (Oslo, Norway) markets 4.5- and 2.8-μm magnetic beads (Dynabeads™) for cell separation. Cells labeled with Dynabeads™ have much higher magnetic susceptibility than cells labeled with superparamagnetic colloids and do not require a high-gradient magnetic field for separation.

The main functional differences between magnetic beads and ferrofluids are (1) rate of cell labeling, (2) removal of particles from cells, and (3) magnetic susceptibility. Adsorption of magnetic colloids at the cell surface is assisted by diffusion and is rapid (around 10 min). Magnetic beads have negligible diffusion and mixing of the cell–bead suspension is required to increase the probability of cell–bead interaction. Dextran-based ferrofluids are biodegradable and do not appear to influence cellular function. A number of reagents have been developed for the removal of cells from Dynabeads™. These include the competitive displacement of antibody using polyclonal antibodies to the Fab region of mouse monoclonal antibody, proteolytic enzymes such as chymopapain, or idiotype peptide analogs (see Table 11.2).

Magnetic beads have a much higher magnetic susceptibility than ferrofluids. Cells labeled with beads will be attracted toward the pole of a standard magnetic at a distance of several millimeters. Magnetized filters (wire or sphere array) are required to create localized high-gradient fields for capture of cells labeled with ferrofluids.

2. High-Gradient Magnetic Separation

In a uniform field, a magnetized particle will be subject to a torque, which results in alignment of its dipole with the magnetic field. There is no net force on the particle. A nonuniform magnetic field (field lines converge or diverge) is required to exert traction force on a magnetized particle and is calculated from the gradient of the magnetic energy ∇U:

$$\mathbf{F} = -\nabla U,$$

where $U = 1-\frac{1}{2}(\chi_p - \chi_m)\int_V H^2\, d\tau$. The integral is taken over V, the particle volume, where H is the magnetic field intensity, and the magnetic susceptibilities of the particle and medium are χ_p and χ_m, respectively (Aharoni, 1976).

Magnetic separators employ stainless steel wire mesh or sphere arrays to create localized high-gradient fields. Figure 11.5 shows calculated magnetic field lines (parallel to **H**) around a magnetized wire. The applied magnetic field is directed left to right and induces a uniform magnetization across the wire cross-section. Field lines surrounding the wire converge (diverge) toward the horizontal (vertical) axis. A magnetic dipole will experience a traction (repulsion) force in regions where field lines converge (diverge). Positively susceptible particles will accumulate on the left and right sides of the magnetized wire.

The flow velocity of the surrounding fluid and the magnetic field gradient in the vicinity of the magnetized wire will influence the trajectory of magnetized particles. Watson (1973) calculated these trajectories and defined a critical radius for cell capture, which depends on the bulk flow rate V_o and the magnetic velocity V_m as defined by

$$V_m = (M_p - M_f)\,\frac{\mu_o b^2 M_s}{9\eta a},$$

where M_p, M_f, and M_s are the magnetization of the particle, fluid, and wire, respectively, μ_o is the permeability of free space, b and a are the radii of the wire and cell, and η is the viscosity of the fluid.

Richards *et al.* performed a semiempirical analysis of the high-gradient magnetic separation of human blood and bone marrow using this definition of magnetic velocity (Richards *et al.*, 1996a, 1996b). A power law relating the

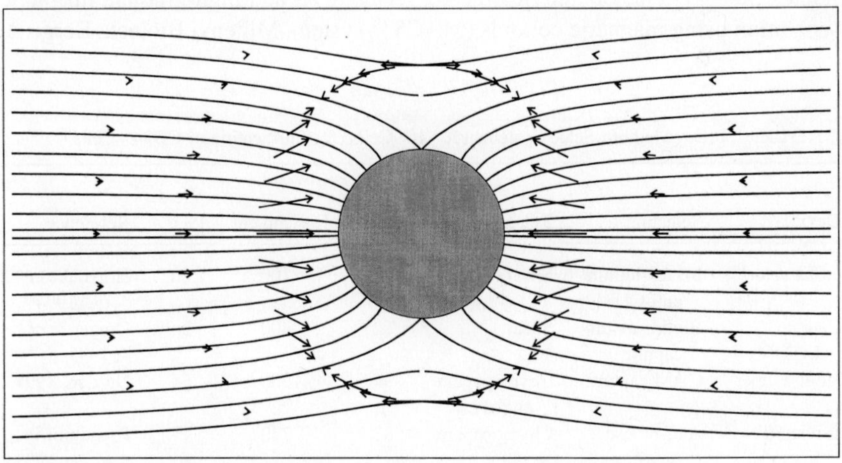

FIGURE 11.5 Magnetic field lines around a magnetized wire. The magnetic field traverses the wire at 90° (left to right). Arrows represent the traction force, which acts on a magnetic dipole moment at various positions around the wire. Positive (negative) susceptibility particle will experience a traction (repulsion) force if field lines converge (diverge).

TABLE 11.4 Factors Influencing Magnetic Separations

Device geometry	Wire array spacing and flow paths (high-gradient magnetic separation) Mixing chamber to promote cell–bead interaction Minimize flow stasis
Magnetic labeling	High binding specificity High magnetic susceptibility Lack of permanent magnetic dipole Cell–bead ratio
Depletion	Low shear stress High magnetic susceptibility
Enrichment	Optimal shear stress for enrichment Selective cleaving reagents (magnetic beads)

characteristic magnetic velocity (V_m/V_o) to the wire capture radius R_{ca} is used to predict the probability of positive-cell capture from operating conditions:

$$R_{ca} = 0.19\left(\frac{V_m}{V_o}\right)^{0.38}.$$

The main physical parameters that influence the performance of magnetic separation are summarized in Table 11.4.

3. Separation Performance of Magnetic Beads and Ferrofluids

Papadimitriou (1995) compared the relative performance of Dynabeads™ (Isolex 50™ system, Baxter Healthcare, Irvine, CA) and high-gradient magnetic separation using magnetic colloids (MACS™ system, Miltenyi Biotech, Bergisch

TABLE 11.5 Medium-Scale Systems (10–10^9 Cells) for Purification of $CD34^+$ Cells

Device	Selection	Release	Purity (%)	Enrichment	Yield (%)	Reference
CELLector™	SB agglutinin/ anti-CD34	Shear	80	100	60	Holyoake *et al.*, 1994[a]
Ceprate LC34™	Biotin–avidin capture	Shear	62	400	26	Dreger *et al.*, 1995
Hollow fiber	CD34 capture	Chymopapain and shear	94	1200	61	Nordon, 1994
Isolex 50™	Anti-mouse Ig affinity	Chymopapain	83	740	43	Papadimitriou *et al.*, 1995
MACS™	Antihapten affinity	—	74	320	60	Papadimitriou *et al.*, 1995

[a] Purification of $CD34^+$ cells from bone marrow.

TABLE 11.6 Large-Scale Systems (10^{10}–10^{11} Cells) for Purification of $CD34^+$ Cells

Trade name	Isolex 300™ (Baxter)	ClinMACS™ (Miltenyi Biotech)	Ceprate SC™ (CellPro)
Method	Dynabeads™	High-gradient magnetic separation	Avidin–biotin immunoadsorption
Purity (%)	75 ± 8	91.5 ± 0.6	57 ± 3
Enrichment	360 ± 90	5000 ± 500	200 ± 60
Yield (%)	65 ± 26	82 ± 6	58 ± 7
Number of separations	8	30	13
Reference	McNiece *et al.*, 1997	McQuaker *et al.*, 1997	Papadimitriou *et al.*, 1995

Gladbach, Germany). $CD34^+$ cells were enriched from mobilized peripheral blood mononuclear cell collections. The enrichment performance of the Isolex 50™ system was marginally better than that of the MACS™ system, with final purities of 83 ± 3% ($n = 12$, input 1.2 ± 2%, 740-fold enrichment) and 74 ± 6% ($n = 5$, input 2.3 ± 1.6%, 320-fold enrichment), respectively. However, the overall $CD34^+$ cell yield was lower for the Isolex 50™ system (43 ± 6%) than the MACS™ system (60 ± 9%).

4. Selection of $CD34^+$ Cells

Purification of $CD34^+$ cells from mobilized peripheral blood mononuclear cell collections has presented a significant challenge. $CD34^+$ cells are at low frequency (0.5–3%) and need to compete with intrinsically sticky cells (*e.g.*, neutrophils and monocytes) that are present in greater numbers. This difficult separation has provided a benchmark test for emerging enrichment technologies. The performance of medium-scale systems (10^8–10^9 cells) is summarized in Table 11.5. Large-scale devices (10^{10}–10^{11} cells) have been developed recently for clinical application. Their performance is summarized in Table 11.6 (Zimmerman *et al.*, 1996; McNiece *et al.*, 1997; McQuaker *et al.*, 1997).

III. RATIONALE FOR CELL SELECTION

The rationale for cell selection procedures is to (1) remove pathogenic cells, (2) retain therapeutic cell subsets, (3) reduce the number of cells to be cultured or cryopreserved, and/or (4) prevent systemic toxicity associated with cell infusion. To date, most of the clinical applications for cell selection technologies have been to remove tumor cells or host-reactive cells or to reduce cryopreservation volume and DMSO toxicity of stem cell grafts.

A. TUMOR CELL DEPLETION OF AUTOLOGOUS STEM CELL GRAFTS

The clinical rationale for tumor purging of autologous grafts is established by an analysis of the source of tumor recurrence. Intensive therapy can only reduce tumor burden by several logarithmic steps, and even though a complete clinical remission is achieved, remaining minimal residual disease can eventually give rise to clinical relapse. Therefore, in some cases, the number of tumor cells which are reinfused with the graft may be insignificant in comparison to the level of minimal residual disease and not responsible for clinical relapse.

The source of relapse has been determined by genetically marking graft-derived tumor cells (Brenner *et al.,* 1993; Rill *et al.,* 1994). A fraction of the autograft is transduced with neomycin-resistance genes, which are detected by polymerase chain reaction (PCR) or resistance to the neomycin analog G418. Coexpression of the neomycin-resistance gene in tumor cells during relapse is taken as direct evidence for graft-derived tumor recurrence. Brenner *et al.* have shown collateral expression of transduced neomycin-resistance genes in acute myeloid leukemia (AML) (Brenner *et al.,* 1993) and neuroblastoma (Rill *et al.,* 1994). The source of relapse is also determined from the rate of tumor recurrence following tumor purging of grafts, although clinical outcome will also depend on the degree of tumor cell depletion by tumor cell purging techniques.

Methods for *ex vivo* depletion of tumor cells can be divided into cytotoxic drugs (Labopin and Gorin, 1992) and antibody-mediated techniques such as complement-mediated lysis (Greenwood *et al.,* 1994; Yerly-Motta *et al.,* 1996), immunotoxins (Uckun and Myers, 1993; Roy *et al.,* 1996), and magnetic-bead depletion (Treleaven *et al.,* 1984; Gee, 1994). Recently, tumor cells were depleted from grafts by antisense oligonucleotides (Bergan, 1997). Most experience has been gained with the marrow-purging drug 4-hydroperoxycyclophosphamide (4-HC). This drug is toxic to normal progenitors as well as tumor cells, which can result in delayed engraftment. Marrow toxicity is reduced by pretreatment with amifostine (Shpall *et al.,* 1994). Antibody-based techniques are applicable where tumor cells have homogeneous expression of surface markers, though the same markers may not be present on tumor stem cells. Purging by monoclonal antibody-based techniques is nontoxic to hematopoietic progenitors and has been applied to acute myeloid leukemia (anti-CD15 and -CD33) (Robertson *et al.,* 1992; Nimgaonkar *et al.,* 1996), acute lymphoblastic leukemia (anti-CD52, -CD9, -CD10, and -CD24) (Uckun *et al.,* 1992; Mehta *et al.,* 1997), multiple myeloma (anti-B-cell and anti-CD6) (Seiden *et al.,* 1995), neuroblastoma, and small-cell lung cancer (anti-CD56) (Roy *et al.,* 1996). An alternate approach is depletion of tumor cells by enrichment of $CD34^+$ cells (Gorin *et al.,* 1995; Bertolini *et al.,* 1997; Mapara *et al.,* 1997).

The clinical role of purging could be best assessed by multicenter randomized studies where enrolled subjects receive identical preparative regimens and are then randomized to receive purged or unpurged autologous marrow. However, most studies are not randomized, nor prospective, with patients receiving a variety of treatments prior to autologous transplantation. The European Cooperative Group for Blood and Marrow Transplantation retrospectively evaluated the clinical outcome of patients with acute leukemia autografted with bone marrow purged by mafosfamide (Gorin *et al.,* 1990, 1991). Purging appears to be most beneficial when patients are autografted in the first complete remission and where patients receive total body irradiation (TBI) as part of the preparative regime. In 671 AML patients, the relapse rate was 35 ± 5% (purged) versus 47 ± 3% (unpurged, $p = 0.006$). For those patients who received TBI, the relapse rates were 29 ± 5 versus 50 ± 4%, respectively ($p < 0.0001$). The beneficial effect of purging was more pronounced in patients who were autografted within 6 months (16 ± 6 versus 60 ± 6%) and in those who had a slow response to chemotherapy before reaching first remission (20 ± 8 versus 61 ± 6%). These findings suggest that efficacy is detectable in situations in which residual tumor persists in marrow or where relapse originating from residual disease is unlikely.

Retrospective analysis of 5-year leukemia-free survival for patients with acute leukemia (both myeloid and lymphocytic) showed a slight beneficial effect of bone marrow purging if patients were grafted after first remission (57 versus 48%, $p = 0.08$), first remission with TBI (47 versus 36%, $p = 0.09$), or second remission (36 versus 28%, $p = 0.053$). Prospective randomized trials will provide a more definitive analysis of the clinical efficacy of bone marrow purging in acute leukemia (Gorin, 1996).

Many patients with lymphomas achieve long-term disease-free survival with first-line chemotherapy or radiotherapy. About 30% of patients with Hodgkin's disease, 50% of patients with high-grade non-Hodgkin's lymphoma, and virtually all patients with low-grade non-Hodgkin's lymphoma will not be cured with conventional treatment. In these cases dose escalation with autologous marrow or peripheral stem cell rescue may lead to an improvement in outcome (Vandenberghe and Goldstone, 1996).

A case-matched retrospective study of purging of autologous bone marrow transplants for non-Hodgkin's lymphoma did not clearly demonstrate a beneficial effect of purging ($n = 224$) (Williams *et al.,* 1996). The main factors which were matched and had been shown to be correlated with disease progression-free survival were disease stage, prior response to therapy at the time of transplantation, patient age, and disease histology. Purging in patients with low-grade lymphoma resulted in a significantly improved 5-year overall survival (84 versus 48%, $p = 0.18$, $n = 50$), though disease-free progression was not significantly different between the two groups (48.4 versus 44.2%, $p = 0.18$). The increase in overall survival for low-grade lymphoma was found to be associated with non-TBI conditioning and also transplantation at large transplant centers, which may partially explain the better outcome for the group receiving purged marrow. A prospective randomized trial run by the European Blood and Marrow Transplant Lymphoma Registry will address the efficacy of purging in low-grade lymphoma (Williams *et al.,* 1996).

High-risk neuroblastoma (stage IV, overexpression of *N-myc* and poorly differentiated) has a poor prognosis, with less than 15% long-term survival using conventional chemotherapy (Brodeur and Gastleberry, 1993). The best long-term survival rates have been achieved with high-dose myeloablative therapy and bone marrow rescue. A total of 147 children with high-risk neuroblastoma who had received myeloablative therapy and autologous purged bone marrow transplantation after completion of induction therapy had a 4-year disease progression-free survival of 38%. Although it is still unclear whether bone marrow purging of autografts will improve clinical outcome (Matthay, 1996), the gene-marking study of Rill (1994) certainly suggests that recurrence of neuroblastoma after high-dose therapy and autologous transplantation is related partially to contamination of the graft with tumor cells. Once again, a prospective randomized clinical trial is required to demonstrate the efficacy of neuroblastoma purging of autografts.

The prognosis of breast cancer patients with extensive axillary lymph node involvement is poor despite standard-dose adjuvant therapy. Between 55 and 87% of patients will relapse by 5 years and 70–90% will relapse by 10 years (Peters, 1991). The use of high-dose chemotherapy with autologous bone marrow support as consolidation after standard-dose adjuvant chemotherapy has been associated with a better clinical outcome. Event-free survival at 2.5 years was between 56 and 82%. Comparison populations which were derived from the clinical trials of adjuvant chemotherapy in stage II breast cancer had a 2.5-year event-free survival of 38–52% (Peters *et al.,* 1993). Ongoing prospective randomized clinical trials are comparing high-dose chemotherapy followed by autologous peripheral stem cell grafts with adjuvant conventional-dose therapy for poor-prognosis breast cancer. Other solid tumors where chemotherapy dose escalation with hematopoietic stem cell rescue may provide a survival advantage include ovarian cancer (Grenman *et al.,* 1996), Ewing's sarcoma, rhabdomyosarcoma, Wilms' tumor, and germ cell tumors (Atra and Pinkerton, 1996).

B. PREVENTION OF GRAFT-VERSUS-HOST DISEASE BY LYMPHOCYTE DEPLETION

Graft-versus-host disease (GVHD) results from reactivity of immunocompetent donor cells, present in the transplanted bone marrow, against the recipient tissues (Billingham, 1966). The GVHD reaction affects the skin, liver, and gastrointestinal tract and is manifest clinically by erythroderma, generalized wasting, diarrhea, and jaundice. Acute GVHD has an onset within 3 months of transplantation and develops in 30–60% of recipients of histocompatible sibling marrow. Mortality of acute GVHD may reach 50% (Weisdorf *et al.,* 1990).

The risk of GVHD is related to the degree of HLA mismatch; the incidence and severity of acute GVHD are higher where donors are matched but unrelated and for histoincompatible related donors (Beatty, 1992). Despite potent immunosuppression, more than half of the patients who receive unmanipulated allogeneic marrow develop GVHD (Lazarus *et al.,* 1997).

The immunopathophysiology of acute GVHD is complex and can be divided into afferent and efferent phases (Ferrara *et al.,* 1996). The afferent phase is initiated by activation of donor T cells. Host-tissue antigens are presented in association with host class I or II major histocompatibility complex (MHC) molecules so that both $CD8^+$ and $CD4^+$ lymphocytes are involved in the initial response. Host MHC expression is up-regulated by secretion of TNF-α and IL-1 following the conditioning regimen (radiation and chemotherapy). Donor T-cell activation is associated with clonal expansion of type 1 helper T cells (Th1), which secrete interleukin-2 (IL-2) and interferon-γ (IFN-γ). These cytokines initiate the effector phase of GVHD, which results in tissue damage by cytotoxic lymphocytes, natural killer cells, and inflammatory cytokines (IL-2, TNF-α) secreted by mononuclear phagocytes. Effector functions of mononuclear phagocytes are triggered by the release of endotoxin (lipopolysaccharide) from the gut during the conditioning regimen. Activation of donor T cells can also result in production of type 2 helper T cells (Th2), which secrete IL-4 and IL-10. Tissue destruction mediated by the inflammatory cytokines (IL-2, TNF-α) is inhibited by the Th2 cytokines. Chronic GVHD is associated with a Th2 response.

Therapeutic interventions that block the afferent phase of GVHD are more successful. The conventional approach to prophylaxis has been posttransplant *in vivo* immunosuppression using glucocorticoids, methotrexate, and cyclosporin. More effective is the removal of donor T cells from transplanted allogeneic grafts.

Pan-T-cell depletion of allogeneic bone marrow grafts effectively prevents GVHD across major or minor histocompatibility differences. With transplantation of unmodified bone marrow, approximately 35% of HLA-identical sibling transplants with conventional posttransplant immunosuppressive therapy develop moderate to severe acute GVHD (Storb *et al.,* 1992). Only 0–15% of patients develop acute GVHD after T-cell-depleted transplants, and if GVHD occurs, it is usually mild in severity (Marmont *et al.,* 1991). Acute GVHD generally does not occur if less than 10^5 T cells/kg are infused (Kernan *et al.,* 1986).

This benefit has been offset by a higher rate of graft rejection and leukemia relapse (Champlin, 1993). Graft rejection appears to be mediated by host T cells and natural killer cells. In addition to host factors, the allogeneic graft possesses cell subsets that facilitate engraftment but do not have the potential to produce GVHD (Kaufman *et al.,* 1994). A mouse model of mixed allogeneic–syngeneic chimeras was used to establish the phenotype of the facilitating cell subset

($CD8^+$, $CD3^+$, $Thy1^+$, HLA class II $^{dim/intermediate}$), which only comprises about 0.4% of mouse marrow. Engraftment is also enhanced by higher transplanted cell doses or intensification of the immunosuppressive conditioning regimen (Truitt and Atasoylu, 1991).

Initially, the graft-versus-leukemia (GVL) effect was suggested based on the lower rate of leukemia recurrence in patients who developed GVHD (Sullivan *et al.,* 1989). This effect was particularly marked for patients with chronic myeloid leukemia (Horowitz *et al.,* 1990). Although donor lymphocyte infusions posttransplant have induced cytogenetic remission, this procedure was also associated with GVHD (Bacigalupo *et al.,* 1997). Considerable effort is being directed toward engineering grafts that preserve or enhance the GVL effect without a significant risk of GVHD.

A number of novel approaches which aim to preserve T-cell function while depleting the host-specific response (GVHD) are under evaluation. These include more selective depletion of T-cell subsets using monoclonal antibody-based techniques (Nimer *et al.,* 1994; Giralt *et al.,* 1995; Fleming *et al.,* 1996; Soiffer *et al.,* 1997), *in vitro* depletion of alloreactive T-cell clones using *Pseudomonas* exotoxin-based immunotoxins (Mavroudis *et al.,* 1996), and retroviral transfection of alloreactive T-cell clones with a "suicide gene" (herpes simplex virus thymidine kinase), which may be depleted *in vivo* by administration of ganciclovir should GVHD develop. An understanding of the molecular basis for recognition of leukemia cell antigens by allogeneic T cells may lead to development of more selective T-cell depletion methods which preserve GVL reactivity (Jiang *et al.,* 1997).

IV. SUMMARY

Monoclonal antibody-based systems for purification or depletion of cell subsets have enabled the development of novel cell therapies. Recently, large-scale systems that are based on magnetic beads, colloids, or avidin–biotin immunoaffinity have been developed for clinical-scale purification of $CD34^+$ cells from bone marrow or mobilized peripheral blood. Although it has been possible to demonstrate relatively high levels of purity for these cell separation methods, performance must be determined in a clinical context. A significant shortcoming of some of these systems has been low and variable $CD34^+$ cell yields (approximately 50%), making it sometimes necessary for patients to have additional apheresis to purify enough $CD34^+$ cells for transplantation.

Despite a reduction in the incidence and severity of GVHD by T-cell depletion of allografts, long-term survival has not improved because of poor graft survival (Hale and Waldmann, 1986) and an increased leukemia recurrence rate (Horowitz *et al.,* 1990). Nevertheless, these studies have led to the identification of GVL reactivity and suggest that long-term survival for CML has an immunological basis.

Tumor cell depletion of autologous grafts may require combined enrichment and depletion methods to reduce tumor cell burden to levels that are clinically insignificant. At this stage it is difficult to assess what are "safe" levels of tumor contamination. In any case, the complexity of tumor–graft–host interactions indicates that carefully controlled and randomized, prospective clinical studies are required for clinical evaluation of these cell separation strategies.

In the future, cell separation will be an essential technology for the continued development of cellular immunotherapy and gene therapy. Furthermore, these multistep procedures will require development of in-line separation procedures to minimize the chance of contamination. As large-scale separation technologies become more efficient and cost-effective, they will be applied in an increasing variety of clinical contexts.

REFERENCES

Aharoni, A. (1976). Traction force on paramagnetic particles in magnetic separators. *IEEE Trans. Magn.* **3,** 234–235.

Anderson, N. A. B., Pamphilon, D. H., Tandy, N. J., *et al.* (1991). Comparison of platelet-rich plasma collection using the Haemonetics PCS and Baxter Autopheresis C. *Vox Sang.* **60,** 155–158.

Atra, A., and Pinkerton, C. R. (1996). Autologous stem cell transplantation in solid tumours of childhood. *Ann. Int. Med.* **28,** 159–164.

AuBuchon, J. P., Carter, S., Adde, M. A., *et al.* (1986). Optimization of parameters of plateletpheresis and lymphocytapheresis yields on the Haemonetics Model V50. *J. Clin. Apheresis* **3,** 103–108.

Bacigalupo, A., Soracco, M., Vassallo, F., *et al.* (1997). Donor lymphocyte infusions (DLI) in patients with chronic myeloid leukemia following allogeneic bone marrow transplantation. *Bone Marrow Transplant* **19,** 927–932.

Bean, C. P., and Livingston, J. D. (1959). Superparamagnetism. *J. Appl. Phys.* **30,** 120s–129s.

Beatty, P. G. (1992). Results of allogeneic bone marrow transplantation with unrelated or mismatched donors. *Semin. Oncol.* **19,** 13–19.

Bell, A. J. (1978). Models for the specific adhesion of cells to cells. *Science* **200,** 618–627.

Berenson, R. J., Andrews, R. G., Bensinger, W. I., *et al.* (1988). Antigen $CD34^+$ marrow cells engraft lethally irradiated baboons. *J. Clin. Invest.* **81,** 951–955.

Bergan, R. C. (1997). *Ex vivo* bone marrow purging with oligonucleotides. *Antisense Nucleic Acid Drug Dev.* **7,** 251–255.

Bertolini, F., Thomas, T., Battaglia, M., *et al.* (1997). A new "two step" procedure for 4.5 log depletion of T and B cells in allogeneic transplantation and of neoplastic cells in autologous transplantation. *Bone Marrow Transplant* **19,** 615–619.

Billingham, R. E. (1968). The biology of graft-versus-host reactions. *In* "The Harvey Lectures, 1966–67," (Zinder, N. D. *et al.*, Eds.), pp. 21–78. Academic Press, New York.

Bird, R. B., Stewart, W. E., and Lightfoot, E. N. (1960a). Diffusivity and the mechanisms of mass transport. *In* "Transport Phenomena" (R. B. Bird, W. E. Stewart, and E. N. Lightfoot, Eds.), pp. 495–518. Wiley, New York.

Bird, R. B., Stewart, W. E., and Lightfoot, E. N. (1960b). Viscosity and the mechanism of momentum transport. *In* "Transport Phenomena" (R. B. Bird, W. E. Stewart, and E. N. Lightfoot, Eds.), pp. 3–33. Wiley, New York.

Braun, R., Teute, H., Kirchner, H., *et al.* (1982). Rapid separation of T cell subpopulations with monoclonal antibodies and affinity chromatography. *J. Immunol. Methods* **54,** 251–258.

Brenner, M. K., Rill, D. R., Moen, R. C., *et al.* (1993). Gene-marking to trace the origin of relapse after autologous marrow transplantation. *Lancet* **341,** 85–86.

Brodeur, G. M., and Gastleberry, R. P. (1993). Neuroblastoma. *In* "Principles and practice of pediatric oncology" (P. A. Pizzo and D. G. Poplack, Eds.), pp. 739–767. Lippincott, Philadelphia.

Brown, R. I. (1989). The physics of continuous flow centrifugal cell separation. *Artif. Organs* **13,** 4 20.

Champlin, R. (1993). T-cell depletion for allogeneic bone marrow transplantation: Impact on graft-versus-host disease, engraftment, and graft-versus-leukemia. *J. Hematother.* **2,** 27–42.

Civin, C. I. (1992). Release of cells from affinity matrices. US Patent 5081030, The Johns Hopkins University, Baltimore, MD.

Cox, R. G., and Brenner, H. (1968). The lateral migration of solid particles in Poiseuille flow: I. Theory. *Chem. Eng. Sci.* **23,** 147–173.

Cozens-Roberts, C., and Lauffenburger, D. A. (1990). Receptor-mediated cell attachment and detachment kinetics: II. Experimental model studies with the radial-flow detachment assay. *Biophys. J.* **58,** 857–872.

Cozens-Roberts, C., Lauffenburger, D. A., and Quinn, D. A. (1990). Receptor-mediated cell attachment and detachment kinetics: I. Probabilistic model and analysis. *Biophys. J.* **58,** 841–856.

Davis, J., Schepers, K. G., Eby, L. L., *et al.* (1993). Comparison of progenitor cell concentration techniques: Continuous flow separation versus density-gradient isolation. *J. Hematother.* **2,** 315–320.

Davis, R. H., and Leighton, D. T. (1987). Shear-induced transport of a particle layer along a porous wall. *Chem. Eng. Sci.* **42,** 275–281.

Dreger, P., Viehmann, K., Steinmann, J., *et al.* (1995). G-CSF-mobilized peripheral blood progenitor cells for allogeneic transplantation: Comparison of T cell depletion strategies using different $CD34^+$ selection systems or CAMPATH-1. *Exp. Hematol.* **23,** 147–154.

Ellingsen, T., Aune, O., Ugelstad, J., *et al.* (1990). Monosized stationary phases for chromatography. *J. Chromatogr.* **535,** 147–161.

Elmore, W. C. (1938). The magnetization of ferromagnetic colloids. *Phys. Rev.* **54,** 1092–1095.

Ferrara, J. L. M., Cooke, K. R., and Pan, L. (1996). The immunopathophysiology of acute graft-versus-host disease. *Stem Cells* **14,** 473–489.

Fleming, D. R., Henslee-Downey, P. J., Romond, E. H., *et al.* (1996). Allogeneic bone marrow transplantation with T cell-depleted partially matched related donors for advanced acute lymphoblastic leukemia in children and adults: A comparative matched cohort study. *Bone Marrow Transplant* **17,** 917–922.

Gee, A. P. (1994). Immunologically based methods for the elimination of tumor cells from autologous stem cell grafts. *Immunomethods* **5,** 232–242.

Giralt, S., Hester, J., Huh, Y., *et al.* (1995). CD8-depleted donor lymphocyte infusion as treatment for relapsed chronic myelogenous leukemia after allogeneic bone marrow transplantation. *Blood* **86,** 4337–4343.

Goldberg, S. L., Mangan, K. F., Klumpp, T. R., *et al.* (1995). Complications of peripheral blood stem cell harvesting: Review of 554 PBSC leukaphereses. *J. Hematother.* **4,** 85–90.

Gorin, N. C. (1996). The potential role of amifostine in haematological malignancies. *Eur. J. Cancer* **32,** S31–S39.

Gorin, N. C., Aegerter, P., Auvert, B., *et al.* (1990). Autologous bone marrow transplantation for acute myeloblastic leukemia in first remission: A European survey of the role of marrow purging. *Blood* **75,** 1606–1614.

Gorin, N. C., Labopin, M., Meloni, G., *et al.* (1991). Autologous bone marrow transplantation for acute myeloblastic leukemia in Europe: Further evidence of the role of marrow purging by mafosfamide. *Leukemia* **5,** 896–904.

Gorin, N. C., Lopez, M., Laporte, J. P., *et al.* (1995). Preparation and successful engraftment of purified CD34$^+$ bone marrow progenitor cells in patients with non-Hodgkin's lymphoma. *Blood* **85,** 1647–1654.

Greenwood, J., Gorman, S. D., Routledge, E. G., *et al.* (1994). Engineering multiple-domain forms of the therapeutic antibody CAMPATH-1H: Effects on complement lysis. *Ther. Immunol.* **1,** 247–255.

Gremba, C., Tarosky, T., and Rosenfeld, C. S. (1994). Inadequate red cell depletion of large-volume marrow harvests processed by the Fenwal CS-3000 plus. *J. Hematother.* **3,** 225–227.

Grenman, S. E., Rantanen, V. T., and Salmi, T. A. (1996). High-dose chemotherapy with autologous stem cell support in advanced ovarian cancer. *Ann. Int. Med.* **28,** 151–158.

Grimsley, P. G., Amos, T. A., Gordon, M. Y., *et al.* (1993). Rapid positive selection of CD34$^+$ cells using magnetic microspheres coated with monoclonal antibody QBEND/10 linked via a cleavable disulphide bond. *Leukemia* **7,** 898–908.

Hale, G., and Waldmann, H. (1988). Depletion of T-cells with CAMPATH-1 and human complement: Analysis of GVHD and graft failure in a multi-centre study. *Bone Marrow Transplant* **3**(Suppl. 1), 11–14.

Hammer, D. A., and Lauffenburger, D. A. (1987). A dynamic model for receptor-mediated cell adhesion to surfaces. *Biophys. J.* **52,** 475–487.

Hester, J. P., Kellog, R. M., and Freireich, E. J. (1983). Mononuclear cell collection by continuous-flow centrifugation. *J. Clin. Apheresis* **1,** 197–201.

Hester, J. P., Rondon, G., Huh, Y. O., *et al.* (1995). Principles of bone marrow processing and progenitor cell/mononuclear cell concentrate collection in a continuous flow blood cell separation system. *J. Hematother.* **4,** 299–306.

Holyoake, T. L., Alcorn, M. J., Richmond, L. J., *et al.* (1994). Efficient isolation of human CD34 positive hematopoietic progenitor cells by immune panning. *Stem Cells* **12,** 114–124.

Horowitz, M. M., Gale, R. P., Sondel, P. M., *et al.* (1990). Graft-versus-leukemia reactions after bone marrow transplantation. *Blood* **75,** 555–562.

Jiang, Y. Z., Mavroudis, D. A., Dermime, S., *et al.* (1997). Preferential usage of T cell receptor (TCR) V beta by allogeneic T cells recognizing myeloid leukemia cells: Implications for separating graft-versus-leukemia effect from graft-versus-host disease. *Bone Marrow Transplant* **19,** 899–903.

Kataoka, K., Sakurai, Y., Hanai, T., *et al.* (1988). Immunoaffinity chromatography of lymphocyte subpopulations using *tert*-amine derived matrices with adsorbed antibodies. *Biomaterials* **9,** 218–224.

Kaufman, C. L., Colson, Y. L., Wren, S. M., *et al.* (1994). Phenotypic characterization of a novel bone marow-derived cell that facilitates engraftment of allogeneic bone marrow stem cells. *Blood* **84,** 2436–2446.

Kernan, N. A., Collins, N. H., Juliano, L., *et al.* (1986). Clonable T lymphocytes in T cell-depleted bone marrow transplants correlate with development of graft-v-host disease. *Blood* **68,** 770–773.

Kiefer, H. (1975). Separation of antigen-specific lymphocytes. A new general method of releasing cells bound to nylon mesh. *Eur. J. Immunol.* **5,** 624–627.

Klumpp, T. R. (1992). Autologous peripheral blood stem cell transplantation versus autologous bone marrow transplantation: Which is better? *Heme/Oncol. Ann.* **2,** 30–38.

Korbling, M., Juttner, C., Henon, P., *et al.* (1994). Blood versus bone marrow transplants. *In* "Blood Stem Cell Transplants" (R. P. Gale, C. A. Juttner, and P. Henon, Eds.), pp. 87–98. Cambridge Univ. Press, Cambridge, UK.

Labopin, M., and Gorin, N. C. (1992). Autologous bone marrow transplantation in 2502 patients with acute leukemia in Europe: A retrospective study. *Leukemia* **6,** S95–S99.

Lazarus, H. M., Vogel, G. B., and Rowe, J. M. (1997). Prevention and treatment of acute graft-versus-host disease: The old and the new. A report from The Eastern Cooperative Oncology Group (ECOG). *Bone Marrow Transplant* **19,** 577–600.

Liberty, P. A., and Feeley, B. P. (1991). Analytical- and process-scale cell separation with bioreceptor ferrofluids and high gradient magnetic separation. *In* "Cell Separation Science and Technology" (D. S. Kompala and P. Todd, Eds.), pp. 268–288. Am. Chem. Soc., Washington, DC.

Mapara, M. Y., Korner, I. J., Hildebrandt, M., *et al.* (1997). Monitoring of tumor cell purging after highly efficient immunomagnetic selection of CD34 cells from leukapheresis products in breast cancer patients: Comparison of immunocytochemical tumor cell staining and reverse transcriptase-polymerase chain reaction. *Blood* **89,** 337–344.

Marmont, A. M., Horowitz, M. M., Gale, R. P., *et al.* (1991). T-cell depletion of HLA-identical transplants in leukemia. *Blood* **78,** 2120–2130.

Maruyama, A., Tsuruta, T., Kataoka, K., *et al.* (1989). Separation of B and T lymphocytes by cellular adsorption chromatography with polyamine graft copolymers as column matrices. II. Recovery of adsorbed B cell enriched populations from the column. *Biomaterials* **10,** 393–399.

Matthay, K. K. (1996). Impact of myeloablative therapy with bone marrow transplantation in advanced neuroblastoma. *Bone Marrow Transplant* **18,** S21–S24.

Mavroudis, D. A., Jiang, Y. Z., Hensel, N., *et al.* (1996). Specific depletion of alloreactivity against haplotype mismatched related individuals by a recombinant immunotoxin: A new approach to graft-versus-host disease prophylaxis in haploidentical bone marrow transplantation. *Bone Marrow Transplant* **17,** 793–799.

McNiece, I., Briddell, R., Stoney, G., *et al.* (1997). Large-scale isolation of $CD34^+$ cells using the Amgen cell selection device results in high levels of purity and recovery. *J. Hematother.* **6,** 5–11.

McQuaker, I. G., Haynes, A. P., Anderson, S., *et al.* (1997). Engraftment and molecular monitoring of $CD34^+$ peripheral-blood stem-cell transplants for follicular lymphoma: A pilot study. *J. Clin. Oncol.* **15,** 2288–2295.

Mehta, J., Powles, R., Treleaven, J., *et al.* (1997). Autologous transplantation with CD52 monoclonal antibody-purged marrow for acute lymphoblastic leukemia: Long-term follow-up. *Leuk. Lymphoma* **25,** 479–486.

Molday, R. S., and Mackenzie, D. (1982). Immunospecific ferromagnetic iron-dextran reagents for the labeling and magnetic separation of cells. *J. Immunol. Methods* **52,** 353–367.

Morecki, S., Topalian, S. L., Myers, W. W., *et al.* (1990). Separation and growth of human $CD4^+$ and $CD8^+$ tumor-infiltrating lymphocytes and peripheral blood mononuclear cells by direct positive panning on covalently attached monoclonal antibody-coated flasks. *J. Biol. Resp. Mod.* **9,** 463–474

Nimer, S. D., Giorgi, J., Gajewski, J. L., *et al.* (1994). Selective depletion of $CD8^+$ cells for prevention of graft-versus-host disease after bone marrow transplantation. A randomized controlled trial. *Transplantation* **57,** 82–87.

Nimgaonkar, M., Kemp, A., Lancia, J., *et al.* (1996). A combination of CD34 selection and complement-mediated immunopurging (anti-CD15 monoclonal antibody) eliminates tumor cells while sparing normal progenitor cells. *J. Hematother.* **5,** 39–48.

Nordon, R. E. (1994). "Uniform shear elution affinity cell separation," Doctoral Thesis, Graduate School of Biomedical Engineering, University of New South Wales.

Nordon, R. E., and Schindhelm, K. (1997). Design of hollow fiber modules for uniform shear elution affinity cell separation. *Artif. Organs* **21,** 107–115.

Nordon, R. E., Haylock, D. N., Gaudry, L., *et al.* (1996). Hollow-fibre affinity cell separation system for $CD34^+$ cell enrichment. *Cytometry* **24,** 340–347.

Nordon, R. E., Milthorpe, B. K., Schindhelm, K., *et al.* (1994). An experimental model of affinity cell separation. *Cytometry* **16,** 25–33.

O'Neill, M. E. (1968). A sphere in contact with a plane wall in a slow linear shear flow. *Chem. Eng. Sci.* **23,** 1293–1298.

Owen, C. S., and Sykes, N. L. (1984). Magnetic labeling and cell sorting. *J. Immunol. Methods* **73,** 41–48.

Owen, C. S., Silvia, J. C., D'Angelo, L. *et al.* (1989). Magnetic-polymer particles. US Patent 4795698, Immunicon Corp., Huntingdon Valley, PA.

Papadimitriou, C. A., Roots, A., Koenigsmann, M., *et al.* (1995). Immunomagnetic selection of $CD34^+$ cells from fresh peripheral blood mononuclear cell preparations using two different separation techniques. *J. Hematother.* **4,** 539–544.

Peters, W. P. (1991). High-dose chemotherapy and autologous bone marrow support for breast cancer. *In* "Important Advances in Oncology" (V.T. DeVita, S. Hellman, and S. A. Rosenberg, Eds.), pp. 135–150. Lippincott, Philadelphia.

Peters, W. P., Ross, M., Vredenburgh, J. J., *et al.* (1993). High-dose chemotherapy and autologous bone marrow support as consolidation after standard-dose adjuvant therapy for high risk primary breast cancer. *J. Clin. Oncol.* **11,** 1132–1143.

Pierelli, L., Menichella, G., Paoloni, A., *et al.* (1993). Evaluation of a novel automated protocol for the collection of peripheral blood stem cells mobilized with chemotherapy or chemotherapy plus G-CSF using the Fresenius AS 104 cell separator. *J. Hematother.* **2,** 145–153.

Rasmussen, A. M., Smeland, E. B., Erikstein, B. K., *et al.* (1992). A new method for detachment of Dynabeads from positively selected B lymphocytes. *J. Immunol. Methods* **146,** 195–202.

Richards, A. J., Roath, S., Smith, R. J. S., *et al.* (1996a). High purity, recovery and selection of human blood cells with a novel high gradient magnetic separator. *J. Hematother.* **5,** 415–426.

Richards, A. J., Roath, S., Smith, R. J. S., *et al.* (1996b). The mechanisms of high gradient magnetic separation of human blood and bone marrow. *IEEE Trans. Magn.* **32,** 459–469.

Rill, D. R., Santana, V. M., Robert, V. M., *et al.* (1994). Direct demonstration that autologous bone marrow transplantation for solid tumors can return a multiplicity of tumorigenic cells. *Blood* **83,** 3068–3076.

Robertson, M. J., Soiffer, R. J., Freedman, A. S., *et al.* (1992). Human bone marrow depleted of CD33-positive cells mediates delayed but durable reconstitution of hematopoiesis: Clinical trial of MY9 monoclonal antibody-purged autografts for the treatment of acute myeloid leukemia. *Blood* **79,** 2229–2236.

Rock, G., Herzig, R., McCombie, N., *et al.* (1983). Automated platelet production during plasmapheresis. *Transfusion* **23,** 290–293.

Rock, G., Herzig, R., McCombie, N., *et al.* (1985). Comparative function of centrifugal apheresis devices. *Vox Sang.* **48,** 136–142.

Roy, D. C., Ouellet, S., Le Houillier, C., *et al.* (1996). Elimination of neuroblastoma and small-cell lung cancer cells with an anti-neural cell adhesion molecule immunotoxin. *J. Natl. Cancer Inst.* **88,** 1136–1145.

Saterbak, A., Kuo, S. C., and Lauffenburger, D. A. (1993). Heterogeneity and probabilistic binding contributions to receptor-mediated cell detachment kinetics. *Biophys. J.* **65,** 243–252.

Schrempf-Decker, G., Baron, D., and Wernet, P. (1980). Helix pomatia agglutinin (HpA) affinity chromatography: The isolation of pure B and T cell populations and their use for routine HLA-DR (Ia) serology. *J. Immunol. Methods* **32,** 285–296.

Seiden, M. V., Schlossman, R., Andersen, J., *et al.* (1995). Monoclonal antibody-purged bone marrow transplantation therapy for multiple myeloma. *Leuk. Lymphoma* **17,** 87–93.

Shapiro, H. M. (1988). How a flow cytometer works. *In* "Practical Flow Cytometry" (H. M. Shapiro, Ed.), pp. 21–86. A. R. Liss, NewYork.

Sharpe, P. T. (1988). Centrifugal elutriation. *In* "Methods of Cell Separation" (R. H. Burdon, and P. H. van Knippenberg, Eds.), pp. 91–105. Elsevier, Amsterdam/New York.

Shpall, E. J., Stemmer, S. M., Hami, L., *et al.* (1994). Amifostine (WR-2721) shortens the engraftment period of 4-hydroperoxycyclophosphamide-purged bone marrow in breast cancer patients receiving high-dose chemotherapy with autologous bone marrow support. *Blood* **83,** 3132–3137.

Soiffer, R. J., Fairclough, D., Robertson, M., *et al.* (1997). CD6-depleted allogeneic bone marrow transplantation for acute leukemia in first complete remission. *Blood* **89,** 3039–3047.

Storb, R., Pepe, M., Deeg, H. J., *et al.* (1992). Long-term follow-up of a controlled trial comparing a combination of methotrexate plus cyclosporine with cyclosporine alone for prophylaxis of graft-versus-host disease in patients administered HLA-identical marrow grafts for leukemia. *Blood* **80,** 560–561.

Sullivan, K. M., Storb, R., Buckner, C. D., *et al.* (1989). Graft-versus-host disease as adoptive immunotherapy in patients with advanced hematologic neoplasms. *N. Engl. J. Med.* **320,** 828–834.

Sutherland, D. R., Marsh, J. C. W., Davidson, J., *et al.* (1992). Differential sensitivity of CD34 epitopes to cleavage by *Pasteurella haemolytica* glycoprotease: Implications for purification of CD34-positive progenitor cells. *Exp. Hematol.* **20,** 590–599.

Thomas, T. E., Sutherland, H. J., and Lansdorp, P. M. (1989). Specific binding and release of cells from beads using cleavable tetrameric antibody complexes. *J. Immunol. Methods* **120,** 221–231.

Treleaven, J. G., Gibson, F. M., Ugelstad, J., *et al.* (1984). Removal of neuroblastoma cells from bone marrow with monoclonal antibodies conjugated to magnetic microspheres. *Lancet* **8368,** 70–73.

Truitt, R. L., and Atasoylu, A. A. (1991). Impact of pretransplant conditioning and donor T cells on chimerism, graft-versus-host disease, graft-versus-leukemia reactivity, and tolerance after bone marrow transplantation. *Blood* **77,** 2515–2523.

Tseng-Law, J., Kobori, J. A., and Al-Abdaly, F. (1995). Positive and positive–negative cell selection mediated by peptide release, Patent PCT/US95/07491.

Uckun, F. M., and Myers, D. E. (1993). Allograft and autograft purging using immunotoxins in clinical bone marrow transplantation for hematologic malignancies. *J. Hematother.* **2,** 155–163.

Uckun, F. M., Kersey, J. H., Haake, R., *et al.* (1992). Autologous bone marrow transplantation in high-risk remission B-lineage acute lymphoblastic leukemia using a cocktail of three monoclonal antibodies (BA-1/CD24, BA-2/CD9, and BA-3/CD10) plus complement and 4-hydroperoxycyclophosphamide for *ex vivo* bone marrow purging. *Blood* **79,** 1094–1104.

Ugelstad, J., Berge, A., Ellingson, T., *et al.* (1992). Preparation and application of new monosized polymer particles. *Prog. Polym. Sci.* **17,** 87–161.

Ugelstad, J., Mork, P. C., Mfutakamba, H. R., *et al.* (1983). Thermodynamics of swelling of polymer, oligomer and polymer–oligomer particles. Preparation and application of monodispersed particles. *In* "Science and Technology of Polymer Colloids". (G. W. Poehlein, R. H. Ottewill, and J. M. Goodwin, Eds.), pp. 51–99. Nijhoff, Boston.

Ugelstad, J., Stenstad, P., Kilaas, L., *et al.* (1993). Monodisperse magnetic polymer particles. *Blood Purif.* **11,** 349–369.

Vandenberghe, E. A., and Goldstone, A. H. (1996). Autologous stem cell transplants in lymphomas. *Ann. Med.* **28,** 137–143.

Van den Engh, G., and Stokdijk, W. (1989). Parallel processing data acquisition system for multi-laser flow cytometry and cell sorting. *Cytometry* **10,** 282–293.

Wagner, J. E. (1992). Isolation of primitive hematopoietic stem cells. *Semin. Hematol.* **29,** S6–S9.

Watson, J. H. P. (1973). Magnetic filtration. *J. Appl. Phys.* **44,** 4209–4213.

Weisdorf, D., Haake, R., Blazar, B., *et al.* (1990). Treatment of moderate/severe acute graft-versus-host disease after allogeneic bone marrow transplantation: An analysis of clinical risk features and outcome. *Blood* **75,** 1024–1030.

Wigzell, H., and Andersson, B. (1969). Cell separation on antigen-coated columns. *J. Exp. Med.* **129,** 23–36.

Williams, C. D., Goldstone, A. H., Pearce, R. M., *et al.* (1996). Purging of bone marrow in autologous bone marrow transplantation for non-Hodgkin's lymphoma: A case-matched comparison with unpurged cases by the European Blood and Marrow Transplant Lymphoma Registry. *J. Clin. Oncol.* **14,** 2454–2464.

Wofsy, L., Kimura, J., and Truffa-Bachi, P. (1971). Cell separation on affinity columns: The preparation of pure populations of anti-hapten specific lymphocytes. *J. Immunol.* **107,** 725–729.

Wysocki, L. J., and Sato, V. L. (1978). "Panning" for lymphocytes: A method of cell selection. *Proc. Natl. Acad. Sci. USA* **75,** 2844–2848.

Yau, J. C., Reading, C. L., Thomas, M. W., *et al.* (1990). Purging of T-lymphocytes with magnetic affinity colloid. *Exp. Hematol.* **18,** 219–222.

Yerly-Motta, V., Racadot, E., Fest, T., *et al.* (1996). Comparative preclinical study of three bone marrow purging methods using PCR evaluation of residual t(14;18) lymphoma cells. *Leuk. Lymphoma* **23,** 313–321.

Zimmerman, T. M., Bender, J. G., Lee, W. J., *et al.* (1996). Large-scale selection of $CD34^+$ peripheral blood progenitors and expansion of neutrophil precursors for clinical applications. *J. Hematother.* **5,** 247–253.

12

ENVIRONMENTAL REQUIREMENTS OF HEMATOPOIETIC PROGENITOR CELLS IN *EX VIVO* EXPANSION SYSTEMS

PETER W. ZANDSTRA,[*,†]
CONNIE J. EAVES,[‡,§] AND
JAMES M. PIRET[*,†]

[*]Biotechnology Laboratory and [†]Department of Chemical and Bio-Resource Engineering, University of British Columbia, Vancouver, British Columbia V6T 1Z3, Canada
[‡]Department of Medical Genetics, University of British Columbia, Vancouver, British Columbia V6T 1Z3, Canada
[§]Terry Fox Laboratory, British Columbia Cancer Agency, Vancouver, British Columbia V5Z 1L3, Canada

I. INTRODUCTION

The continuity of blood formation throughout adult life is vested in a small population of pluripotent hematopoietic stem cells. These cells have the unique ability to undergo dozens of divisions without any apparent change in their undifferentiated state or ability to meet the daily need to generate billions of new blood cells of multiple types. These remarkable properties distinguish hematopoietic stem cells from their lineage-restricted progeny, although the latter may still give rise to thousands, even millions, of mature progeny. Nevertheless, no known lineage-restricted hematopoietic progenitor appears able to sustain the *in vivo* production of blood cells for more than a few weeks. Such considerations have highlighted the importance of hematopoietic stem cell production as the ultimate goal of most *ex vivo* cell expansion systems and hence the need to identify *in vitro* conditions that will support not only the proliferation and differentiation of these cells but also the amplification of their numbers.

The ability of hematopoietic cells from genetically identifiable clones regenerated in primary mice to engraft and repopulate the blood-forming systems of multiple secondary recipients has provided rigorous experimental evidence that adult murine hematopoietic stem cells remain capable of further expansion (Wu *et al.,* 1968; Dick *et al.,* 1985; Lemischka *et al.,* 1986; Keller and Snodgrass, 1990; Pawliuk *et al.,* 1996). A similar approach has shown that murine hematopoietic stem cells can be stimulated to divide in culture without loss of their original properties (Fraser *et al.,* 1992). Recently, conditions that support a net expansion of both murine and human stem cells *in vitro* have been identified (Bhatia *et al.,* 1997; Conneally *et al.,* 1997; Miller and Eaves, 1997). Such findings, together with previous studies demonstrating the rapid and marked increases *in vitro* of more differentiated cell types that appear more readily obtainable, have greatly spurred interest in the potential of bioreactor systems to provide optimized and reproducible expansions of hematopoietic cells *ex vivo* in clinically useful numbers. This chapter reviews some of the progress that has been made in this evolving area, including an analysis of bioreactor operating parameters and design principles investigated thus far, as well as some of the experimental culture results which advances in bioreactor technology should be able to realize on the larger scales required for clinical applications.

II. POTENTIAL APPLICATIONS OF CELL EXPANSION AS A CELLULAR THERAPY

There are many clinical situations where *ex vivo* expanded hematopoietic cells could be useful. Initial interest in this area focused on the potential use of such cells to further abrogate the early phase of neutropenia and thrombocytopenia, which can compromise patient recovery from high-dose therapy (To *et al.*, 1992; Williams *et al.*, 1996). Specific growth factors are required for the *in vitro* amplification of different types of mature cells and, in theory, these could allow the production of very large numbers of granulocytes, macrophages, erythrocytes, platelets, natural killer (NK) cells, and other myeloid or lymphoid cells from the sorts of primitive populations that can be readily accessed by current clinical procedures. For example, the recent identification of *in vitro* conditions for generating large numbers of dendritic cells from $CD34^+$ cells (Szabolcs *et al.*, 1995) has focused interest on the use of this technology to develop novel immunotherapy protocols requiring large numbers of these efficient antigen-presenting cells to induce the production *ex vivo* of cytotoxic T lymphocytes as specific effectors against foreign or cancer-specific antigens.

The bone marrow and peripheral blood of patients with many types of leukemia typically contain both normal and leukemic stem cells and, in some leukemias, the normal and leukemic stem cells appear to differ in their ability to be maintained under certain conditions *in vitro* (Coulombel *et al.*, 1983b, 1985; Udomsakdi *et al.*, 1992). One approach to the treatment of these diseases exploits this differential behavior of normal and leukemic stem cells *in vitro* to remove residual leukemic stem cells from blood or marrow autografts prior to their reinfusion (Barnett *et al.*, 1997). Addition of specific cytokines to the medium in which the autografts are incubated may allow an even greater therapeutic benefit to be achieved by causing a selective amplification of the normal stem cell population (Petzer *et al.*, 1997) or by initiating an immunotherapeutic reaction within the autograft that can be magnified by low-dose immunostimulation *in vivo* immediately posttransplant (Klingemann *et al.*, 1993, 1994; Wong *et al.*, 1996).

Hematopoietic stem cells are of interest for many gene therapy applications because of their ability to produce large numbers of progeny for the lifetime of the individual. However, retroviral infection requires that the target cells be stimulated to divide (Miller *et al.*, 1995). Thus, the design of a culture system in which hematopoietic stem cells can be rapidly and efficiently activated without alteration of their other biological properties is an important requirement for clinical situations where the genetic marking or modification of these cells is desired.

There is also increasing interest in the possibility of using expanded hematopoietic stem cell transplants to overcome histocompatibility barriers in allogeneic hosts or engraftment barriers in nonmyeloablated-syngeneic hosts. In both of these situations, the effectiveness of simply increasing the number of stem cells transplanted (Brecher *et al.*, 1982; Stewart *et al.*, 1993; Shizuru *et al.*, 1996) has created new clinical opportunities for the use of expanded stem cell populations pending the development of methods suitable for achieving this goal on a routine basis. The extent of this challenge is brought home by the knowledge that current transplant protocols already require the infusion of large numbers of cells to ensure rapid and sustained engraftment. Thus, even for current protocols, *ex vivo* expansion technology could be useful to reduce the number of cells that would need to be harvested and hence permit less invasive procedures to be used to obtain them.

III. DEVELOPMENT OF HEMATOPOIETIC CELL CULTURE SYSTEMS SUITABLE FOR CLINICAL SCALE-UP

A. STATIC CULTURE SYSTEMS

The first culture system in which the production of hematopoietic cells could be reproducibly sustained for many weeks was described for murine bone marrow cells by Dexter and colleagues in 1977 (Dexter *et al.,* 1977). Following the recognition of the importance of adding corticosteroids to the medium (Greenberger, 1978), this system was also shown to be applicable to human bone marrow cells (Gartner and Kaplan, 1980). Critical for the establishment of durable hematopoiesis in these cultures, which require no addition of growth factors, was the rapid formation (or simple provision) of a supportive adherent feeder layer containing cells of the fibroblast, endothelial, and adipocyte lineages (Dexter *et al.,* 1980). Subsequent studies showed that the more primitive hematopoietic cells in such cultures actually become concentrated in the adherent layer (Mauch *et al.,* 1980; Coulombel *et al.,* 1983a) and their maturing granulopoietic progeny are then continuously released into the nonadherent fraction (Slovick *et al.,* 1984). Many aspects of this system, including some of the cellular interactions that take place between the hematopoietic cells and the stroma, have been shown to mimic hematopoiesis *in vivo* (Dexter *et al.,* 1980; Eaves *et al.,* 1993a). However, because of the complex cellular composition of the *in vitro* system, the critical molecular mediators responsible for sustaining and stimulating the most primitive hematopoietic cells cultured in the presence of stroma have not yet been fully characterized.

In stromal cell-containing cultures, the pH and concentrations of nutrients and metabolic byproducts (e.g., lactic acid) to which the cells are exposed fluctuate greatly outside physiological ranges (Koller and Palsson, 1993). These fluctuations may be responsible, at least in part, for the eventual decrease observed within 2–3 months in the numbers of total and progenitor hematopoietic cells (Eaves *et al.,* 1991a). Nevertheless, this type of culture system has attracted considerable interest because, until recently, it provided the only conditions which allowed repopulating stem cells to proliferate and remain detectable for more than a few days (Fraser *et al.,* 1992).

A number of empirical approaches have been tested to improve the productivity and longevity of stromal cell-containing hematopoietic culture systems. One has been to simply increase the frequency of medium exchange. This alone can significantly increase the output of hematopoietic cells in such cultures (Slovick *et al.,* 1984; Schwartz *et al.,* 1991a, 1991b; Koller *et al.,* 1993a, 1993b). These findings, together with the knowledge that the addition of fresh, serum-containing medium activates the cycling of intermediate progenitor cell types (Toksoz *et al.,* 1980; Cashman *et al.,* 1985), led to the development of several continuous perfusion culture systems.

The first of these were either flat-plate bioreactors consisting of 25-cm^2 rectangular glass inserts that served to promote adherent cell attachment (Koller *et al.,* 1992b) or membrane bioreactors in which the hematopoietic cells were inoculated into a collagen porous microcarrier matrix in a compartment enclosed by ultrafiltration membranes (Wang and Wu, 1992). In the flat-plate bioreactors, medium was continuously recirculated from a pH- and O_2-controlled reservoir with weekly exchange of half the bioreactor medium (Koller *et al.,* 1992a, 1992c). In the porous microcarrier bioreactor, serum-free medium (replaced daily) was continuously recirculated unidirectionally above and below the unperfused serum- and collagen bead-containing center chamber (Wang and Wu, 1992). The center chamber was fed serum-containing medium periodically upon cell sampling.

Each of these systems was reported to increase both total cell and CFU-GM production from bone marrow cells compared to stromal layer-containing cultures established in standard plastic tissue culture flasks and fed by weekly exchange of half the medium (Koller *et al.,* 1992b; Wang and Wu, 1992). The collagen-chamber bioreactor has subsequently been shown to support multilineage (lymphoid, erythroid, granulocyte, macrophage, and megakaryocyte) murine hematopoiesis when maintained at 37°C for periods of up to 60 days without any exogenously added cytokines (Wang *et al.,* 1995). In the flat-plate bioreactor, it has been possible to obtain significant net expansions of LTC-IC (long-term culture initiating cells, approximately 3-fold over input) from human bone marrow inocula using cytokine-supplemented media (Koller *et al.,* 1993a). In both, oxygenation of the medium occurred in the reservoir. Thus oxygen gradients would have been likely to develop in the direction of medium flow (Piret and Cooney, 1991). Moreover, since the medium in the reservoir was replaced intermittently, abrupt changes in the concentrations of nutrients, cytokines, and inhibitory metabolites to which the cells were exposed would also have occurred on a recurring basis.

Palsson *et al.* (1993) developed a modified bioreactor system to reduce oxygen gradients in the direction of medium flow. In this system, a gas-permeable membrane along the entire length of the culture bed provided independent oxygenation of the medium. Their modified design was operated with continuous, single-pass medium perfusion to minimize temporal gradients, although axial spatial gradients of soluble nutrients and inhibitory metabolites would still be present. This system was subsequently scaled-up 10-fold to enable inoculation of approximately 3×10^7 light-density human bone marrow cells per bioreactor, and in this system 10-, 21-, and 7.5-fold net expansions of total cells, CFU-GM, and LTC-IC, respectively, were observed (Koller *et al.,* 1993b). Cell proliferation, however, may have still been limited by the surface area available since an overall 2- to 3-fold further increase in cell and progenitor expansions could be achieved by intermittent harvesting or by using the expanded cells to initiate new reactors (Oh *et al.,* 1994; Palsson *et al.,* 1995).

Highfill *et al.* (1996) also reported results for murine bone marrow cells using a packed glass fiber matrix bed bioreactor. They found that the long-term (greater than 10 weeks) generation of cells achieved in this system (in the absence of exogenously added cytokines) was dependent on the marrow cell inoculum density, regardless of whether a stromal cell layer was already established on the matrix before a source of primitive hematopoietic cells was added or whether the conditioned medium was recirculated.

B. STIRRED SUSPENSION CULTURES

Appreciation of the undesirable features and practical difficulties of designing static, clinical-scale bioreactors containing fixed stromal adherent layers, plus the discovery that contact is not required for stromal cells to support the most primitive hematopoietic cells (Verfaillie, 1992), encouraged the exploration of stirred suspension culture systems. This alternative also offered the attractive advantages of ready scalability and relative simplicity. Moreover, because of their relatively homogeneous nature, stirred suspension cultures are uniquely suited for investigations of different culture parameters (e.g., O_2 tension, cytokine concentration, serum components, medium exchange rates, etc.) that may influence the viability and turnover of specific stages and types of hematopoietic cells.

Sardonini and Wu (1993) were the first to document the behavior of hematopoiesis in a spinner culture system inoculated with human marrow cells. In their studies, total cell numbers could be increased 10- to 20-fold over input values over a 4-week period with maintenance of CFU-GM numbers. By increasing the initial marrow cell density and addition to the medium of a more complex mixture of more concentrated cytokines (FL, Flt-3 ligand; SF, stem cell factor; IL-3, interleukin-3; IL-6, interleukin-6; IL-11, interleukin-11), we were able to demonstrate reproducible CFC (colony-forming cell) and LTC-IC expansions of 66- and 9-fold, respectively, in a similar stirred suspension system (Zandstra *et al.,* 1994, 1997). Such outputs have been shown to be dependent on input cell density, vessel configuration (specifically, the ratio of impeller diameter to vessel diameter), and medium composition (Zandstra *et al.,* 1994; Collins *et al.,* 1996).

C. STROMA-FREE CULTURES

Recent studies (see later) have identified cytokine cocktails that, *in vitro,* can stimulate a significant net amplification of both murine (Miller and Eaves, 1997) and human (Bhatia *et al.,* 1997; Conneally *et al.,* 1997) hematopoietic stem cells in the absence of any stromal elements. Thus, there is a potential role for *in vitro* bioreactor systems in which all medium components would be molecularly defined. Interest in the development of such systems is enhanced by the knowledge that cord blood and mobilized peripheral blood represent clinically relevant sources of hematopoietic stem cells but may be devoid of stromal cells and their precursors. One recently described example is a flat-plate, perfusion culture system (Sandstrom *et al.,* 1996). In this bioreactor the culture chamber contains multiple microgrooves perpendicular to the direction of medium flow. This design facilitates the diffusion of nutrients to multiple layers of cells while retaining them in the culture chamber. The dimensions of the grooves (200 × 200 μm) have been chosen based on the calculated constraint of providing oxygen above 60% of air saturation at the bottom of the last groove. In tests of this system, significant expansions of total cells (10- to 15-fold) and CFU-GM (15- to 25-fold) from mobilized peripheral blood precursors have been observed (Sandstrom *et al.,* 1996). Spinner cultures have also been used to amplify progenitor populations from cord blood inocula. For example, Kogler *et al.* reported 50- to 100-fold expansions of CFC and 5- to 15-fold expansions of LTC-IC in stirred suspension cultures initiated with enriched populations (44–72% purity) of $CD34^+$ cord blood cells using FL-supplemented medium (Kogler *et al.,* 1996a, 1996b).

Gas-permeable, tissue culture bags, in which the cells generally grow undisturbed until harvested, have also been used successfully for the culture of large numbers of human hematopoietic cells (Purdy *et al.,* 1995; Williams *et al.,* 1996). In these studies, the size of the $CD34^+$ cell population isolated from the marrow or blood of normal individuals could be expanded 30- to 120-fold when appropriate cytokines were added to the culture medium. The inability of cells to become adherent in plastic bags makes it easier to harvest the cells under aseptic conditions, although it also makes it more cumbersome to exchange the medium without losing cells or increasing the volume. In addition, due to the static nature of these cultures, the cells settle onto the lower surface of the bag in multilayers and are, therefore, likely to be exposed to undesirable gradients of oxygen and medium components.

These initial investigations of bioreactor configurations—or potential configurations—have provided insights into potential approaches to optimize the growth of hematopoietic progenitors *ex vivo.* They have also provided information about their various limitations. In the following section, we will focus on a review of what has been learned about general bioreactor operating parameters and their specific effects on the output of functionally and phenotypically defined populations of hematopoietic cells.

IV. REFINEMENT OF BIOREACTOR OPERATING PARAMETERS

A. OXYGEN TENSION

Oxygen tension is an important culture parameter because of its low solubility at 37°C (~0.2 mM) when in equilibrium with air (21% O_2 v/v). As a result, the oxygen in medium can be readily depleted and become limiting, especially in cultures containing multiple layers of cells. Numerous studies have documented effects of different oxygen levels on the proliferation of hematopoietic cells *in vitro,* both in semisolid and in liquid cultures. These studies have been motivated primarily by the observation that, *in vivo,* oxygen tensions in the marrow are on the order of 2–7% (Lindop and Rotblat, 1960) as well as reports that oxygen tensions ranging from 1 to 10% can enhance both the size and number of hematopoietic colonies obtained in semisolid media in comparison to parallel cultures maintained in atmospheric oxygen (Bradley *et al.,* 1978; Rich and Kubanek, 1982; Broxmeyer *et al.,* 1990). Increased growth of hematopoietic cells at low oxygen concentrations has been attributed, in part, to an increased responsiveness of cells to growth factors (Rich and Kubanek, 1982), to increased fibroblast cell proliferation (Bradley *et al.,* 1978), and to a decreased generation of hydrogen peroxide, oxygen radicals, and other oxygen intermediates produced by mature granulocytes (Pick and Mizel, 1981; Darfler and Insel, 1983; Meagher *et al.,* 1988). In some cases, these conclusions have been further supported by studies of the effects of adding various oxygen radical scavengers such as catalase or mannitol.

Probably because of difficulties in measuring the oxygen tension in the culture environment, attempts to extend these results to liquid culture systems have given conflicting results (Table 12.1). Thus, although the oxygen tension in the bulk medium has been controlled in the studies shown in Table 12.1, the concentration of oxygen in the immediate microenvironment of the cells was not directly determined. Inconsistencies in the results obtained could, therefore, simply reflect differences in oxygen depletion due to differences in cell density (Koller *et al.,* 1992a, 1992b, 1992c; Palsson *et al.,* 1993). The effects of oxygen tension reported in these latter studies may be further confounded by the different cytokines used, since these may differently influence cell-specific oxygen radical production as shown by studies with M-CSF (Broxmeyer *et al.,* 1990).

In addition, more mature cells may be less sensitive to changes in their oxygen environment than their immediate precursors (Table 12.1) (Cipolleschi *et al.,* 1993), suggesting that the definition of an "optimal" oxygen tension may depend not only on the cytokines used to stimulate the cells but also on the particular cells being stimulated. Thus, under *in vitro* conditions where there was little or no net expansion of very primitive cells (LTC-IC), low oxygen conditions resulted in the maintenance of higher numbers of these cells (Koller *et al.,* 1992a), whereas greater LTC-IC expansions (Table 12.2) (Koller *et al.,* 1993b) were subsequently shown to occur under conditions that included higher oxygen concentrations (Table 12.1) (Palsson *et al.,* 1993). Overall, these results indicate that oxygen tension is likely an important, although still incompletely understood, determinant of hematopoietic cell production. Additional investigations of the effects of controlled manipulations of oxygen tension in parallel with measurements of oxygen concentrations in the cellular environment should be informative, particularly in relation to differences in the types of cells present and/or being stimulated.

TABLE 12.1 Apparent Effects of Differences in Bulk Medium Oxygen Tension on Progenitor Cell Expansion

Oxygen bulk levels (% v/v)	Culture system	Species	Cell type assessed	Optimal O_2 concentration	Reference
5 20	Tissue culture flask	Human	CFU-GM	5	Koller *et al.*, 1992a
5 20	Tissue culture flask	Human	CFU-GM, BFU-E and CFU-GEMM numers	5	Koller *et al.*, 1992c
5 20	Perfusion bioreactor	Human	CFC	5	Koller *et al.*, 1992b
5 20 40 60	Perfusion bioreactor	Human	CFU-GM and BFU-E	20	Palsson *et al.*, 1993
1 20	Tissue culture flask	Mouse	Total cells, CFC, marrow repopulating ability (MRA)	20 1	Cipolleschi *et al.*, 1993

TABLE 12.2 Summary of Cytokine-Stimulated Expansion of Human Hematopoietic Cell Types in Cultures Initiated with Mononuclear or $CD34^+$ Cells

	Culture parameters		Maximum expansions					Growth factors (ng/ml)										
					CFC													
Cell type[a]	Size (ml)	Innoculum density (ml^{-1})	Vessel	TC	E	GM	LTC-IC	Epo	G-CSF	GM-CSF	IL-1	IL-3	IL-6	IL-11	SF	Tpo	FL	Reference
BM MNC	100	3×10^6	Bag	10	9	6				50*	5							Lemoli *et al.*, 1992
PB $CD34^+$	50	5×10^3	T-flask	28	n.r.	26		10	10	10	10	10		10				Haylock *et al.*, 1992
PB $CD34^+$	100	1.5×10^4	T-flask	205		170	n.r.	1*			100	100*	100*		100			Brugger *et al.*, 1993
BM MNC	360	3×10^7/vessel	Flat-bed bioreactor	10	12	21	7.5	0.1*		5		2			10			Koller *et al.*, 1993b
BM MNC	n.r.	$10^5/cm^2$	Flat-bed bioreactor	25	5	23	n.r.	0.1*		5		5			10			Palsson *et al.*, 1993
BM MNC	250	5×10^5	Spinner	21		3	n.r.			0.3		0.5			100			Sardonini and Wu, 1993
PB MNC	n.r.	1×10^6	Flat-bed bioreactor	22	n.r.	25	16	0.1*		5		2			10			Van Zant *et al.*, 1994
BM MNC	40	10^6	Spinner	16		35	7					2			10			Zandstra *et al.*, 1994

PB $CD34^+$	100	5×10^4	T-flask	64	50	1.5	1		3	100	100		10			Henschler *et al.*, 1994
CB $CD34^+$	300	4×10^4	Bag	75	60	17	5*	10^3*		50	20		20			Moore and Hoskins, 1994
$CD34^+$ PB	100	3×10^4	T-flask	62	50	n.r.										Brugger *et al.*, 1995
BM $CD34^+$	250	2×10^4	Bag	n.r.	120	3		100		100	100		100			Purdy *et al.*, 1995
PB $CD34^+$	120	5×10^3	Grooved bioreactor	113	18	0.76		150*		150*	40		50			Sandstrom *et al.*, 1996
PB $CD34^+$	100	$(0.5-1) \times 10^5$	Bag	26	5	n.r.		100[b]		100[b]						Williams *et al.*, 1996
CB $CD34^+$	40	1×10^4	Spinner	150	60	8				100			100		100	Kogler *et al.*, 1996b
CD $CD34^+$	1	$(1-3) \times 10^3$	24-well plate	n.r.	197 60	n.r.				20		40	10	100		Ohmizono *et al.*, 1997
CD $CD34^+$	1	$(0.2-1) \times 10^4$	24-well plate	$>10^5$	$>10^5$	$>10^4$								10*	50	Piacibello *et al.*, 1997
BM MNC	40	1×10^6	Spinner	14	66	9				10	10	10	50		50	Zandstra *et al.*, 1997b

[a]BM, bone marrow; PB, peripheral blood; CB, cord blood.
[b]PIXY321 (100 ng/ml of the fusion product of IL-3 and GM-CSF) was used in this study.
Note: *indicates U/ml; n.r. = not reported.

B. PERFUSION/MEDIUM EXCHANGE RATE

Traditional static marrow cultures initiated with large inocula of unseparated cells and fed by periodic replacement of part of the culture medium with fresh medium can be limited by nutrient and cytokine depletion and by exposure to inhibitory compounds (Cashman *et al.,* 1985; Caldwell *et al.,* 1991b; Zandstra *et al.,* 1997b). Medium exchanges have been shown to stimulate the proliferation of quiescent, primitive clonogenic progenitor cells present in the adherent layer of long-term marrow cultures likely as a result of the activation of endogenous sources of cytokines following the addition of fresh serum (Caldwell *et al.,* 1991b; Eaves *et al.,* 1993a). Increasing the medium exchange rate in static cultures (Cashman *et al.,* 1985; Schwartz *et al.,* 1991) and in several intermitXtently fed perfusion culture systems (Koller *et al.,* 1992; Wang and Wu, 1992; Sandstrom *et al.,* 1995, 1996) has also been found to increase total cell and progenitor cell output. Cell-specific glucose consumption and lactate production both increase with faster medium exchange rates and increased serum supplementation (Caldwell *et al.,* 1991a). On the other hand, the faster introduction of fresh medium also dilutes metabolic byproducts and inhibitory factors more rapidly (Koller *et al.,* 1995b; Lauffenburger *et al.,* 1996; Zandstra *et al.,* 1997b). Thus several mechanisms may operate to enhance hematopoietic progenitor production *in vitro* when increased medium exchange rates are used.

In vivo perfusion rates have been estimated to be 0.1 ml per cm^3 of bone marrow tissue per minute (Martial *et al.,* 1987). Assuming *in vivo* cell densities on the order of 5×10^8 cells per cm^3, this perfusion rate corresponds to $\sim$0.2 ml of serum per 10^6 cells per day or, for 20% serum-containing medium, one medium exchange per day (Koller and Palsson, 1993). However, the applicability of these numbers to *in vitro* cell expansion systems may be questionable since the microenvironmental conditions and extracellular fluid composition in each case may be very different.

C. CULTURE INOCULUM, CELL DENSITY, AND ACCESSORY CELLS OF STROMAL ORIGIN

The role *in vitro* of mature hematopoietic and nonhematopoietic cells that are constituents of bone marrow tissue *in vivo* (i.e., macrophages, adipocytes, endothelial cells, and adventitial reticular cells) has been a subject of ongoing controversy. It is now clear that soluble hematopoietic growth factors can replace, at least for short periods of time, the supportive, stimulatory function of marrow stromal cells (Brandt *et al.,* 1990; Sutherland *et al.,* 1993; Simmons *et al.,* 1994; Levesque *et al.,* 1996; Miller and Eaves, 1997; Zandstra *et al.,* 1997a). However, cultures containing a "feeder" layer of stromal cells have generally been found to sustain stem cells longer than those carried out in their absence (Dexter *et al.,* 1980; Rowley *et al.,* 1993; Wineman *et al.,* 1993; Koller *et al.,* 1995b) although the precise role of the stromal cells in these cultures is still not known. Stromal cells produce a variety of soluble cytokines and extracellular matrix components, as well as presenting cellular adhesion molecules and membrane-bound cytokines that are recognized by specific types of hematopoietic cells (Eaves *et al.*, 1991b; Guba *et al.*, 1992; Long *et al.*, 1992; Quesenberry *et al.*, 1993; Verfaillie *et al.*, 1994). Consequently, the presence of such a substratum can deliver highly localized and potent signals, both positive and negative, even when these are produced at low levels (Cashman *et al.*, 1990; Otsuka *et al.*, 1991). However, there may also be other important molecular interactions in such cultures that have not yet been defined or characterized.

Studies showing enhanced primitive progenitor cell production when hematopoietic cells are separated from the stromal layer suggest that direct contact between LTC-IC and the supportive cells of the adherent layer is not an essential part of the mechanism by which the LTC-IC are stimulated under these conditions (Verfaillie, 1992). Our results showing LTC-IC expansion in stirred suspension culture supplemented with cocktails of cytokines also support these observations (Zandstra *et al.*, 1994, 1997b).

As discussed earlier, increased medium supplementation rates have yielded significant LTC-IC expansions from light-density normal marrow cells when the cells were inoculated either at a high density (Zandstra *et al.*, 1994; Koller *et al.*, 1996) or onto a preformed stroma (Koller *et al.*, 1993a, 1996). In contrast, LTC-IC numbers were just maintained in cultures initiated with purified $CD34^+lin^-$ (lineage negative) cells, even in the presence of preformed stroma and with rapid medium exchange rates (Koller *et al.*, 1996). In the absence of stroma, LTC-IC numbers from highly purified cells decreased regardless of the rate of medium exchange (Koller *et al.*, 1995). This further supports the hypothesis that increased medium exchange rates enhance factor secretion by stromal cells. When purified cells were remixed with increasing numbers of $CD34^-$ cells, LTC-IC expansion ratios were inversely correlated with input $CD34^+lin^-$ purity (regardless of the presence of stroma), such that maximum LTC-IC expansions were obtained at $CD34^+lin^-$ purity typical of unfractionated bone marrow cells (Koller *et al.*, 1995b). This suggested that $CD34^-$ cells are an important source of stimulatory factors. Similarly, we found that hematopoietic progenitor expansions were cell concentration dependent in stirred suspension cultures initiated with unseparated human marrow cells, again supporting the concept that endogenously produced factor(s) can play an important stimulatory role (Zandstra *et al.*, 1994). Moreover, bioassays of medium from such 1-week-old stirred cultures supplemented only with SF and IL-3 revealed levels of IL-6 and G-CSF on the order of 1–10 ng/ml (Zandstra *et al.*, 1994).

Taken together, these results support a model whereby factors secreted by *both* stromal and more differentiated hematopoietic (i.e., $CD34^-$) cells can affect the maintenance and proliferation of primitive hematopoietic cells. It is also important at this point to emphasize that none of the foregoing bioreactor expansion studies were performed in the presence of FL or TPO, two subsequently isolated factors found to have direct stimulatory effects on primitive hematopoietic cells (see next subsection).

D. CYTOKINE SUPPLEMENTS

The ability of particular growth factor supplements to stimulate an expansion of primitive hematopoietic cell populations *in vitro* has been extensively investigated. Not surprisingly, these studies have mirrored the cloning and production by recombinant technology over the past 15 years of a large number of factors that can stimulate primitive hematopoietic cells, either alone or in combination. For example, Haylock *et al.* (1992) showed that significant CFU-GM expansion (26-fold) could be obtained from purified CD34$^+$ cells isolated from mobilized peripheral blood and maintained in static cultures in the presence of IL-1, IL-3, IL-6, G-CSF, and SF. Using similar protocols, many other groups have reproduced these results, attesting to their generality (Brugger *et al.,* 1993; Sato *et al.,* 1993; Srour *et al.,* 1993; Fel *et al.,* 1994; Henschler *et al.,* 1994). Such studies revealed that the extent of hematopoietic progenitor cell expansion was greater when a purified starting population of CD34$^+$ cells was used, suggesting that the presence of large numbers of mature cells, or their secreted metabolites, can exert strong negative effects on hematopoietic cell proliferation (Emerson, 1996). Moreover, the duration as well as the magnitude of progenitor expansion could be increased by the adoption of suitable medium dilution protocols (Moore and Hoskins, 1994; Van Zant *et al.,* 1994). As discussed earlier, the continuous or semicontinuous addition of medium can even reduce the need for CD34$^+$ cell purification of the input inoculum (Koller *et al.,* 1993a; Palsson *et al.,* 1993), presumably by providing an adequate rate of dilution of inhibitory cells or metabolites.

A summary of many of these expansion reports is presented in Table 12.2. The studies of progenitor expansion in cultures of CD34$^+$ cells isolated from mobilized peripheral blood have repeatedly shown cytokine combinations containing SF, IL-3, IL-1, IL-6, and G-CSF to be the most effective. The extent and longevity of both total cell and CFC expansion in these cultures are also correlated with the concentration of the cytokines. For example, Haylock *et al.* (1992), who added these cytokines at 10 ng/ml, obtained 28- and 26-fold total cell and CFU-GM expansions, whereas Brugger *et al.* (1993), who used 100 ng/ml concentrations, obtained 200- and 170-fold CFC expansions. Henschler *et al.* (1994) and Sandstrom *et al.* (1996) added similar cytokines at intermediate concentrations (10–100 ng/ml) and obtained intermediate total cell (60- to 100-fold) and CFC (20- to 50-fold) expansions. However, LTC-IC numbers appeared, at best, to be just maintained in these studies (although they were not always monitored). In other reports, similar expansions of total cells and CFC with lower but, nevertheless, significant expansions of LTC-IC were obtained when light-density marrow cells were cultured in a medium that was supplemented with only SF and IL-3, and in some studies GM-CSF (granulocyte–macrophage colony-stimulating factor) also, all at low concentrations (i.e., ≤ 10 ng/ml) (Koller *et al.,* 1993a; Zandstra *et al.,* 1994). The lack of a requirement for IL-1, IL-6, and G-CSF supplementation in these latter cultures may be attributed to the substantial quantities of these factors produced endogenously by the stromal and accessory cells also present (Zsebo *et al.,* 1988; Eaves *et al.,* 1991b; Guba *et al.,* 1992). On the other hand, factorial analysis of results obtained with FL-containing cytokine combinations showed that FL, SF, and IL-3 were sufficient to obtain 30-fold amplification of LTC-IC in 10-day cultures initiated with highly purified inocula of $CD34^+CD38^-$ marrow cells (Petzer *et al.,* 1996b).

Investigations of the cytokine concentration dependence of LTC-IC amplification from $CD34^+CD38^-$ marrow cells (Zandstra *et al.,* 1997a) have revealed that 30-fold higher cytokine concentrations are required for maximal expansion of LTC-IC (30- to 80-fold over input) than are sufficient to obtain near-maximal expansions of CFC. These latter studies have also demonstrated the importance of the *relative* concentration of the cytokines in a given "optimal" combination. In particular, a higher concentration of FL was found to be critical for enhancing the amplification of LTC-IC. In addition, negative concentration dependent effects of IL-3 on LTC-IC numbers were observed when the concentrations of both FL and SF were reduced (Zandstra *et al.,* 1997a). Interestingly, our previous studies had shown that IL-3 is the only factor of many that on its own can expand the CFC population from a starting population of $CD34^+CD38^-$ human bone marrow cells (Petzer *et al.,* 1996b). These results, along with the demonstrated ability of IL-3 and IL-1 to block the cytokine-stimulated generation of lymphoid precursors from primitive murine hematopoietic cells (Hirayama *et al.,* 1994) and to impair their self-renewal *in vitro* (Yonemura *et al.,* 1996), underscore the possibility of detrimental consequences of the indiscriminate inclusion of high concentrations of IL-3 (or IL-1) in cytokine cocktails used for the activation or expansion of human stem cell transplants for clinical applications.

A possible role of FL in LTC-IC stimulation was first suggested by the long-term negative effects on hematopoiesis obtained in experiments in which antisense FL was added to LTC (Small *et al.,* 1994) followed by the long-term positive effects obtained in experiments in which exogenous FL was added repeatedly to such cultures (Gabbianelli *et al.,* 1995). Addition of FL in combination with various other cytokines to both human marrow and cord blood cultures was subsequently found to allow significant (>10-fold) LTC-IC expansions from both of these progenitor sources (Kogler *et al.,* 1996b; Petzer *et al.,* 1996a; Zandstra *et al.,* 1997a). In fact, FL is the only factor thus far found to stimulate a net increase in LTC-IC numbers when used alone (Petzer *et al.,* 1996b). Taken together, these results suggest a unique role for FL in activating events that allow primitive hematopoietic cells to maintain their ability to be detected as LTC-IC. Additional studies have shown that this effect of FL is specifically associated with the mitogenic activation of a subset of CD34$^+$ cells that express low levels of CD38 (Nordon *et al.,* 1997) and contain only a portion of all initially detectable LTC-IC (Sauvageau *et al.,* 1994; Conneally *et al.,* 1997). It has also been reported that FL in combination with SF, Pixy-123, IL-3, GM-CSF, and Epo can enhance the plating efficiency of high-proliferative potential CFC (Broxmeyer *et al.,* 1995; Hirayama *et al.,* 1995; McKenna *et al.,* 1995; Muench *et al.,* 1995).

Recently, the addition of just FL and TPO to suspension cultures of CD34$^+$ cord blood cells was found to result in a sustained (>20 week) and marked expansion of many types of primitive cells, including LTC-IC (Piacibello *et al.,* 1997). Evidence of other factors that may have the ability to stimulate or enhance the production of very primitive hematopoietic cells is also emerging, although these have not yet been examined in comparable detail. Included among this group are basic fibroblast growth factor (FGF) (Gabbianelli *et al.,* 1990), other factors such as IL-11 and IL-12 that activate members of the family of gp130-like signal transducers (Jacobsen *et al.,* 1993; Miller and Eaves, 1997), and products of the notch ligand (Moore *et al.,* 1997) and Wnt gene families (Austin *et al.,* 1997).

It should be noted that, although qualitatively similar progenitor expansions have been obtained in cytokine-supplemented cultures of adult marrow and newborn (cord) blood cells, some interesting differences have also been reported. For example, 30- to 250-fold higher expansions of CD34$^+$ cells were achieved over a 5-week period with inputs of cord blood versus adult marrow cells, even when the same number of CD34$^+$CD45RA^{lo}CD71^{lo} cells were initially present (Lansdorp *et al.,* 1993; Moore, 1993). The extent to which the total cell and CFC populations were expanded was also higher in the cord blood experiments. Although some of this differential expansion may be attributable to a higher frequency of functionally defined primitive cells in cord blood, even within phenotypically identified subsets of CD34$^+$ cells (Conneally *et al.,* 1997), the higher rates of cell accumulation in the cord blood cultures suggest additional, ontogeny-related changes in cell cycle times.

Differences in the cytokine responsiveness of phenotypically similar cell populations isolated from cord blood and adult marrow have also been reported. Thus, both total and CD34$^+$ cell numbers were found to indicate synergistic responses of cord blood cells to SF, IL-11, and G- or GM-CSF (van de Ven *et al.,* 1995), whereas only additive (at best) responses of adult marrow cells to these cytokines were detected. Conversely, Tsujino *et al.* (1993) have shown significant increases in the short-term generation of clonogenic cells from CD34$^+$lin$^-$HLA-DR$^-$ marrow cells with the addition of IL-3 to SF-containing cultures, whereas no differences were reported between SF alone and SF plus IL-3-containing medium for similarly maintained cultures of cord blood cells. Factorial analysis of 10-day cultures initiated with CD34$^+$CD38$^-$ marrow cells has shown that a combination of FL, SF, and IL-3 is necessary and sufficient to stimulate maximal production of LTC-IC, whereas in similar cultures of cord blood cells, this required a combination of FL and IL-6/sIL-6R (Zandstra *et al.,* 1998). These results, like those of Deutsch *et al.,* suggest that the magnitude of the cytokine-inducible proliferative and differentiative responses of specific progenitor subclasses is subject to changes that occur during development (Deutsch *et al.,* 1995). Signaling through gp130 may be particularly important in this regard (Deutsch *et al.,* 1995; Sui *et al.,* 1996; Miller and Eaves, 1997; Zandstra *et al.,* 1998). Given the potential impact of successful *ex vivo* expansion of primitive progenitors from cord blood sources, the need to understand their unique cytokine requirements may be of particular importance.

Numerous other cytokines of both hematopoietic and nonhematopoietic origin have been reported to influence the proliferative activity and differentiation behavior of primitive hematopoietic cells. For example, transforming growth factor-β (TGF-β) and macrophage inflammatory protein-1α (MIP-1α) have both been shown to inhibit the cell cycle progression of primitive normal hematopoietic cells (Keller *et al.,* 1988; Sing *et al.,* 1988; Cashman *et al.,* 1990; Eaves *et al.,* 1991b; Graham and Pragnell, 1992; Eaves *et al.,* 1993b). A negative effect of added TNF-α (tumor necrosis factor α) on LTC-IC expansion *in vitro* has also been observed (Petzer *et al.,* 1996b). Recent studies suggest that this may result from a direct nontoxic but deterministic action of the added TNF-α factors (Maguer-Satta *et al.*, 1998).

V. RECEPTOR STIMULATION AND MEDIUM CYTOKINE DEPLETION

Cellular responses to cytokines are initiated by the specific high-affinity binding of the cytokines to soluble and/or cell membrane-bound receptors and the resultant aggregation on the cell surface of the binding protein with associated transmembrane signaling molecules, often themselves having partial or total ligand-binding activity. The formation of this ligand–receptor complex results in the activation of an intrinsic (or intracytoplasmically associated) tyrosine kinase and the subsequent triggering of a cascade of phosphorylation events that are passed from one intracellular signaling intermediate to the next. On the basis of studies of the mechanism by which IL-2 stimulates T cells to proliferate, Smith has proposed a "ligand–receptor model" which accounts for the observed effects of changes in IL-2 concentration, IL-2 receptor density, and the duration of receptor activation on the cell cycle progression of T cells (Smith, 1995). Particularly noteworthy is the finding that Gaussian distributions in cycle progression times closely correlate with parallel differences in IL-2 receptor expression even within otherwise identical clonal T-cell populations. This suggests that the rate-limiting step in the IL-2-stimulated expansion of T-cell populations is the interaction of IL-2 with its receptor (Cantrell and Smith, 1984). From studies of the responses of separated T-cell subpopulations isolated on the basis of their individual IL-2 receptor densities, evidence has been obtained to indicate that some finite number of ligand–receptor interactions must occur before the cell replicates its DNA (Smith, 1995). This threshold may be reached earlier in cells expressing high numbers of receptors than in cells expressing low numbers of receptors. A similar mechanism was, in fact, postulated many years ago to explain the increasing sensitivity of differentiating erythroid progenitors to erythropoietin (Epo) (Eaves and Eaves, 1978; Eaves *et al.,* 1979), which was subsequently supported by documented increases in Epo receptor density during erythroid progenitor cell differentiation (Fraser *et al.,* 1988; Wognum *et al.,* 1990). Data showing that overexpression of cell surface receptors potentiates a mitogenic response have also been obtained for the EGF (epidermal growth factor) (Wells *et al.,* 1990) and insulin ligand–receptor systems (de Meyts, 1994; Martial *et al.,* 1994).

The kinetics of EGF binding to its receptor have been extensively investigated by Wiley and Lauffenburger (Wiley and Cunningham, 1982; Knauer *et al.,* 1984; Starbuck *et al.,* 1990; Reddy *et al.,* 1994; French *et al.,* 1995; Reddy *et al.,* 1996). Their results are consistent with the concept that the information essential for cell regulation can be found in both the magnitude and persistence of the cytokine signal instead of simply the presence of the signal per se (Marshall, 1995). To investigate mechanisms for the concentration-dependent stimulation of cytokine-responsive cells, the human EGF receptor (EGFR) system was used to compare the kinetics of the wild-type receptor with that of a carboxy-terminal truncated mutant deficient in receptor-mediated internalization (Chen *et al.,* 1989). When EGF depletion was minimized by medium replenishment, the EGF concentration dependencies of the proliferation responses of the two cell types were similar, whereas when EGF depletion was not prevented, maximal proliferation of cells expressing wild-type receptors required initial EGF concentrations approximately 10-fold higher than those required by the cells expressing the mutant EGFR (Reddy *et al.,* 1994). This suggested that diminished depletion of EGF from the extracellular medium was the major reason for the increased mitogenic activity of EGF by cells possessing internalization-deficient receptors.

These results are further supported by differences observed in the mitogenic potency and factor depletion rates of EGF and a human EGF mutein (EGF_{Y13G}) that has a 50-fold lower EGFR-binding affinity (Tadaki and Niyogi, 1993). The reduced binding affinity of the mutant cytokine to the EGFR results in a lower rate of EGFR down-regulation and a lower rate of depletion of the mutant cytokine. However, when medium containing one or the other cytokines was continuously replaced, EGF provided the more potent stimulus due to its higher binding affinity. In contrast, when the medium was not replaced, EGF_{Y13G} was depleted less rapidly than EGF and thus provided the more potent stimulus (Reddy *et al.,* 1996).

Effects of cytokine exposure on cytokine receptor internalization and reexpression on factor-dependent cell lines and terminally differentiated primary cells have been extensively investigated (Murthy *et al.,* 1989; Pietsch *et al.,* 1992; Yee *et al.,* 1993; Korpelainen *et al.,* 1995; Miyazawa *et al.,* 1995; Shimizu *et al.,* 1996). However, relatively few systematic studies of the rate of cytokine depletion in liquid cultures of primary hematopoietic cells have been reported. A twofold decrease in the concentrations of IL-1, IL-3, and IL-6 in the fluid phase of continuously perfused, flat-bed cultures of primitive primary hematopoietic cells maintained on irradiated stromal cell feeder layers was reported (Koller *et al.,* 1995a). However, calculation of cell-specific rates of cytokine depletion in static bioreactors is understandably limited by the difficulty of recovering cell samples and accurately measuring cytokine levels within the layers of adherent cells in which the most primitive hematopoietic cells are concentrated.

Stirred suspension and dilute static cell suspension cultures have thus been more useful for deriving cell-specific cytokine depletion rates. In such studies, we found that supplementation of stirred suspension cultures with increasing concentrations of a cytokine cocktail resulted in increasing rates of cell-specific cytokine depletion (Zandstra *et al.,* 1997). Cytokine depletion rates of cultures initiated with $CD34^{+}CD38^{-}$ adult marrow cells reduced SF and IL-3 concentrations at approximately 35-fold faster rates than low-density unseparated marrow cells (Zandstra *et al.,* 1997). This is consistent with previous reports of higher densities of receptors for some cytokines on more primitive cell types (Metcalf and Nicola 1991; Wagner *et al.,* 1995).

Taken together, these results suggest the existence of a self-limiting mechanism that may regulate primitive hematopoietic cell stimulation by cytokines in a transport-limited environment. More primitive cells (i.e., LTC-IC) appear to have both higher cytokine requirements for their maximal amplification than later types of hematopoietic cells (Petzer *et al.,* 1996b; Zandstra *et al.,* 1997a) and greater capacities to decrease the extracellular concentration of cytokines to which they are responsive (Zandstra *et al.,* 1997b). In the diffusion-limited microenvironment thought to prevail in the adult marrow (Naeim *et al.,* 1996), the operation of these two principles may have a profound influence on constraining the recruitment of very primitive cells into self-renewal divisions. These results also have implications for the design of clinical-scale hematopoietic cell culture systems where optimization of cytokine delivery will need to take into account the changing rates of depletion of each cytokine required. Because of the homogeneity of the microenvironment to which each cell is exposed in stirred suspension cultures, bioreactors of this design may be particularly well suited for the development of culture systems in which the continuous control of that environment could prove to be an important parameter in determining the rate, extent, and longevity of expansion achievable.

VI. SUMMARY AND FUTURE DEVELOPMENTS

Multiparameter analysis of factors that can affect hematopoietic cell expansion *in vitro* has yielded important insights into the regulation of these cells. Continuing investigations of these mechanisms are likely to be critical to the development of *ex vivo* systems that will eventually allow specific numbers and types of hematopoietic cells to be produced routinely on a clinically useful scale. In particular, cytokine concentrations appear to be of major importance. Thus, continuous monitoring of cytokine depletion may be essential to maintain cytokine concentrations at levels that optimize the production of specific types of hematopoietic cells. Perfusion systems for medium replacement may also be necessary to optimize cytokine concentrations over time. An additional consequence of stimulating cells with high concentrations of cytokines is to temporally modulate their capacity to continue to respond to the same cytokines (e.g., through changes in cell surface receptor density). The design of soluble or surface-bound ligands that prevent or limit receptor internalization, along with investigations of the biological consequences of stimulation with such molecules, may help achieve sustained expansions of self-renewing stem cells. Finally, the observations that different types and concentrations of cytokines may influence the production of functionally distinct progenitors, along with the possibility that the identity of these cytokines may change with the ontological source of the responding cells, underscore the need for more detailed information about how these different responses are elicited.

ACKNOWLEDGMENTS

The work performed in the authors' laboratories was supported by grants from StemCell Technologies and the British Columbia Science Council, Novartis Pharmaceuticals Canada, and the National Cancer Institute of Canada, with funds from the Terry Fox Run. P.Z. held Studentships from the Natural Sciences and Engineering Research Council of Canada and the BC Science Council, and C.J.E. is a Terry Fox Cancer Research Scientist of the National Cancer Institute of Canada.

REFERENCES

Austin, T. W., Solar, G. P., Ziegler, F. C., *et al.* (1997). A role for the Wnt gene family in hematopoiesis: Expansion of multilineage progenitor cells. *Blood* **89,** 3624–3635.

Barnett, M. J., Klingemann, H. G., Eaves, C. J., *et al.* (1997). Autografting with cultured marrow for the myeloid leukemias. The Vancouver experience. *In* "Autologous Stem Cell Transplantations. Biological and Clinical Results in Malignancies" (A. M. Carella, Ed.), pp. 287–298. Harwood Academic, Reading, UK.

Bhatia, M., Bonnet, D., Kapp, U., *et al.* (1997). Quantitative analysis reveals expansion of human hematopoietic repopulating cells after short-term *ex vivo* culture. *J. Exp. Med.* **186,** 619–624.

Bradley, T. R., Hodgson, G. S., and Rosendaal, M. (1978). The effect of oxygen tension on haemopoietic and fibroblast cell proliferation *in vitro*. *J. Cell. Physiol.* **97,** 517–522.

Brandt, J., Srour, E. F., van Besien, K., *et al.* (1990). Cytokine-dependent long-term culture of highly enriched precursors of hematopoietic progenitor cells from human bone marrow. *J. Clin. Invest.* **86,** 932–941.

Brecher, G., Ansel, J. D., Micklem, H. S., *et al.* (1982). Special proliferative sites are not needed for seeding and proliferation of transfused bone marrow cells in normal syngeneic mice. *Proc. Natl. Acad. Sci. USA* **79,** 5085–5087.

Broxmeyer, H. E., Cooper, S., Lu, L., *et al.* (1990). Enhanced stimulation of human bone marrow macrophage colony formation *in vitro* by recombinant human macrophage colony-stimulating factor in agarose medium and at low oxygen tension. *Blood* **76,** 323–329.

Broxmeyer, H. E., Lu, L., Cooper, S., *et al.* (1995). Flt3 ligand stimulates/costimulates the growth of myeloid stem/progenitor cells. *Exp. Hematol.* **23,** 1121–1129.

Brugger, W., Heimfeld, S., Berenson, R. J., *et al.* (1995). Reconstitution of hematopoiesis after high-dose chemotherapy by autologous progenitor cells generated *ex vivo*. *N. Engl. J. Med.* **333,** 283–287.

Brugger, W., Mocklin, W., Heimfeld, S., *et al.* (1993). *Ex vivo* expansion of enriched peripheral blood $CD34^+$ progenitor cells by stem cell factor, interleukin-1β (IL-1β), IL-6, IL-3, interferon-γ, and erythropoietin. *Blood* **81,** 2579–2584.

Caldwell, J., Locey, B., Clarke, M. F., *et al.* (1991a). Influence of medium exchange schedules on metabolic, growth, and GM-CSF secretion rates of genetically engineered NIH-3T3 cells. *Biotechnol. Prog.* **7,** 1–8.

Caldwell, J., Palsson, B. O., Locey, B., *et al.* (1991b). Culture perfusion schedules influence the metabolic activity and granulocyte-macrophage colony-stimulating factor production rates of human bone marrow stromal cells. *J. Cell. Physiol.* **147,** 344–353.

Cantrell, D. A., and Smith, K. A. (1984). The interleukin-2 T-cell system: A new cell growth model. *Science* **224,** 1312–1316.

Cashman, J., Eaves, A. C., and Eaves, C. J. (1985). Regulated proliferation of primitive hematopoietic progenitor cells in long-term human marrow cultures. *Blood* **66,** 1002–1005.

Cashman, J. D., Eaves, A. C., Raines, E. W., *et al.* (1990). Mechanisms that regulate the cell cycle status of very primitive hematopoietic cells in long-term human marrow cultures. I. Stimulatory role of a variety of mesenchymal cell activators and inhibitory role of TGF-β. *Blood* **75,** 96–101.

Chen, W. S., Lazar, C. S., Lund, K. A., *et al.* (1989). Functional independence of the epidermal growth factor receptor from a domain required for ligand-induced internalization and calcium regulation. *Cell* **59,** 33–43.

Cipolleschi, M. G., Sbarba, P. D., and Olivotto, M. (1993). The role of hypoxia in the maintenance of hematopoietic stem cells. *Blood* **82,** 2031–2037.

Collins, P. C., Papoutsakis, E. T., and Miller, W. M. (1996). *Ex vivo* culture systems for hematopoietic cells. *Curr. Opin. Biotechnol.* **7,** 223–230.

Conneally, E., Cashman, J., Petzer, A., *et al.* (1997). Expansion *in vitro* of transplantable human cord blood stem cells demonstrated using a quantitative assay of their lympho-myeloid repopulating activity in nonobese diabetic-scid/scid mice. *Proc. Natl. Acad. Sci. USA* **94,** 9836–9841.

Coulombel, L., Eaves, A. C., and Eaves, C. J. (1983a). Enzymatic treatment of long-term human marrow cultures reveals the preferential location of primitive hemopoietic progenitors in the adherent layer. *Blood* **62,** 291–297.

Coulombel, L., Eaves, C., Kalousek, D., *et al.* (1985). Long-term marrow culture of cells from patients with acute myelogenous leukemia. Selection in favor of normal phenotypes in some but not all cases. *J. Clin. Invest.* **75,** 961–969.

Coulombel, L., Kalousek, D. K., Eaves, C. J., *et al.* (1983b). Long-term marrow culture reveals chromosomally normal hematopoietic progenitor cells in patients with Philadelphia chromosome-positive chronic myelogenous leukemia. *N. Engl. J. Med.* **308,** 1493–1498.

Darfler, F. J., and Insel, P. A. (1983). Clonal growth of lymphoid cells in serum-free media requires elimination of H_2O_2 toxicity. *J. Cell. Physiol.* **116,** 31–36.

de Meyts, P. (1994). The structural basis of insulin and insulin-like growth factor-1 receptor binding and negative co-operativity, and its relevance to mitogenic versus metabolic signaling. *Diabetologia* **37,** S135–S148.

Deutsch, V. R., Olson, T. A., Nagler, A., *et al.* (1995). The response of cord blood megakaryocyte progenitors to IL-3, IL-6 and aplastic canine serum varies with gestational age. *Br. J. Haematol.* **89,** 8–16.

Dexter, T. M., Allen, T. D., and Lajtha, L. G. (1977). Conditions controlling the proliferation of haemopoietic stem cells *in vitro. J. Cell. Physiol.* **91,** 335–344.

Dexter, T. M., Spooncer, E., Toksoz, D., *et al.* (1980). The role of cells and their products in the regulation of *in vitro* stem cell proliferation and granulocyte development. *J. Supramol. Struct.* **13,** 513–524.

Dick, J. E., Magli, M. C., Huszar, D., *et al.* (1985). Introduction of a selectable gene into primitive stem cells capable of long-term reconstitution of the hemopoietic system of W/Wv mice. *Cell* **42,** 71–79.

Eaves, C. J., and Eaves, A. C. (1978). Erythropoietin (Ep) dose–response curves for three classes of erythroid progenitors in normal human marrow and in patients with polycythemia vera. *Blood* **52,** 1196–1210.

Eaves, C. J., Cashman, J. D., and Eaves, A. C. (1993a). Control mechanisms for primitive hemopoietic cells. *In* "Colloques INSERM" (M. Guigon, F. M. Lemoine, N. Dainaik, *et al.,* Eds.), pp. 123–131. John Libbey Eurotext Ltd., Montrouge, France.

Eaves, C. J., Cashman, J. D., and Eaves, A. C. (1991a). Methodology of long-term culture of human hemopoietic cells. *J. Tissue Cult. Methods* **13,** 55–62.

Eaves, C. J., Cashman, J. D., Kay, R. J., *et al.* (1991b). Mechanisms that regulate the cell cycle status of very primitive hematopoietic cells in long-term human marrow cultures. II. Analysis of positive and negative regulators produced by stromal cells within the adherent layer. *Blood* **78,** 110–117.

Eaves, C. J., Cashman, J. D., Wolpe, S. D., *et al.* (1993b). Unresponsiveness of primitive chronic myeloid leukemia cells to macrophage inflammatory protein 1α, an inhibitor of primitive normal hematopoietic cells. *Proc. Natl. Acad. Sci. USA* **90,** 12015–12019.

Eaves, C. J., Humphries, R. K., and Eaves, A. C. (1979). *In vitro* characterization of erythroid precursor cells and the erythropoietic differentiation process. *In* "Cellular and Molecular Regulation of Hemoglobin Switching" (G. Stamatoyannopoulos and A. W. Nienhuis, Eds.), pp. 251–273. Grune and Stratton, New York.

Emerson, S. G. (1996). *Ex vivo* expansion of hematopoietic precursors, progenitors, and stem cells. The next generation of cellular therapeutics. *Blood* **87,** 3082–3088.

Fel, R., Heimfeld, S., Xu, Z., *et al.* (1994). *Ex-vivo* expansion of human $CD34^+$ hematopoietic progenitor cells. *Exp. Hematol.* **22,** 726.

Fraser, C. C., Szilvassy, S. J., Eaves, C. J., *et al.* (1992). Proliferation of totipotent hematopoietic stem cells *in vitro* with retention of long-term competitive *in vivo* reconstituting ability. *Proc. Natl. Acad. Sci. USA* **89,** 1968–1972.

Fraser, J. K., Nicholls, J., Coffey, C., *et al.* (1988). Down-modulation of high-affinity receptors for erythropoietin on murine erythroblasts by interleukin 3. *Exp. Hematol.* **16,** 769–773.

French, A. R., Tadaki, D. K., Niyogi, S. K., *et al.* (1995). Intracellular trafficking of epidermal growth factor family ligands is directly influenced by the pH sensitivity of the receptor/ligand interaction. *J. Biol. Chem.* **270,** 4334–4340.

Gabbianelli, M., Pelosi, E., Montesoro, E., *et al.* (1995). Multi-level effects of flt3 ligand on human hematopoiesis. Expansion of putative stem cells and proliferation of granulomonocytic progenitors/monocytic precursors. *Blood* **86,** 1661–1670.

Gabbianelli, M., Sargiacomo, M., Pelosi, E., *et al.* (1990). "Pure" human hematopoietic progenitors. Permissive action of basic fibroblast growth factor. *Science* **249,** 1561–1564.

Gartner, S., and Kaplan, H. S. (1980). Long-term culture of human bone marrow cells. *Proc. Natl. Acad. Sci. USA* **77,** 4756–4759.

Graham, G. J., and Pragnell, I. B. (1992). SCI/MIP-1α: A potent stem cell inhibitor with potential roles in development. *Dev. Biol.* **151,** 377–381.

Greenberger, J. S. (1978). Sensitivity of corticosteroid-dependent insulin-resistant lipogenesis in marrow preadipocytes of obese-diabetic (db/db) mice. *Nature* **275,** 752–754.

Guba, S. C., Sartor, C. I., Gottschalk, L. R., *et al.* (1992). Bone marrow stromal fibroblasts secrete interleukin-6 and granulocyte-macrophage colony-stimulating factor in the absence of inflammatory stimulation. Demonstration by serum-free bioassay, enzyme-linked immunosorbent assay, and reverse transcriptase polymerase chain reaction. *Blood* **80,** 1190–1198.

Haylock, D. N., To, L. B., Dowse, T. L., *et al.* (1992). *Ex vivo* expansion and maturation of peripheral blood CD34$^+$ cells into the myeloid lineage. *Blood* **80,** 1405–1412.

Henschler, R., Brugger, W., Luft, T., *et al.* (1994). Maintenance of transplantation potential in *ex vivo* expanded CD34$^+$-selected human peripheral blood progenitor cells. *Blood* **84,** 2898–2903.

Highfill, J. G., Haley, S. D., and Kompala, D. S. (1996). Large-scale production of murine bone marrow cells in an airlift packed bed bioreactor. *Biotechnol. Bioeng.* **50,** 514–520.

Hirayama, F., Clark, S. C., and Ogawa, M. (1994). Negative regulation of early B lymphopoiesis by interleukin 3 and interleukin 1α. *Proc. Natl. Acad. Sci. USA* **91,** 469–473.

Hirayama, F., Lyman, S. D., Clark, S. C., *et al.* (1995). The flt3 ligand supports proliferation of lymphohematopoietic progenitors and early B-lymphoid progenitors. *Blood* **85,** 1762–1768.

Jacobsen, S. E. W., Veiby, O. P., and Smeland, E. B. (1993). Cytotoxic lymphocyte maturation factor (interleukin 12) is a synergistic growth factor for hematopoietic stem cells. *J. Exp. Med.* **178,** 413–418.

Keller, G., and Snodgrass, R. (1990). Life span of multipotential hematopoietic stem cells *in vivo. J. Exp. Med.* **171,** 1407–1418.

Keller, J. R., Mantel, C., Sing, G. K., *et al.* (1988). Transforming growth factor β_1 selectively regulates early murine hematopoietic progenitors and inhibits the growth of IL-3-dependent myeloid leukemia cell lines. *J. Exp. Med.* **168,** 737–750.

Klingemann, H. G., Deal, H., Reid, D., *et al.* (1993). Design and validation of a clinically applicable culture procedure for the generation of interleukin-2 activated natural killer cells in human bone marrow autografts. *Exp. Hematol.* **21,** 1263–1270.

Klingemann, H. G., Eaves, C. J., Barnett, M. J., *et al.* (1994). Transplantation of patients with high risk acute myeloid leukemia in first remission with autologous marrow cultured in interleukin-2 followed by interleukin-2 administration. *Bone Marrow Transplant.* **14,** 389–396.

Knauer, D. J., Wiley, H. S., and Cunningham, D. D. (1984). Relationship between epidermal growth factor receptor occupancy and mitogenic response. *J. Biol. Chem.* **259,** 5623–5631.

Kogler, G., Callejas, J., Hakenberg, P., *et al.* (1996a). Hematopoietic transplant potential of unrelated cord blood. Critical issues. *J. Hematother.* **5,** 105–116.

Kogler, G., Callejas, J., Miggliaccio, A., *et al.* (1996b). The effect of different growth factor combinations and medium on the *ex vivo* expansion of umbilical cord blood committed and primitive progenitors in "steady state" or stirred suspension cultures. *Blood* **88,** 603a.

Koller, M. R., and Palsson, B. O. (1993). Tissue engineering. Reconstitution of human hematopoiesis *ex vivo. Biotechnol. Bioeng.* **42,** 909–930.

Koller, M. R., Bender, J. G., Miller, W. M., *et al.* (1992a). Reduced oxygen tension increases hematopoiesis in long-term culture of human stem and progenitor cells from cord blood and bone marrow. *Exp. Hematol.* **20,** 264–270.

Koller, M. R., Bender, J. G., Miller, W. M., *et al.* (1993a). Expansion of primitive human hematopoietic progenitors in a perfusion bioreactor system with IL-3, IL-6, and stem cell factor. *Biotechnology* **11,** 358–363.

Koller, M. R., Bender, J. G., Papoutsakis, E. T., *et al.* (1992b). Beneficial effects of reduced oxygen tension and perfusion in long-term hematopoietic cultures. *Ann. N.Y. Acad. Sci.* **665,** 105–116.

Koller, M. R., Bender, J. G., Papoutsakis, E. T., *et al.* (1992c). Effects of synergistic cytokine combinations, low oxygen, and irradiated stroma on the expansion of human cord blood progenitors. *Blood* **80,** 403–411.

Koller, M. R., Bradley, M. S., and Palsson, B. O. (1995a). Growth factor consumption and production in perfusion cultures of human bone marrow correlate with specific cell production. *Exp. Hematol.* **23,** 1275–1283.

Koller, M. R., Emerson, S. G., and Palsson, B. O. (1993b). Large-scale expansion of human stem and progenitor cells from bone marrow mononuclear cells in continuous perfusion cultures. *Blood* **82,** 378–384.

Koller, M. R., Manchel, I., Palsson, M. A., *et al.* (1996). Different measures of human hematopoietic cell culture performance are optimized under vastly different conditions. *Biotechnol. Bioeng.* **50,** 505–513.

Koller, M. R., Palsson, M. A., Manchel, I., *et al.* (1995b). Long-term culture-initiating cell expansion is dependent on frequent medium exchange combined with stromal and other accessory cell effects. *Blood* **86,** 1784–1793.

Korpelainen, E. I., Gamble, J. R., Smith, W. B., *et al.* (1995). Interferon-γ upregulates interleukin-3 (IL-3) receptor expression in human endothelial cells and synergizes with IL-3 in stimulating major histocompatibility complex class II expression and cytokine production. *Blood* **86,** 176–182.

Lansdorp, P. M., Dragowska, W., and Mayani, H. (1993). Ontogeny-related changes in proliferative potential of human hematopoietic cells. *J. Exp. Med.* **178,** 787–791.

Lauffenburger, D. A., Chu, L., French, A., *et al.* (1996). Engineering dynamics of growth factors and other therapeutic ligands. *Biotechnol. Bioeng.* **52,** 61–80.

Lemischka, I. R., Raulet, D. H., and Mulligan, R. C. (1986). Developmental potential and dynamic behavior of hematopoietic stem cells. *Cell* **45,** 917–927.

Lemoli, R. M., Tafuri, A., Strife, A., *et al.* (1992). Proliferation of human hematopoietic progenitors in long-term bone marrow cultures in gas-permeable plastic bags is enhanced by colony-stimulating factors. *Exp. Hematol.* **20,** 569–575.

Levesque, J. P., Haylock, D. N., and Simmons, P. J. (1996). Cytokine regulation of proliferation and cell adhesion are correlated events in human $CD34^+$ hemopoietic progenitors. *Blood* **88,** 1168–1176.

Lindop, P. J., and Rotblat, J. (1960). Protection against acute effects of radiation by hypoxia. *Nature* **185,** 593–594.

Long, M. W., Briddell, R., Walter, A. W., *et al.* (1992). Human hematopoietic stem cell adherence to cytokines and matrix molecules. *J. Clin. Invest.* **90,** 251–255.

Maguer-Satta, V., Burl, S., Liu, L., *et al.* (1998). BCR-ABL accelerates C2-ceramide-induced apoptosis. *Oncogene* Jan 15;**16**(2), 237–248.

Marshall, C. J. (1995). Specificity of receptor tyrosine kinase signaling. Transient versus sustained extracellular signal-regulated kinase activation. *Cell* **80,** 179–185.

Martial, A., Dousset, B., Dardenne, M., *et al.* (1994). Insulin utilization and kinetic effect on hybridoma metabolism in batch and continuous cultures. *Biotechnology* **34,** 195–203.

Martiat, P., Ferrant, A., Cogneau, M., *et al.* (1987). Assessment of bone marrow blood flow using positron emission tomography. No relationship with bone marrow cellularity. *Br. J. Haematol.* **66,** 307–310.

Mauch, P., Greenberger, J. S., Botnik, L., *et al.* (1980). Evidence for structured variation in self-renewal capacity within long-term bone marrow cultures. *Proc. Natl. Acad. Sci. USA* **77,** 2927–2930.

McKenna, H. J., deVries, P., Brasel, K., *et al.* (1995). Effect of flt3 ligand on the *ex vivo* expansion of human $CD34^+$ hematopoietic progenitor cells. *Blood* **86,** 3413–3420.

Meagher, R. C., Salvado, A. J., and Wright, D. G. (1988). An analysis of the multilineage production of human hematopoietic progenitors in long-term bone marrow culture. Evidence that reactive oxygen intermediates derived from mature phagocytic cells have a role in limiting progenitor cell self-renewal. *Blood* **72,** 273–281.

Metcalf, D., and Nicola, N. A. (1991). Direct proliferative actions of stem cell factor on murine bone marrow cells *in vitro.* Effects of combination with colony-stimulating factors. *Proc. Natl. Acad. Sci. USA* **88,** 6239–6243.

Miller, C. L., and Eaves, C. J. (1997). Expansion *in vitro* of adult murine hematopoietic stem cells with transplantable lympho-myeloid reconstituting ability. *Proc. Natl. Acad. Sci. USA* **94,** 13648–13653.

Miller, D. G., Adam, M. A., and Miller, A. D. (1995). Gene transfer by retrovirus vectors occurs only in cells that are actively replicating at the time of infection. *Mol. Cell. Biol.* **10,** 4239–4242.

Miyazawa, K., Williams, D. A., Gotoh, A., *et al.* (1995). Membrane-bound Steel factor induces more persistent tyrosine kinase activation and longer life span of c-kit gene-encoded protein than its soluble form. *Blood* **85,** 641–649.

Moore, K. A., Pytowski, B., Witte, L., *et al.* (1997). Hematopoietic activity of a stromal cell transmembrane protein containing epidermal growth factor-like repeat motifs. *Proc. Natl. Acad. Sci. USA* **94,** 4011–4016.

Moore, M. A. S. (1993). *Ex vivo* expansion and gene therapy using cord blood CD34$^+$ cells. *J. Hematother.* **2,** 221–224.

Moore, M. A. S., and Hoskins, I. (1994). *Ex vivo* expansion of cord blood-derived stem cells and progenitors. *Blood Cells* **20,** 468–481.

Muench, M. O., Roncarolo, M. G., Menon, S., *et al.* (1995). FLK-2/FLT-3 ligand regulates the growth of early myeloid progenitors isolated from human fetal liver. *Blood* **85,** 963–972.

Murthy, S. C., Sorensen, P. H. B., Mui, A.-F., *et al.* (1989). Interleukin-3 down-regulates its own receptor. *Blood* **73,** 1180–1187.

Naeim, F., Moatamed, F., and Sahimi, M. (1996). Morphogenesis of the bone marrow. Fractal structures and diffusion-limited growth. *Blood* **87,** 5027–5031.

Nordon, R. E., Ginsberg, S. S., and Eaves, C. J. (1997). High resolution cell division tracking demonstrates the Flt3-ligand-dependence of human marrow CD34$^+$CD38$^-$ cell production *in vitro*. *Br. J. Haematol.* **98,** 528–539.

Oh, D. J., Koller, M. R., and Palsson, B. O. (1994). Frequent harvesting from perfused bone marrow cultures results in increased overall cell and progenitor expansion. *Biotechnol. Bioeng.* **44,** 609–616.

Ohmizono, Y., Sakabe, H., Kimura, T., *et al.* (1997). Thrombopoietin augments *ex vivo* expansion of human cord blood-derived hematopoietic progenitors in combination with stem cell factor and flt3 ligand. *Leukemia* **11,** 524–530.

Otsuka, T., Thacker, J. D., Eaves, C. J., *et al.* (1991). Differential effects of microenvironmentally presented interleukin 3 versus soluble growth factor on primitive human hematopoietic cells. *J. Clin. Invest.* **88,** 417–422.

Palsson, B. O., Oh, D. J., and Koller, M. R. (1995). Replating of bioreactor expanded human bone marrow results in extended growth of primitive and mature cells. *Cytotechnology* **18,** 125–131.

Palsson, B. O., Paek, S. H., Schwartz, R. M., *et al.* (1993). Expansion of human bone marrow progenitor cells in a high cell density continuous perfusion system. *Biotechnology* **11,** 368–372.

Pawliuk, R., Eaves, C., and Humphries, R. K. (1996). Evidence of both ontogeny and transplant dose-regulated expansion of hematopoietic stem cells *in vivo*. *Blood* **88,** 2852–2858.

Petzer, A. L., Eaves, C. J., Barnett, M. J., *et al.* (1997). Selective expansion of primitive normal hematopoietic cells in cytokine-supplemented cultures of purified cells from patients with chronic myeloid leukemia. *Blood* **90,** 64–69.

Petzer, A. L., Hogge, D. E., Lansdorp, P. M., *et al.* (1996a). Self-renewal of primitive human hematopoietic cells (long-term-culture-initiating cells) *in vitro* and their expansion in defined medium. *Proc. Natl. Acad. Sci. USA* **93,** 1470–1474.

Petzer, A. L., Zandstra, P. W., Piret, J. M., *et al.* (1996b). Differential cytokine effects on primitive (CD34$^+$CD38$^-$) human hematopoietic cells. Novel responses to flt3-ligand and thrombopoietin. *J. Exp. Med.* **183,** 2551–2558.

Piacibello, W., Sanavio, F., Garetto, L., *et al.* (1997). Extensive amplification and self-renewal of human primitive hematopoietic stem cells from cord blood. *Blood* **89,** 2644–2653.

Pick, E., and Mizel, D. (1981). Rapid microassays for the measurement of superoxide and hydrogen peroxide production by macrophages in culture using an automatic enzyme immunoassay reader. *J. Immunol. Methods* **46,** 211–226.

Pietsch, T., Kyas, U., Steffens, U., *et al.* (1992). Effects of human stem cell factor (c-kit ligand) on proliferation of myeloid leukemia cells. Heterogeneity in response and synergy with other hematopoietic growth factors. *Blood* **80,** 1199–1206.

Piret, J. M., and Cooney, C. L. (1991). Model of oxygen transport limitations in hollow fiber bioreactors. *Biotechnol. Bioeng.* **37,** 80–92.

Purdy, M. H., Hogan, C. J., Hami, L., *et al.* (1995). Large volume *ex vivo* expansion of CD34$^-$ positive hematopoietic progenitor cells for transplantation. *J. Hematother.* **4,** 515–525.

Quesenberry, P. J., Kittler, E., Lowry, P., *et al.* (1993). Stromal regulation of hemopoiesis and the role of subliminal growth factor production. *Leukemia* **7,** S3–S4.

Reddy, C. C., Wells, A., and Lauffenburger, D. A. (1994). Proliferative response of fibroblasts expressing internalization-deficient epidermal growth factor (EGF) receptors is altered via differential EGF depletion effect. *Biotechnol. Prog.* **10,** 377–384.

Reddy, C. C., Wells, A., and Lauffenburger, D. A. (1996). Receptor-mediated effects on ligand availability influence relative mitogenic potencies of epidermal growth factor and transforming growth factor α. *J. Cell. Physiol.* **166,** 512–522.

Rich, I. N., and Kubanek, B. (1982). The effect of reduced oxygen tension on colony formation of erythropoietic cells *in vitro. Br. J. Haematol.* **52,** 579–588.

Rowley, S. D., Brashem-Stein, C., Andrews, R., *et al.* (1993). Hematopoietic precursors resistant to treatment with 4-hydroperoxycyclophosphamide. Requirement for an interaction with marrow stroma in addition to hematopoietic growth factors for maximal generation of colony-forming activity. *Blood* **82,** 60–65.

Sandstrom, C. E., Bender, J. G., Miller, W. M., *et al.* (1996). Development of novel perfusion chamber to retain nonadherent cells and its use for comparison of human "mobilized" peripheral blood mononuclear cell cultures with and without irradiated bone marrow stroma. *Biotechnol. Bioeng.* **50,** 493–504.

Sandstrom, C. E., Bender, J. G., Papoutsakis, E. T., *et al.* (1995). Effects on $CD34^+$ cell selection and perfusion on *ex vivo* expansion of peripheral blood mononuclear cells. *Blood* **86,** 958–970.

Sardonini, C. A., and Wu, Y. J. (1993). Expansion and differentiation of human hematopoietic cells from static cultures through small-scale bioreactors. *Biotechnol. Prog.* **9,** 131–137.

Sato, N., Sawada, K., Koizumi, K., *et al.* (1993). *In vitro* expansion of human peripheral blood $CD34^+$ cells. *Blood* **82,** 3600–3609.

Sauvageau, G., Lansdorp, P. M., Eaves, C. J., *et al.* (1994). Differential expression of homeobox genes in functionally distinct $CD34^+$ subpopulations of human bone marrow cells. *Proc. Natl. Acad. Sci. USA* **91,** 12223–12227.

Schwartz, R. M., Emerson, S. G., Clarke, M. F., *et al.* (1991a). *In vitro* myelopoiesis stimulated by rapid medium exchange and supplementation with hematopoietic growth factors. *Blood* **78,** 3155–3161.

Schwartz, R. M., Palsson, B. O., and Emerson, S. G. (1991b). Rapid medium perfusion rate significantly increases the productivity and longevity of human bone marrow cultures. *Proc. Natl. Acad. Sci. USA* **88,** 6760–6764.

Shimizu, Y., Ashman, L. K., Du, Z., *et al.* (1996). Internalization of kit together with stem cell factor on human fetal liver-derived mast cells. *J. Immunol.* **156,** 3443–3449.

Shizuru, J. A., Jerabek, L., Edwards, C. T., *et al.* (1996). Transplantation of purified hematopoietic stem cells. Requirements for overcoming the barriers of allogeneic engraftment. *Biol. Blood Marrow Transplant.* **2,** 3–14.

Simmons, P. J., Aylett, G. W., Niutta, S., *et al.* (1994). c-kit is expressed by primitive human hematopoietic cells that give rise to colony-forming cells in stroma-dependent or cytokine-supplemented culture. *Exp. Hematol.* **22,** 157–165.

Sing, G. K., Keller, J. R., Ellingsworth, L. R., *et al.* (1988). Transforming growth factor β selectively inhibits normal and leukemic human bone marrow cell growth *in vitro. Blood* **72,** 1504–1511.

Slovick, F. T., Abboud, C. N., Brennan, J. K., *et al.* (1984). Survival of granulocytic progenitors in the nonadherent and adherent compartments of human long-term marrow cultures. *Exp. Hematol.* **12,** 327–338.

Small, D., Levenstein, M., Kim, E., *et al.* (1994). STK-1, the human homolog of Flk-2/Flt-3, is selectively expressed in $CD34^+$ human bone marrow cells and is involved in the proliferation of early progenitor/stem cells. *Proc. Natl. Acad. Sci. USA* **91,** 459–463.

Smith, K. A. (1995). Cell growth signal transduction is quantal. *Ann. N.Y. Acad. Sci.* **766,** 263–271.

Srour, E. F., Brandt, J. E., Briddell, R. A., *et al.* (1993). Long-term generation and expansion of primitive hematopoietic progenitor cells *in vitro. Blood* **81,** 661–669.

Starbuck, C., Wiley, H. S., and Lauffenburger, D. A. (1990). Epidermal growth factor binding and trafficking dynamics in fibroblasts. Relationship to cell proliferation. *Chem. Eng. Sci.* **45,** 2367–2373.

Stewart, F. M., Crittenden, R. B., Lowry, P. A., *et al.* (1993). Long-term engraftment of normal and post-5-fluorouracil murine marrow into normal nonmyeloablated mice. *Blood* **81,** 2566–2571.

Sui, X., Tsuji, K., Tajima, S., *et al.* (1996). Erythropoietin-independent erythrocyte production. Signals through gp130 and c-kit dramatically promote erythropoiesis from human CD34$^+$ cells. *J. Exp. Med.* **183,** 837–845.

Sutherland, H. J., Hogge, D. E., Cook, D., *et al.* (1993). Alternative mechanisms with and without Steel factor support primitive human hematopoiesis. *Blood* **81,** 1465–1470.

Szabolcs, P., Moore, M. A. S., and Young, J. W. (1995). Expansion of immunostimulatory dendritic cells among the myeloid progeny of human CD34$^+$ bone marrow precursors cultured with c-kit ligand, granulocyte-macrophage colony-stimulating factor, and TNF-α. *J. Immunol.* **154,** 5851–5861.

Tadaki, D. K., and Niyogi, S. K. (1993). The functional importance of hydrophobicity of the tyrosine at position 13 of human epidermal growth factor in receptor binding. *J. Biol. Chem.* **268,** 10114–10119.

To, L. B., Roberts, M. M., Haylock, D. N., *et al.* (1992). Comparison of haematological recovery times and supportive care requirements of autologous recovery phase peripheral blood stem cell transplants, autologous bone marrow transplants and allogeneic bone marrow transplants. *Bone Marrow Transplant.* **9,** 277–284.

Toksoz, D., Dexter, T. M., Lord, B. I., *et al.* (1980). The regulation of hemopoiesis in long-term bone marrow cultures. II. Stimulation and inhibition of stem cell proliferation. *Blood* **55,** 931–936.

Tsujino, Y., Wada, H., Misawa, M., *et al.* (1993). Effects of mast cell growth factor, interleukin-3, and interleukin-6 on human primitive hematopoietic progenitors from bone marrow and cord blood. *Exp. Hematol.* **21,** 1379–1386.

Udomsakdi, C., Eaves, C. J., Swolin, B., *et al.* (1992). Rapid decline of chronic myeloid leukemic cells in long-term culture due to a defect at the leukemic stem cell level. *Proc. Natl. Acad. Sci. USA* **89,** 6192–6196.

van de Ven, C., Ishizawa, L., Law, P., *et al.* (1995). IL-11 in combination with SLF and G-CSF or GM-CSF significantly increases expansion of isolated CD34$^+$ cell population from cord blood vs. adult bone marrow. *Exp. Hematol.* **23,** 1289–1295.

Van Zant, G., Rummel, S. A., Koller, M. R., *et al.* (1994). Expansion in bioreactors of human progenitor populations from cord blood and mobilized peripheral blood. *Blood Cells* **20,** 482–491.

Verfaillie, C. M. (1992). Direct contact between human primitive hematopoietic progenitors and bone marrow stroma is not required for long-term *in vitro* hematopoiesis. *Blood* **79,** 2821–2826.

Verfaillie, C., Hurley, R., Bhatia, R., *et al.* (1994). Role of bone marrow matrix in normal and abnormal hematopoiesis. *Crit. Rev. Oncol. Hematol.* **16,** 201–224.

Wagner, J. E., Collins, D., Fuller, S., *et al.* (1995). Isolation of small, primitive human hematopoietic stem cells. Distribution of cell surface cytokine receptors and growth in SCID-Hu mice. *Blood* **86,** 512–523.

Wang, T. Y., and Wu, J. H. D. (1992). A continuous perfusion bioreactor for long-term bone marrow culture. *Ann. N.Y. Acad. Sci.* **665,** 274–284.

Wang, T. Y., Brennan, J. K., and Wu, J. H. (1995). Multilineal hematopoiesis in a three-dimensional murine long-term bone marrow culture. *Exp. Hematol.* **23,** 26–32.

Wells, A., Welsh, J. B., Lazar, C. S., *et al.* (1990). Ligand-induced transformation by a noninternalizing epidermal growth factor receptor. *Science* **247,** 962–964.

Wiley, H. S., and Cunningham, D. D. (1982). The endocytotic rate constant: A cellular parameter for quantitating receptor-mediated endocytosis. *J. Biol. Chem.* **257,** 4222–4229.

Williams, S. F., Lee, W. J., Bender, J. G., *et al.* (1996). Selection and expansion of peripheral blood CD34$^+$ cells in autologous stem cell transplantation for breast cancer. *Blood* **87,** 1687–1691.

Wineman, J. P., Nishikawa, S. I., and Muller-Sieburg, C. E. (1993). Maintenance of high levels of pluripotent hematopoietic stem cells *in vitro.* Effect of stromal cells and c-kit. *Blood* **81,** 365–372.

Wognum, A. W., Lansdorp, P. M., Humphries, R. K., *et al.* (1990). Detection and isolation of the erythropoietin receptor using biotinylated erythropoietin. *Blood* **76,** 697–705.

Wong, E. K., Eaves, C., and Klingemann, H. G. (1996). Comparison of natural killer activity of human bone marrow and blood cells in cultures containing IL-2, IL-7 and IL-12. *Bone Marrow Transplant.* **18,** 63–71.

Wu, A. M., Till, J. E., Siminovitch, L., *et al.* (1968). Cytological evidence for a relationship between normal hematopoietic colony-forming cells and cells of the lymphoid system. *J. Exp. Med.* **127,** 455–464.

Yee, N. S., Langen, H., and Besmer, P. (1993). Mechanism of kit ligand, phorbol ester, and calcium-induced down-regulation of c-kit receptors in mast cells. *J. Biol. Chem.* **268,** 14189–14201.

Yonemura, Y., Ku, H., Hirayama, F., *et al.* (1996). Interleukin 3 or interleukin 1 abrogates the reconstituting ability of hematopoietic stem cells. *Proc. Natl. Acad. Sci. USA* **93,** 4040–4044.

Zandstra, P. W., Conneally, E., Petzer, A. L., *et al.* (1997a). Cytokine manipulation of primitive human hematopoietic cell self-renewal. *Proc. Natl. Acad. Sci. USA* **94,** 4698–4703.

Zandstra, P. W., Conneally, E., Piret, J. M., *et al.* (1998). Ontogeny associated changes in the cytokine responses of primitive human haematopoietic cells. *Br. J. Haematol.* **101,** 770–778.

Zandstra, P. W., Eaves, C. J., and Piret, J. M. (1994). Expansion of hematopoietic progenitor cell populations in stirred suspension bioreactors of normal human bone marrow cells. *Biotechnology* **12,** 909–914.

Zandstra, P. W., Petzer, A. L., Eaves, C. J., *et al.* (1997b). Cellular determinants affecting the rate of cytokine depletion in cultures of human hematopoietic cells. *Biotechnol. Bioeng.* **54,** 58–66.

Zsebo, K. M., Yuschenkoff, V. N., Schiffer, S., *et al.* (1988). Vascular endothelial cells and granulopoiesis. Interleukin-1 stimulates release of G-CSF and GM-CSF. *Blood* **71,** 99–103.

13

Ex Vivo HEMATOPOIETIC CELL EXPANSION FOR BONE MARROW TRANSPLANTATION

RAMKUMAR MANDALAM,
MANFRED KOLLER,[1] AND ALAN SMITH

Aastrom Biosciences, Inc. P.O. Box 376, Ann Arbor, Michigan 48106

I. INTRODUCTION

During the past few decades major advances have been made in the field of *ex vivo* manipulation and generation of cells for patient care. Some areas in

[1]Current address: Oncosis, 6199 Cornerstone Ct., Suite 111, San Diego, CA 92121.

which *ex vivo* generated cells have found therapeutic use include adoptive immunotherapy, skin care, and, more recently, hematopoietic stem cell transplantation. The development of a new *ex vivo* cell therapy requires a series of issues to be successfully addressed to achieve the ultimate goal of patient care. These issues include (i) development of an optimized biological process that results in the generation of a potentially therapeutic cell population, (ii) implementation of the biological process in a reproducible manner in a single-use disposable sterile device at clinical scale, (iii) integration of the device into a system which facilitates Good Manufacturing Practice (GMP) quality control, and (iv) clinical trials to establish safety and efficacy of the cell therapy. Only when each of these issues has been addressed will the new cell therapy become viable for widespread clinical practice. This chapter describes developments addressing each of these issues, focusing on hematopoietic stem cell (HSC) transplantation as an example.

II. HEMATOPOIETIC STEM CELL TRANSPLANTATION

HSC transplantation is perhaps the most widely practiced form of cell therapy today, with over 30,000 procedures performed worldwide in 1995 and the number of procedures increasing every year (Rowlings, 1996). Although current HSC transplantation therapies have been very successful for treating a number of diseases, issues of morbidity, mortality, and cost persist (Huan *et al.,* 1992; Hénon, 1993; Stroncek *et al.,* 1993; Peters and Rogers, 1994). In response to these problems a number of academic and industrial investigators have been developing methods to improve HSC transplantation therapies. A significant amount of this effort has focused on the *ex vivo* expansion and/or genetic modification of hematopoietic cells to either augment or replace current HSC transplantation methods. The advantages of *ex vivo* expansion include use of a smaller initial cell sample, potential acceleration of engraftment by producing the appropriate therapeutic mixture of cells, and passive purging of tumor cells during the *ex vivo* expansion process (Lundell *et al.,* 1997). Most of the developed cell expansion protocols essentially conform to one of two methods: the culture of native unpurified cell populations at high density in perfused systems, or the culture of enriched HSC populations at low density in static systems.

Within the body HSCs are surrounded by a microenvironment that influences their growth and development (Fig. 13.1). This microenvironment consists of neighboring cells that communicate with HSCs through the secretion of soluble growth factors, extracellular matrix (ECM), and direct cell–cell contact. The microenvironment is served by local blood circulation, which brings in nutrients and oxygen as it removes waste products and CO_2. The microenvironment is also connected to the other major organ systems in the body through the systemic circulation. For example, erythropoietin (Epo), which is produced in the

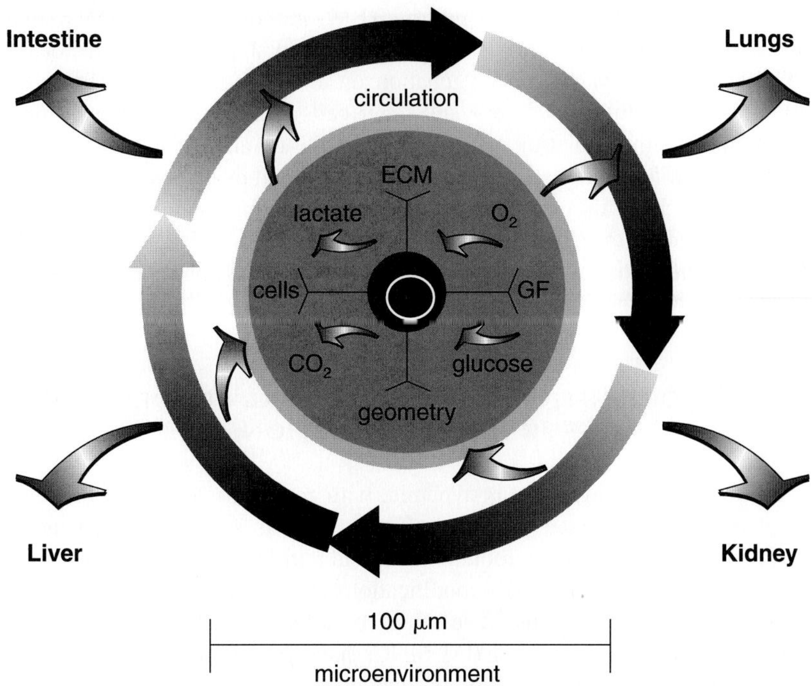

FIGURE 13.1 Tissue engineering model of the *in vivo* cellular microenvironment.

kidney, can be carried to the bone marrow to act on cells in their microenvironment. It is reasonable to assume that re-creation of this environment *ex vivo* will allow HSCs to behave *in vitro* as they do *in vivo.* Because it is currently impossible to completely re-create this environment *ex vivo,* it is important to understand which components are critical for HSC growth and development.

The approach of expanding enriched HSCs in low-density culture assumes that the critical component(s) of the hematopoietic microenvironment can be re-created in a plastic dish or bag by the use of serum-free culture medium and addition of soluble growth factors. Although significant cell expansion can be achieved through this technique, currently available preclinical (Knobel *et al.,* 1994; Koller *et al.,* 1995c) and clinical (Brugger *et al.,* 1995; Alcorn *et al.,* 1996; Williams *et al.,* 1996) data suggest that repopulating HSCs are not present after culture of CD34-enriched cells. Also, the expansion potential of CD34-enriched cells between different donors is quite variable (Koller *et al.,* 1996a).

In contrast, the approach of expanding unpurified cell populations in high-density perfused cultures assumes that accessory cells are capable of providing an environment conducive to HSC growth and development, as they do *in vivo.* The specific molecular mechanisms through which this occurs have not yet been defined well enough to provide the critical components of the microenvironment through other means, although that may be preferred once the knowledge becomes available. In human bone marrow (BM) mononuclear cell (MNC) cultures, perfusion of an appropriate medium with recombinant growth factor supplementation and physiologic oxygenation on an appropriately treated surface facilitate the growth of accessory and other cells along with the expansion of parenchymal (hematopoietic) cells. These conditions allow for HSCs to exist within layers of accessory and other cells, as evidenced by extensive cobblestone formation, much as they exist *in vivo,* resulting in high densities (exceeding 10^7 cells/ml) of stem cells, early and late progenitors, and mature cells. The influence of different components of the biological process on culture productivity is described in the following section.

III. DEFINITION OF A BIOLOGICAL PROCESS FOR HSC PRODUCTION

The BM microenvironment is dynamic, with constant interplay of biological and physicochemical rate processes. These processes include exchange of growth factors, oxygen and carbon dioxide, and nutrients and inhibitory byproducts. In an *ex vivo* system certain modifications are required to optimize production of the desired cell product. The effects of some of the foregoing factors on *ex vivo* expansion of unpurified BM MNCs have been examined to determine optimal process conditions for HSC production.

A. EFFECT OF PERFUSION

The *in vivo* BM microenvironment is continuously perfused with blood. To simulate blood circulation, *ex vivo* cultures can be perfused with medium either continuously or discontinuously (periodic medium exchange) at an appropriate rate. It has been repeatedly demonstrated that frequent medium exchange/perfusion in human hematopoietic MNC cultures results in enhanced expansion of nucleated cells, CFU-GM (colony forming units-granulocyte, macrophage), CFU-F (colony forming units-fibroblast), and LTC-IC (long-term culture initiating cells) (Koller *et al.,* 1993a, 1993b; Zandstra *et al.,* 1994; Sandstrom *et al.,* 1995). The effect of frequency of medium exchange on MNC culture expansion is illustrated in the example shown in Fig. 13.2. It was observed that the feeding interval has a significant impact on the culture output, with higher nucleated cell, CFU-GM, and LTC-IC production if medium is exchanged every other day. The optimal medium perfusion rate for a culture process also depends upon the desired cell product and other variables such as inoculum type and inoculum cell density (Koller *et al.,* 1996b).

B. EFFECT OF MEDIUM COMPOSITION

Traditionally, human hematopoietic cell cultures have utilized medium containing relatively high concentrations of animal sera (Savinell *et al.,* 1989).

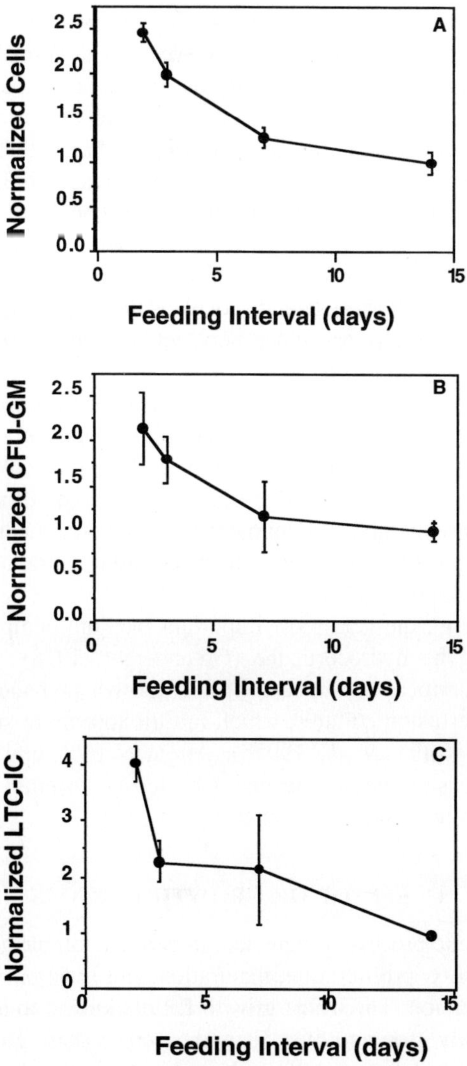

FIGURE 13.2 Effect of feeding interval on MNC culture output. MNC cultures were set up at 0.5×10^6 cells/well in a 24-well plate system. The experiments were conducted in growth medium containing IL-3, GM-CSF, SCF (c-kit ligand), and Epo. The cultures were fed with a 50% medium exchange at various intervals. Averaged normalized (A) total cells, (B) CFU-GM, and (C) LTC-IC numbers are shown.

Recent studies in perfused human BM MNC cultures have shown that a serum level of 20% was optimal, and this could be provided by fetal bovine serum (FBS) alone or a mixture of horse serum and FBS, but not by horse serum alone (Koller *et al.,* 1998a). However, the increasing use of cultured human cells in clinical trials is highlighting the need for alternatives to animal sera.

Human plasma and serum, either from autologous or from allogeneic sources, were found to support less than 50% of the culture output as compared with animal sera containing medium (Koller *et al.,* 1998a). Although this reduced level of performance may be offset by the use of correspondingly larger cultures, the use of human plasma and serum still carries considerable concerns with respect to the transmission of known and unknown human infectious agents. Although autologous human serum overcomes this issue, the generation of autologous serum for cell expansion protocols is undesirable from a manufacturing, regulatory, and quality assurance perspective, particularly as the scope of cell expansion therapy increases.

Although a number of commercial serum-free media have been developed for the purpose of CD34-enriched cell expansion (Lebkowski *et al.,* 1995), these media were not supportive of accessory cells in BM MNC cultures (particularly CFU-F) and therefore resulted in significantly decreased culture output (Koller *et al.,* 1998a). Consequently, development of serum-free medium for accessory cell containing cultures is underway, with encouraging initial reports (Sandstrom *et al.,* 1996; Koller *et al.,* 1998a).

Aside from serum, the next most important medium component in BM cell culture appears to be hydrocortisone (Greenberger, 1978). The provision of hydrocortisone is critical to support CFU-F growth and the development of dense stroma in perfusion cultures, which in turn appears to support the expansion of LTC-IC (Koller *et al.,* 1995b). Attempts to optimize other medium components, such as amino acid levels or buffering systems, have not resulted in significant gains in culture performance.

C. EFFECT OF GROWTH FACTORS

The hematopoietic process is regulated, in part, by soluble growth factors that influence the viability, proliferation, maturation, and function of cells at various stages of differentiation. The list of growth factors known to act on hematopoiesis has very quickly grown to over 50, and together, these growth factors form a regulatory network, which has been the subject of considerable research. As additional recombinant growth factors have become available, they have been examined for their utility in *ex vivo* hematopoietic cell expansion.

In the absence of exogenous growth factors, CD34-enriched cells *in vitro* rapidly lose viability and disappear within several days (Takaue *et al.,* 1992; Sato *et al.,* 1993; Henschler *et al.,* 1994). A number of studies have shown that many (four or more) growth factors must be added to CD34-enriched cell cultures at relatively high concentrations to achieve maximal proliferation (Haylock *et al.,* 1992; Brugger *et al.,* 1993; Koller *et al.,* 1995c). In contrast, perfused BM MNC cultures support a high-density stromal and accessory cell population which endogenously produces a significant number of growth factors (Koller *et al.,* 1995a, 1995c, 1995d). Consequently, perfused BM MNC cultures can be supported for at least 19 weeks in the absence of exogenous growth factors (Schwartz *et al.,* 1991b). However, the production of the greatest cell number within a clinically relevant time frame requires the addition of some exogenous growth factors at relatively low concentrations. This was originally demonstrated in this culture system with the use of IL (interleukin)-3, GM-CSF (granulocyte, macrophage-colony stimulating factor), and Epo (erythropoietin) (Schwartz *et al.,* 1991a). Since that time, many other growth factor combinations have been tested in perfused BM MNC culture, and the mature cell composition can be greatly altered. For example, the addition of Epo (in the presence of c-kit ligand) generates a mature cell population that is greater than 50% erythroid, as compared with less than 5% in the absence of Epo (Koller *et al.,* 1995a). The output of more primitive cells such as progenitors, and particularly LTC-IC, is more difficult to influence in this manner. However, several growth factors have been shown to be effective for increasing primitive cell output whereas their effect on mature cell output is smaller (Table 13.1). Therefore, significant advances have been made through the use of novel growth factors, and this is likely to continue as new growth factors are identified and become available.

D. EFFECT OF OXYGENATION

Extreme levels (both high and low) of oxygen can be inhibitory to cell growth. The appropriate level of oxygen is a function of cell density, cell

TABLE 13.1 Relative Effect[a] of Exogenous Growth Factors on BM MNC Expansion

Growth factor	Cells	CFU-GM	LTC-IC
IL-3/GM-CSF/Epo	1.0	1.0	1.0
+KL	1.7	2.4	1.0
+FL	1.3	4.4	2.1
+KL[b]/FL	1.8	5.1	2.1
+FL/IL-11	1.3	4.4	3.2
+FL/Tpo[c]	1.3	6.4	5.3
+FL/IL-11/Tpo	1.8	7.0	4.2

[a]The effect of the addition of several exogenous growth factors on BM MNC expansion is shown relative to the originally described IL-3/GM-CSF/Epo combination (Schwartz *et al.,* 1991a) (defined as 1.0).

[b]c-kit ligand.

[c]Thrombopoietin.

composition, oxygen consumption rate, and culture depth. The expansion potential and cell composition are in turn influenced by growth factor combination and other culture conditions. There have been numerous studies on the benefit of hypoxic concentrations (~5% O_2 gas phase) on progenitor cell growth and development in low-density static cultures (Smith and Broxmeyer, 1986; Pennathur-Das and Levitt, 1987; Ghezzi *et al.,* 1991; Koller *et al.,* 1992; Cipolleschi *et al.,* 1993). In high-cell-density continuous-perfusion cultures, human BM cell expansion was optimal using a higher oxygen concentration (20% O_2 gas phase) (Palsson *et al.,* 1993), although the actual oxygen concentration at the cell bed is certainly lower than that measured in the gas phase.

The oxygen uptake rate for rat BM has been determined to be 0.02 μmol/10^6 cells/h (Gesinski *et al.,* 1968; Olander, 1972). Specific oxygen uptake rates of human BM MNCs have been measured, using methods described previously (Peng and Palsson, 1996b), to be in the range of 0.004–0.035 μmol/10^6 cells/h over the period of 12 days (Mandalam *et al.,* 1995). The maximum specific oxygen uptake rate was found to occur on day 4 to day 5 of culture, when the percentage of myeloid cells was greatest.

The maximum possible oxygen flux at the cell bed with a culture depth of 3 mm and a 20% oxygen gas phase is calculated to be 0.07 μmol/(cm^2.h). At the maximum specific oxygen uptake rate measured, the theoretical maximum achievable cell density would be 2×10^6 cells/cm^2 (6.7×10^6/cells/ml). Thus, a constant optimal flux of oxygen to the cell bed is required to maintain high-density cultures.

E. DONOR-TO-DONOR VARIABILITY

The potential utility of hematopoietic cell cultures in clinical applications will depend upon the ability to reliably produce a consistent therapeutic cell product from different donors. In fact, even if cultured cells are found to be capable of durable engraftment, the use of a backup hematopoietic cell source can only be eliminated if the expansion process is robust and yields a predictable cell product. Furthermore, variable culture behavior due to the use of primary human cells can be a confounding factor in basic scientific studies. Variability in hematopoietic culture output is therefore of major concern, so a study was undertaken to assess the donor-to-donor variability in the expansion of human BM CD34-enriched cells and MNCs in culture.

Primary human BM cells from a large number of donors were expanded *in vitro* under controlled culture conditions. The inoculum density in the expansion culture was always adjusted by the measured cell purity to give an initial density of 3000 CD34$^+$lin$^-$ (lineage negative) cells per culture. After a 2-week culture period, the different CD34-enriched cell samples exhibited a wide range of cell output ($(0.02-5.07) \times 10^6$) with a high coefficient of variation ($C_v = 0.69$, $n = 52$) (Koller *et al.,* 1996a). The range in CFU-GM output was even greater (12–9455, $C_v = 0.90$). Addition of a preformed stromal layer to these cultures had a significant positive effect and resulted in narrower ranges of cell ((0.19–8.27) × 10^6, C_v = 0.41) and CFU-GM (218–17586, C_v = 0.54) output. MNC cultures, which provide an accessory cell environment including endogenous stroma, were then compared with CD34-enriched cell cultures from an additional 25 donors. MNC cultures (containing 3000 $CD34^+lin^-$cells) were found to give the greatest and most consistent cell ((2.51–5.20) × 10^6, C_v = 0.17) and CFU-GM (2618–14745, C_v = 0.46) output (Fig. 13.3) (Koller *et al.,* 1996b). In contrast to the accessory cell environment, the type of growth factor supplementation did not significantly affect donor-to-donor variability

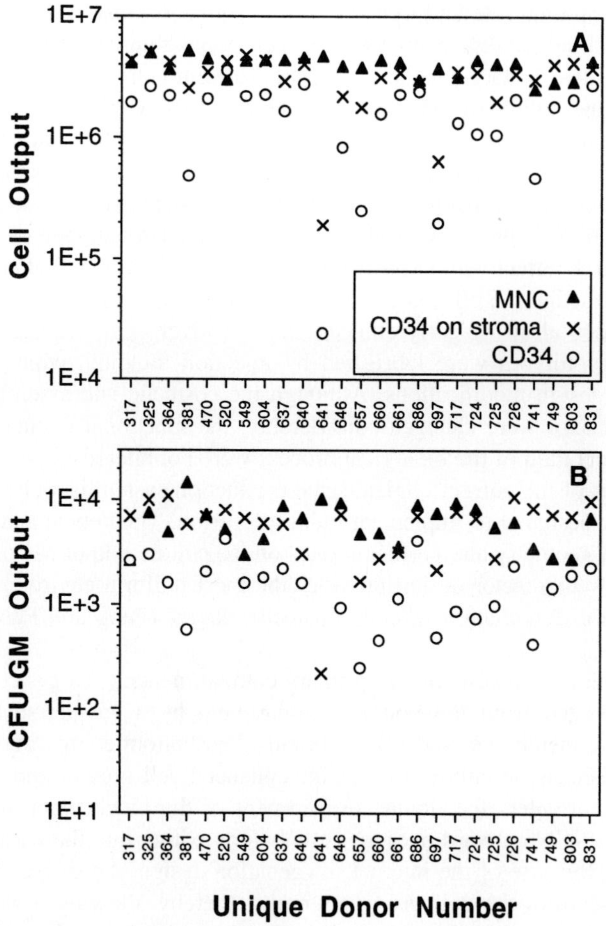

FIGURE 13.3 Variability from parallel cultures of human BM CD34-enriched cells, CD34-enriched cells on preformed stroma, and MNCs. (A) Cell and (B) CFU-GM output from 3000 $CD34^+lin^-$ cells within each type of culture inoculum is shown for 25 donors.

because all samples were similarly affected by changes in growth factor addition (Koller *et al.,* 1996a). These results have significant implications for the design of clinical trials of cultured hematopoietic cells.

IV. DESIGN OF A DEVICE TO IMPLEMENT THE BIOLOGICAL PROCESS

The main parameters defining the optimized biological process for HSC production have been described. Implementation of the process at the clinical scale required a device to be developed. The device needed to enable implementation of the optimized process conditions, such as providing perfusion of medium at an optimal rate and uniform oxygenation. Because BM MNC cultures result in both adherent and nonadherent cell populations, the device must also provide an ideal surface for the growth of adherent cells (Koller *et al.,* 1998b) as well as the ability to maintain nonadherent cells within the system and in close contact with the adherent accessory cell layer. As no existing bioreactor designs were able to provide all of these features, a novel perfusion-based, internally oxygenated parallel-plate bioreactor system was developed (Koller *et al.,* 1993b; Palsson *et al.,* 1993). Early machined prototypes of this design were shown to effectively expand human BM MNCs at small (Palsson *et al.,* 1993), intermediate (Koller *et al.,* 1993b), and clinical scale (Armstrong *et al.,* 1993), thereby allowing early clinical studies (Silver *et al.,* 1993). Subsequent bioreactor prototypes were fabricated by injection molding, which required a number of design modifications. Although the cosmetic and assembly features of the bioreactor were changed considerably, the functional features necessary for implementation of the biological process were not altered.

The heart of the current clinical-scale production system is a circular-shaped biochamber designed to implement radial perfusion. Theoretical analysis of the influence of geometry and perfusion rates on the culture output has revealed that a radial flow bioreactor design provides the most uniform environment for the growth and differentiation of cells on feeder layers (Peng and Palsson, 1995, 1996a).

The bioreactor consists of two primary compartments: (1) a gas compartment that is separated from the bottom compartment by a gas-permeable, liquid-impermeable membrane; and (2) the liquid-filled bottom compartment with the plastic surface tissue culture treated for enhanced cell growth and attachment. Fresh medium enters the liquid compartment of the biochamber at the center and flows radially outward over the cells before exiting into the waste container (Fig. 13.4). Because of the internal oxygenation design, the delivery of oxygen is decoupled from the delivery of nutrients, thereby allowing a slower, more physiological rate of medium perfusion without recirculation. The bioreactor is accompanied by a closed fluid pathway that includes a pumping mechanism for enabling continuous perfusion.

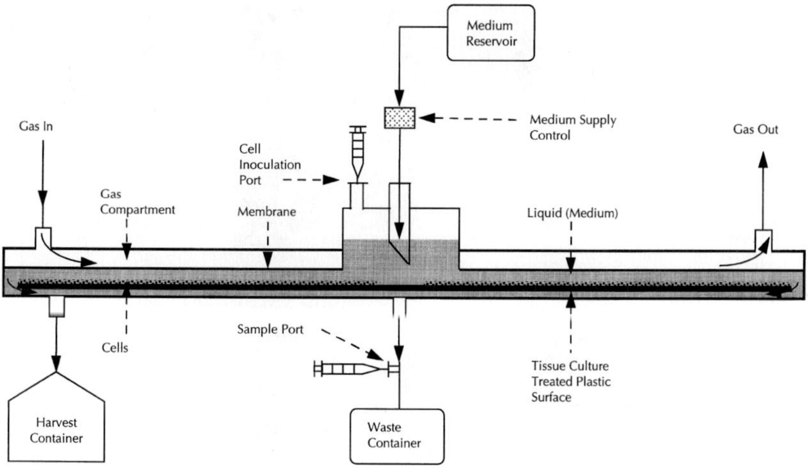

FIGURE 13.4 Schematic of the Aastrom perfusion bioreactor system showing various compartments, components, and fluid pathway.

V. INTEGRATION OF DEVICE INTO A NOVEL GMP CELL PRODUCTION SYSTEM

The bioreactor device required integration into a GMP cell production system for controlled and reproducible production of cells for cell therapy. A fluid pathway, an incubator unit, a processor unit to conduct inoculation and harvest procedures, and a computer-based system manager have been developed to support the system to achieve reliable cell production under GMP conditions. The entire system, called the AastromReplicell™ Cell Production System, is automated by incorporating microprocessor-controlled hardware and software to implement and control the *ex vivo* expansion processes (Fig. 13.5). A brief description of the different components of the AastromReplicell System follows.

The closed, sterile bioreactor and fluid pathway are contained in a "cell cassette." Specifically, the cell cassette includes a bioreactor (cell growth chamber), a medium supply container, a pump for delivery of medium, a container for waste medium collection, a collection container for harvested cells, tubing between the components, and sterile barrier elements throughout. Each cell cassette is designed for single use and is disposable.

A dedicated incubator controls the biological and physical environment and operations of each cell cassette necessary to support the cell growth process. The incubator receives and self-engages the disposable cell cassette. The incubator controls the flow of medium to the growth chamber, the temperature of the growth medium supply compartment (4°C), the temperature of the cell growth chamber compartment (37°C), and the concentration and flow rate of

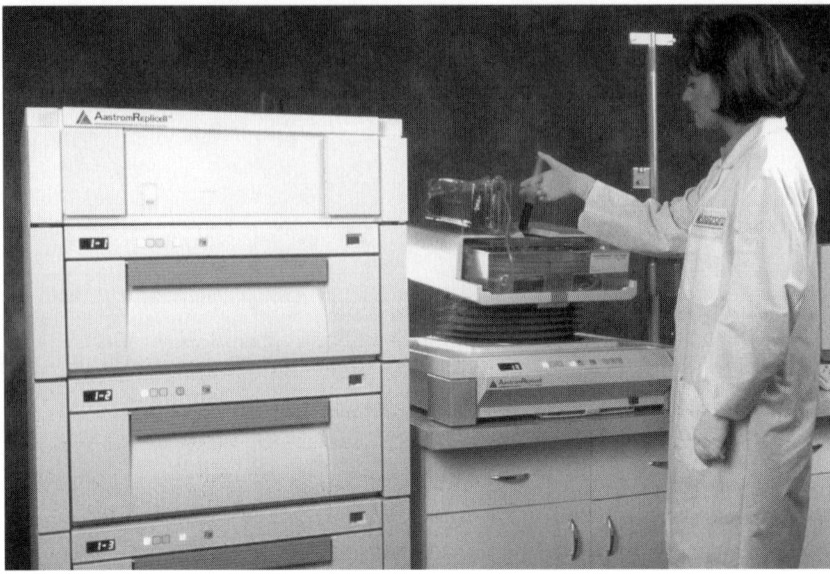

FIGURE 13.5 The AastromReplicell™ Cell Production System consists of modular incubators (two shown stacked on the left), system manager (center), and processor (right). A cell cassette is being inoculated with cells as it is engaged in the processor.

gases delivered to the gas compartment of the culture chamber. The incubator also monitors various safety/alarm parameters to ensure that the cell production process is proceeding as prescribed.

The processor performs the initial priming of the cell cassette with growth medium and inoculation and controlled distribution of cells. The same unit also performs the removal (harvest) of the cells from the growth chamber at the completion of the cell production process, sterilely transferring the cells to a preattached harvest container (analogous to a blood transfusion bag).

The system manager employs a user-friendly graphical interface. The system manager provides a convenient central user interface and provides for redundant monitoring of each incubator in the network as well as a real-time display of current status and conditions for the process. An operator can perform scheduling tasks for up to 50 incubators using the system manager, which provides a daily or weekly printed record of process conditions and alarm events for quality control and record-keeping purposes.

The ID key contains a semiconductor memory device and clock and is affixed to each cell cassette at the beginning of a cell production procedure. The ID key provides reliable identification of the cell product, instructs the instruments for the cell production process, prevents mix-ups and operator error, and stores the primary data for a complete process history record, effectively a "manufacturing batch record" for the cell product.

The entire system allows for easy and reproducible production of cells under GMP conditions. The automation and data recording minimizes multiple manual steps involved in a cell culture process. The closed, sterile, and single-use fluid system prevents microbial contamination during the process and also prevents any potential cross-contamination between the samples.

VI BIOLOGICAL PERFORMANCE OF THE AUTOMATED AASTROMREPLICELL SYSTEM

The biological process for expansion of BM MNCs has been implemented in the AastromReplicell System to evaluate its performance with a large number of donor samples. The growth medium, including serum and the growth factors PIXY (GM-CSF and IL-3 fusion molecule), Epo, and flt3-ligand, was continuously perfused starting on day 3, with perfusion rates increasing toward the end of the culture. The cells were cultured for 12 days and flow cytometric and functional assays were performed on the expanded cell product. The performance of the expansion process is shown in the Table 13.2. As is evident from the data, the procedure resulted in an expansion of both committed progenitor and stem/primitive multipotent cells. The average nucleated and myeloid progenitor cell expansion was 4.7- and 22.6-fold, respectively. The primitive cells, as measured by LTC-IC and $CD34^+lin^-$, were also expanded. The expanded product is typically composed of about 80% myeloid and 12% erythroid cells. The culture and environmental conditions promote the formation of a dense stromal layer as described previously and the final product contains stromal cells with $Thy\text{-}1^+$ and $CD14^+$ phenotypes. The $Thy\text{-}1^+$ cells and clonogenic CFU-F cells

TABLE 13.2 Average Cell Output from MNC[a] Expansion in AastromReplicell System ($n = 44$)

	In	Out
Total nucleated cells	225×10^6	1065×10^6
CFU-GM	4.2×10^5	95.7×10^5
LTC-IC	5.4×10^3	12.0×10^3
$CD34^+lin^-$	4.9×10^6	6.0×10^6
CFU-F	4.7×10^4	1.7×10^6

[a]MNCs were obtained from normal human BM by Ficoll separation. The assay conditions used in this study have been described previously (Koller *et al.,* 1996c).

expand 48- and 46-fold, respectively. These data indicate that the biological process has been successfully implemented in the automated AastromReplicell System, and the cell product appears equivalent to that produced in the earlier prototypes.

VII. IMPLEMENTATION OF THE PROCESS AND AASTROMREPLICELL SYSTEM IN CLINICAL TRIALS

Preliminary studies at MD Anderson Cancer Center using transplantation of cells produced *ex vivo* in the AastromReplicell System, augmenting a standard autologous BM transplant, indicated that *ex vivo* expansion can be performed reliably and reproducibly, and no toxicity occurred with intravenous infusion of the expanded cells (Champlin *et al.,* 1995). Ten patients with high-risk metastatic breast carcinoma, aged 18–60 years, were entered into this study transplanting unmanipulated BM cells plus *ex vivo* produced cells. BM was harvested, collecting greater than 2×10^8 nucleated cells/kg and greater than 0.5×10^6 $CD34^+$ cells/kg. Twelve days prior to the planned BM transplant, 2.25×10^8 MNC cells were inoculated into each of three biochambers, part of the AastromReplicell System, and continuously perfused with medium containing PIXY (5 ng/ml) and Epo (0.1 U/ml). The expansion reproducibly increased total nucleated cells, CFU-GM, and LTC-IC. Patients received cyclophosphamide (2.0 g/m^2/day), thiotepa (240 mg/m^2/day), and BCNU (150 mg/m^2/day) on days 7, 6, and 5, with reinfusion of the cryopreserved BM on day 0 and the *ex vivo* produced cells 4 h later. The white blood cell (WBC) nadir was less than 0.1 cells/mm^3 in all patients.

All patients engrafted within narrow time ranges, with median recovery of WBC greater than 200 cells/mm^3 on day 8 (range 7–8) and an ANC (automated nucleated cell count) greater than 500 cells/mm^3 on day 11 (range 10–13). Platelet levels were greater than 25,000 platelets/mm^3 on day 16 (range 13–21) and greater than 50,000 on day 20 (range 18–27). The toxicity associated with chemotherapy or cell infusions was less than grade two. Four patients had infections unrelated to the infusion of cells produced in the AastromReplicell System. These data compare favorably with 29 historical controls evaluated at the same institution who received the same chemotherapy and autologous BM transplantation without expanded cells, in which ANC recovered to greater than 500 on day 11 (range 7–29) and platelets to greater than 25,000 and 50,000 on days 24 (range 9–78) and 28 (range 9–147), respectively.

This initial augmentation feasibility trial was followed by a replacement feasibility trial at Loyola University Medical Center (LUMC). The patients received an ablative STAMP V chemotherapy regimen known to be lethal without appropriate cell rescue therapy. A principal objective of this study was

TABLE 13.3 Cell Dose Characteristics for the Loyola University Clinical Trial

Cell type (n = 6)	Median values infused per kg of patient weight[a]
Nucleated cells	66×10^6
Lin^-CD34^+ cells	2.2×10^5
CFU-GM	5.4×10^5

[a]Median cell doses harvested, washed, and infused into patients post-*ex vivo* expansion.

to achieve the first clinical demonstration that cells produced *ex vivo* using the AastromReplicell System could indeed restore the hematopoietic system in myeloablated patients. In the absence of any such previous data, a conservative cell backup strategy was employed to assure patient recovery and safety.

Data from the first six patients accrued in the replacement trial at LUMC have been presented (Stiff *et al.,* 1997). Briefly, the patients underwent a BM harvest prior to chemotherapy. A portion of the cells harvested were expanded in the AastromReplicell System in media supplemented with PIXY, Epo, and flt3-L for 12 days while the majority of the harvested marrow was cryopreserved for safety backup. The median volume of BM expanded for the first six patients was 38.6 ml. Typically, four cell cassettes were used per patient, with each cassette inoculated with 225×10^6 MNCs. The patients received AastromReplicell System-expanded cells alone 72 h after cessation of chemotherapy. The median quantity of cells infused into patients is shown in Table 13.3. Three general measures of hematopoietic recovery were used in this trial. These were the first day of WBC recovery to greater than 200 cells/mm^3, ANC recovery to greater than 500 cells/mm^3, and platelet recovery to greater than 20,000 platelets/mm^3. The mean recovery results for the first six patients are summarized in Table 13.4. The data demonstrate the first successful engraftment using only *ex vivo* expanded cells as the source of rescue following ablative

TABLE 13.4 Days to Engraftment for the Six Patients in the Loyola University Trial

Cell type	Median days (range)
WBC > 200/μl	8 (7–9)
ANC > 500/μl	16 (13–22)
Platelet > 20,000/μl	23 (22–27)

chemotherapy. Hematological results were similar to those for historical control patients treated at LUMC using the same chemotherapy regimen and standard autologous BM transplantation. No patient experienced any significant infusion toxicity from infusion of the *ex vivo* expanded cell product and these patients have demonstrated stable engraftment, even with the use of radiation therapy posttransplant (followup data available up to 18 months). The trial is ongoing, with continued objectives of demonstrating (i) safety, operational feasibility, and reliability of the AastromReplicell System in a hospital setting and (ii) added evidence that *ex vivo* expanded cells from small-volume BM samples provide stable engraftment in STAMP V ablated breast cancer patients.

VIII. SUMMARY

With an increasing number of HSC transplants being performed yearly, there is a clear need for an easier and more efficient method for generation of cells for HSC rescue after myeloablative therapy. One approach is *ex vivo* expansion of hematopoietic cells to either augment or replace current HSC transplantation methods. Advantages of *ex vivo* expansion include graft optimization allowing generation of an appropriate mixture of cells leading to more rapid engraftment time, use of smaller initial sample volume, thus reducing the tumor burden, and passive purging during expansion. Expansion of unpurified BM cells in an *in vivo* mimicking stromal-based environment results in high-density cultures containing stem, early and late progenitor, and mature cells, in contrast to expansion of enriched HSCs that result in expansion of progenitor and mature cells only. It has also been established that the donor-to-donor variability is significantly reduced if unpurified BM cells are used for expansion. The biological process for the expansion of unpurified BM cells was defined by optimizing critical factors such as medium composition, growth factor supplementation, medium perfusion rates, culture surface chemistry, and oxygenation. The optimized conditions were then implemented in a single-use disposable sterile device for clinical-scale production of cells. The novel cell culture device was integrated into an automated GMP cell production system called the AastromReplicell™ Cell Production System. The culture of BM cells in the AastromReplicell System resulted in expansion of $lin^{-}CD34^{+}$ cells, LTC-IC, CFU-GM, CFU-F, and nucleated cells. Clinically, the transplantation of AastromReplicell System-produced cells derived from autologous BM cells, augmenting a standard autologous transplant in the treatment of breast cancer, indicated safe, reliable, and reproducible production of cells with no toxicity in patients. Furthermore, in a subsequent clinical trial, it was demonstrated that the infusion of *ex vivo* expanded cells derived from less than 40 ml of BM as the sole source of rescue resulted in hematopoietic reconstitution in myeloablated patients who had received a STAMP V chemotherapy regimen. All six patients receiving AastromReplicell System expanded cells as the sole source of hematopoietic rescue engrafted within the period historically observed with patients receiving the same chemotherapy regimen and standard autologous BM transplantation. These results show that the *ex vivo* expansion of a small volume of BM generated under GMP conditions in the AastromReplicell System is capable of producing a mixture of stem and progenitor cells that can durably reconstitute the hematopoietic system in patients who have undergone myeloablative chemotherapy.

ACKNOWLEDGMENTS

The authors thank Drs. R. D. Champlin and P. J. Stiff for providing the clinical data and Elizabeth Oram for editorial assistance.

REFERENCES

Alcorn, M. J., Holyoake, T. L., Richmond, L., *et al.* (1996). CD34-positive cells isolated from cryopreserved peripheral-blood progenitor cells can be expanded *ex vivo* and used for transplantation with little or no toxicity. *J. Clin. Oncol.* **14,** 1839–1847.

Armstrong, R. D., Koller, M. R., Paul, L. A., *et al.* (1993). Clinical scale production of stem and hematopoietic cells *ex vivo. Blood* **82,** 296a.

Brugger, W., Heimfeld, S., Berenson, R. J., *et al.* (1995). Reconstitution of hematopoiesis after high-dose chemotherapy by autologous progenitor cells generated *ex vivo. N. Engl. J. Med.* **333,** 283–287.

Brugger, W., Möcklin, W., Heimfeld, S., *et al.* (1993). *Ex vivo* expansion of enriched peripheral blood CD34$^+$ progenitor cells by stem cell factor, interleukin-1β (IL-1β), IL-6, IL-3, interferon-γ, and erythropoietin. *Blood* **81,** 2579–2584.

Champlin, R., Mehra, R., Gajewski, J., *et al.* (1995). *Ex vivo* expanded progenitor cell transplantation in patients with breast cancer. *Blood* **86,** 295a.

Cipolleschi, M. G., Dello Sbarba, P., and Olivotto, M. (1993). The role of hypoxia in the maintenance of hematopoietic stem cells. *Blood* **82,** 2031–2037.

Gesinski, R. M., Morrison, J. H., and Toepfer, J. R. (1968). Measurement of oxygen consumption of rat bone marrow cells by a polarographic method. *J. Appl. Physiol.* **24,** 751–754.

Ghezzi, P., Dinarello, C. A., Bianchi, M., *et al.* (1991). Hypoxia increases production of interleukin-1 and tumor necrosis factor by human mononuclear cells. *Cytokine* **3,** 189–194.

Greenberger, J. S. (1978). Sensitivity of corticosteroid-dependent insulin-resistant lipogenesis in marrow preadipocytes of obese-diabetic (db/db) mice. *Nature* **275,** 752–754.

Haylock, D. N., To, L. B., Dowse, T. L., *et al.* (1992). *Ex vivo* expansion and maturation of peripheral blood CD34$^+$ cells into the myeloid lineage. *Blood* **80,** 1405–1412.

Hénon, P. R. (1993). Peripheral blood stem cell transplantations: Past, present and future. *Stem Cells* **11,** 154–172.

Henschler, R., Brugger, W., Luft, T., *et al.* (1994). Maintenance of transplantation potential in *ex vivo* expanded CD34$^+$-selected human peripheral blood progenitor cells. *Blood* **84,** 2898–2903.

Huan, S. D., Hester, J., Spitzer, G., *et al.* (1992). Influence of mobilized peripheral blood cells on the hematopoietic recovery by autologous marrow and recombinant human granulocyte-macrophage colony-stimulating factor after high-dose cyclophosphamide, etoposide, and cisplatin. *Blood* **79,** 3388–3393.

Knobel, K. M., McNally, M. A., Berson, A. E., *et al.* (1994). Long-term reconstitution of mice after *ex vivo* expansion of bone marrow cells: Differential activity of cultured bone marrow and enriched stem cell populations. *Exp. Hematol.* **22,** 1227–1235.

Koller, M. R., Bender, J. G., Miller, W. M., *et al.* (1993a). Expansion of human hematopoietic progenitors in a perfusion bioreactor system with IL-3, IL-6, and stem cell factor. *Bio/Technology* **11,** 358–363.

Koller, M. R., Bender, J. G., Papoutsakis, E. T., *et al.* (1992). Beneficial effects of reduced oxygen tension and perfusion in long-term hematopoietic cultures. *Ann. N.Y. Acad. Sci.* **665,** 105–116.

Koller, M. R., Bradley, M. S., and Palsson, B. (1995a). Growth factor consumption and production in perfusion cultures of human bone marrow correlates with specific cell production. *Exp. Hematol.* **23,** 1275–1283.

Koller, M. R., Emerson, S. G., and Palsson, B. (1993b). Large-scale expansion of human stem and progenitor cells from bone marrow mononuclear cells in continuous perfusion culture. *Blood* **82,** 378–384.

Koller, M. R., Maher, R. J., Manchel, I., *et al.* (1998a). Alternatives to animal sera for human bone marrow cell expansion: Human serum and serum-free media. *J. Hematother.* **7,** 413–423.

Koller, M. R., Maher, R. J., and Palsson, B. (1995b). Hydrocortisone influences hematopoiesis through modulation of the growth factor network in the bone marrow microenvironment. *Exp. Hematol.* **23,** 768.

Koller, M. R., Manchel, I., Brott, D. A., *et al.* (1996a). Donor-to-donor variability in the expansion potential of human bone marrow cells is reduced by accessory cells but not by soluble growth factors. *Exp. Hematol.* **24,** 1484–1493.

Koller, M. R., Manchel, I., Newsom, B. S., *et al.* (1995c). Bioreactor expansion of human bone marrow: Comparison of unprocessed, density-separated, and CD34-enriched cells. *J. Hematother.* **4,** 159–169.

Koller, M. R., Manchel, I., Palsson, M. A., *et al.* (1996b). Different measures of human hematopoietic cell culture performance are optimized under vastly different conditions. *Biotechnol. Bioeng.* **50,** 505–513.

Koller, M. R., Oxender, M., Brott, D. A., *et al.* (1996c). Flt-3 ligand is more potent than c-kit ligand for the synergistic stimulation of *ex vivo* hematopoietic cell expansion. *J. Hematother.* **5,** 449–459.

Koller, M. R., Palsson, M. A., Manchel, I., *et al.* (1998b). Tissue culture surface characteristics influence the expansion of human bone marrow cells. *Biomaterials* **19,** 1963–1972.

Koller, M. R., Palsson, M. A., Manchel, I., *et al.* (1995d). LTC-IC expansion is dependent on frequent medium exchange combined with stromal and other accessory cell effects. *Blood* **86,** 1784–1793.

Lebkowski, J. S., Schain, L. R., and Okarma, T. B. (1995). Serum-free culture of hematopoietic stem cells: A review. *Stem Cells* **13,** 607–612.

Lundell, B. I., Tyer, C., DeSombre, K., *et al.* (1997). *Ex-vivo* perfusion culture expansion of tumor positive bone marrow from breast cancer patients results in passive purging during the culture period. *Blood* **90,** 216a.

Mandalam, R. K., Palsson, M., Vento, C. A., *et al.* (1995). Oxygen requirements of bone marrow mononuclear cells. *Am. Inst. Chem. Eng. Annu. Meet.* 231a.

Olander, C. P. (1972). Respiration and erythropoiesis of bone marrow of normal and hypoxically stimulated rats. *Am. J. Physiol.* **222,** 45–48.

Palsson, B., Paek, S. H., Schwartz, R. M., *et al.* (1993). Expansion of human bone marrow progenitor cells in a high cell density continuous perfusion system. *Bio/Technology* **11,** 368–371.

Peng, C. A., and Palsson, B. (1995). Importance of nonhomogeneous concentration distributions near walls in bioreactors for primary cell cultures. *Ind. Eng. Chem. Res.* **34,** 3239–3245.

Peng, C. A., and Palsson, B. (1996a). Cell growth and differentiation on feeder layers is predicted to be influenced by bioreactor geometry. *Biotechnol. Bioeng.* **50,** 479–492.

Peng, C. A., and Palsson, B. (1996b). Determination of specific oxygen uptake rates in human hematopoietic cultures and implications for bioreactor design. *Ann. Biomed. Eng.* **24,** 373–381.

Pennathur-Das, R., and Levitt, L. (1987). Augmentation of *in vitro* human marrow erythropoiesis under physiological oxygen tensions is mediated by monocytes and T lymphocytes. *Blood* **69,** 899–907.

Peters, W. P., and Rogers, M. C. (1994). Variation in approval by insurance companies of coverage for autologous bone marrow transplantation for breast cancer. *N. Engl. J. Med.* **330,** 473–477.

Rowlings, P. A. (1996). Summary slides show current use and outcome of blood and marrow transplantation. *Am. Bone Marrow Transpl. Registry Newsl.* November, 6–12.

Sandstrom, C. E., Bender, J. G., Papoutsakis, E. T., *et al.* (1995). Effects of CD34$^+$ cell selection and perfusion on *ex vivo* expansion of peripheral blood mononuclear cells. *Blood* **86,** 958–970.

Sandstrom, C. E., Collins, P. C., McAdams, T. A., *et al.* (1996). Comparison of whole serum-deprived media for *ex vivo* expansion of hematopoietic progenitor cells from cord blood and mobilized peripheral blood mononuclear cells. *J. Hematother.* **5,** 461–473.

Sato, N., Sawada, K., Koizumi, K., *et al.* (1993). *In vitro* expansion of human peripheral blood CD34$^+$ cells. *Blood* **82,** 3600–3609.

Savinell, J. M., Lee, G. M., and Palsson, B. (1989). On the orders of magnitude of epigenic dynamics and monoclonal antibody production. *Bioprocess Eng.* **4,** 231–234.

Schwartz, R. M., Emerson, S. G., Clarke, M. F., *et al.* (1991a). *In vitro* myelopoiesis stimulated by rapid medium exchange and supplementation with hematopoietic growth factors. *Blood* **78,** 3155–3161.

Schwartz, R. M., Palsson, B., and Emerson, S. G. (1991b). Rapid medium perfusion rate significantly increases the productivity and longevity of human bone marrow cultures. *Proc. Natl. Acad. Sci. USA* **88,** 6760–6764.

Silver, S. M., Adams, P. T., Hutchinson, R. J., *et al.* (1993). Phase I evaluation of *ex vivo* expanded hematopoietic cells produced by perfusion cultures in autologous bone marrow transplantation. *Blood* **82,** 296a.

Smith, S., and Broxmeyer, H. E. (1986). The influence of oxygen tension on the long-term growth *in vitro* of haematopoietic progenitor cells from human cord blood. *Br. J. Haematol.* **63,** 29–34.

Stiff, P. J., Oldenberg, D., Hsi, E., *et al.* (1997). Successful hematopoietic engraftment following high dose chemotherapy using only *ex-vivo* expanded bone marrow grown in Aastrom (stromal-based) bioreactors. *Proc. Am. Soc. Clin. Oncol.* **16,** 88a.

Stroncek, D. F., Holland, P. V., Bartch, G., *et al.* (1993). Experiences of the first 493 unrelated marrow donors in the National Marrow Donor Program. *Blood* **81,** 1940–1946.

Takaue, Y., Abe, Y., Kawano, A., *et al.* (1992). Combination of recombinant cytokines fails to produce *ex vivo* expansion of human blood hematopoietic progenitor cells. *Ann. Hematol.* **64,** 217–220.

Williams, S. F., Lee, W. J., Bender, J. G., *et al.* (1996). Selection and expansion of peripheral blood CD34$^+$ cells in autologous stem cell transplantation for breast cancer. *Blood* **87,** 1687–1691.

Zandstra, P. W., Eaves, C. J., and Piret, J. M. (1994). Expansion of hematopoietic progenitor cell populations in stirred suspension bioreactors of normal human bone marrow cells. *Bio/Technology* **12,** 909–914.

14

Gene Delivery Technology: Nonviral and Viral Vector Systems

Mitch Raponi and Geoff Symonds

Johnson & Johnson Research
74 McLachlan Avenue, Rushcutters Bay
Sydney, New South Wales 2001, Australia

I. Introduction

The first gene therapy clinical protocol was initiated in September 1990 (Blaese and Anderson, 1990). Since then, 177 gene marker/gene therapy clinical protocols have been approved or are pending approval in the United States alone (Table 14.1). Although there are numerous different cell types being targeted for the treatment of a variety of diseases, all gene therapy strategies have two essential technical requirements. These are the efficient introduction of the foreign therapeutic gene into the target cell and the expression of the transgene at levels that are of therapeutic benefit. The technical problems involved with these requirements are still the object of active research and, although there are significant conceptual and technical hurdles to overcome, progress is being made.

TABLE 14.1 RAC-Approved Gene Therapy Clinical Protocols

Gene transfer strategy	Number of protocols[a]
Viral vector	
Retrovirus	103
Adenovirus	26
Adeno-associated virus	2
Vaccinia virus	4
Fowlpox virus	3
Canarypox virus	1
Nonviral vector	
Liposome	29
Particle mediated	3
Direct DNA injection	3
Electroporation	1
Total	175

[a] NIH-listed protocols as of March 25, 1997.

Gene therapy can be defined as the introduction of genetic material (DNA or RNA) into a target tissue for therapeutic benefit (Morgan and Anderson, 1993). This can be aimed at (i) delivering DNA to encode for a substance that is lacking in the cell, thereby providing the cell with a new function (genetic augmentation), or (ii) selectively replacing a defective gene with a normal one (Capecchi, 1994; Huber, 1994). The more established gene therapy strategies are those employing gene augmentation techniques.

The constitutive production of a therapeutic molecule within the target cell of an individual offers the advantage of eliminating the need to continuously administer drugs to the patient. In this respect gene therapy (Friedmann, 1989; Anderson, 1992; Miller, 1992; Morgan and Anderson, 1993; Mulligan, 1993; Crystal, 1995) can be applied to a variety of complex and acquired genetic diseases, including cancer (Avalosse *et al.,* 1995; Culver *et al.,* 1995), cardiovascular disease (Nabel *et al.,* 1990; Chowdhury *et al.,* 1991; Wilson *et al.,* 1992), AIDS (Yu *et al.,* 1994; Bridges and Sarver, 1995; Sun *et al.,* 1995a), and arthritis (Evans and Robbins, 1994; Robbins *et al.,* 1994), as well as single-gene (monogenic) disorders such as cystic fibrosis (Colladge and Evans, 1995), lysosomal storage disease (Beutler, 1992; Salvetti *et al.,* 1995), hemophilia (Hseuh *et al.,* 1992; Hoeben *et al.,* 1993; Brownlee, 1995), Parkinson's disease (Wolff *et al.,* 1989; Freese *et al.,* 1990; Jiao *et al.,* 1993), and severe combined immunodeficiency (SCID) (Hoogerbrugge *et al.,* 1995). Examples of gene therapy strategies targeted against these diseases are summarized in Table 14.2.

Two of the most important advances in gene therapy have been (i) the ability to transfer genes into target tissues and (ii) the ability to maintain their stable

TABLE 14.2 Selection of Diseases Targeted by Gene Therapy

Disease	Target tissue	Vector	Gene insert
Monogenic disease			
Severe combined immunodeficiency	Hematopoietic cells	Retrovirus	Adenosine deaminase
Cystic fibrosis	Airway epithelium	Liposome Retrovirus Adenovirus Adeno-associated virus	Cystic fibrosis transmembrane regulator
Hemophilia	Hematopoietic cells	Retrovirus	Clotting factor gene
Lysosomal storage disease	Hematopoietic cells Fibroblasts	Retrovirus	α-Iduronidase β-Glucuronidase
Complex genetic disease			
Cardiovascular disease	Hepatocytes Smooth muscle	Retrovirus Liposome Direct DNA injection	Low-density lipoprotein receptor β-Galactosidase
Cancer	Hematopoietic cells Tumor cells	Retrovirus Liposome Adenovirus	Cytokines HLA-B7
Arthritis	Synovial cells	Retrovirus Liposome	Cytokines
Parkinson's disease	Neurons Fibroblasts Muscle	Herpes simplex virus Retrovirus Liposome	Tyrosine hydroxylase
Acquired genetic disease			
Human immunodeficiency virus	Hematopoietic cells	Retrovirus Particle bombardment	Antisense Ribozyme RNA decoy Transdominant mutant

expression. A variety of chemical, physical, and biological gene transfer methods exist for this purpose (see Tables 14.3 and 14.4). These techniques range from the relatively simple forms of *in vitro* transformation such as calcium phosphate coprecipitation (Graham and van Der Eb, 1973) and electroporation (Potter *et al.,* 1984) to the exploitation of the natural infectious properties of recombinant viral vectors.

II. NONVIRAL GENE DELIVERY

Nonviral methods of introducing DNA into a target cell and consequently altering the cell's genotype and phenotype are known as cellular transformation (Pellicer *et al.,* 1980). Advantages and disadvantages of the following nonviral gene transfer techniques are summarized in Table 14.3.

TABLE 14.3 Summary of Nonviral Gene Transfer Techniques

Gene transfer method	Advantages	Disadvantages
Calcium phosphate coprecipitation	Simple Stable expression	Random integration into host genome Low transfection efficiency
Lipoplexes	*In vivo* applications Accepts large DNA inserts	Transient expression Low transfection efficiency
Receptor-mediated endocytosis	High transfection efficiency Tissue specific	Some lysosomal degradation Transient expression
Electroporation	Simple Stable expression	Transient expression Low transfection efficiency
Direct DNA injection	No cotransfer of unwanted genetic material Gene transfer into specific target cell	Transient expression Low transfection efficiency
Particle bombardment	High transfection efficiency Tissue specific	Transient expression Low transfection efficiency
Extrachromosomal replicating vectors	Stable, long-term expression Nonintegrating Large DNA inserts	Largely theoretical vectors

A. CALCIUM PHOSPHATE COPRECIPITATION

Introduction of DNA by calcium phosphate coprecipitation was one of the first methods of cellular transformation utilized (Graham and Van Der Eb, 1973). Advantages include its simplicity, toleration of a variety of DNA preparations, and its efficacy in a variety of cell types. Disadvantages are its toxicity to certain cell types and generally low transformation efficiency (Chang, 1994). The mechanisms involved in transfection by the calcium phosphate method are (i) coprecipitation of DNA with calcium phosphate, (ii) adsorption of this complex by the target cell, (iii) endocytosis, and (iv) transfer of the calcium phosphate–DNA complex from the endosome to the nucleus (Chang, 1994). Integration within the cellular genome occurs at very low frequency and is random, with no apparent preferential chromosomal locations. Although the simplicity of this method maintains its wide acceptance at the research level, its use as a gene transfer method in gene therapy approaches has not been significant.

B. ELECTROPORATION

Electrotransfection is commonly employed to achieve transformation in a wide variety of cell types (Potter *et al.,* 1984). In this technique, cells are exposed to a high-voltage electrical discharge in the presence of exogenous DNA. The consequent formation of hydrophilic pores in the cell membrane (electroporation) allows uptake of exogenous DNA (Neumann *et al.,* 1982). Although this strategy generally achieves transient gene expression of the introduced DNA *in vitro,* permanent introduction of the genetic material into a small proportion of the cell population does occur. *In vivo* electrotransfection techniques have also been investigated (Sukharev *et al.,* 1994). Sukharev *et al.* (1994) have demonstrated that *in vivo* gene transfer by electrotransfection is feasible by subcutaneously injecting plasmid DNA (containing a selectable marker) into mice and then applying two high-voltage pulses across a pleat of skin. A few days after electroporation, skin samples were excised and analyzed for stable transformants. Although transformation efficiency was very low, stable transformants were isolated (Titomirov *et al.,* 1991).

Electroporation has recently been combined with direct DNA injection *in vivo* to obtain high-efficiency gene transfer into a specific tissue. This "electrogene therapy" was described by Nishi *et al.* (1996), who injected a mammalian expression plasmid containing the *Eschericia coli lacZ* gene into the internal carotid artery of rats whose brain tumors had been subjected to an electric pulse just prior to injection of the plasmid. In addition to avoiding the disadvantages of viral-mediated systems of gene transfer (see later), this method demonstrated that genes could be expressed in particular target organs while avoiding any toxic effects. One gene therapy clinical trial has been approved using electroporation as a means of gene transfer (Black and Fakhrai, 1995). This strategy involves injecting malignant glioma patients with irradiated TGF-J2 antisense gene modified autologous tumor cells.

C. PARTICLE BOMBARDMENT

Also known as the ballistic microprojectile or gene gun method, this gene delivery technology is a physical means of transferring exogenous DNA into the target cell nucleus (Klein *et al.,* 1987). Although initially developed for the transformation of plants, particle bombardment has found its application in a number of gene therapy strategies (Burkholder, 1993; Woffendin *et al.,* 1996). The technique involves coating microparticles (usually 1- to 3- μm gold beads) with plasmid DNA and projecting them onto a target tissue by an electrical discharge or gas pulse device. The physical force involved in this bombardment allows the coated particles to penetrate the cell. The DNA construct is then gradually released from the microparticles and expressed within the cell. In most cases, the transferred genes exist episomally within the host cell nucleus (Schofield and Caskey, 1995); however, it has recently been demonstrated that stable integration can be achieved (Woffendin *et al.,* 1996). In that study T lymphocytes were transfected with an anti-HIV-containing expression plasmid and infused into HIV-positive individuals. Maintenance of a subpopulation of transfected cells was demonstrated by Southern analysis 10 days following infusion.

Particle bombardment has been successful in a range of somatic tissues *in vivo* and *ex vivo* and has achieved up to 100-fold higher transgene expression levels when compared to other nonviral-mediated gene delivery systems (Yang *et al.,* 1990; Williams *et al.,* 1991; Cheng *et al.,* 1993; Yang *et al.,* 1994; Sun *et al.,* 1995b). Particle bombardment of internal organs or tissues requires surgery to expose the target site. Clearly, the most accessible somatic tissue *in vivo* is the skin. Particle bombardment is being explored for the application of gene vaccines (Tang *et al.,* 1992; Fynan *et al.,* 1993; Wang *et al.,* 1993). Skin is an excellent target for immunization as epidermal tissue contains antigen-presenting cells (e.g., Langerhans cells) as well as cytotoxic T lymphocytes, the latter being involved in cellular immune responses. Tang *et al.* have demonstrated a protective immune response in mice following particle-mediated gene delivery of human growth hormone (hGH) or human α_1 antitrypsin (hAAT) DNA into the ear epidermis (Tang *et al.,* 1992). Only two gene therapy protocols currently employ particle-mediated gene transfer, both being *ex vivo* approaches. The first is the HIV protocol previously mentioned (Woffendin *et al.,* 1996) and the second involves the transfection of autologous tumor cells with the GM-CSF gene (Mahvi, 1996).

The advantages of particle-mediated gene transfer include (i) the delivery of large amounts of DNA to specific target tissues, (ii) the ability to deliver large DNA fragments (e.g., cosmids or artificial chromosomes), and (iii) the lack of cytopathic responses (Yang *et al.,* 1994). Compared to viral-mediated gene transfer, no potentially deleterious material is cotransfected with the therapeutic DNA. The major technical disadvantage of this new technology is the apparent lack of long-term gene expression.

D. DIRECT DNA INJECTION

Transformation of mammalian cells has been achieved through direct microinjection of "naked" DNA (Anderson *et al.,* 1980; Capecchi, 1980). Wolff *et al.* (1990) reported gene expression for at least 2 months following microinjection of RNA and DNA expression vectors directly into mouse muscle *in vivo.* Southern analysis determined that the DNA expression vector existed episomally; however, myoblasts are nondividing cells and therefore gene expression can be maintained in the transfected tissue for years. Gene transfer using this method is relatively inefficient as demonstrated by the limited number of cells expressing reporter genes. Davis *et al.* (1993) enhanced muscle uptake and expression of DNA by injecting the DNA in hypertonic sucrose. Injecting muscles with the anesthetic bupivacaine prior to DNA transfer has also led to substantial increase in DNA expression (Danko *et al.,* 1994).

DNA has been administered *in vivo* via a number of different routes. These include intravenous, intraperitoneal, subcutaneous, intramuscular, dermal, nasal, ocular, oral, and catheter delivery (Wickstrom, 1992). Three clinical trials have been approved in the United States which employ direct DNA injection as the mode of gene transfer (Isner and Walsh, 1994; Curiel, 1995; Isner, 1995). One of these involves injection of autologous tumor cells with DNA for augmentation of an antitumor immune response (Curiel, 1995). The other two protocols deliver therapeutic genes to vascular endothelial cells intra-arterially via an angioplasty catheter to treat peripheral artery disease (Isner and Walsh, 1994; Isner, 1995). As for particle-mediated gene transfer, direct DNA injection has been used to express viral antigens *in vivo* to achieve immune responses (Davis *et al.,* 1993; Ulmer *et al.,* 1993). Boyer *et al.* recently demonstrated that a specific cellular and humoral immune response could be evoked in chimpanzees following inoculation with the *env, rev,* and *gag/pol* genes (Boyer *et al.,* 1997).

The advantage of direct DNA injection is its simplicity. As with other nonviral delivery systems, potential risks such as producing replication-competent virus are avoided. Furthermore, the technical procedure is not as demanding as using viral-based vectors. However, direct DNA injection does have the disadvantage of poor gene transfer efficiency and a low level of stable gene integration.

E. LIPOPLEXES

Gene transfer can be achieved by membrane fusion techniques. These include the use of liposomes (Fraley and Papahadjopoulos, 1982) and erythrocyte ghosts (Straus and Raskas, 1980). DNA entrapped in human erythrocyte ghosts has been transferred to target cells; however, even compared to calcium phosphate coprecipitation and electroporation, the technique is relatively inefficient and has therefore not been pursued in recent years.

Liposomes have been used to encapsulate or complex nucleic acid (a lipoplex, as defined by Felgner *et al.* (1997)). Nucleic acid can then be transferred into target cells by fusion with the plasma membrane and consequent endocytosis (Fraley and Papahadjopoulos, 1982; Singhal and Huang, 1994; Zabner *et al.,* 1995). By avoiding normal physiological transfer processes, this method allows introduction of molecules that cannot normally enter the cell. Cationic liposomes (or cationic amphiphiles) have been developed which facilitate fusion and uptake by the cell (Felgner *et al.,* 1987; Behr, 1994). The net positive charge associated with cationic liposomes results in increased transfection efficiency by reducing electrostatic interference between the cell and lipoplex (Farhood *et al.,* 1994). Following endocytosis, the lipoplex is surrounded by an endosome and either degraded by lysosomal action or released into the cytoplasm (Singhal and Huang, 1994). The DNA or lipoplex then migrates to the nucleus, where gene expression occurs (Fig. 14.1). It has been shown that lipid and DNA must dissociate before transcription can take place (Zabner *et al.,* 1995). Twenty-nine of the currently approved gene therapy protocols use liposome technology (Table 14.1).

Liposomes are widely used as an efficient method of delivering DNA to a variety of target tissues *in vitro* and *in vivo.* Zhu *et al.* demonstrated gene expression in virtually all tissues in a mouse model following intravenous injection of a lipoplex (Zhu *et al.,* 1993). Importantly, gene transfer has not been

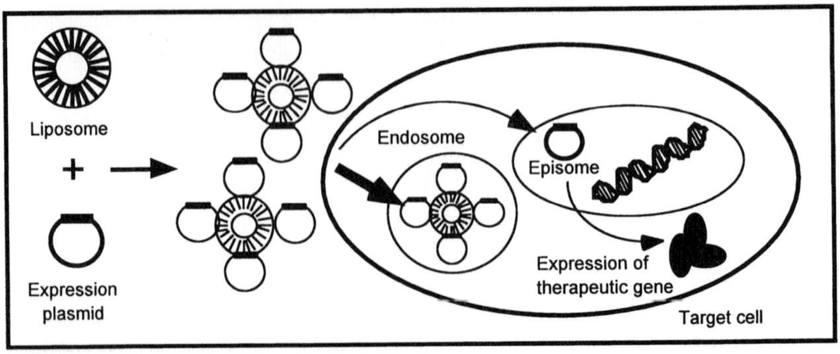

FIGURE 14.1 Lipoplex gene transfer. The negatively charged expression plasmid is complexed with the positively charged liposome to produce a lipoplex. Following fusion with the target cell membrane, the lipoplex is transferred to the endosome. Although the majority of lipoplex is consequently degraded by lysosomal action, some is shunted to the nucleus, where the plasmid dissociates from the liposome. Existing as an episome, the plasmid is then capable of expressing the therapeutic gene.

detected in the germline cells following intravenous injection or intra-arterial transfection (Nabel *et al.,* 1992). The most common cell types transfected by lipoplexes following intravenous delivery are vascular endothelial cells, monocytes, and macrophages (Liu *et al.,* 1997). Introduction of the p53 gene via liposomes has been shown to significantly reduce tumor volume, relapse, and metastases in an animal model system (Lesoon-Wood *et al.,* 1995). Nabel *et al.* (1990) used liposomes to transfer the β-galactosidase reporter gene into specific sites in the arterial wall. This method may therefore have implications for the treatment of diseases such as cancer and artherosclerosis *in vivo.* Direct transfer of the HLA-B7 gene into established melanoma cells *in vivo* has been achieved using lipoplexes and appears to augment specific immune response to the melanoma cells (Nabel *et al.,* 1993). Nabel *et al.* (1996) recently reported tumor-specific immune responses in patients receiving this gene therapy, one patient experiencing complete tumor remission. In these studies no toxicity or anti-DNA antibodies was demonstrated (Nabel, 1993). The cystic fibrosis transmembrane conductance regulator (CFTR) gene has been delivered to the airway epithelium in cystic fibrosis (CF) patients using liposomes (Caplen *et al.,* 1995). In this study, lipoplexes were delivered by nasal aerosol (Stribling *et al.,* 1992). No evidence of toxicity was observed with the quantity of lipoplex administered. CF is a monogenetic disorder which results in defective chloride ion transport across lung epithelial cells. A partial restoration of chloride perfusion was seen following CFTR cDNA gene transfer; however, this lasted for no longer than 7 days. The primary reasons for limited correction of the CF deficit were probably variation in lipoplex deposition and transfection efficiency.

As with other chemical transfer mechanisms, a major disadvantage of liposome-based gene transfer is its low efficiency of gene transfer and toxicity both *in vitro* and *in vivo.* Schuele *et al.* (1997) reported that using abundant quantities of lipoplex in an attempt to achieve higher transfection efficiencies *in vivo* also leads to toxic effects. In that case, pulmonary inflammation was observed in a dose-dependent manner when lipoplexes were delivered to the lung. Recent advances in liposome-mediated gene delivery have improved the efficiency of DNA transfer by employing receptor-mediated techniques (see Receptor-Mediated Gene Transfer).

F. RECEPTOR-MEDIATED GENE TRANSFER

Receptor-mediated techniques employ polycation complexes, or polyplexes (Felgner *et al.,* 1997), to transfer DNA into target cells (Curiel, 1994). DNA plasmids have been conjugated with transferrin and polylysine to take advantage of the physiological transfer processes of the cell (Cotten *et al.,* 1990). This avoids perturbation of the cell membrane and consequent cellular toxicity. Polylysine has also been conjugated to an asialoglycoprotein, which, after complexing with plasmid DNA, specifically targets the asialoglycoprotein receptors on hepatocytes (Wu *et al.,* 1991). Similarly, hepatic cells have been targeted using an α_2-macroglobulin–polylysine–DNA conjugate *in vitro* (Schneider *et al.,* 1996). Transfection efficiency has been increased by partially gluconoylating polylysine to reduce the positive charge associated with the polyplex (Erbacher *et al.,* 1997). Buschle *et al.* have conjugated CD3 antibodies to polylysine–DNA complexes to transfer DNA into activated T lymphocytes (Buschle *et al.,* 1995) whereas macrophages were transfected using a mannosylated polyplex (Ferkol *et al.,* 1996). Liposomes have also been targeted specifically to $CD4^+$ T lymphocytes by employing a CDR2 (complementarity-determining region 2 of the CD4 epitope) derived peptide (Slepushkin *et al.,* 1996). This gene transfer process is, however, restricted by entrapment of the DNA complex within cellular endosomes and ultimate lysosomal degradation (Cotten *et al.,* 1990).

Adenovirus possesses a mechanism allowing it to disrupt the endosome and escape into the cytosol. Coupling adenovirus to conjugate DNA complexes has been shown to facilitate lysosomal disruption and consequently augment the transfer of DNA using receptor-mediated techniques (Curiel *et al.,* 1991, 1992) (Fig. 14.2). Adenoviral–polyplex conjugates can increase gene transfer efficiency 100- to 1000-fold (Curiel *et al.,* 1991, 1992; Cristiano *et al.,* 1993; Curiel, 1994). Furthermore, large amounts of DNA can be transferred into a wide range of replicating or nonreplicating cells by this method (Cristiano *et al.,* 1993; Curiel, 1994). Adenovirus and adeno-associated virus have also been conjugated to liposomes to increase transfection efficiencies *in vitro* and *in vivo* (Philip *et al.,* 1994; Raja-Walia *et al.,* 1995; Vieweg *et al.,* 1995; Fasbender *et al.,*

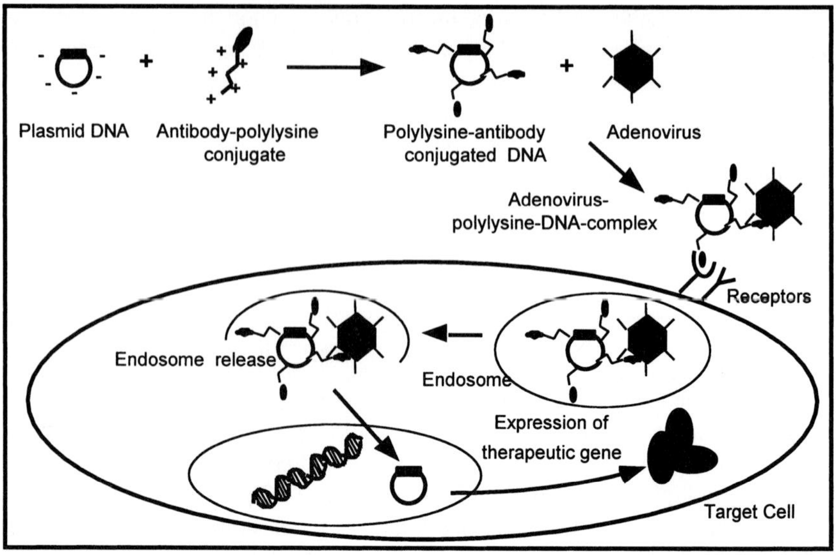

FIGURE 14.2 Receptor-mediated gene transfer. Expression plasmids are conjugated to an antibody–polycation molecule (e.g., polylysine) via electrostatic interaction. In most cases these conjugates are then complexed with an adenoviral vector. Vector uptake is mediated by interaction with membrane receptors specific for either the antibody or adenovirus. Transfer to the cytoplasmic endosome is followed by adenoviral-facilitated escape and transfer to the nucleus. The DNA plasmid expresses its therapeutic gene in an episomal fashion.

1997). A similar technique employing influenza virus peptide conjugates has been used to augment the transfer of marker genes into a number of cell lines (Wagner *et al.,* 1992). Recently, polylysine has been replaced with low molecular weight polyethylenimine (PEI) to conjugate large bacterial artificial chromosomes (up to 170-kb DNA) with inactivated adenovirus (Baker and Cotten, 1997). The use of PEI increased DNA transfer into a mammalian cell line with a 10-fold greater efficiency compared to polylysine conjugates.

HMG1 (high-mobility-group 1) is an abundant chromosomal protein which has been shown to bind DNA and efficiently transfect cells (Bottger *et al.,* 1988). The advantages of using this system for gene transfer are twofold. First, it has demonstrated a lack of cell toxicity as compared to other nonviral gene delivery agents and second, HMG1 seems to facilitate DNA transport to the nucleus of nondividing cells (Mistry *et al.,* 1997). Viral proteins have been incorporated into liposomes in conjunction with HMG1 to facilitate gene transfer by receptor-mediated fusion. These fusigenic viral liposomes, or "viriosomes," include the hemagglutinating virus of Japan (HVJ) (Kaneda *et al.,* 1989; Dzau *et al.,* 1996). HVJ is nonpathogenic to humans and its envelope protein allows fusion with the target cell membrane. The properties of this viriosome facilitate

DNA delivery directly into the cytoplasm, bypassing the lysosomal degradation pathway. HVJ-based viriosomes have been used to transfect a variety of tissues *in vivo* (Morishita *et al.,* 1993; Dzau *et al.,* 1996; Nakamura *et al.,* 1996; Aoki *et al.,* 1997). The advantages of this system include the ability of the HVJ–liposome to encapsulate up to 100 kb of DNA and to transfer it to target cells with high efficiency when compared to lipoplex or naked DNA (Dzau *et al.,* 1996). To increase the length of transgene expression, the HVJ–liposome system has been incorporated with elements of the Epstein–Barr virus (Sugden *et al.,* 1995) (see Extrachromosomal Replicating Vectors). This viriosome-based system is an attractive tool for gene transfer; however, as it is a viral hybrid, safety issues must be addressed. For example, low-level antigenicity against HVJ has been reported following viriosome injection *in vivo* (Dzau *et al.,* 1996).

The bacterial protein perfringolysin O (PFO) has also been used to efficiently deliver DNA into murine myoblasts (Gottschalk *et al.,* 1995). PFO was bound to a *lac Z*-containing plasmid using a biotin–streptavidin poly(L-lysine)bridge (similar to polyplex conjugation). DNA entry was mediated by permeabilizing the cell membrane in a receptor ligand independent manner. β-Galactosidase activity was demonstrated in 15–20% of cells under optimal conditions. A nonlipid, water-soluble polyamine has recently been reported to achieve high-efficiency gene transfer *in vivo* when compared to the liposome carrier Lipofectamine. The transfer efficiency was also comparable to a recombinant adenoviral vector. In preliminary studies the novel polyplex has also shown no toxic effects in mice nor was any serum inhibition observed (Goldman *et al.,* 1997).

G. EXTRACHROMOSOMAL REPLICATING VECTORS

Although integrating vectors can maintain stable and long-term gene expression, there are potential risks involved in using these vectors (see Retroviral Vectors). Furthermore, current viral vectors which do not integrate cannot replicate extrachromosomally and therefore have a limited lifetime in proliferating cells. Vectors that can replicate extrachromosomally include mammalian artificial chromosomes (MACS), Epstein–Barr viral (EBV) vectors, and human *ori* vectors (Calos, 1996). MACS contain the minimum chromosomal components required for DNA replication and segregation, namely, the origin of replication, telomeres, and centromeres. For gene therapy, MACS would also need to contain sequences for driving controlled expression of the therapeutic gene (Huxley, 1994). The first complete construction of a human artificial chromosome was recently demonstrated by Harrington *et al.* (1997). Although this is a significant step forward in the production of a MAC-based vector, many problems still exist. Currently, the largest size DNA that can be transferred to a target tissue is around a few hundred kilobases. The human artificial chromosomes described (Harrington *et al.,* 1997) were between 6 and 10 megabases, which is an impractical size for delivering into cells. For MACS to become a reality in gene therapy, gene delivery systems will have to accommodate larger DNA fragments or the size of artificial chromosomes will have to be reduced. The advantages of using MAC-based vectors include the ability to deliver large fragments of DNA while maintaining the ability to replicate and segregate in parallel with the cells' chromosomes. Although the DNA would not integrate into the natural genome, there would be the possibility of recombination with cellular chromosomes and potential oncogenesis due to the introduced genetic instability. No artificial chromosomes have yet been developed for gene therapy.

EBV vectors, like other viral vectors employed in gene therapy, contain elements derived from many sources. These include bacterial fragments (drug-resistance genes, replication origins, promoters) and eukaryotic genes (replication origins, promoters, splice sites, polyadenylation signals). It is the EBV replication origin and *EBNA-1* gene that confer the unique ability of these vectors to be maintained as stable, freely replicating plasmids (Margolskee, 1992). Although EBV vectors are retained in the nucleus, they have a random segregation pattern and they do gradually disappear from rapidly dividing cells after a few months. It has been demonstrated that EBV vectors can accommodate up to 160 kb of DNA and are B-cell specific (Banerjee *et al.,* 1995). These cells are attractive targets for hematopoietic diseases such as the immunoglobulin immunodeficiency, agammaglobulinemia. Furthermore, the EBNA-1 protein which is required for replication and retention has been shown to overcome immune surveillance, allowing efficient *in vivo* persistence (Levitskaya *et al.,* 1995; Miyashita *et al.,* 1995). Human *ori* vectors have been developed that rely on native human sequences to mediate vector replication (Krysan *et al.,* 1989; Calos, 1996). It has been shown that these vectors can express a gene of interest for approximately 2 months, depending on the frequency of cell division (Wohlgemuth *et al.,* 1996). Replication and retention of both EBV and human *ori* based vectors have not yet been demonstrated in primary cells.

III. VIRAL DELIVERY SYSTEMS

Biological gene transfer methods employ modified DNA and RNA viral vectors to infect the cell, thereby introducing and expressing its genome which contains the foreign gene (Morgan and Anderson, 1993). This is known as transduction. The most commonly used viral vectors are herpes virus (Freese *et al.,* 1990), adenovirus (Berkner, 1992; Wilkinson and Akrigg, 1992; Neve, 1993; Brody and Crystal, 1994), adeno-associated virus (Muzyczka, 1992; Kotin, 1994), and various retroviruses (Drumm *et al.,* 1990; Chowdhury *et al.,* 1991; Freeman *et al.,* 1993; Hoeben *et al.,* 1993; Miller *et al.,* 1993; Mulligan, 1993). Many other viruses are also being investigated as potential gene delivery systems but cannot be reviewed comprehensively here. In each case, viruses have had the genes encoding essential replicative/packaging proteins replaced with the therapeutic gene of interest. These modified viruses are termed recombinant viral

TABLE 14.4 Summary of Viral Gene Transfer Techniques

Gene transfer method	Advantages	Disadvantages
Adenoviral vectors	Broad host range Infects nondividing cells Higher titer Minimal target cell damage No possibility of insertional mutagenesis High transduction efficiency	<8-kb DNA can be incorporated Possible RCA production Transient expression Lack of tissue specificity Immune response
Adeno-associated viral vectors	Broad host range Infects nondividing cells High titer Chromosomal integration Wild-type integrates at specific site Stable expression	<5-kb DNA can be incorporated Vectors lose specific site integration Difficult to purify vector Low titer
Herpes simplex viral vectors	Neurotropic No possibility of insertional mutagenesis Infects nondividing cells Accepts large DNA inserts (~15 kb)	Lytic infection Transient expression Low titer
Retroviral vectors	Chromosomal integration Can be tissue specific Stable expression	Possible insertional mutagenesis Possible RCR production Transduces dividing cells only
HIV vectors	Infects nondividing cells Tissue specific Stable expression	Possible insertional mutagenesis Possible RCR production Low titer Ethical considerations

vectors. Advantages and disadvantages of recombinant viral mediated gene transfer techniques are summarized in Table 14.4.

A. HERPES SIMPLEX VIRUS VECTORS

The herpes simplex virus (HSV) is a neurotropic vector (Freese *et al.,* 1990; Efstathiou and Minson, 1995; Blomer *et al.,* 1996). It has therefore been developed for therapeutic gene transfer in the treatment of neurodegenerative disorders such as Parkinson's disease (Freese *et al.,* 1990; During *et al.,* 1994). Neurons have been transduced with a recombinant HSV vector containing the tyrosine hydroxylase (TH) gene. This resulted in increased production of the neurotransmitter, dopamine. Alzheimer's disease, epilepsy, Lesch–Nyhan syndrome, and multiple sclerosis may become targets for similar gene therapy techniques (Culver, 1994). Strategies for treating brain tumors with HSV-based vectors are also being developed (Martuza *et al.,* 1991; Blomer *et al.,* 1996). HSV may remain latent or cause cell lysis following cell infection. This is detrimental for gene therapy strategies so it will be necessary to develop HSV vectors in which the genes required for lytic replication are removed. This has proved difficult due to the size of the vector (~ 150 kb) and the number of genes it harbors (~ 70). However, in one study, where a recombinant HSV vector containing the TH gene was injected into the brains of Parkinsonian rats (During *et al.,* 1994), an improved disease status was observed for 1 year without significant toxic effects.

B. ADENOVIRAL VECTORS

The adenovirus is a DNA virus which can produce upper respiratory tract infections. Adenoviral vectors cannot integrate into the host cell's genome and have been demonstrated to be safe when administered to humans. Millions of American army recruits have been vaccinated with unattenuated adenovirus without major adverse reactions (Kremer and Perricaudet, 1995). Although its natural tropism for respiratory epithelium has prompted its use in cystic fibrosis gene therapies, it has also been employed for gene delivery to most other human tissues (Stratford-Perricaudet and Perricaudet, 1994). Twenty-six of the gene therapy protocols approved by the Recombinant DNA Advisory Committee (RAC) employ adenovirus-based strategies, nine being for cystic fibrosis, sixteen for cancer, and one for partial ornithine transcarbamylase deficiency. Although gene transfer into airway epithelial cells has also been mediated via liposomes (Hyde *et al.,* 1993) and retroviruses (Drumm *et al.,* 1990), the most widely used vector for cystic fibrosis therapy has been the adenovirus (Engelhardt *et al.,* 1993; Zabner *et al.,* 1993). Transient rectification of the chloride transport defect has been achieved by adenovirus-mediated transfer of the CFTR gene (Zabner *et al.,* 1993). However, long-term correction of associated pulmonary disease is yet to be demonstrated. Knowles *et al.* (1995) failed to show correction of the CFTR deficit in an adenoviral-based clinical trial, although increasing the contact time between the vector and target epithelial cells may improve clinical efficacy (Jiang *et al.,* 1997). This clinical study also demonstrated the deleterious inflammatory responses involved when using adenoviruses.

Adenoviral-specific inflammatory responses are a major disadvantage in adenoviral-based gene transfer strategies. To avoid antigenicity, DeMatteo *et al.* (1995) transferred an adenoviral vector to mouse thymus *in vivo,* achieving immunologic unresponsiveness to the recombinant virus. Like other viral systems, adenoviruses pose a risk of infectious recombination. To avoid the manifestation of replication-competent adenovirus (RCA), most adenoviral vectors carry deletions in the E1a, E1b, and E3 genes. These defective vectors must be grown in the presence of a helper virus for it to be packaged into an infectious

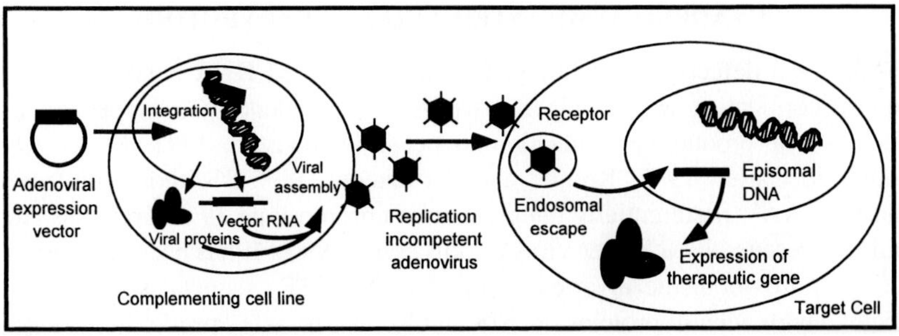

FIGURE 14.3 Adenoviral gene transfer. The adenoviral vector is produced by transfecting an expression plasmid, containing the therapeutic gene, into a complementing cell line. This cell line provides the elements necessary for replication in trans. Due to this genomic complementarity, the adenoviral vectors produced are replication incompetent. The vector binds to the target cell via specific receptors and is transferred to the cytoplasmic endosome. Following endosomal release, the adenoviral DNA is delivered to the nucleus, where it expresses the therapeutic gene in an episomal fashion.

particle (Fig. 14.3). Other viral elements (e.g., E2a, E2b, and E4 genes) have also been mutated or deleted to decrease RCA and host immune responses (Gorziglia *et al.,* 1996; Guangping and Wilson, 1997). However, Ilan *et al.* (1997) have demonstrated that the insertion of the E3 region into recombinant adenovirus inhibits antiviral antibody and adenovirus-specific cellular immune responses. An adenoviral vector was recently constructed which lacks all viral coding sequences (Kochanek *et al.,* 1996). Fender *et al.* have also developed a "dodecahedron" adenovirus which consists of only one or two of the eleven proteins contained in the adenovirus virion (Fender *et al.,* 1997). These recombinant vectors, which lack the viral genome, have similar gene transfer efficacy and yet seem to be a safer alternative to other adenovirus-based systems.

Adenovirus infects most cell types regardless of the cell's mitotic state (Kremer and Perricaudet, 1995). To avoid gene expression in nontarget cells, tissue-specific enhancers and promoters have been employed. Specific gene expression was recently demonstrated in a gastric carcinoma model *in vivo* using the carcinoembryonic antigen (CEA) promoter (Tanaka *et al.,* 1997). Selective cell targeting has also been achieved by modifying the protein required for attachment to cell surface receptors (Stevenson *et al.,* 1997). This approach could be developed further by using specific ligands for targeted gene transfer.

An advantage of the adenovirus-based system is its high transduction efficiency. Almost 100% of myocytes were transduced when a rabbit heart was perfused with adenovirus containing the *lacZ* gene (Donahue *et al.,* 1997). Transduction efficiency was increased to this level by using higher numbers of viral particles, longer duration of viral exposure, specific culture media composition, and physiological temperatures. This was the first report of complete transduction of an intact organ *ex vivo.*

C. ADENO-ASSOCIATED VIRAL (AAV) VECTORS

AAV is a defective, nonpathogenic, single-stranded DNA parvovirus which requires coinfection with a helper virus (usually an adenovirus, herpes virus, or vaccinia) for productive replication to occur (Muzyczka, 1992; Kotin, 1994; Flotte and Carter, 1995; Kremer and Perricaudet, 1995). Without coinfection, wild-type AAV preferentially integrates into chromosome 19 and remains latent until it is rescued by a helper virus (Kotin *et al.,* 1990). It has been reported that AAV requires both the inverted terminal repeat (ITR) sequences and the site-specific endonuclease proteins, Rep78 and Rep68, for site-specific integration to occur (Weitzman *et al.,* 1994). Although some reports have suggested recombinant constructs may partially maintain site-specific integration (Giraud *et al.,* 1994, 1995), no AAV vectors have yet been manufactured which maintain this capability completely (Ponnazhagan *et al.,* 1997).

AAV has a broad host range and does not require target cells to proliferate for successful infection (Kotin, 1994). Respiratory epithelium, fibroblasts, hematopoietic stem cells, cells of the central nervous system (CNS), and tumor cells have been efficiently transduced with AAV vectors (Flotte and Carter, 1995). As a gene transfer vector, AAV can accommodate about 4.7 kb of DNA insert (Muzyczka, 1992). This size restriction limits the potential cDNA that can be employed as a therapeutic. Another disadvantage of this delivery system includes difficulty in purifying the vector from contaminating wild-type AAV and helper virus during vector preparation. Two RAC-approved gene therapy protocols are employing AAV-mediated gene transfer. Both of these aim to treat cystic fibrosis patients (Flotte, 1994; Gardner, 1995). Preclinical data have demonstrated detection of the CFTR protein up to 6 months after intrabronchial transduction of an AAV vector in a rabbit model (Flotte *et al.,* 1993). No cytotoxicity was observed in this study.

D. RETROVIRAL VECTORS

Retroviruses have been employed in the majority of clinical trials to date (Table 14.1). It is the unique life cycle of the retrovirus that has made it an attractive vector for gene transfer (Varmus, 1988; Boris-Lawrie and Temin, 1994; Vile and Russell, 1995). Retroviral transduction is usually performed in the presence of a polycation (e.g., Polybrene™ or protamine sulfate) to reduce electrostatic interference between the virus and its target cell. After fusion with cell membrane glycoproteins, the virus enters the cell, where its RNA genome is reverse transcribed into DNA. This DNA copy, termed cDNA, becomes double-stranded and is transported to the nucleus, where it is integrated into the host genome to form the provirus (Miller, 1992). The provirus is then transcribed by the host cell transcriptional machinery into RNA that becomes genomic viral RNA or is multiply spliced to produce subgenomic messages from which viral

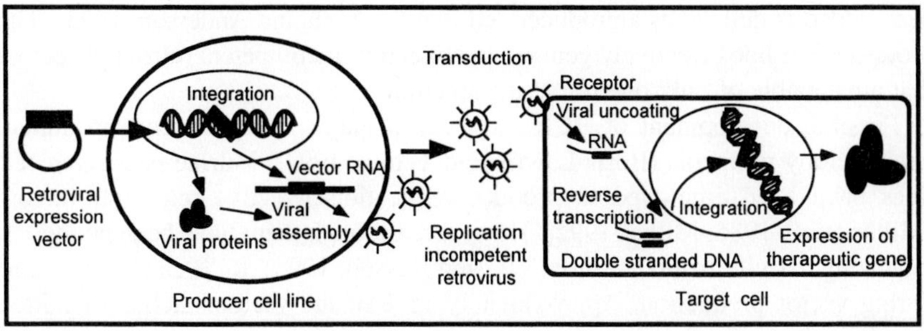

FIGURE 14.4 Retroviral gene transfer. Expression plasmids lacking the essential genes for replication are transfected into a packaging cell line. The packaging cell line contains a complementary integrated provirus which lacks a packaging sequence but provides the structural proteins and reverse transcriptase in *trans.* The producer cell line thereby provides replication-incompetent retroviral particles. The retroviral vector transduces a target cell by interaction with specific receptors present on the cell surface. After entering the cell, the viral RNA is reverse transcribed, and the proviral DNA randomly integrates into the target cell genomes, where the therapeutic gene is expressed.

proteins are produced. Figure 14.4 outlines the steps involved in retroviral-mediated gene transfer. The advantages of retroviral vectors are the high transfer efficiencies, precise and stable integration into the host cell genome, and, by using replication-incompetent retroviruses, the lack of further retroviral production after infection. Disadvantages include the inability to infect nondividing cells (Springett *et al.,* 1989; Miller *et al.,* 1990), potential insertional mutagenesis, which may cause activation of cellular oncogenes or gene inactivation (Boris-Lawrie and Temin, 1994), and the possibility of recombination to produce replication-competent retrovirus (RCR). Targeting retroviral integration to specific chromosomal positions may reduce the possibility of insertional mutagenesis. It may also allow integration into transcriptionally active sites. Targeted integration of retroviral sequences has been achieved by employing a fusion of HIV-1 integrase and the DNA-binding domain of lambda repressor (Bushman, 1994). The size of the gene insert is also limited as retroviruses are only capable of packaging around 10 kb of genetic material (Naviaux and Verma, 1992).

Retroviral vectors are packaged into infectious particles after introduction into a retroviral packaging cell line (Miller *et al.,* 1993; Morgan and Anderson, 1993). A typical packaging cell is a cell that contains a modified retroviral genome that has the structural genes (*gag, pol,* and *env*) but lacks the packaging signal (psi). Psi is required for efficient encapsidation of the vector RNA into virions (Bender *et al.,* 1987). The engineered retroviral vector, which is introduced into the packaging cell, is in effect the complement to the packaging cell genome and retains the psi encapsidation and replication signals but is devoid of the structural genes. A packaging cell line to which a retroviral vector has been added is defined as a producer cell line (Morgan and Anderson, 1993). The producer cell line effectively generates replication-incompetent retroviral vectors that are capable of only one round of infection.

Possible development of RCR has been a major safety concern in human gene therapy protocols (Boris-Lawrie and Temin, 1994). Murine packaging cell lines have been employed to produce replication-defective retroviral vectors (Miller *et al.,* 1986; Miller, 1992). However, although there have been no reports indicating toxicity from their use (Vile and Russell, 1995), RCR can be produced during vector production. Approximately 15% of the vector batches submitted for clinical certification have tested positive for RCR (Hodgson *et al.,* 1997). This is due to recombination with endogenous retroviral elements (ERE) in the packaging cell line. To eliminate RCR manifestation and to increase vector titer, human packaging cell lines are being employed which do not express murine-like ERE (Finer *et al.,* 1994; Cosset *et al.,* 1995). This should reduce recombination during vector production. Russ *et al.* (1996) have recently developed retroviral vectors which have the ability to excise themselves after chromosomal integration, leaving the gene of interest intact. Deletion of viral sequences, which is mediated by the *Cre/LoxP* site-specific recombinase system, may improve long-term gene expression as well as reduce possible recombination with endogenous retroviruses present in human target cells.

The host range of a retroviral particle is primarily determined by the type of envelope gene—in the case of retroviral vectors, the *env* gene—contained within the packaging cell line (Miller, 1992). Although murine amphotropic virus-based retroviral vectors are capable of infecting human cells, transduction generally results in moderate-level gene transfer (Friedmann and Yee, 1995). Retroviral vectors employing the gibbon ape leukemia virus (GALV) *env* gene, however, have been shown to be more efficient at gene transduction into mammalian cells as compared with vectors with amphotropic Env glycoproteins (Miller *et al.,* 1991; Bayle *et al.,* 1993; Von Kalle *et al.,* 1994; Bunnell *et al.,* 1995). These pseudotyped retroviruses have been designed by replacing the amphotropic Moloney murine leukemia virus (MoMLV) *env* product by a vesicular stomatitis virus G (VSV-G) protein that interacts with ubiquitous membrane receptors such as phosphatidylserine. The vast distribution of these receptors appears to increase the vector host range (Friedmann and Yee, 1995). The stability of VSV-G also allows pseudotyped particles to be concentrated by ultracentrifugation to titers that are higher than those of amphotropic vectors (Friedmann and Yee, 1995).

To increase the cellular tropism of retroviral vectors, the Env protein has been modified by incorporation of polypeptides to alter its binding properties for targeted gene delivery (Salmons *et al.,* 1995; Cosset and Russell, 1996; Schnierle and Groner, 1996). Vector host range has been altered by the addition of an antibody-envelope fusion protein (Chu and Dornburg, 1995). These vectors efficiently infect cells that express the corresponding antigen. Retroviral vectors have also been designed for tissue-specific gene expression *in vivo* by adding tissue-specific enhancers (Naviaux and Verma, 1992; Kasahara *et al.,* 1994).

A number of methods have been investigated to increase retroviral transduction efficiency. Retroviral titer and transduction have been increased by harvesting virus-containing medium at 32°C compared to 37°C (Kotani *et al.,* 1994; Bunnell *et al.,* 1995; Kaptein *et al.,* 1997). This is probably due to increased vector stability at this temperature. Furthermore, it has been established that retroviral titer is not predictive of gene transfer efficiency (Forestell *et al.,* 1995). A high-titer virus may not transduce as efficiently as virus with a longer half-life. Increasing transduction efficiency has also been achieved by increasing the rate of virus–cell association by fluid "flow-through" techniques (Palsson *et al.,* 1995; Chuck and Palsson, 1996) and by centrifugation of target cells during transduction (Kotani *et al.,* 1994; Bahnson *et al.,* 1995; Bunnell *et al.,* 1995). By providing a net flow of virus-containing medium to the target cells, retroviruses located distally are brought into contact with the cells, increasing the probability of viral adsorption (Palsson and Andreadis, 1997). Flow-through methods also eliminate the problems associated with viral titer, allowing transduction to be performed with reduced viral volumes even without the use of polycations (Chuck and Palsson, 1996). Transduction efficiency has also been enhanced by coprecipitation with calcium chloride (Morling and Russell, 1995) and by cotransfection with liposomes (Hodgson *et al.,* 1997). These techniques also bring the virus closer to target cells, thereby increasing the rate of viral uptake.

E. HUMAN IMMUNODEFICIENCY VIRUS (HIV) VECTORS

Recombinant human immunodeficiency virus type 2 (HIV-2) (Garzino-Demo *et al.,* 1995) and HIV-1-based vectors (Shimada *et al.,* 1991; Buchschacher and Panganiban, 1992; Reiser *et al.,* 1996; Mammano *et al.,* 1997) have been investigated for cell-specific gene transfer. Therapeutic gene transfer into $CD4^+$ cells, the cells which wild-type HIV naturally infects, would be ideal for HIV gene therapy strategies. Furthermore, unlike other retroviruses, HIV can infect nondividing cells (Lewis and Emerman, 1994). This would allow HIV-based vectors to transduce noncycling macrophages and microglial cells (the other cells targeted by wild-type HIV) as well as hematopoietic stem cells. The ability of lentiviruses, such as HIV, to infect noncycling cells is due to the preintegration complex traversing the intact nuclear envelope. This nuclear-targeting property stems from the p17 Gag protein and Vpr accessory protein (Heinzinger *et al.,* 1994). Inducible gene expression using Tat- and Rev-inducible promoters is also an advantage of the HIV-based delivery system. Inducible expression of anti-HIV therapeutic genes will only occur in HIV-infected cells. Such strategies employing "toxin" genes have been suggested. In these cases, HIV-infected cells will be selectively destroyed by the expressed toxin (Buchschacher and Panganiban, 1992; Caruso and Klatzmann, 1992).

The viral elements allowing $CD4^+$ cell-targeted gene transfer have been exploited in other viral-based systems. For example, recombinant rabies virus pseudotyped with the HIV-1 envelope spike protein has demonstrated specific infection of the $CD4^+$ HeLa cell line (Mebatsion and Conzelmann, 1996). Mammano *et al.* (1997) have recently demonstrated targeted gene transfer into human $CD4^+$ cells using MoMLV pseudotyped to a truncated version of the HIV-1 envelope glycoprotein. Progress has been made with vesicular stomatitis virus G protein (VSV-G) pseudotyped HIV-1 vectors (Naldini *et al.,* 1996). In this case, the HIV-based vector was capable of transducing nondividing neuronal cells *in vivo.* The producer cell lines used to generate this virus were, however, based on a transient transfection system. Stable HIV packaging cell lines have recently been produced (Corbeau *et al.,* 1996). The viral titer from these producer cell lines was up to 10^5 transducing units per milliliter, and the vector was replication-incompetent.

Although HIV vectors have been made replication-incompetent (Richardson *et al.,* 1995), there is a risk of recombination with wild-type virus leading to new forms of virulent HIV. Exposing patients to this risk is a major concern. Precautions are therefore being examined to prevent this possibility. The suicide gene herpes simplex thymidine kinase (HS-tk) could be incorporated into HIV-based vectors, allowing infected cells to be selectively destroyed by the drug ganciclovir. Due to the ethical considerations involved with HIV-based vectors, it may be several years before they enter clinical trials.

F. NEO-ORGANS AND ENCAPSULATED CELLS

A novel approach for the long-term delivery of a therapeutic molecule has been demonstrated by implanting genetically modified autologous fibroblasts into animal models (Moullier *et al.,* 1993, 1995; Salvetti *et al.,* 1995). In this strategy, skin biopsies are taken from the subject and the fibroblasts are genetically altered with a retroviral vector containing the therapeutic gene. The cells are grown *in vitro* with collagen to form a solid matrix. These modified fibroblasts (also known as "organoides" or "neo-organs") are then implanted into the intraperitoneal cavity, where they display continual delivery of the therapeutic molecule. This strategy is also known as "transkaryotic" therapy (Selden *et al.,* 1997). Fibroblasts have been used to deliver factor VIII to immunodeficient mice (Hoeben *et al.,* 1993). Myoblasts have also been used as the cell type in transkaryotic-based therapy (Stratford-Perricaudet *et al.,* 1992). Human growth hormone has been delivered systemically in a mouse model by transducing cultured myoblasts with a retroviral vector and injecting the modified myoblasts into mouse muscle (Dhawan *et al.,* 1991). This strategy has also been employed to deliver factor IX *in vivo* (Dai *et al.,* 1992).

Another strategy for delivering therapeutic molecules systemically is to inject microcapsules containing genetically engineered cells *in vivo* (Lanza *et al.*, 1996). These capsules can protect allogeneic cells from destruction by the immune system. Emerich (1997) has recently shown a protective effect in a monkey model of Huntington's disease after delivering encapsulated cells producing ciliary neurotrophic factor to the striatum.

IV. CONCLUSIONS

Gene therapy is the introduction of genetic material (DNA or RNA) into target cells for therapeutic benefit. The key to gene therapy is appropriate and efficient delivery systems. Both nonviral and viral systems are reviewed here, with assessments of their efficacy, utility, advantages, and disadvantages. As gene transfer mechanisms become more efficient and long-term gene expression is enhanced, the variety of diseases that can be treated by gene therapy will continue to expand. However, until the problem of delivery and subsequent expression is adequately resolved, gene therapy will not realize its potential.

ACKNOWLEDGMENTS

We thank Allison Arndt, Alison Knop, and Josh Stern for comments on the manuscript.

REFERENCES

Anderson, W. (1992). Human gene therapy. *Science* **256,** 808–813.

Anderson, W. F., Killos, L., Sanders-Haigh, L., *et al.* (1980). Replication and expression of thymidine kinase and human globin genes microinjected into mouse fibroblasts. *Proc. Natl. Acad. Sci. USA* **77,** 5399–5403.

Aoki, M., Morishita, R., Higaki, J., *et al.* (1997). *In vivo* transfer efficiency of antisense oligonucleotides into the myocardium using HVJ-liposome method. *Biochem. Biophys. Res. Commun.* **231,** 540–545.

Avalosse, B., Dupont, F., and Burny, A. (1995). Gene therapy for cancer. *Curr. Opin. Oncol.* **7,** 94–100.

Bahnson, A. B., Dunigan, J. T., Baysal, B. E., *et al.* (1995). Centrifugal enhancement of retroviral mediated gene transfer. *J. Virol. Methods* **54,** 131–143.

Baker, A., and Cotten, M. (1997). Delivery of bacterial artificial chromosomes into mammalian cells with psoralen-inactivated adenovirus carrier. *Nucleic Acids Res.* **25,** 1950–1956.

Banerjee, S., Livanos, E., and Vos, J. M. (1995). Therapeutic gene delivery in human B-lymphoblastoid cells by engineered non-transforming infectious Epstein–Barr virus. *Nat. Med.* **1,** 1303–1308.

Bayle, J. Y., Johnson, L., St. George, J., *et al.* (1993). High-efficiency gene transfer to primary monkey airway epithelial cells with retrovirus vectors using the gibbon ape leukemia virus receptor. *Hum. Gene Ther.* **4,** 161–170.

Behr, J. P. (1994). Gene transfer with synthetic cationic amphiphiles: Prospects for gene therapy. *Bioconjug. Chem.* **5,** 382–389.

Bender, M., Palmer, T., Gelinas, R., *et al.* (1987). Evidence that the packaging signal of Moloney murine leukemia virus extends into the gag region. *J. Virol.* **61,** 1639–1646.

Berkner, K. L. (1992). Expression of heterologous sequences in adenoviral vectors. *Curr. Top. Microbiol. Immunol.* **158,** 39–66.

Beutler, E. (1992). Gaucher disease: New molecular approaches to diagnosis and treatment. *Science* **256,** 794–799.

Black, K., and Fakhrai, H. (1995). A phase I study of the safety of injecting malignant glioma patients with irradiated TGF-J2 antisense gene modified autologous tumor cells. University of California, Los Angeles School of Medicine, Los Angeles, California. *RAC Report,* Office of Recombinant DNA Activities, NIH No. 9512-138.

Blaese, R. M., and Anderson, W. F. (1990). The ADA human gene therapy clinical protocol. *Hum. Gene Ther.* **1,** 327–362.

Blomer, U., Naldini, L., Verma, I., *et al.* (1996). Applications of gene therapy to the CNS. *Hum. Mol. Genet.* **5,** 1397–1404.

Boris-Lawrie, K., and Temin, H. M. (1994). The retroviral vector: Replication cycle and safety considerations for retrovirus-mediated gene therapy. *Ann. N.Y. Acad. Sci.* **716,** 59–71.

Bottger, M., Vogel, F., Platzer, M., *et al.* (1988). Condensation of vector DNA by the chromosomal protein HMG1 results in efficient transfection. *Biochim. Biophys. Acta* **950,** 221–228.

Boyer, J., Ugen, K., Wang, B., *et al.* (1997). Protection of chimpanzees from high-dose heterologous HIV-1 challenge by DNA vaccination. *Nat. Med.* **3,** 526–532.

Bridges, S. H., and Sarver, N. (1995). Gene therapy and immune restoration for HIV disease. *Lancet* **345,** 427–432.

Brody, S. L., and Crystal, R. G. (1994). Adenovirus-mediated *in vivo* gene transfer. *Ann. N.Y. Acad. Sci.* **716,** 90–103.

Brownlee, S. G. (1995). Prospects for gene therapy of haemophilia A and B. *Br. Med. Bull.* **51,** 91–105.

Buchschacher, G. L. J., and Panganiban, A. T. (1992). Human immunodeficiency virus vectors for inducible expression of foreign genes. *J. Virol.* **66,** 2731–2739.

Bunnell, B. A., Muul, L. M., Donahue, R. E., *et al.* (1995). High-efficiency retroviral-mediated gene transfer into human and onohuman primate peripheral blood lymphocytes. *Proc. Natl. Acad. Sci. USA* **92,** 7739–7743.

Burkholder, J. (1993). Rapid transgene expression in lymphocyte and macrophage primary cultures after particle bombardment-mediated gene transfer. *J. Immunol. Methods* **165,** 149–156.

Buschle, M., Cotten, M., Kirlappos, H., *et al.* (1995). Receptor-mediated gene transfer into human T lymphocytes via binding of DNA/CD3 antibody particles to the CD3 T cell receptor complex. *Hum. Gene Ther.* **6,** 753–761.

Bushman, F. D. (1994). Tethering human immunodeficiency virus 1 integrase to a DNA site directs integration to nearby sequences. *Proc. Natl. Acad. Sci. USA* **91,** 9233–9237.

Calos, M. (1996). The potential of extrachromosomal replicating vectors for gene therapy. *Trends Genet.* **12,** 463–466.

Capecchi, M. (1994). Targeted gene replacement. *Sci. Am.* March, 34–41.

Capecchi, M. R. (1980). High efficiency transformation by direct microinjection of DNA into cultured mammalian cells. *Cell* **22,** 479–488.

Caplen, N., Alton, E., Middleton, P., *et al.* (1995). Liposome-mediated CFTR gene transfer to the nasal epithelium of patients with cystic fibrosis. *Nat. Med.* **1,** 39–46.

Caruso, M., and Klatzmann, D. (1992). Selective killing of $CD4^+$ cells harboring a human immunodeficiency virus-inducible suicide gene prevents viral spread in an infected cell population. *Proc. Natl. Acad. Sci. USA* **89,** 182–186.

Chang, P. (1994). Calcium phosphate-mediated DNA transfection. *In* "Gene Therapeutics: Methods and Applications of Direct Gene Transfer" (J. Wolff, Ed.), pp. 157–179. Birkhauser, Boston, Cambridge, MA.

Cheng, L., Ziegelhoffer, P., and Yang, N. S. (1993). *In vivo* promoter activity and transgene expression in mammalian somatic tissues evaluated by using particle bombardment. *Proc. Natl. Acad. Sci. USA* **90,** 4455–4459.

Chowdhury, J. R., Grossman, M., Gupta, S., *et al.* (1991). Long-term improvement of hypercholesterolemia after *ex vivo* gene therapy in LDLR-deficient rabbits. *Science* **254,** 1802–1805.

Chu, T. H., and Dornburg, T. (1995). Retroviral vector particles displaying the antigen-binding site of an antibody enable cell-type-specific gene transfer. *J. Virol.* **69,** 2659–2663.

Chuck, A. S., and Palsson, B. O. (1996). Consistent and high rates of gene transfer can be obtained using flow-through transduction over a wide range of retroviral titers. *Hum. Gene Ther.* **7,** 743–750.

Colladge, W. H., and Evans, M. J. (1995). Cystic fibrosis gene therapy. *Br. Med. Bull.* **51,** 82–90.

Corbeau, P., Kraus, G., Wong-Staal, F., *et al.* (1996). Efficient gene transfer by a human immunodeficiency virus type 1 (HIV-1)-derived vector utilizing a stable HIV packaging cell line. *Proc. Natl. Acad. Sci. USA* **93,** 14070–14075.

Cosset, F., and Russell, S. (1996). Targeting retrovirus entry. *Gene Ther.* **3,** 946–956.

Cosset, F. L., Takeuchi, Y., Battini, J. L., *et al.* (1995). High-titer packaging cells producing recombinant retroviruses resistant to human serum. *J. Virol.* **69,** 7430–7436.

Cotten, M., Langle-Rouault, F., Kirlappos, H., *et al.* (1990). Transferrin-polycation-mediated introduction of DNA into human leukemic cells: Stimulation by agents that affect the survival of transfected DNA or modulate transferrin receptor levels. *Proc. Natl. Acad. Sci. USA* **87,** 4033–4037.

Cristiano, R., Smith, L., Kay, M., *et al.* (1993). Hepatic gene therapy: Efficient gene delivery and expression in primary hepatocytes utilizing a conjugated adenovirus-DNA complex. *Proc. Natl. Acad. Sci. USA* **90,** 11548–11552.

Crystal, R. (1995). Transfer of genes to humans: Early lessons and obstacles to success. *Science* **270,** 404–410.

Culver, K. W. (1994). "Gene Therapy: A Handbook for Physicians." Mary Ann Liebert, New York.

Culver, K. W., Vickers, T. M., Lamsam, J. L., *et al.* (1995). Gene therapy for solid tumors. *Br. Med. Bull.* **51,** 192–204.

Curiel, D. (1995). Phase I trial of a polynucleotide augmented anti-tumor immunization to human carcinoembryonic antigen in patients with metastatic colorectal cancer. *RAC Report,* Office of Recombinant DNA Activities, NIH No. 9406-073.

Curiel, D. T. (1994). High efficiency gene transfer mediated by adenovirus-polylysine-DNA complexes. *Ann. N.Y. Acad. Sci.* **716,** 36–58.

Curiel, D. T., Agarwal, S., Wagner, E., *et al.* (1991). Adenovirus enhancement of transferrin-polysine-mediated gene delivery. *Proc. Natl. Acad. Sci. USA* **88,** 8850–8854.

Curiel, D. T., Wagner, E., Cotten, M., *et al.* (1992). High-efficiency gene transfer mediated by adenovirus coupled to DNA-polylysine complexes. *Hum. Gene Ther.* **3,** 147–154.

Dai, Y., Roman, M., Naviaux, R. K., *et al.* (1992). Gene therapy via primary myoblasts: Long-term expression of factor IX protein following transplantation *in vivo. Proc. Natl. Acad. Sci. USA* **89,** 10892–10895.

Danko, I., Fritz, J., Jiao, S., *et al.* (1994). Pharmacological enhancement of *in vivo* foreign gene expression in muscle. *Gene Ther.* **1,** 114–121.

Davis, H., Whalen, R., and Demeneix, B. (1993). Direct gene transfer into skeletal muscle *in vivo:* Factors affecting efficiency of transfer and stability of expression. *Hum. Gene Ther.* **4,** 151–159.

DeMatteo, R., Raper, S., Ahn, M., *et al.* (1995). Gene transfer to the thymus. A means of abrogating the immune response to recombinant adenovirus. *Ann. Surg.* **222,** 229–242.

Dhawan, J., Pan, L. C., Pavlath, G. K., *et al.* (1991). Systemic delivery of human growth hormone by injection of genetically engineered myoblasts. *Science* **254,** 1509–1512.

Donahue, J., Kikkawa, K., Johns, D., *et al.* (1997). Ultrarapid, highly efficient gene transfer to the heart. *Proc. Natl. Acad. Sci. USA* **94,** 4664–4668.

Drumm, M. L., Pope, H. A., Clif, W. H., *et al.* (1990). Correction of the cystic fibrosis defect *in vitro* by retrovirus-mediated gene transfer. *Cell* **62,** 1227–1233.

During, M., Naegele, J., O'Malley, K., *et al.* (1994). Long-term behavioral recovery in Parkinsonian rats by an HSV vector expressing tyrosine hydroxylase. *Science* **266,** 1399–1403.

Dzau, V., Mann, M., Morishita, R., *et al.* (1996). Fusigenic viral liposomes for gene therapy in cardiovascular diseases. *Proc. Natl. Acad. Sci. USA* **93,** 11421–11425.

Efstathiou, S., and Minson, A. (1995). Herpes virus-based vectors. *Brit. Med. Bull.* **51,** 45–55.

Emerich, D., Winn, S., Hantraye, P., *et al.* (1997). Protective effect of encapsulated cells producing neurotrophic factor CNTF in a monkey model of Huntington's disease. *Nature* **386,** 395–399.

Engelhardt, J. F., Yang, Y., Stratford-Perricaudet, L. D., *et al.* (1993). Direct gene transfer of human CFTR into human bronchial epithelia of xenografts with E1-deleted adenovirus. *Nat. Genet.* **4,** 27–34.

Erbacher, P., Roche, A. C., Monsigny, M., *et al.* (1997). The reduction of the positive charges of polylysine by partial gluconylation increases the transfection efficiency of polylysine/DNA complexes. *Biochim. Biophys. Acta* **1324,** 27–36.

Evans, C., and Robbins, P. (1994). Gene therapy for arthritis. *In* "Gene Therapeutics: Methods and Applications of Direct Gene Transfer" (J. Wolff, Ed.), pp. 320–343. Birkhauser, Boston, Cambridge, MA.

Farhood, H., Gao, X., Son, K., *et al.* (1994). Cationic liposomes for direct gene transfer in therapy of cancer and other diseases. *Ann. N.Y. Acad. Sci.* **716,** 23–35.

Fasbender, A., Zabner, J., Chillon, M., *et al.* (1997). Complexes of adenovirus with polycationic polymers and cationic lipids increase the efficiency of gene transfer *in vitro* and *in vivo. J. Biol. Chem.* **272,** 6479–6489.

Felgner, P. L., Barenholz, Y., Behr, J. P., *et al.* (1997). Nomenclature for synthetic gene delivery systems. *Hum. Gene Ther.* **8,** 511–512.

Felgner, P. L., Gadek, T. R., Holm, M., *et al.* (1987). Lipofection: A highly efficient, lipid-mediated DNA-transfection procedure. *Proc. Natl. Acad. Sci. USA* **84,** 7413–7417.

Fender, P., Ruigrok, R., Gout, E., *et al.* (1997). Adenovirus dodecahedron, a new vector for human gene transfer. *Nat. Biotechnol.* **15,** 52–56.

Ferkol, T., Perales, J., Mularo, F., *et al.* (1996). Receptor-mediated gene transfer into macrophages. *Proc. Natl. Acad. Sci. USA* **93,** 101–105.

Finer, M., Dull, T., Qin, L., *et al.* (1994). kat: A high-efficiency retroviral transduction system for primary human T lymphocytes. *Blood* **83,** 43–50.

Flotte, T. (1994). A phase I study of an adeno-associated virus-CFTR gene vector in adult CF patients with mild lung disease. *RAC Report,* Office of Recombinant DNA Activities, NIH No. 9409-083.

Flotte, T., and Carter, B. (1995). Adeno-associated virus vectors for gene therapy. *Gene Ther.* **2,** 357–362.

Flotte, T., Afione, S., Conrad, C., *et al.* (1993). Stable *in vivo* expression of the cystic fibrosis transmembrane conductance regulator with an adeno-associated virus vector. *Proc. Natl. Acad. Sci. USA* **90,** 10613–10617.

Forestell, S., Bohnlein, E., and Rigg, R. (1995). Retroviral end-point titer is not predictive of gene transfer efficiency: Implications for vector production. *Gene Ther.* **2,** 723–730.

Fraley, R., Papahadjopoulos, D. (1982). Liposomes: The development of a new carrier system for introducing nucleic acid into plant and animal cells. *Curr. Top. Microbiol. Immunol.* **96,** 171–191.

Freeman, S. M., Abboud, C. N., Whartenby, K. A., *et al.* (1993). The "bystander effect": Tumor regression when a fraction of the tumor mass is genetically modified. *Cancer Res.* **53,** 5274–5283.

Freese, A., Geller, A. I., and Neve, R. (1990). HSV-1 vector mediated neuronal gene delivery. *Biochem. Pharmacol.* **40,** 2189–2199.

Friedmann, T. (1989). Progress toward human gene therapy. *Science* **244,** 1275–1281.

Friedmann, T., and Yee, J. K. (1995). Pseudotyped retroviral vectors for studies of human gene therapy. *Nat. Med.* **1,** 275–277.

Fynan, E., Webster, R., Fuller, D., *et al.* (1993). DNA vaccines: Protective immunizations by parenteral, mucosal, and gene-gun inoculations. *Proc. Natl. Acad. Sci. USA* **90,** 11478–11482.

Gardner, P. (1995). A phase I/II study of tgAAV-CF for the treatment of chronic sinusitis in patients with cystic fibrosis. *RAC Report,* Office of Recombinant DNA Activities, NIH No. 9507-114.

Garzino-Demo, A., Gallo, R., and Arya, S. (1995). Human immunodeficiency virus type 2 (HIV-2): Packaging signal and associated negative regulatory element. *Hum. Gene Ther.* **6,** 177–184.

Giraud, C., Winocour, E., and Berns, K. (1994). Site-specific integration by adeno-associated virus is directed by a cellular DNA sequence. *Proc. Natl. Acad. Sci. USA* **91,** 10039–10043.

Giraud, C., Winocour, E., and Berns, K. (1995). Recombinant junctions formed by site-specific integration of adeno-associated virus into an episome. *J. Virol.* **69,** 6917–6924.

Goldman, C., Soroceanu, L., Smith, N., *et al.* (1997). *In vitro* and *in vivo* gene delivery mediated by a synthetic polycationic amino polymer. *Nat. Biotechnol.* **15,** 462–466.

Gorziglia, M., Kadan, M., Yei, S., *et al.* (1996). Elimination of both E1 and E2a from adenovirus vectors further improves prospects for *in vivo* human gene therapy. *J. Virol.* **70,** 4173–4178.

Gottschalk, S., Tweten, R., Smith, L., *et al.* (1995). Efficient gene delivery and expression in mammalian cells using DNA coupled with perfringolysin O. *Gene Ther.* **2,** 498–503.

Graham, F. L., and Van Der Eb, A. J. (1973). A new technique for the assay of infectivity of human adenovirus 5 DNA. *Virology* **52,** 456–467.

Guangping, G., and Wilson, J. (1997). Recombinant adenoviral vector with a ts125 mutation in E2a and deletions in both E1 and E4 regions prolongs transgene expression and reduces toxicity in liver-directed gene transfer. Keystone Symposia on Molecular and Cellular Biology of Gene Therapy, Snowbird, UT.

Harrington, J., Van Bokkelen, G., Mays, R., *et al.* (1997). Formation of *de novo* centromeres and construction of first-generation human artificial chromosomes. *Nat. Genet.* **15,** 345–355.

Heinzinger, N., Bukrinsky, M., Haggerty, S., *et al.* (1994). The Vpr protein of human immunodeficiency virus type 1 influences nuclear localization of viral nucleic acids in nondividing host cells. *Proc. Natl. Acad. Sci. USA* **91,** 7311–7315.

Hodgson, C., Xu, G., Solaiman, F., *et al.* (1997). Biosynthetic retrovectoring systems for gene therapy. *J. Mol. Med.* **75,** 249–258.

Hoeben, R. C., Fallaux, F. J., Van Tilburg, N. H., *et al.* (1993). Toward gene therapy for hemophilia A: Long-term persistence of factor VIII-secreting fibroblasts after transplantation into immunodeficient mice. *Hum. Gene Ther.* **4,** 179–186.

Hoogerbrugge, P. M., Vonbeusachem, V. W., Kaptein, L. C. M., *et al.* (1995). Gene therapy for adenosine deaminase deficiency. *Br. Med. Bull.* **51,** 72–81.

Hseuh, J. L., Qiu, X., Zhou, J., *et al.* (1992). Treatment of hemophilia B with autologous skin fibroblasts transduced with a human clotting factor IX cDNA. *Hum. Gene Ther.* **3,** 543–552.

Huber, B. E. (1994). Gene therapy strategies for treating neoplastic disease. *Ann. N.Y. Acad. Sci.* **716,** 6–11.

Huxley, C. (1994). Mammalian artificial chromosomes: A new tool for gene therapy. *Gene Ther.* **1,** 7–12.

Hyde, S. C., Gill, D. R., Higgins, C. F., *et al.* (1993). Correction of the ion transport defect in cystic fibrosis transgenic mice by gene therapy. *Nature* **362,** 250–255.

Ilan, Y., Drougett, G., Chowdhury, N., *et al.* (1997). Insertion of the adenoviral E3 region into recombinant viral vector prevents antiviral humoral and cellular immune responses and permits long-term gene expression. *Proc. Natl. Acad. Sci. USA* **94,** 2587–2592.

Isner, J. (1995). Accelerated re-endothelialization and reduced neointimal thickening following catheter transfer of phVEGF165. *RAC Report,* Office of Recombinant DNA Activities, NIH No. 9508-118.

Isner, J., and Walsh, K. (1994). Arterial gene transfer for therapeutic angiogenesis in patients with peripheral artery disease. *RAC Report,* Office of Recombinant DNA Activities, NIH No. 9409-088.

Jiang, C., Akita, G., Colledge, W., *et al.* (1997). Increased contact time improves adenovirus-mediated CFTR gene transfer to nasal epithelium of CF mice. *Hum. Gene Ther.* **8,** 671–680.

Jiao, S., Gurevich, V., and Wolff, J. A. (1993). Long-term correction of rat model of Parkinson's disease by gene therapy. *Nature* **362,** 450–453.

Kaneda, Y., Iwai, K., and Uchida, T. (1989). Increased expression of DNA cointroduced with nuclear protein in adult rat liver. *Science* **243,** 375–378.

Kaptein, L., Greijer, A., Valerio, D., *et al.* (1997). Optimized conditions for the production of recombinant amphotropic retroviral vector preparations. *Gene Ther.* **4,** 172–176.

Kasahara, N., Dozy, A., and Kan, Y. (1994). Tissue-specific targeting of retroviral vectors through ligand-receptor interactions. *Science* **266,** 1373–1376.

Klein, T. M., Wolf, E. D., Wu, R., *et al.* (1987). High-velocity microprojectiles for delivering nucleic acids into living cells. *Nature* **327,** 70–73.

Knowles, M., Hohneker, K., Zhou, Z., *et al.* (1995). A controlled study of adenoviral-vector-mediated gene transfer in the nasal epithelium of patients with cystic fibrosis. *N. Engl. J. Med.* **333,** 823–831.

Kochanek, S., Clemens, P. R., Mitani, K., *et al.* (1996). A new adenoviral vector: Replacemnet of all viral coding sequences with 28 kb of DNA independently expressing both full-length dystrophin and β-galactosidase. *Proc. Natl. Acad. Sci. USA* **93,** 5731–5736.

Kotani, H., Newton, P., Zhang, S., *et al.* (1994). Improved methods of retroviral vector transduction and production for gene therapy. *Hum. Gene Ther.* **5,** 19–28.

Kotin, R. (1994). Prospects for the use of adeno-associated virus as a vector for human gene therapy. *Hum. Gene Ther.* **5,** 793–801.

Kotin, R., Siniscalo, M., Samulski, R., *et al.* (1990). Site-specific integration by adeno-associated virus. *Proc. Natl. Acad. Sci. USA* **87,** 2211–2215.

Kremer, E. J., and Perricaudet, M. (1995). Adenovirus and adeno-associated virus mediated gene transfer. *Br. Med. Bull.* **51,** 31–44.

Krysan, P., Haase, S., and Calos, M. (1989). Isolation of human sequences that replicate autonomously in human cells. *Mol. Cell. Biol.* **9,** 1026–1033.

Lanza, R. P., Hayes, J. L., and Chick, W. L. (1996). Encapsulated cell technology. *Nat. Biotechnol.* **14,** 1107–1111.

Lesoon-Wood, L. A., Kim, W. H., Kleinman, H. K., *et al.* (1995). Systemic gene therapy with p53 reduces growth and metastases of a malignant human breast cancer in nude mice. *Hum. Gene Ther.* **6,** 395–405.

Levitskaya, J., Coram, M., Levitsky, V., *et al.* (1995). Inhibition of antigen processing by the internal repeat region of the Epstein–Barr virus nuclear antigen-1. *Nature* **375,** 685–688.

Lewis, P. F., and Emerman, M. (1994). Passage through mitosis is required for oncoretroviruses but not for the human immunodeficiency virus. *J. Virol.* **68,** 510–516.

Liu, Y., Mounkes, L., Liggit, H., *et al.* (1997). Factors influencing the efficiency of cationic liposome-mediated intravenous gene delivery. *Nat. Biotechnol.* **15,** 167–173.

Mahvi, D. (1996). Phase I/IB study of immunization with autologous tumor cells transfected with the GM-CSF gene by particle-mediated transfer in patients with melanoma or sarcoma. *RAC Report,* Office of Recombinant DNA Activities, NIH No. 9608-158.

Mammano, F., Salvatori, F., Indraccolo, S., *et al.* (1997). Truncation of the human immunodeficiency virus type 1 envelope glycoprotein allows efficient pseudotyping of Moloney murine leukemia virus particles and gene transfer into $CD4^+$ cells. *J. Virol.* **71,** 3341–3345.

Margolskee, R. F. (1992). Epstein–Barr virus based expression vectors. *Curr. Top. Microbiol. Immunol.* **158,** 67–95.

Martuza, R, Malick, A., Markert, J., *et al.* (1991). Experimental therapy of human glioma by means of a genetically engineered virus mutant. *Science* **252,** 854–856.

Mebatsion, T., and Conzelmann, K. K. (1996). Specific infection of $CD4^+$ target cells by recombinant rabies virus pseudotypes carrying the HIV-1 envelope spike protein. *Proc. Natl. Acad. Sci. USA* **93,** 11366–11370.

Miller, A., Garcia, J., Von Suhr, N., *et al.* (1991). Construction and properties of retrovirus packaging cells based on gibbon ape leukemia virus. *J. Virol.* **65,** 2220–2224.

Miller, A., Miller, D., Garcia, J., *et al.* (1993). Use of retroviral vectors for gene transfer and expression. *Methods Enzymol.* **217,** 581–599.

Miller, A., Trauber, D., and Buttimore, C. (1986). Factors involved in production of helper virus-free retrovirus vectors. *Somat. Cell Mol. Genet.* **12,** 175–183.

Miller, A. D. (1992). Human gene therapy comes of age. *Nature* **357,** 455–460.

Miller, D. G., Adam, M. A., and Miller, A. D. (1990). Gene transfer by retrovirus vectors occurs only in cells that are actively replicating at the time of infection. *Mol. Cell. Biol.* **10,** 4239–4242.

Mistry, A., Falciola, L., Monaco, L., *et al.* (1997). Recombinant HMG1 protein produced in *Pichia pastoris:* A nonviral gene delivery agent. *BioTechniques* **22,** 718–729.

Miyashita, E., Yang, B., Lam, K., *et al.* (1995). A novel form of Epstein–Barr virus latency in normal B cells *in vivo. Cell* **80,** 593–601.

Morgan, R., and Anderson, F. (1993). Human gene therapy. *Annu. Rev. Biochem.* **62,** 191–217.

Morishita, R., Gibbons, G. H., Ellison, K. F., *et al.* (1993). Single intraluminal delivery of antisense cdc2 kinase and proliferating-cell nuclear antigen oligonucleotides results in chronic inhibition of neointimal hyperplasia. *Proc. Natl. Acad. Sci. USA* **90,** 8474–8478.

Morling, F., and Russell, S. (1995). Enhanced transduction efficiency of retroviral vectors coprecipitated with calcium phosphate. *Gene Ther.* **2,** 504–508.

Moullier, P., Bohl, D., Cardoso, J., *et al.* (1995). Long-term delivery of a lysosomal enzyme by genetically modified fibroblasts in dogs. *Nat. Med.* **1,** 353–357.

Moullier, P., Bohl, D., Heard, J. M., *et al.* (1993). Correction of lysosomal storage in the liver and spleen of MPS VII mice by implantation of genetically modified skin fibroblasts. *Nat. Genet.* **4,** 154–159.

Mulligan, R. C. (1993). The basic science of gene therapy. *Science* **260,** 926–931.

Muzyczka, N. (1992). Use of adeno-associated virus as a general transduction vector for mammalian cells. *Curr. Top. Microbiol. Immunol.* **158,** 97–129.

Nabel, E., Gordon, D., Yang, Z. Y., *et al.* (1992). Gene transfer *in vivo* with DNA-liposome complexes: Lack of autoimmunity and gonadal localization. *Hum. Gene Ther.* **3,** 649–656.

Nabel, E. G., Plautz, G., and Nabel, G. J. (1990). Site-specific gene expression *in vivo* by direct gene transfer into the arterial wall. *Science* **249,** 1285–1288.

Nabel, G. (1993). Immunotherapy for cancer by direct gene transfer. *RAC Report,* Office of Recombinant DNA Activities, NIH No. 9306-045.

Nabel, G., Gordon, D., Bishop, D., *et al.* (1996). Immune response in human melanoma after transfer of an allogenic class I major histocompatibility complex gene with DNA-liposome complexes. *Proc. Natl. Acad. Sci. USA* **93,** 15388–15393.

Nabel, G. J., Nabel, E. G., Yang, Z. Y., *et al.* (1993). Direct gene transfer with DNA-liposome complexes in melanoma: Expression, biologic activity, and lack of toxicity in humans. *Proc. Natl. Acad. Sci. USA* **90,** 11307–11311.

Nakamura, N., Horibe, S., Matsumoto, N., *et al.* (1996). Transient introduction of a foreign gene into healing rat patellar ligament. *J. Clin. Invest.* **97,** 226–231.

Naldini, L., Blomer, U., Gallay, P., *et al.* (1996). *In vivo* gene delivery and stable transduction of nondividing cells by a lentiviral vector. *Science* **272,** 263–267.

Naviaux, R. K., and Verma, I. M. (1992). Retroviral vectors for persistent expression *in vivo. Curr. Opin. Cell. Biotechnol.* **3,** 540–547.

Neumann, E., Schaefer-Ridder, M., Wang, Y., *et al.* (1982). Gene transfer into mouse lyoma cells by electroporation in high electric fields. *EMBO J.* **1,** 841–845.

Neve, R. L. (1993). Adenovirus vectors enter the brain. *Trends Neurosci.* **16,** 251–253.

Nishi, T., Yoshizato, K., Yamashiro, S., *et al.* (1996). High-efficiency *in vivo* gene transfer using intra-arterial plasmid DNA injection following *in vivo* electroporation. *Cancer Res.* **56,** 1050–1055.

Palsson, B., and Andreadis, S. (1997). The physico-chemical factors that govern retrovirus-mediated gene transfer. *Exp. Hematol.* **25,** 94–102.

Palsson, B. O., Chuck, A. S., and Huang, G. (1995). Retroviral infection is limited by random Brownian motion. *Blood* **86,** 237a.

Pellicer, A., Robins, D., Wold, B., *et al.* (1980). Altering genotype and phenotype by DNA-mediated gene transfer. *Science* **209,** 1414–1422.

Philip, R., Brunette, E., Kilinski, L., *et al.* (1994). Efficient and sustained gene expression in primary T lymphocytes and primary and cultured tumor cells mediated by adeno-associated virus plasmid DNA complexed to cationic liposomes. *Mol. Cell. Biol.* **14,** 2411–2418.

Ponnazhagan, S., Erikson, D., Kearns, W., *et al.* (1997). Lack of site-specific integration of the recombinant adeno-associated virus 2 genomes in human cells. *Hum. Gene Ther.* **8,** 275–284.

Potter, H., Weir, L., and Leder, P. (1984). Enhancer-dependent expression of human κ-immunoglobulin genes introduced into mouse pre-B lymphocytes by electropotation. *Proc. Natl. Acad. Sci. USA* **81,** 7161–7165.

Raja-Walia, R., Webber, J., Naftilan, J., *et al.* (1995). Enhancement of liposome-mediated gene transfer into vascular tissue by replication deficient adenovirus. *Gene Ther.* **2,** 521–530.

Reiser, J., Harminson, G., Kluepfel-Stahl, S., *et al.* (1996). Transduction of nondividing cells using pseudotyped defective high-titer HIV type 1 particles. *Proc. Natl. Acad. Sci. USA* **93,** 15266–15271.

Richardson, J., Kaye, J., Child, L., *et al.* (1995). Helper virus-free transfer of human immunodeficiency virus type 1 vectors. *J. Gen. Virol.* **76,** 691–696.

Robbins, P. D., Tahara, H., Mueller, G., *et al.* (1994). Retroviral vectors for use in human gene therapy for cancer, Gaucher disease, and arthritis. *Ann. N. Y. Acad. Sci.* **716,** 72–89.

Russ, A. P., Friedel, C., Grez, M., *et al.* (1996). Self-deleting retrovirus vectors for gene therapy. *J. Virol.* **70,** 4927–4932.

Salmons, B., Saller, R., Baumann, J., *et al.* (1995). Construction of retroviral vectors for targeted delivery and expression of therapeutic genes. *Leukemia* **9,** S53–S60.

Salvetti, A., Heard, J. M., and Danos, O. (1995). Gene therapy of lysosomal storage disease. *Br. Med. Bull.* **51,** 106–122.

Schneider, H., Huse, K., Birkenmeier, G., *et al.* (1996). Gene transfer mediated by α-macroglobulin. *Nucleic Acids Res.* **24,** 3873–3874.

Schnierle, B., and Groner, B. (1996). Retroviral targeted delivery. *Gene Ther.* **3,** 1069–1073.

Schofield, J. P., and Caskey, C. T. (1995). Non-viral approaches to gene therapy. *Br. Med. Bull.* **51,** 56–71.

Schuele, R., St. George, J., Bagley, R., *et al.* (1997). Basis of pulmonary toxicity associated with cationic lipid-mediated gene transfer to the mammalian lung. *Hum. Gene Ther.* **8,** 689–707.

Selden, R., Heartlein, M., and Treco, D. (1997). Transkaryotic therapy and homologous recombination. Keystone Symposia on Molecular and Cellular Biology of Gene Therapy, Snowbird, UT.

Shimada, T., Fujii, H., Mitsuya, H., *et al.* (1991). Targeted and highly efficient gene transfer into $CD4^+$ cells by a recombinant human immunodeficiency virus retroviral vector. *J. Clin. Invest.* **88,** 1043–1047.

Singhal, A., and Huang, L. (1994). Gene transfer in mammalian cells using liposomes as carriers. *In* "Gene Therapeutics: Methods and Applications of Direct Gene Transfer" (J. Wolff, Ed.), pp. 118–142. Birkhauser, Boston, Cambridge, MA.

Slepushkin, V., Salem, I., Andreev, S., *et al.* (1996). Targeting of liposomes to HIV-1-infected cells by peptides derived from the CD4 receptor. *Biochem. Biophys. Res. Commun.* **227,** 827–833.

Springett, G. M., Moen, R. C., Anderson, S., *et al.* (1989). Infection efficiency of T lymphocytes with amphotropic retroviral vectors is cell cycle dependent. *J. Virol.* **63,** 3865–3869.

Stevenson, S., Rollience, M., Marshall-Neff, J., *et al.* (1997). Selective targeting of human cells by a chimeric adenovirus vector containing a modified fiber protein. *J. Virol.* **71,** 4782–4790.

Stratford-Perricaudet, L., and Perricaudet, M. (1994). Gene therapy: The advent of the adenovirus. *In* "Gene Therapeutics: Methods and Applications of Direct Gene Transfer" (J. Wolff, Ed.), pp. 344–362. Birkhauser, Boston, Cambridge, MA.

Stratford-Perricaudet, L., Makeh, I., Perricaudet, M., *et al.* (1992). Widespread long-term gene transfer to mouse skeletal muscles and heart. *J. Clin. Invest.* **90,** 626–630.

Straus, S. E., and Raskas, H. J. (1980). Transfection of KB cells by polyethylene glycol-induced fusion with erythrocyte ghosts containing adenovirus type 2 DNA. *J. Gen. Virol.* **48,** 241–245.

Stribling, R., Brunette, E., Liggitt, D., *et al.* (1992). Aerosol gene delivery *in vivo. Proc. Natl. Acad. Sci. USA* **89,** 11277–11281.

Sugden, B., Marsh, K., and Yates, B. (1995). A vector that replicates as a plasmid and can be efficiently selected in B-lymphoblasts transformed by Epstein–Barr virus. *Mol. Cell. Biol.* **5,** 410–413.

Sukharev, S., Titomirov, A., and Klenchin, V. (1994). Electrically-induced DNA transfer into cells. Electrotransfection *in vivo. In* "Gene Therapeutics: Methods and Applications of Direct Gene Transfer" (J. Wolff, Ed.), pp. 210–232. Birkhauser, Boston, Cambridge, MA.

Sun, L. Q., Gerlach, W., and Symonds, G. (1995a). Antisense and ribozyme strategies for HIV-1 infection. *Drugs New Perspect.* **8,** 325–331.

Sun, W., Burkholder, J., Sun, J., *et al.* (1995b). *In vivo* cytokine gene transfer by gene gun reduces tumor growth in mice. *Proc. Natl. Acad. Sci. USA* **92,** 2889–2893.

Tanaka, T., Kanai, F., Lan, K. H., *et al.* (1997). Adenovirus-mediated gene therapy of gastric carcinoma using cancer-specific gene expression *in vivo. Biochem. Biophys. Res. Commun.* **231,** 775–779.

Tang, D. C., DeVit, M., and Johnston, S. (1992). Genetic immunization is a simple method for eliciting an immune response. *Nature* **356,** 152–154.

Titomirov, A., Sukharev, S., and Kistanova, E. (1991). *In vivo* electroporation and stable transformation of skin cells of newborn mice by plasmid DNA. *Biochim. Biophys. Acta* **1088,** 131–134.

Ulmer, J. B., Donnelly, J. J., Parker, S. E., *et al.* (1993). Heterologous protection against influenza by injection of DNA encoding a viral protein. *Science* **259,** 1745–1749.

Varmus, H. (1988). Retroviruses. *Science* **240,** 1427–1435.

Vieweg, J., Boczkowski, D., Roberson, K., *et al.* (1995). Efficient gene transfer with adeno-associated virus-based plasmids complexed to cationic liposomes for gene therapy of human prostate cancer. *Cancer Res.* **55,** 2366–2372.

Vile, R. G., and Russell, S. J. (1995). Retroviruses as vectors. *Br. Med. Bull.* **51,** 12–30.

Von Kalle, C., Kiem, H. P., Goehle, S., *et al.* (1994). Increased gene transfer into human hematopoietic progenitor cells by extended *in vitro* exposure to a pseudotyped retroviral vector. *Blood* **84,** 2890–2897.

Wagner, E., Plank, C., Zatloukal, K., *et al.* (1992). Influenza virus hemagglutinin HA-2 N-terminal fusogenic peptides augment gene transfer by transferrin-polylysine-DNA complexes: Toward a synthetic virus-like gene-transfer vehicle. *Proc. Natl. Acad. Sci. USA* **89,** 7934–7938.

Wang, B., Ugen, K., Srikantan, V., *et al.* (1993). Gene inoculation generates immune responses against human immunodeficiency virus type 1. *Proc. Natl. Acad. Sci. USA* **90,** 4156–4160.

Weitzman, M., Kyostio, S., Kotin, R., *et al.* (1994). Adeno-associated virus (AAV) Rep proteins mediate complex formation between AAV DNA and its integration site in human DNA. *Proc. Natl. Acad. Sci. USA* **91,** 5808–5812.

Wickstrom, E. (1992). Strategies for administering targeted therapeutic oligodeoxynucleotides. *Trends Biotechnol.* **10,** 281–287.

Wilkinson, G. W. G., and Akrigg, A. (1992). Constitutive and enhanced expression from the CMV major IE promoter in a defective adenovirus vector. *Nucleic Acids Res.* **20,** 2233–2239.

Williams, R., Johnston, S., Riedy, M., *et al.* (1991). Introduction of foreign genes into tissues of living mice by DNA-coated microprojectiles. *Proc. Natl. Acad. Sci. USA* **88,** 2726–2730.

Wilson, J. M., Grossman, M., Raper, S. E., *et al.* (1992). *Ex vivo* gene therapy of familial hypercholesterolemia. *Hum. Gene Ther.* **3,** 179–222.

Woffendin, C., Ranga, U., Yang, Z. Y., *et al.* (1996). Expression of a protective gene prolongs survival of T-cells in human immunodeficiency virus-infected patients. *Proc. Natl. Acad. Sci. USA* **93,** 2889–2894.

Wohlgemuth, J., Kang, S., Bulboaca, G., *et al.* (1996). Long-term gene expression from autonomously replicating vectors in mammalian cells. *Gene Ther.* **3,** 503–512.

Wolff, J. A., Fisher, L. J., Xu, L., *et al.* (1989). Grafting fibroblasts genetically modified to produce L-dopa in a rat model of Parkinson disease. *Proc. Natl. Acad. Sci. USA* **86,** 9011–9014.

Wolff, J. A., Malone, R. W., Williams, P., *et al.* (1990). Direct gene transfer into mouse muscle *in vivo. Science* **247,** 1465–1468.

Wu, G. Y., Wilson, J. M., Shalaby, F., *et al.* (1991). Receptor-mediated gene delivery *in vivo. J. Biol. Chem.* **266,** 14338–14342.

Yang, N. S., Burkholder, J., Roberts, B., *et al.* (1990). *In vivo* and *in vitro* gene transfer to mammalian somatic cells by particle bombardment. *Proc. Natl. Acad. Sci. USA* **87,** 9568–9572.

Yang, N. S., De Luna, C., and Cheng, L. (1994). Gene transfer via particle bombardment: Applications of the Accell gene gun. *In* "Gene Therapeutics: Methods and Applications of Direct Gene Transfer" (J. Wolff, Ed.), pp. 193–209. Birkhauser, Boston, Cambridge, MA.

Yu, M., Poeschla, E., and Wong-Staal, F. (1994). Progress towards gene therapy for HIV infection. *Gene Ther.* **1,** 13–26.

Zabner, J., Couture, L. A., Gregory, R. J., *et al.* (1993). Adenovirus-mediated gene transfer transiently corrects the chloride transport defect in nasal epithelia of patients with cystic fibrosis. *Cell* **75,** 207–216.

Zabner, J., Fasbender, A., Moninger, T., *et al.* (1995). Cellular and molecular barriers to gene transfer by a cationic lipid. *J. Biol. Chem.* **270,** 18997–19007.

Zhu, N., Liggitt, D., Liu, Y., *et al.* (1993). Systemic gene expression after intravenous DNA delivery into adult mice. *Science* **261,** 209–211.

15

Summary and Future Directions

Robert E. Nordon and Klaus Schindhelm

Graduate School of Biomedical Engineering
University of New South Wales
Sydney, New South Wales 2052, Australia

I. INTRODUCTION

Advances in medical science have led to treatments that are based on an understanding of disease pathogenesis. The steps in this process have been (a) study of the molecular mechanisms governing normal cell growth, development, and function of differentiated cells, (b) characterization of the cellular or molecular lesions that cause the disease state, and (c) development of strategies that repair the cellular system.

Cell therapy may assist in this process by providing an exogenous source of cells with the potential to eliminate diseased cells and regenerate the cellular system. Therefore the aims of cell therapy will be (a) development of *in vitro* methods for production of cells with the potential to eliminate diseased cell types and restore normal function, (b) refinement of *ex vivo* processes so that they may be implemented under pharmaceutical or blood bank standards of Good Manufacturing Practice (GMP), and (c) evaluation of clinical safety and efficacy.

Graft engineering commences with harvest of bone marrow, peripheral mononuclear cells, or cord blood, followed by modification of the transplant to facilitate clinical efficacy. This may involve removal of cell subsets that have detrimental effects on the host and supplementation with cells that can restore hematopoietic or immune function. The transfer of *in vitro* laboratory techniques to a cell production facility is not a trivial task and requires continued development of technologies that will improve the safety and efficiency of the process.

II. EXTRINSIC OR INTRINSIC MANIPULATION OF HEMATOPOIETIC CELL GROWTH AND DIFFERENTIATION

In theory it should be possible to modify cell growth and differentiation by modulation of the cell microenvironment or by altering transcriptional and signaling pathways using gene transfer techniques. Our current knowledge of the molecular mechanisms for transcriptional control of cell development is rapidly growing, though our understanding of these systems is not advanced enough to develop *in vitro* strategies for cell therapy. Nevertheless, in the future a more complete picture of these complex pathways may lead to *in vitro* methods for control of differentiation at a transcriptional level.

A. TRANSCRIPTIONAL CONTROL OF HEMATOPOIESIS

The advent of molecular cloning and gene "knockout" technology has led to the discovery and functional characterization of transcription factors, the regulators of gene transcription. In mice with a null mutation of a transcription factor, hematopoietic development will be arrested at an embryological stage where the transcription factor is first required for development. For example, knockout of SCL results in block of primitive (yolk sac) hematopoiesis at day 9.5 of mouse embryonic development (Robb *et al.,* 1995; Shivdasani *et al.,* 1995) whereas definitive hematopoiesis (fetal liver, day 12.5) is blocked by knockout of AML-1 (Wang *et al.,* 1996). It is possible to study the role of hematopoietic transcription factors later in development by creation of mouse chimeras that are derived from doubly targeted embryonal stem cells (homozygous-null mutation).

It is likely that multiple transcriptional molecules interact in an orchestrated fashion to regulate gene expression during hematopoietic and lymphoid development. Although gene-targeting techniques have defined the functions for some transcription factors, in many cases the target genes that they regulate are unknown. In the future more specific techniques will help to define the role of transcription factors. For example, the use of developmental- or tissue-restricted knockout of gene expression may be accomplished by promoter-driven expression of specific recombinases that delete genes inserted between specific recombination sites (e.g., *Cre* recombinase). Definition of the genetic program for hematopoietic development will provide techniques that allow control of cell differentiation at a transcriptional level.

B. CELL SIGNALING BY HEMATOPOIETIC GROWTH FACTOR RECEPTORS

Signal transduction pathways are a cascade of cytoplasmic reactions that link signals generated by receptor–ligand engagement at the cell surface to changes in gene expression or cellular metabolism. Hematopoietic growth factors (HGF), when bound to their cognate cell membrane receptor, deliver important signals that regulate cell cycle status, cell survival, and differentiation. The response to HGF stimulation depends on the cell type and developmental stage, as well as the presence of other costimulatory or inhibitory molecules. Therefore it appears that these signaling pathways interpret inputs differently depending on the transcriptional context as well as the presence of other exogenous signals. The mechanism for integration of multiple signals may be by cross-talk between interrelated signaling pathways or receptor types.

The JAK–STAT signal transduction pathways are utilized by many different HGF receptors (Darnell *et al.,* 1994), in addition to certain members of the receptor tyrosine kinase family (Novak *et al.,* 1995; Vignais *et al.,* 1996). The specificity of the STAT family is conferred at many levels, including their restricted pattern of expression in specific tissues and the specific affinities they have for growth factor receptors, SH2 domains, and nucleotide sequences (Rothman *et al.,* 1994; Heim *et al.,* 1995; Stahl *et al.,* 1995).

Similarly, the Ras–Raf pathway mediates signaling from a variety of receptor types, including cytokine receptors, receptor tyrosine kinases, and G protein-coupled receptors (Duronio *et al.,* 1992; van Corven *et al.,* 1993). Furthermore, this pathway may be activated by Raf-1-independent pathways (Czar *et al.,* 1997). The response to activation of the Ras–Raf pathway includes mitogenesis and differentiation.

Inhibitors of cytokine signaling include SHP-1 (Schulz *et al.,* 1993; Tsui *et al.,* 1993) and the SOCS family of proteins (Starr *et al.,* 1997) that act by tyrosine dephosphorylation of activated proteins or by binding and inhibiting specific tyrosine kinases. Apoptosis may be induced by withdrawal of cytokine stimulation or by direct induction by proapoptotic cytokines (tumor necrosis factor and FAS ligand) (Nicholson and Thornberry, 1997). Growth factor stimulation or withdrawal influences cell survival and apoptosis by interaction with the Bcl family of proteins, which form homo- or heterodimers that have either pro- or antiapoptotic activity (Jacobson, 1997; Kroemer, 1997).

It is likely that few biological responses are mediated by a single, linear-signaling pathway, but rather by a network of parallel and branching pathways. Therefore defining the configuration and kinetic regulation of these interacting pathways will present a major challenge.

C. INFLUENCE OF CYTOKINES AND ADHESION MOLECULES ON HEMATOPOIETIC STEM CELL DEVELOPMENT

Hematopoiesis can be considered as a process regulated by signals provided to developing hematopoietic cells by their surrounding microenvironment. These signals are provided by stromal cell–hematopoietic cell interactions mediated by various cell adhesion molecules and also through the action of specific HGF following binding to their cognate cell surface receptors. *In vitro,* combinations of synergistic HGF and colony-stimulating factors (CSFs) are required for hematopoietic proliferation and differentiation (Haylock *et al.,* 1992). It appears that early-acting growth factors such as FLT3 ligand (FL), stem cell factor (SCF), and thrombopoietin (TPO) stimulate the production of pluripotent hematopoietic cells. Initial studies have demonstrated that cells generated *in vitro* with these cytokines regenerate hematopoiesis in the NOD–SCID mouse model of human engraftment (Kusadasi *et al.,* 1998), though the engraftment potential of cytokine-generated cells has not been demonstrated in the human transplantation setting.

The marrow stroma presents a diversity of adhesive ligands (cell surface molecules, membrane or extracellular matrix-anchored HGF, and extracellular matrix components) which, in addition to modulating growth and survival, appear to contribute to the lodgment of hematopoietic progenitor cells within the bone marrow microenvironment. Hematopoietic cells present a variety of cell adhesion molecules representing at least five superfamilies, including integrin, immunoglobulin, selectin, sialomucin, and the CD44 family of adhesion molecules (Kincade *et al.,* 1989; Clark *et al.,* 1992; Long, 1992; Simmons *et al.,* 1994, 1997). It is likely that these cell adhesion molecules will perform hematopoietic trafficking roles that are analogous to the multistep paradigm recently described for leukocyte emigration and lymphocyte recirculation (Springer 1994).

Integrin-knockout studies in mice have demonstrated a role for VLA-4 (Hirsch *et al.,* 1996) and possibly VLA-5 (Traycoff *et al.,* 1997; van der Loo *et al.,* 1997) in homing to sites of secondary hematopoiesis (liver, spleen, and bone marrow) during embryological development. In addition to a role in cell homing, there is likely to be significant interaction between adhesion molecules, mitogenic pathways (Lévesque *et al.,* 1995, 1996), and the cytoskeleton (Bockholt and Burridge, 1993; Takahira *et al.,* 1997). For example, integrins require a conformational change for functional activation (inside-out signaling), and once in an active form can signal following ligand engagement (outside-in signaling). More specifically, VLA-4 and VLA-5 are nonactive, nonligand binding on quiescent hematopoietic stem cells but are transiently activated by cytokines and bind to their counterligands, fibronectin and VCAM-1 (Lévesque *et al.,* 1996).

Despite the definition of stroma-based and cytokine-driven systems for hematopoietic cell production *in vitro,* the contribution of adhesive interactions on stem cell survival and the modulation of mitogenesis and differentiation remain largely unknown. It is possible that culture systems for *ex vivo* production of hematopoietic cells will require both cytokine and cell adhesion molecule interactions to promote hematopoietic cell renewal *in vitro.*

III. HEMATOPOIETIC CELL GRAFT ENGINEERING

High-dose chemo- and radiotherapy followed by hematopoietic reconstitution using stem cell transplant currently provides the best chance of survival for some hematological malignancies and advanced solid tumors. The cytotoxicity of conditioning and immunosuppressive drugs is a significant drawback associated with high-dose therapy. Patient survival depends on availability of a suitable stem cell source, the rate of hematopoietic reconstitution following transplant, and the incidence of potentially fatal complications such as graft-versus-host disease, opportunistic infections, and tumor recurrence.

The goal of hematopoietic graft engineering is to improve the safety and efficacy of stem cell transplant by manipulation of the constituent cell types present in the graft. This would include depletion of harmful cell types such as tumor cells and *ex vivo* selection and expansion procedures to generate cells that improve the rate of engraftment and graft-versus-tumor reactivity. Since only about 30% of transplant candidates have a human leukocyte antigen (HLA)-identical sibling, there is considerable interest in alternative sources of stem cells such as HLA-matched unrelated donors, umbilical cord blood, or autologous stem cells mobilized into peripheral blood by chemotherapy and growth factors.

A. ALLOGENEIC HEMATOPOIETIC STEM CELL TRANSPLANTATION

Most allogeneic hematopoietic stem cell transplants require sibling donors matched with the recipients for HLA. Allogeneic transplantation remains the treatment of choice for young patients with chronic myeloid leukemia with an HLA-identical sibling donor available. Three-year survival depends on disease phase: 66% in first chronic phase, 44% in accelerated phase, and 19% if transplanted during blast phase. Since most patients with acute lymphoblastic leukemia are cured with conventional therapy, transplants are usually reserved for patients failing chemotherapy or having poor prognostic factors such as older age, difficulty obtaining first remission or patients that are Philadelphia-chromosome positive. The probabilities of 3-year leukemia-free survival are 54% (first remission), 40% (greater than or second remission), and 20% (not in remission). As for acute lymphoblastic leukemia, 3-year survival for acute myelogenous leukemia is related to remission state: 59% for transplants done in first remission, 35% in second remission or greater, and 26% if not in remission (Rowlings, 1996). Other diseases where allogeneic transplants may be indicated include severe aplastic anemia, non-Hodgkin's lymphoma, myelodysplastic syndromes, and multiple myeloma.

Complications associated with allogeneic stem cell transplantation relate to high-dose conditioning regimes and pancytopenia, graft failure, acute and chronic graft-versus-host disease, reactivation of latent viral infections, and long-term complications such as late infections, chronic lung disease, and an increased incidence of leukemia, myelodysplasia, and solid tumors.

Recent therapeutic advances include the use of donor lymphocyte infusions that promote a graft-versus-leukemia effect most predictably in patients who relapse into the chronic phase of chronic myeloid leukemia (Kolb *et al.,* 1990; Drobyski *et al.,* 1993; Collins *et al.,* 1997). Lymphoma associated with Epstein–Barr virus has been treated by transfer of reactive cytotoxic T lymphocytes (Papadopoulos *et al.,* 1994; Bonini *et al.,* 1997). Similarly, infusion of cytomegalovirus-reactive T lymphocytes has been an effective way to reconstitute cellular immunity after allogeneic stem cell transplantation (Walter *et al.,* 1995).

If future research demonstrates that graft-versus-host and graft-versus-tumor reactivity are separable, then the graft could be modified so that cytotoxic T cells that cause the clinical manifestations of graft-versus-host disease are depleted while T cells with graft-versus-tumor reactivity are retained or expanded *ex vivo.* Another *ex vivo* procedure that may have a significant impact on the safety and efficacy of allogeneic transplantation would be generation of cells that facilitate more rapid engraftment.

B. AUTOLOGOUS STEM CELL TRANSPLANTATION

The virtual complete absence of graft-versus-host disease, a low incidence of engraftment failure, and the significantly lower incidence of cytomegalovirus infection make autologous stem cell transplantation a much safer procedure in comparison to allogeneic transplant. Malignant contamination of the autograft and the lack of graft-versus-tumor effect comprise the most important drawbacks of this procedure.

A significant clinical advance has been the development of mobilized peripheral stem cell transplantation (Sheridan *et al.,* 1994; Körbling *et al.,* 1995) and techniques for increasing the yield of hematopoietic progenitors in peripheral blood (Glaspy *et al.,* 1995). Mobilized peripheral stem cell grafts have largely superseded autologous bone marrow grafts; enhanced granulocyte and platelet recovery is translated into less infection and hemorrhage, with shorter hospital stays and lower antibiotic and transfusion requirements (Henon *et al.*, 1992; To *et al.*, 1992; Peters *et al.*, 1993a). This enhanced margin for safety has also led to wider application of high-dose chemotherapy followed by autologous stem cell transplant, particularly for advanced breast cancer and other solid tumors.

The potential indications for high-dose chemotherapy followed by autologous transplantation are similar to those for allogeneic transplantation where a matched-sibling donor or mismatched HLA donor is not available or where older age precludes allogeneic stem cell transplantation. At this stage autologous stem cell transplantation is not recommended for chronic myeloid leukemia because of a lack of graft-versus-tumor effect and contamination of the graft with Philadelphia chromosome positive cells.

Autologous stem cell transplant provides long-term disease-free survival for relapsed patients with Hodgkin's disease (Reece *et al.*, 1991; Schmitz *et al.*, 1993) and possibly non-Hodgkin's lymphoma with a poor prognosis (Haioun *et al.*, 1994; Gianni *et al.*, 1996). Without high-dose chemotherapy followed by autologous stem cell transplant, less than 10% of patients with stage III multiple myeloma will survive longer than 5 years. Complete remission can be achieved in about 50% of patients using high-dose therapy and autologous stem cell transplant (Barlogie, 1991; Cunningham *et al.*, 1994).

Autologous stem cell transplant may be warranted in women with primary breast cancer in the high-risk setting (spread to greater than 10 lymph nodes) since the 10- year disease-free survival with conventional adjuvant chemotherapy is only 10–20% (Bonadonna *et al.*, 1995). Nonrandomized clinical trials have suggested a superior outcome with high-dose chemotherapy (Gianni *et al.*, 1992; Peters *et al.*, 1993b). A number of randomized clinical studies are now examining the role of dose-escalated chemotherapy in the high-risk setting.

Ex vivo tumor cell purging technologies will play a prominent role in autologous transplantation since residual tumor cells in the graft may contribute to disease relapse. Furthermore, *ex vivo* generated hematopoietic progenitors may potentially shorten the period of severe neutropenia and thrombocytopenia following myeloablation. In addition, expansion of long-term repopulating cells from harvested material may allow multiple cycles of high-dose chemotherapy. Finally, *ex vivo* generation of autologous cell subsets which amplify the graft-versus-tumor reactivity of autologous grafts will be an important research objective.

C. CORD BLOOD STEM CELL TRANSPLANTATION

For those 70% of patients with no HLA-matched sibling donor, registries of unrelated donors established worldwide only provide a further 25% of patients with an HLA-matched transplant, the remaining patients requiring an alternate or more rapidly accessible source of stem cells. As a by product of childbirth, placental cord blood is seen as a readily available source of stem cells. Cord blood has a high content of primitive hematopoietic progenitor cells (Broxmeyer *et al.,* 1989) and transplantation studies have shown that graft-versus-host disease might be less frequent and severe compared to that seen with the use of bone marrow. HLA-identical related cord blood transplants engraft in 87% of children, and the incidence of graft-versus-host disease (4–8%) is less than in HLA-matched sibling transplants (20%) (Wagner *et al.,* 1995; Gluckman *et al.,* 1997). Unrelated cord blood transplants engraft in 90% of children, and 50% of patients developed acute graft-versus-host disease grades II–IV, a low incidence considering that 90% of these patients were mismatched for one to three HLA (Kurtzberg *et al.,* 1996; Wagner *et al.,* 1997).

Cord blood banks are being established worldwide. The number and antigenic diversity of donations to be banked should provide a majority of patients with units that are matched for 5/6 or 6/6 HLA. For example, the New York cord blood bank could provide matches for 58% of 2800 patients searching a bank containing 6000 donations (Beatty *et al.,* 1995). Other issues related to cord blood banking include informed consent and anonymity for collection and use of cord blood. Quality control procedures to prevent the transmission of genetic disease or infectious agents, accurate HLA typing, and adequate stem cell yields are also important considerations. The cost of storage is significant so volume reduction and bank size are critical economic issues.

Risk factors for engraftment failure include low cell dose and CFU-GM, and graft-versus-host disease correlates with increasing degree of HLA mismatch (Kurtzberg *et al.,* 1996; Migliaccio *et al.,* 1996; Rubinstein *et al.,* 1996; Wagner *et al.,* 1997). Decreased survival is also correlated to body weight and age. Delayed engraftment accounts for significant morbidity and mortality in children, neutrophils reaching safe levels in 3 weeks ($>0.5 \times 10^9$/L) and platelets 2–3 months ($>50 \times 10^9$/L). Low rates of survival in cord blood transplants in adults might be related to low cell number or degree of mismatching. Until these issues are resolved, cord blood transplant will be restricted to small recipients.

Ex vivo expansion of cord blood stem cells may assist engraftment by reducing the rate of graft failure and shortening engraftment times, leading to a potential role in larger patients. Thus, development of *ex vivo* stem cell expansion technologies will enable the use of cord blood for both small and large recipients.

D. RECONSTITUTION OF IMMUNITY BY ADOPTIVE IMMUNOTHERAPY WITH T CELLS

Cellular adoptive immunotherapy is defined as the transfer of effector cells of the immune system to augment or restore immune responses for the treatment of malignant or infectious disease. Animal models support the principle that immune effector cells such as lymphokine-activated killer cells, major histocompatibility complex (MHC)-restricted T cells, and activated monocytes can be used therapeutically to prevent progression of virus infections or virus-induced malignancies (Bukowski *et al.*, 1985; Mule *et al.*, 1985; Ada and Jones, 1986; Shinomiya *et al.*, 1989). Furthermore, these models have shown that it is possible for the host immune system to identify and eliminate cells that express tumor antigens. The techniques that have facilitated initial clinical investigation of these approaches have been (a) isolation of effector cells with reactivity to the relevant tumor or pathogen, (b) expansion of effector cells to numbers that are sufficient for clinical efficacy, (c) infusion of effector cells without toxicity to the host, and (d) maintenance of effector cell function following eradication or control of the tumor or pathogen. Clinical studies have now demonstrated the feasibility of preventing or treating infections caused by Epstein–Barr virus and cytomegalovirus in immunocompromised bone marrow transplant patients by transfer of T cells specific for target antigens encoded by these viruses (Riddell *et al.*, 1992a; Walter *et al.*, 1995; Heslop *et al.*, 1996).

Specific reactivity of T cells is conferred by the $\alpha\beta$ T-cell receptor. The two major subsets of mature T cells, CD4 (T helper) and CD8 (cytotoxic), are activated upon binding to a peptide fragment of a protein antigen presented in association with class II and class I MHC, respectively (Morrison *et al.*, 1986; Neefjes and Ploegh, 1992; Yewdell and Bennink, 1992). Peptides in association with class II MHC molecules are presented by antigen-presenting cells such as dendritic cells or monocytes. Class I MHC molecules are expressed on most cell types. The activation process is amplified by binding of costimulatory molecules (CD28/B7) (Linsley *et al.*, 1992; Linsley and Ledbetter, 1993), interleukin-2 (IL-2) autocrine loops, and T helper cell cytokines (IL-2, γ-interferon, GM-CSF) (Mosmann *et al.*, 1986; Maggi *et al.*, 1988). Essentially, the T-cell cytotoxic response mediates direct lysis of the target cell, while T helper cells amplify or maintain long-term T-cell-mediated cytotoxicity and B-cell immunoglobulin production.

Viruses have evolved mechanisms that subvert antigen processing and presentation so that they may survive long term in the host in a latent state. These evasive tactics include reduction in the transcription and/or translation of host proteins including class I MHC and other cellular proteins involved in antigen processing (Schrier *et al.*, 1983; Paabo *et al.*, 1989; Read *et al.*, 1993). The viral genome may also encode proteins that interfere with antigenic peptide transport to the endoplasmic reticulum for assembly with MHC I molecules (Korner and Burgert, 1994). Viruses also escape T-cell recognition by mutation, particularly where viral replication is error prone (e.g., reverse transcriptase) (Phillips *et al.*, 1991; Klenerman *et al.*, 1994; Koup, 1994; Goulder *et al.*, 1997).

There are many possible mechanisms whereby tumors might evade immunological surveillance. These include down-regulation of class I MHC on tumor cells (Zeff *et al.*, 1990), outgrowth of antigen loss variants (Lehmann *et al.*, 1995; Dudley and Roopenian, 1996), tumor secretion of immunosuppressive cytokines such as TGFβ (Torre-Amione *et al.*, 1990), tumor expression of Fas ligand (Villunger *et al.*, 1997), or an absence of costimulatory molecules (Schultze *et al.*, 1995; Cardoso *et al.*, 1996). Graft-versus-tumor reactivity in acute myeloid leukemia and lymphoma may be the result of minor histocompatibility antigens presented as peptides by class I MHC (Horowitz *et al.,* 1990; Jones *et al.,* 1991).

Despite the evolution of mechanisms that allow virus-infected cells to escape recognition by host immune cells, techniques have been developed to select antigen-specific T-cell clones that have reactivity against virus-infected cells. The principal methodology for isolating T cells specific for individual viruses relies on the presentation of viral antigen by an autologous antigen-presenting cell to activate virus-specific T cells, followed by culture in IL-2 to expand reactive T cells (Riddell *et al.,* 1992a, 1992b; Walter *et al.,* 1995). Fibroblasts that express class I MHC are used to stimulate production of $CD8^+$-specific cytotoxic lymphocyte clones whereas cells that present class II MHC and peptide (monocytes and dendritic cells) can be used to select and expand $CD4^+$ T helper clones.

Various strategies for isolation of tumor-reactive cytotoxic T cells are developing. A central issue to be addressed will be the identification of immunodominant peptides that associate with the patient's MHC and are responsible for initiating antitumor T-cell response. These peptides have been isolated by a number of techniques, including the screening of cDNA expression libraries of tumor cells with tumor-reactive T-cell clones (Van Pel *et al.,* 1995). Dendritic cells obtained from the patient's blood after culture with GM-CSF and IL-4 are pulsed with previously defined immunogenic peptides (Alexander-Miller *et al.,* 1996) and are then used to stimulate peripheral blood T cells. An alternate approach is to express the gene encoding the tumor antigen in an autologous antigen-presenting cell using recombinant vectors (Yang *et al.,* 1995; Boczkowski *et al.,* 1996; Yee *et al.,* 1996).

Minor H antigens that are possibly restricted to hematopoietic cells, including leukemic cells, have been used as targets for T-cell immunotherapy to enhance graft-versus-leukemia activity without inducing graft-versus-host disease. *In vivo,* stimulation of allogeneic minor H antigen specific T cells occurs after bone marrow transplant, and further amplification of donor derived specific T-cell lines is obtained by stimulation with γ-irradiated PBMC obtained and stored from the recipient pretransplant (Goulmy *et al.,* 1983; de Bueger *et al.,* 1992). Enhancement of graft-versus-tumor reactivity by expansion of cytotoxic T-cell lymphocytes will rely on the identification of minor H antigens or other immunogenic peptides.

The safety and efficacy of cellular immunotherapy may be improved by using methods that intrinsically alter T-cell function. Recent gene therapy strategies include retroviral-mediated gene transfer of marker genes that allow the *in vivo* persistence of transferred T cells to be tracked or the introduction of genes that confer sensitivity to gancyclovir so that cells may be eliminated should harmful side effects develop (Bonini *et al.,* 1997). Immune cell effector function may be enhanced by modification of signaling pathways. For example, persistent $CD8^+$ cytotoxic activity *in vivo* may be accomplished by transfer of genes that encode a chimeric IL-2 receptor that retains signaling domains but is responsive to GM-CSF that is normally produced by $CD8^+$ cytotoxic lymphocytes after antigen stimulation (Nelson *et al.,* 1994).

E. MODIFICATION OF LYMPHOID AND HEMATOPOIETIC CELL FUNCTION BY EXOGENOUS GENE TRANSFER

The biological properties of hematopoietic stem cells that make them a suitable cellular target for delivery of somatic gene therapy are their long-term persistence in the host following transplant and their potential to differentiate into all of the hematopoietic lineages including cells of the immune system. Replication-incompetent retroviral vectors have been adopted most commonly for gene transfer studies since they have a high transduction efficiency for proliferating cells.

Because most hematopoietic stem cells are quiescent (out of cycle) and retroviral integration requires cells to enter cycle, it has been difficult to attain high levels of stem cell transduction. HGF stimulation has been used to recruit hematopoietic stem cells into the cell cycle; however, this has also been associated with differentiation and hematopoietic lineage commitment. Therefore it is critical to develop *in vitro* systems that stimulate self-renewal divisions of stem cells. Lentiviral-based vectors may not require cell division for long-term gene expression (Crooks *et al.,* 1998) and require further evaluation.

Lymphoid cells are more easily manipulated *in vitro* and were used in the first gene therapy trial for adenosine deaminase deficiency (Blaese *et al.,* 1995). T cells are stimulated to divide using IL-2 and T-cell receptor cross-linking antibodies. Therapeutic levels of retroviral-mediated gene transfer were achieved and transduced lymphocytes persisted *in vivo* for at least 2 years. This study demonstrated initial proof of principle for gene therapy of the immune systems and has provided a strong impetus for development of this new form of therapy.

The number of potential applications for somatic cell gene therapy continues to grow, though the low efficiency of gene transfer continues to limit this development. At this stage, gene therapy approaches are confined to disorders where unregulated expression of a single gene will have therapeutic efficacy. Lymphoid cells have been used as targets for gene transfer for hereditary diseases such as defects in purine metabolism and leukocyte adhesion or acquired diseases such as acquired immunodeficiency syndrome (AIDS).

Resistance to HIV infection can be conferred to T-cell subsets by a number of different approaches. These include expression of ribozymes that catalytically cleave and inactivate HIV-1 RNA molecules (Sun *et al.,* 1995), antisense RNAs (Chatterjee *et al.,* 1992; Crisell *et al.,* 1993) or RNA decoys that inhibit reverse transcription or translation of HIV-1 RNA molecules (Lisziewicz *et al.,* 1993), and mutant-dominant repressor proteins that inhibit HIV structural or regulatory genes (Woffendin *et al.,* 1994).

A number of gene therapy strategies are currently under investigation for the treatment of cancer. These include the development of tumor vaccines by expression of immunogenic peptides on antigen-presenting cells, "suicide" genes that provide a method for eliminating transfected lymphocytes should toxicity occur (e.g., graft-versus-host disease), and gene-marking studies to determine the source of tumor recurrence following autologous transplantation or to track the long-term persistence of transplanted cells (Bonini *et al.,* 1997).

Ex vivo genetic modification of hematopoietic and lymphoid cells will require high efficiency and safe vectors, an understanding of the biology of stem cell self-renewal, and development of the various cell-processing technologies required to bring this therapy to the clinic.

IV. CORE TECHNOLOGIES REQUIRED FOR DELIVERY OF *EX VIVO* CELL THERAPY

The core technologies that have enabled initial development of processes for *ex vivo* cell therapy include methods for cell selection, cell growth, and gene transfer. Although these methods are suitable for initial investigations in a research laboratory setting, more robust processes are required for safe and efficacious delivery of therapy.

A. PROCESS DEVELOPMENT FOR *EX VIVO* CELL THERAPY

The regulation of *ex vivo* cell processing poses a challenge to regulators, who are required to simultaneously support innovation of new therapeutic modalities and to protect public health and safety. The advent of cellular therapies has led the U.S. Food and Drug Administration (FDA) to reassess these therapeutic products and to align them with the regulatory framework already in place for biotechnology and pharmaceutical products. Therefore a new cell therapy product would require an IND (Investigational New Drug) approval for the initiation of clinical trials, followed by licensing of the manufacturing process (Product License Application (PLA)) and facilities (Establishment License Application (ELA)) prior to marketing. More recently, in a draft proposal, the FDA has relaxed the requirement for cellular manipulations to be conducted in an approved site but still requires the approval of devices and processes (FDA, 1997).

The main emphasis for regulation of cell therapy processes relates to adequate standards of current Good Manufacturing Practice (cGMP) and quality assurance of the final cellular product using approved specifications for purity, identity, potency, and sterility. Guidelines for what constitutes cGMP for new and emerging technologies are established using precedent established in the pharmaceutical setting. This would include appropriate control of facilities, personnel, equipment, labels, and records. Because quality control (QC) testing is limited by the quantity of the final cellular product, a quality assurance program must be implemented to ensure that the product will be safe and efficacious on a lot-to-lot basis. For example, to effectively argue that products are sterile, the organization would need to establish an appropriate process in a proper facility, with testing of environment, equipment, and raw materials, and programs for personnel training, documentation, and internal auditing.

Regulation of these processes might be fully centralized by a government authority or partially vested with internal institutional review boards and government-recognized professional organizations so that the system becomes self-regulating. In the United States, transplant centers have created their own organization (Foundation for the Accreditation of Hematopoietic Cell Therapy) to develop QC standards and an infrastructure to accredit the practice of stem cell transplantation.

Replacing animal-derived products used for growth of cells with autologous plasma, serum, or recombinant proteins can reduce the risk of contamination with adventitious pathogens. Furthermore, development of a closed-system manufacturing process will have a major impact on product and worker safety. Therefore the development of automated and functionally closed cell processing devices should be a goal for technologists working in this field.

B. CELL SEPARATION

The goal of cell separation processes is to debulk harvested cells for further downstream processing such as cryopreservation or gene transfer and to remove potentially pathogenic cell subsets. The performance of cell separation technologies is dictated by clinical and GMP requirements. Ideally, an automated closed system that has high yields with adequate enrichment or depletion factors is required.

Centrifugal elutriation has been developed for the extracorporeal harvesting of leukocytes from peripheral blood (apheresis). Blood is continuously (cross-flow elutriation) or semicontinuously (counterflow elutriation) drawn from a central venous catheter or peripheral line, leukocyte-rich plasma is removed from the blood, and the remaining red mass is returned to the patient (Wagner, 1992). Apheresis has found great utility in the development of peripheral stem cell transplant; bone marrow stem cells are mobilized into the peripheral circulation by growth factors and chemotherapy and harvested using centrifugal elutriation.

Monoclonal antibody based separation methods are required to purify or deplete cell subsets from leukapheresis products, bone marrow, or cord blood. Fluorescent-activated cell sorting is one of the most selective cell separation methods, allowing a cell-by-cell sort based on multiparameter discriminators such as two surface antigens and forward and side light scatter. Cellular throughput is limited to less than 30,000 cells/s (Van den Engh and Stokdijk, 1989) so a debulking procedure is still required before cell processing. SyStemix Inc. (Palo Alto, CA) has developed a clinical high-speed sorting facility (Chapter 10).

Immunoaffinity separation methods are based on an immunoaffinity substrate that is created by immobilization of specific ligands such as monoclonal antibody to a material with little intrinsic cell affinity. The geometry of the substrate may be flat plates (panning) (Wysocki and Sato, 1978), hollow fibers (Nordon and Schindhelm, 1997) or spherical beads (Berenson *et al.,* 1988; Maruyama *et al.,* 1989). Once bound to the substrate, cells are recovered by shear stress and chemical agents that disrupt antigen–ligand bonds. Magnetic cell separation is accomplished by labeling cells with beads (Ugelstad *et al.,* 1983, 1992, 1993) or colloids (Molday and Mackenzie, 1982) that contain ferromagnetic material. A high-gradient magnetic field usually created by a magnetizable wire or sphere array is required for separation of cells that are labeled with magnetic colloids (Richards *et al.,* 1996a, 1996b). Magnetic colloids are small enough to be left on cells without interfering significantly with biological function or flow cytometry (Papadimitriou *et al.,* 1995).

Clinical devices at various stages of regulatory approval that are either semiautomated or fully automated have been developed for $CD34^+$ cell enrichment from bone marrow or leukapheresis collections. These are based on avidin–biotin immunoaffinity (Ceprate SC™, CellPro Inc., Bothel, WA), magnetic beads (Isolex 300™, Baxter Inc., Irvine, CA), or magnetic colloids (CliniMACS™, Miltenyi Biotec GmbH, Bergisch Gladbach, Germany). The recovery of $CD34^+$ cells for these systems is between 50 and 80%, with the final purity of the cell product usually greater than 80%, provided that the input frequency of $CD34^+$ cells is greater than 1% (Papadimitriou *et al.,* 1995; McNiece *et al.,* 1997; McQuaker *et al.,* 1997).

In a clinical context, the performance of these systems may have some shortcomings. Multiple apheresis collection procedures are sometimes required if $CD34^+$ cell yields are low. Reproducibility of enrichment yield has been a problem for some systems. Tumor cell depletion of autologous grafts may require combined enrichment and depletion methods to reduce tumor cell burden to levels that will not lead to tumor recurrence. Long-term and prospective randomized trials are also required to determine whether tumor cell depletion of autografts is warranted for specific tumor types.

T-cell depletion of allogeneic grafts has reduced the incidence of graft-versus-host disease (Kernan *et al.,* 1986; Marmont *et al.,* 1991); however, benefits have been offset by a greater rate of graft failure or recurrence of leukemia (Champlin, 1993). Cell selection procedures and cellular immunotherapy are likely to play an important role in selectively augmenting graft-versus-leukemia reactivity without graft-versus-host disease.

C. ENVIRONMENTAL REQUIREMENTS OF HEMATOPOIETIC PROGENITOR CELLS IN *EX VIVO* EXPANSION SYSTEMS

Recent demonstration of the conditions that support a net expansion of stem cells in both murine and human systems has stimulated interest in the potential of bioreactor systems to provide hematopoietic cells *ex vivo* in clinically useful number. The clinical applications that would require bioreactor technologies include the *ex vivo* production of neutrophil and platelet precursors to abrogate neutropenia and thrombocytopenia following high-dose myelotoxic or ablative therapy and the generation of specific cell types for immunotherapy such as dendritic cells and cytotoxic T cells. Retroviral transduction of lymphoid or hematopoietic cells requires that cells be stimulated to enter the cell cycle. Therefore clinical-scale transduction processes may also require bioreactor systems for high-density cell culture.

The Dexter culture system, which requires no addition of growth factors, but the provision of a supportive adherent feeder layer containing fibroblasts, endothelial cells, and adipocytes (stroma), can support long-term hematopoiesis *in vitro* (Dexter *et al.,* 1977). Frequent medium exchanges have improved the productivity and longevity of stromal cell containing hematopoietic culture systems, which has spurred interest in the development of continuous perfusion bioreactors for support of the Dexter culture system (Slovick *et al.,* 1984; Schwartz *et al.,* 1991a, 1991b; Koller *et al.,* 1993a, 1993b).

The first perfusion systems were flat-plate or membrane bioreactors where cells were inoculated onto glass plates or collagen microcarriers, respectively (Koller *et al.,* 1992; Wang and Wu, 1992). A gas-permeable membrane along the entire length of the culture bed has been used to reduce oxygen gradients in the direction of media perfusion (Palsson *et al.,* 1993). This perfusion bioreactor was the forerunner of the first clinical device specifically designed for expansion of bone marrow (Aastrom Biosciences, Inc., Chapter 13).

The discovery of human HGF has led to stroma-free suspension culture systems which can apparently stimulate net amplification of both murine and human hematopoietic cells. Potential advantages of a growth factor driven system are that all medium components are molecularly defined and differentiation can be "fine tuned" by specification of optimal growth factor combinations. Furthermore, stem cell sources that do not provide stroma (cord blood and mobilized peripheral blood) can also be expanded. The culture devices that have been evaluated for growth of hematopoietic cell suspension culture are gas-permeable bags (Purdy *et al.,* 1995; Williams *et al.,* 1996), grooved parallel-plate perfusion bioreactors (Sandstrom *et al.,* 1996), and spinner flasks (Zandstra *et al.,* 1994).

The bioreactor operating parameters that influence the phenotype and number of cells generated from hematopoietic stem cells include oxygen tension, media perfusion and exchange rates, and the nature of the culture inoculum. The combination and concentration of human HGF are central control parameters since they govern the nature of the proliferative stimulus. Many studies have shown that a cytokine combination containing SCF, IL-3, IL-1, IL-6, and G-CSF is able to stimulate production of colony-forming cells. On the other hand, factorial analysis experiments have revealed that that FL, SCF, IL-3, and TPO are required for production of long-term culture-initiating cells (LTC-IC) (Petzer *et al.,* 1996a, 1996b; Zandstra *et al.,* 1997). Maximal production of LTC-IC from adult marrow $CD34^{+}CD38^{-}$ cells was obtained using cytokine levels that were 30- to 80-fold greater than those required for production of colony-forming cells (Zandstra *et al.,* 1997a).

Cytokines, once bound to their growth factor receptor, are internalized, and therefore cell-specific, cytokine-consumption rates would determine the rate at which cytokines are depleted from media. For example, $CD34^{+}CD38^{-}$ cells consume SCF and IL-3 approximately 35-fold faster than unseparated marrow cells (Zandstra *et al.,* 1997b).

Given that multiple factors influence the quality and quantity of cellular output, further research is required to determine the molecular mechanisms responsible for hematopoietic cell growth and development. Bioreactor device development should examine new methods for exchange of media and control of the concentration of cytokines, oxygen, and other metabolites.

D. *EX VIVO* HEMATOPOIETIC CELL EXPANSION FOR BONE MARROW TRANSPLANTATION

The role of a device for *in vitro* manipulation and generation of cells is to perform an optimized biological process at clinical scale in a system that facilitates GMP quality control. Clinical trials are then required to establish safety and efficacy of the system. These issues have been addressed for development of the Aastrom Cell Production System (CPS, Aastrom Biosciences, Inc., Ann Arbor, MI), a plate-perfusion bioreactor for expansion of bone marrow using a Dexter culture system.

The factors that were addressed in defining the biological process of hematopoietic stem cell production were the effect of media perfusion rate and composition, growth factors, oxygenation, and the degree of donor-to-donor variability. The provision of hydrocortisone was required to support outgrowth of a dense stroma for hematopoietic cell production. A serum level of 20% was found to be optimal and animal sera supported a greater cellular output compared to autologous and allogeneic human sera, though it was acknowledged that a fully defined, serum-free system is more desirable from a manufacturing, regulatory, and quality assurance perspective. Commercially available serum-free media for $CD34^{+}$-enriched cell expansion were not supportive of accessory cell cultures, and development of serum-free media for support of stromal cells is underway.

Although endogenous production of growth factors by stroma can support hematopoietic cell production, the greatest cell output was achieved by addition of growth factors, albeit at relatively low concentrations (IL-3, GM-CSF, and Epo). The combination of growth factors affected the nature of the progenitor cell output. For example, addition of FL, TPO, or SCF increased the output of LTC-IC. It may also be feasible to expand cord blood using this system without the use of a preformed stroma (Koller *et al.,* 1998).

Ten patients with high-risk metastatic breast carcinoma were transplanted with unmanipulated and *ex vivo* produced cells with no toxicity and a rate of hematopoietic reconstitution that was similar to that of historical controls. A subsequent clinical trial examined the use of expanded cells alone in patients who received myeloablative therapy known to be lethal without appropriate cell rescue therapy. The first four patients had a hematological recovery that was similar to that of historic controls who had received the same chemotherapy regime and autologous bone marrow transplantation.

E. GENE DELIVERY TECHNOLOGY: NONVIRAL AND VIRAL VECTOR SYSTEMS

The introduction of genetic material into a target tissue can be aimed at delivering DNA to encode for a substance that is lacking in the cell, thereby augmenting cell function, or to selectively replace a defective gene with a normal one. Long-term expression of the transgene offers the advantage of eliminating the need to continuously administer drugs to the patient. Therefore gene delivery technology offers an unparalleled facility to specifically target a wide range of acquired and genetic disease, including cancer, cardiovascular disease, AIDS, and rarer single-gene disorders such as severe combined immunodeficiency disease, cystic fibrosis, lysosomal storage disease, and hemophilia.

Gene transfer techniques are based on purely physicochemical interactions such as coprecipitation, electroporation, and particle bombardment or exploit the natural infectious properties of recombinant viral vectors. Calcium phosphate–DNA complexes, or lipoplexes, are physically adsorbed onto the cell membrane and endocytosed, with transfer of the complex from the endosome to the nucleus (Graham and Van Der Eb, 1973; Fraley and Papahadjopoulos, 1982). For this process to occur, the complex should be resistant to lysosomal action and released into the cytoplasm. Most nonspecific physicochemical techniques rely on selection of stable transfectants since genomic integration occurs at very low frequency.

Receptor-mediated gene transfer techniques take advantage of physiological transfer processes of the cell. Ligands that facilitate uptake include polylysine, transferrin, glycoproteins, or immunoglobulin domains (CDR) that bind to specific cell surface receptors (Cotten *et al.,* 1990; Wu *et al.,* 1991; Slepushkin *et al.,* 1996). Endocytosed receptor–ligand–DNA complexes are subject to lysosomal degradation, and mechanisms are required to disrupt this pathway. For example, coupling adenovirus to conjugate DNA has been shown to facilitate lysosomal disruption and increase gene transfer efficiency (Curiel, 1994).

Other proteins when introduced into the DNA complex facilitate DNA transport to the nucleus (Mistry *et al.,* 1997) or endosome-independent uptake (Sendai virus hemagglutinin) (Kaneda *et al.,* 1989; Dzau *et al.,* 1996).

Host genomic integration of the transgene has the theoretical risk of insertional mutagenesis and possible malignant transformation. Furthermore, cell cycle transit is required for genomic integration of some vectors. These disadvantages could be overcome by development of extrachromosomal vectors that include sites for DNA replication and other genes to facilitate chromosomal-independent replication. Mammalian artificial chromosomes contain the minimum chromosomal components required for DNA replication and segregation (telomeres and centromeres). Epstein–Barr virus (EBV) based vectors include the EBV replication and the *EBNA-1* genes, which confer the ability of these vectors to maintain stable, freely replicating plasmids. Human *ori* vectors incorporate native human sequences to mediate vector replication (Calos, 1996).

Because the frequency of stable genomic integration is low for nonviral delivery systems, multiple passages of cell lines and selection steps are required to obtain high levels of gene expression. Primary hematopoietic and lymphoid cells have limited capacity for self-renewal divisions *in vitro* and therefore require vectors that integrate with high efficiency.

The process whereby gene transfer methods employ modified DNA or RNA viral vectors is known as transduction. The most commonly used viral vectors are herpes virus (Freese *et al.,* 1990), adenovirus (Brody and Crystal, 1994), adeno-associated virus (Muzyczka, 1992; Kotin, 1994), and various retroviruses (Drumm *et al.*, 1990; Chowdhury *et al.,* 1991; Freeman *et al.,* 1993; Hoeben *et al.,* 1993; Miller *et al.,* 1993; Mulligan, 1993). These modified viruses are termed recombinant viral vectors since they have had genes encoding essential replicative/packaging proteins replaced with the therapeutic gene. The relatively high transduction efficiency of these agents has led to their use in gene therapy protocols.

Development of recombinant viral vectors for clinical therapy requires consideration of many factors. Because these vectors are rendered noninfective by deletion of replicative/packaging proteins, production and packaging of these recombinant viruses require development of packaging and producer cell lines. The vector and cell-line constructs should have little homology to ensure that recombination events leading to production of replication-competent virus are unlikely to occur. Screening for production of replication-competent virus in viral supernatants is an important safety and quality control consideration for production of therapeutic vectors.

The uptake of recombinant viruses is limited to a range of host-cell types and, in most cases, is mediated by binding to cell surface receptors. Retroviruses require cell division for stable integration, whereas herpes simplex viruses, adenoviruses, and adeno-associated viruses deliver genes to noncycling cells. Depending on the recombinant virus, the size of the transgene insert is limited to between 5 and 15 kb. For vectors other than wild-type adeno-associated virus, integration of the transgene is generally not site specific, so there is the possibility of insertional mutagenesis. Herpes simplex virus infection may be lytic, and development of this vector would require removal of genes required for cell lysis.

Latent viral infections induce immune tolerance by preventing presentation of immunogenic peptides and recognition by T-cell immunity. Therefore long-term expression of the transgene requires induction of a latent infection by using viral vectors that are derived from latent viruses and express few, if any, foreign proteins. Adenoviral vectors are particularly immunogenic whereas latent viruses such as retroviruses and herpes simplex virus escape recognition by T-cell immunity.

At the present time, retroviral vectors are most commonly used in gene therapy protocols. Advantages of retroviruses are low immunogenicity and high transduction efficiency, provided that cells are in cycle. For the benefits of gene therapy to be fully realized, a number of critical areas relating to gene delivery require further development. These include more efficient agents that can transfect quiescent cells, development of culture systems and bioreactors that facilitate transduction of hematopoietic stem cells, and delivery systems that have a low risk of generating replication-competent vectors.

V. FUTURE DIRECTIONS

The success of *ex vivo* cell therapy will depend on the creation of a dialogue between the various professionals that are involved basic research, technology development, and therapy. In part, this convergence will be facilitated by a cross-disciplinary presentation of research developments. It is important that research groups developing these therapies are conversant with biological, technical, and clinical issues.

In the past, cancer therapy has been based on surgery, chemotherapy, and radiotherapy. The low therapeutic index for high-dose therapy has stimulated the search for safer and more targeted therapies. Restoration and enhancement of cellular function by infusion of therapeutic cell subsets will no doubt play an increasingly important role in the multimodal treatment of cancer.

Adoptive cellular immunotherapy and gene therapy are treatment modalities where our knowledge of the cellular and molecular basis of disease has partnered with new *in vitro* techniques for the manipulation of cells. Even though these approaches may seem close to clinical implementation, a significant research effort is required to fully understand the biology of these systems. Cell selection, expansion, and gene transfer technologies will enable clinical implementation of these *in vitro* processes.

REFERENCES

Ada, G. L., and Jones, P. D. (1986). The immune response to influenza infection. *Curr. Top. Microbiol. Immunol.* **128,** 1–54.

Alexander-Miller, M. A., Leggatt, G. R., and Berzofsky, J. A. (1996). Selective expansion of high- or low-avidity cytotoxic T lymphocytes and efficacy for adoptive immunotherapy. *Proc. Natl. Acad. Sci. USA* **93,** 4102–4107.

Barlogie, B. (1991). Toward a cure for multiple myeloma? *N. Engl. J. Med.* **325,** 1304–1306.

Beatty, P. G., Mori, M., and Milford, E. (1995). Impact of racial genetic polymorphism on the probability of finding an HLA-matched donor. *Transplantation* **60,** 778–783.

Berenson, R. J., Andrews, R. G., Bensinger, W. I., *et al.* (1988). Antigen $CD34^+$ marrow cells engraft lethally irradiated baboons. *J. Clin. Invest.* **81,** 951–955.

Blaese, R. M., Culver, K. W., Miller, A. D., *et al.* (1995). T lymphocyte-directed gene transfer for ADA-SCID: Initial trial results after 4 years. *Science* **270,** 475–480.

Bockholt, S. M., and Burridge, K. (1993). Cell spreading on extracellular matrix proteins induces tyrosine phosphorylation of tensin. *J. Biol. Chem.* **268,** 14565–14567.

Boczkowski, D., Nair, S. K., Snyder, D., *et al.* (1996). Dendritic cells pulsed with RNA are potent antigen-presenting cells *in vitro* and *in vivo. J. Exp. Med.* **184,** 465–472.

Bonadonna, G., Zambetti, M., and Valagussa, P. (1995). Sequential or alternating doxorubicin and CMF regimens in breast cancer with more than three positive nodes. Ten-year results. *JAMA* **273,** 542–547.

Bonini, C., Ferrari, G., Verzeletti, S., *et al.* (1997). HSV-TK gene transfer into donor lymphocytes for control of allogeneic graft-versus-leukemia. *Science* **276,** 1719–1724.

Brody, S. L., and Crystal, R. G. (1994). Adenovirus-mediated *in vivo* gene transfer. *Ann. N.Y. Acad. Sci.* **716,** 90–103.

Broxmeyer, H. E., Douglas, G. W., Hangoc, G., *et al.* (1989). Human umbilical cord blood as a source of transplantable hematopoietic stem/progenitor cells. *Proc. Natl. Acad. Sci. USA* **86,** 3828–3832.

Bukowski, J. F., Warner, J. F., Dennert, G., *et al.* (1985). Adoptive transfer studies demonstrating the antiviral effect of natural killer cells *in vivo. J. Exp. Med.* **161,** 40–52.

Calos, M. (1996). The potential of extrachromosomal replicating vectors for gene therapy. *Trends Genet.* **12,** 463–466.

Cardoso, A. A., Schultze, J. L., Boussiotis, V. A., *et al.* (1996). Pre-B acute lymphoblastic leukemia cells may induce T-cell anergy to alloantigen. *Blood* **88,** 41–48.

Champlin, R. (1993). T-cell depletion for allogeneic bone marrow transplantation: Impact on graft-versus-host disease, engraftment, and graft-versus-leukemia. *J. Hematother.* **2,** 27–42.

Chatterjee, S., Johnson, P. R., and Wong, K. K., Jr. (1992). Dual-target inhibition of HIV-1 *in vitro* by means of an adeno-associated virus antisense vector. *Science* **258,** 1485–1488.

Chowdhury, J. R., Grossman, M., Gupta, S., *et al.* (1991). Long-term improvement of hypercholesterolemia after *ex vivo* gene therapy in LDLR-deficient rabbits. *Science* **254,** 1802–1805.

Clark, B. R., Gallager, J. T., and Dexter, T. M. (1992). Cell adhesion in the stromal regulation of adhesion. *Baillère's Clin. Hematol.* **10,** 21–53.

Collins, R. H., Jr., Shpilberg, O., Drobyski, W. R., *et al.* (1997). Donor leukocyte infusions in 140 patients with relapsed malignancy after allogeneic bone marrow transplantation. *J. Clin. Oncol.* **15,** 433–444.

Cotten, M., Langle-Rouault, F., Kirlappos, H., *et al.* (1990). Transferrin-polycation-mediated introduction of DNA into human leukemic cells: Stimulation by agents that affect the survival of transfected DNA or modulate transferrin receptor levels. *Proc. Natl. Acad. Sci. USA* **87,** 4033–4037.

Crisell, P., Thompson, S., and James, W. (1993). Inhibition of HIV-1 replication by ribozymes that show poor activity *in vitro. Nucleic Acids Res.* **21,** 5251–5255.

Crooks, G. M., Case, S. S., Price, M. A., *et al.* (1998). Efficient transduction of $CD34^+CD38^-$ cells with lentiviral vectors. 27th Annual Meeting of the International Society for Experimental Hematology, Vancouver, BC, Canada.

Cunningham, D., Paz-Ares, L., Milan, S., *et al.* (1994). High-dose melphalan and autologous bone marrow transplantation as consolidation in previously untreated myeloma. *J. Clin. Oncol.* **12,** 759–763.

Curiel, D. T. (1994). High efficiency gene transfer mediated by adenovirus–polylysine–DNA complexes. *Ann. N.Y. Acad. Sci.* **716,** 36–58.

Czar, X. F., Ward, A. C., Hoffman, B. W., *et al.* (1997). cAMP suppresses p21ras and Raf-1 responses but not the Erk-1 response to granulocyte-colony-stimulating factor: Possible Raf-1-independent activation of Erk-1. *Biochem. J.* **322,** 79–87.

Darnell, J., Kerr, I. M., and Stark, G. R. (1994). Jak–STAT pathways and transcriptional activation in response to IFNs and other extracellular signaling proteins. *Science* **264,** 821–825.

de Bueger, M., Bakker, A., Van Rood, J. J., *et al.* (1992). Tissue distribution of human minor histocompatibility antigens. Ubiquitous versus restricted tissue distribution indicates heterogeneity among human cytotoxic T lymphocyte-defined non-MHC antigens. *J. Immunol.* **149,** 1788–1794.

Dexter, T. M., Allen, T. D., and Lajtha, L. G. (1977). Conditions controlling the proliferation of haemopoietic stem cells *in vitro. J. Cell. Physiol.* **91,** 335–344.

Drobyski, W. R., Keever, C. A., Roth, M. S., *et al.* (1993). Salvage immunotherapy using donor leukocyte infusions as treatment for relapsed chronic myelogenous leukemia after allogeneic bone marrow transplantation: Efficacy and toxicity of a defined T-cell dose. *Blood* **82,** 2310–2318.

Drumm, M. L., Pope, H. A., Clif, W. H., *et al.* (1990). Correction of the cystic fibrosis defect *in vitro* by retrovirus-mediated gene transfer. *Cell* **62,** 1227–1233.

Dudley, M. E., and Roopenian, D. C. (1996). Loss of a unique tumor antigen by cytotoxic T lymphocyte immunoselection from a 3-methylcholanthrene-induced mouse sarcoma reveals secondary unique and shared antigens. *J. Exp. Med.* **184,** 441–447.

Duronio, V., Welham, M. J., Abraham, S., *et al.* (1992). p21ras activation via hemopoietin receptors and c-kit requires tyrosine kinase activity but not tyrosine phosphorylation of p21ras-GTPase-activating protein. *Proc. Natl. Acad. Sci. USA* **89,** 1587–1591.

Dzau, V., Mann, M., Morishita, R., *et al.* (1996). Fusigenic viral liposomes for gene therapy in cardiovascular diseases. *Proc. Natl. Acad. Sci. USA* **93,** 11421–11425.

FDA. (1997). Proposed approach to regulation of cellular and tissue-based products (The Food and Drug Administration). *J. Hematother.* **6,** 195–212.

Fraley, R., and Papahadjopoulos, D. (1982). Liposomes: The development of a new carrier system for introducing nucleic acid into plant and animal cells. *Curr. Top. Microbiol. Immunol.* **96,** 171–191.

Freeman, S. M., Abboud, C. N., Whartenby, K. A., *et al.* (1993). The "bystander effect": Tumor regression when a fraction of the tumor mass is genetically modified. *Cancer Res.* **53,** 5274–5283.

Freese, A., Geller, A. I., and Neve, R. (1990). HSV-1 vector mediated neuronal gene delivery. *Biochem. Pharmacol.* **40,** 2189–2199.

Gianni, A. M., Bregni, M., Siena, S., *et al.* (1996). Is high dose better than standard-dose chemotherapy as initial treatment of poor-risk large cell lymphomas? A criticial analysis from available randomized trials. *Ann. Oncol.* **7,**S12.

Gianni, A. M., Siena, S., Bregni, M., *et al.* (1992). Growth-factor supported high-dose sequential (HDS) adjuvant chemotherapy in breast cancer with at least 10 positive nodes. *Proc. Am. Soc. Clin. Oncol.* **11,** 68.

Glaspy, J., LeMaistre, C. F., Lill, M., *et al.* (1995). Dose-response of 7 day administration of recombinant methionyl human stem cell factor (SCF) in combination with filgrastim (G-CSF) for progenitor cell mobilization in patients with stage II–IV breast cancer. *Blood* **86,** 463a.

Gluckman, E., Rocha, V., Boyer-Chammard, A., *et al.* (1997). Outcome of cord-blood transplantation from related and unrelated donors. *N. Engl. J. Med.* **337,** 373–381.

Goulder, P. J., Phillips, R. E., Colbert, R. A., *et al.* (1997). Late escape from an immunodominant cytotoxic T-lymphocyte response associated with progression to AIDS. *Nat. Med.* **3,** 212–217.

Goulmy, E., Gratama, J. W., Blokland, E., *et al.* (1983). A minor transplantation antigen detected by MHC-restricted cytotoxic T lymphocytes during graft-versus-host disease. *Nature* **302,** 159–161.

Graham, F. L., and Van Der Eb, A. J. (1973). A new technique for the assay of infectivity of human adenovirus 5 DNA. *Virology* **52,** 456–467.

Haioun, C., Lepage, E., Gisselbrecht, C., *et al.* (1994). Comparison of autologous bone marrow transplantation with sequential chemotherapy for intermediate-grade and high-grade non-Hodgkin's lymphoma in first complete remission: A study of 464 patients. *J. Clin. Oncol.* **12,** 2543–2550.

Haylock, D. N., To, L. B., Dowse, T. L., *et al.* (1992). *Ex vivo* expansion and maturation of peripheral blood CD34$^+$ cells into the myeloid lineage. *Blood* **80,** 1405–1412.

Heim, M. H., Kerr, I. M., Stark, G. R., *et al.* (1995). Contribution of STAT SH2 groups to specific interferon signaling by the Jak–STAT pathway. *Science* **267,** 1347–1349.

Henon, P. H. R., Liang, H., Beck-Wirth, G., *et al.* (1992). Comparison of hematopoietic and immune recovery after autologous bone marrow or blood stem cell transplants. *Bone Marrow Transplant.* **9,** 285–291.

Heslop, H. E., Ng, C. Y., Li, C., *et al.* (1996). Long-term restoration of immunity against Epstein–Barr virus infection by adoptive transfer of gene-modified virus-specific T lymphocytes. *Nat. Med.* **2,** 551–555.

Hirsch, E., Iglesias, A., Potocnik, A. J., *et al.* (1996). Impaired migration but not differentiation of hemopoietic stem cells in the absence of β_1-integrins. *Nature* **380,** 171–175.

Hoeben, R. C., Fallaux, F. J., Van Tilburg, N. H., *et al.* (1993). Toward gene therapy for hemophilia A: Long-term persistence of factor VIII-secreting fibroblasts after transplantation into immunodeficient mice. *Hum. Gene Ther.* **4,** 179–186.

Horowitz, M. M., Gale, R. P., Sondel, P. M., *et al.* (1990). Graft-versus-leukemia reactions after bone marrow transplantation. *Blood* **75,** 555–562.

Jacobson, M. D. (1997). Apoptosis: Bcl-2-related proteins get connected. *Curr. Biol.* **7,** R277–R281.

Jones, R. J., Ambinder, R. F., Piantadosi, S., *et al.* (1991). Evidence of a graft-versus-lymphoma effect associated with allogeneic bone marrow transplantation. *Blood* **77,** 649–653.

Kaneda, Y., Iwai, K., and Uchida, T. (1989). Increased expression of DNA cointroduced with nuclear protein in adult rat liver. *Science* **243,** 375–378.

Kernan, N. A., Collins, N. H., Juliano, L., *et al.* (1986). Clonable T lymphocytes in T cell-depleted bone marrow transplants correlate with development of graft-v-host disease. *Blood* **68,** 770–773.

Kincade, P. W., Lee, G., Pietrangeli, C. E., *et al.* (1989). Cells and molecules that regulate B-lymphopoiesis in the bone marrow. *Annu. Rev. Immunol.* **7,** 111–143.

Klenerman, P., Rowland-Jones, S., McAdam, S., *et al.* (1994). Cytotoxic T-cell activity antagonized by naturally occurring HIV-1 Gag variants. *Nature* **369,** 403–407.

Kolb, H. J., Mittermuller, J., Clemm, C., *et al.* (1990). Donor leukocyte transfusions for treatment of recurrent chronic myelogenous leukemia in marrow transplant patients. *Blood* **76,** 2462–2465.

Koller, M. R., Bender, J. G., Miller, W. M., *et al.* (1992). Reduced oxygen tension increases hematopoiesis in long-term culture of human stem and progenitor cells from cord blood and bone marrow. *Exp. Hematol.* **20,** 264–270.

Koller, M. R., Bender, J. G., Miller, W. M., *et al.* (1993a). Expansion of primitive human hematopoietic progenitors in a perfusion bioreactor system with IL-3, IL-6, and stem cell factor. *Biotechnology* **11,** 358–363.

Koller, M. R., Emerson, S. G., and Palsson, B. O. (1993b). Large-scale expansion of human stem and progenitor cells from bone marrow mononuclear cells in continuous perfusion cultures. *Blood* **82,** 378–384.

Koller, M. R., Manchel, I., Maher, R. J., *et al.* (1998). Clinical-scale human umbilical cord blood cell expansion in a novel automated perfusion culture system. *Bone Marrow Transplant.* **21,** 653–663.

Körbling, M., Przepiorka, D., Huh, Y. O., *et al.* (1995). Allogeneic blood stem cell transplantation for refractory leukemia and lymphoma: Potential advantage of blood over marrow allografts. *Blood* **85,** 1659–1665.

Korner, H., and Burgert, H. G. (1994). Down-regulation of HLA antigens by the adenovirus type 2 E3/19K protein in a T-lymphoma cell line. *J. Virol.* **68,** 1442–1448.

Kotin, R. (1994). Prospects for the use of adeno-associated virus as a vector for human gene therapy. *Hum. Gene Ther.* **5,** 793–801.

Koup, R. A. (1994). Virus escape from CTL recognition. *J. Exp. Med.* **180,** 779–782.

Kroemer, G. (1997). The proto-oncogene Bcl-2 and its role in regulating apoptosis. *Nat. Med.* **3,** 614–620.

Kurtzberg, J., Laughlin, M., Graham, M. L., *et al.* (1996). Placental blood as a source of hematopoietic stem cells for transplantation into unrelated recipients. *N. Engl. J. Med.* **335,** 157–166.

Kusadasi, N., van Soest, P. L., and Ploemacher, R. E. (1998). *Ex vivo* expansion of NOD/SCID-repopulating cells from umbilical cord blood. 27th Annual Meeting of the International Society of Experimental Hematology, Vancouver, BC, Canada.

Lehmann, F., Marchand, M., Hainaut, P., *et al.* (1995). Differences in the antigens recognized by cytolytic T cells on two successive metastases of a melanoma patient are consistent with immune selection. *Eur. J. Immunol.* **25,** 340–347.

Lévesque, J., Haylock, D., and Simmons, P. (1996). Cytokine regulation of proliferation and cell adhesion are correlated events in human $CD34^+$ hemopoietic progenitors. *Blood* **88,** 1168–1176.

Lévesque, J. P., Leavesley, D. I., Niutta, S., *et al.* (1995). Cytokines increase human hemopoietic cell adhesiveness by activation of very late antigen (VLA)-4 and (VLA)-5 integrins. *J. Exp. Med.* **181,** 1805–1815.

Linsley, P. S., Greene, J. L., Tan, P., *et al.* (1992). Coexpression and functional cooperation of CTLA-4 and CD28 on activated T lymphocytes. *J. Exp. Med.* **176,** 1595–1604.

Linsley, P. S., and Ledbetter, J. A. (1993). The role of the CD28 receptor during T cell responses to antigen. *Annu. Rev. Immunol.* **11,** 191–212.

Lisziewicz, J., Sun, D., Smythe, J., *et al.* (1993). Inhibition of human immunodeficiency virus type 1 replication by regulated expression of a polymeric Tat activation response RNA decoy as a strategy for gene therapy in AIDS. *Proc. Natl. Acad. Sci. USA* **90,** 8000–8004.

Long, M. W. (1992). Blood cell cytoadhesion molecules. *Exp. Hematol.* **20,** 288–301.

Maggi, E., Del-Prete, G., Macchia, D., *et al.* (1988). Profiles of lymphokine activities and helper function for IgE in human T cell clones. *Eur. J. Immunol.* **18,** 1045–1050.

Marmont, A. M., Horowitz, M. M., Gale, R. P., *et al.* (1991). T-cell depletion of HLA-identical transplants in leukemia. *Blood* **78,** 2120–2130.

Maruyama, A., Tsuruta, T., Kataoka, K., *et al.* (1989). Separation of B and T lymphocytes by cellular adsorption chromatography with polyamine graft copolymers as column matrices. II. Recovery of adsorbed B cell enriched populations from the column. *Biomaterials* **10,** 393–399.

McNiece, I., Briddell, R., Stoney, G., *et al.* (1997). Large-scale isolation of $CD34^+$ cells using the Amgen cell selection device results in high levels of purity and recovery. *J. Hematother.* **6,** 5–11.

McQuaker, I. G., Haynes, A. P., Anderson, S., *et al.* (1997). Engraftment and molecular monitoring of $CD34^+$ peripheral-blood stem-cell transplants for follicular lymphoma: A pilot study. *J. Clin. Oncol.* **15,** 2288–2295.

Migliaccio, A. R., Adamson, J. W., Rubinstein, P., *et al.* (1996). Placental/cord blood stem cell transplantation: Correlation between the progenitor cell dose and time to myeloid engraftment in 55 unrelated transplants. *Blood* **88,** 286a.

Miller, A., Miller, D., Garcia, J., *et al.* (1993). Use of retroviral vectors for gene transfer and expression. *Methods Enzymol.* **217,** 581–599.

Mistry, A., Falciola, L., Monaco, L., *et al.* (1997). Recombinant HMG1 protein produced in *Pichia pastoris:* A nonviral gene delivery agent. *BioTechniques* **22,** 718–729.

Molday, R. S., and Mackenzie, D. (1982). Immunospecific ferromagnetic iron-dextran reagents for the labeling and magnetic separation of cells. *J. Immunol. Methods* **52,** 353–367.

Morrison, L. A., Lukacher, A. E., Braciale, V. L., *et al.* (1986). Differences in antigen presentation to MHC class I- and class II-restricted influenza virus-specific cytolytic T lymphocyte clones. *J. Exp. Med.* **163,** 903–921.

Mosmann, T. R., Cherwinski, H., Bond, M. W., *et al.* (1986). Two types of murine helper T cell clone. I. Definition according to profiles of lymphokine activities and secreted proteins. *J. Immunol.* **136,** 2348–2357.

Mule, J. J., Shu, S., and Rosenberg, S. A. (1985). The anti-tumor efficacy of lymphokine-activated killer cells and recombinant interleukin 2 *in vivo. J. Immunol.* **135,** 646–652.

Mulligan, R. C. (1993). The basic science of gene therapy. *Science* **260,** 926–931.

Muzyczka, N. (1992). Use of adeno-associated virus as a general transduction vector for mammalian cells. *Curr. Top. Microbiol. Immunol.* **158,** 97–129.

Neefjes, J. J., and Ploegh, H. L. (1992). Intracellular transport of MHC class II molecules. *Immunol. Today* **13,** 179–184.

Nelson, B. H., Lord, J. D., and Greenberg, P. D. (1994). Cytoplasmic domains of the interleukin-2 receptor beta and gamma chains mediate the signal for T-cell proliferation. *Nature* **369,** 333–336.

Nicholson, D. W., and Thornberry, N. A. (1997). Caspases: Killer proteases. *Trends Biochem. Sci.* **22,** 299–306.

Nordon, R. E., and Schindhelm, K. (1997). Design of hollow fiber modules for uniform shear elution affinity cell separation. *Artif. Organs* **21,** 107–115.

Novak, U., Harpur, A. G., Paradiso, L., *et al.* (1995). Colony-stimulating factor 1-induced STAT1 and STAT3 activation is accompanied by phosphorylation of Tyk2 in macrophages and Tyk2 and JAK1 in fibroblasts. *Blood* **86,** 2948–2956.

Paabo, S., Severinsson, L., Andersson, M., *et al.* (1989). Adenovirus proteins and MHC expression. *Adv. Cancer Res.* **52,** 151–163.

Palsson, B. O., Paek, S. H., Schwartz, R. M., *et al.* (1993). Expansion of human bone marrow progenitor cells in a high cell density continuous perfusion system. *Biotechnology* **11,** 368–372.

Papadimitriou, C. A., Roots, A., Koenigsmann, M., *et al.* (1995). Immunomagnetic selection of $CD34^+$ cells from fresh peripheral blood mononuclear cell preparations using two different separation techniques. *J. Hematother.* **4,** 539–544.

Papadopoulos, E. B., Ladanyi, M., Emanuel, D., *et al.* (1994). Infusions of donor leukocytes to treat Epstein–Barr virus-associated lymphoproliferative disorders after allogeneic bone marrow transplantation. *N. Engl. J. Med.* **330,** 1185–1191.

Peters, W. P., Rosner, G., Ross, M., *et al.* (1993a). Comparative effects of granulocyte-macrophage colony-stimulating factor (GM-CSF) and granulocyte colony-stimulating factor (G-CSF) on priming peripheral blood progenitor cells for use with autologous bone marrow after high-dose chemotherapy. *Blood* **81,** 1709–1719.

Peters, W. P., Ross, M., Vredenburgh, J. J., *et al.* (1993b). High-dose chemotherapy with autologous bone marrow support after standard-dose adjuvant therapy for high-risk breast cancer. *J. Clin. Oncol.* **11,** 1132–1143.

Petzer, A. L., Hogge, D. E., Lansdorp, P. M., *et al.* (1996a). Self-renewal of primitive human hematopoietic cells (long-term-culture-initiating cells) *in vitro* and their expansion in defined medium. *Proc. Natl. Acad. Sci. USA* **93,** 1470–1474.

Petzer, A. L., Zandstra, P. W., Piret, J. M., *et al.* (1996b). Differential cytokine effects on primitive ($CD34^+CD38^-$) human hematopoietic cells: Novel responses to Flt3-ligand and thrombopoietin. *J. Exp. Med.* **183,** 2551–2558.

Phillips, R. E., Rowland-Jones, S., Nixon, D. F., *et al.* (1991). Human immunodeficiency virus genetic variation that can escape cytotoxic T cell recognition. *Nature* **354,** 453–459.

Purdy, M. H., Hogan, C. J., Hami, L., *et al.* (1995). Large volume *ex vivo* expansion of CD34-positive hematopoietic progenitor cells for transplantation. *J. Hematother.* **4,** 515–525.

Read, G. S., Karr, B. M., and Knight, K. (1993). Isolation of a herpes simplex virus type 1 mutant with a deletion in the virion host shutoff gene and identification of multiple forms of the vhs (UL41) polypeptide. *J. Virol.* **67,** 7149–7160.

Reece, D. E., Barnett, M. J., Connors, J. M., *et al.* (1991). Intensive chemotherapy with cyclophosphamide, carmustine, and etoposide followed by autologous bone marrow transplantation for relapsed Hodgkin's disease. *J. Clin. Oncol.* **9,** 1871–1879.

Richards, A. J., Roath, S., Smith, R. J. S., *et al.* (1996a). High purity, recovery and selection of human blood cells with a novel high gradient magnetic separator. *J. Hematother.* **5,** 415–426.

Richards, A. J., Roath, S., Smith, R. J. S., *et al.* (1996b). The mechanisms of high gradient magnetic separation of human blood and bone marrow. *IEEE Trans. Magn.* **32,** 459–469.

Riddell, S. R., Greenberg, P. D., Overell, R. W., *et al.* (1992a). Phase I study of cellular adoptive immunotherapy using genetically modified $CD8^+$ HIV-specific T cells for HIV seropositive patients undergoing allogeneic bone marrow transplant. *Hum. Gene Ther.* **3,** 319–338.

Riddell, S. R., Watanabe, K. S., Goodrich, J. M., *et al.* (1992b). Restoration of viral immunity in immunodeficient humans by the adoptive transfer of T cell clones. *Science* **257,** 238–241.

Robb, L., Lyons, I., Li, R., *et al.* (1995). Absence of yolk sac hematopoiesis from mice with a targeted disruption of the scl gene. *Proc. Natl. Acad. Sci. USA* **92,** 7075–7079.

Rothman, P., Kreider, B., Azam, M., *et al.* (1994). Cytokines and growth factors signal through tyrosine phosphorylation of a family of related transcription factors. *Immunity* **1,** 457–468.

Rowlings, P. A. (1996). Current use and outcome of blood and marrow transplantation. *ABMTR Newsl. 3,* 6–12.

Rubinstein, P., Carrier, C., Adamson, J., *et al.* (1996). New York Blood Center's program for unrelated placental/umbilical cord blood (PCB) transplantation: 243 transplants in the first 3 years. *Blood* **88,** 142a.

Sandstrom, C. E., Bender, J. G., Miller, W. M., *et al.* (1996). Development of novel perfusion chamber to retain nonadherent cells and its use for comparison of human "mobilized" peripheral blood mononuclear cell cultures with and without irradiated bone marrow stroma. *Biotechnol. Bioeng.* **50,** 493–504.

Schmitz, N., Glass, B., Dreger, P., *et al.* (1993). High-dose chemotherapy and hematopoietic stem cell rescue in patients with relapsed Hodgkin's disease. *Ann. Hematol.* **66,** 251–256.

Schrier, P. I., Bernards, R., Vaessen, R. T., *et al.* (1983). Expression of class I major histocompatibility antigens switched off by highly oncogenic adenovirus 12 in transformed rat cells. *Nature* **305,** 771–775.

Schultze, J. L., Cardoso, A. A., Freeman, G. J., *et al.* (1995). Follicular lymphomas can be induced to present alloantigen efficiently: A conceptual model to improve their tumor immunogenicity. *Proc. Natl. Acad. Sci. USA* **92,** 8200–8204.

Schulz, L. D., Schweitzer, P. A., Rajan, T. V., *et al.* (1993). Mutations at the murine motheaten locus are within the hematopoietic cell protein tyrosine phosphatase (Hcph) gene. *Cell* **73,** 1445–1454.

Schwartz, R. M., Emerson, S. G., Clarke, M. F., *et al.* (1991a). *In vitro* myelopoiesis stimulated by rapid medium exchange and supplementation with hematopoietic growth factors. *Blood* **78,** 3155–3161.

Schwartz, R. M., Palsson, B. O., and Emerson, S. G. (1991b). Rapid medium perfusion rate significantly increases the productivity and longevity of human bone marrow cultures. *Proc. Natl. Acad. Sci. USA* **88,** 6760–6764.

Sheridan, W. P., Begley, C. G., To, L. B., *et al.* (1994). Phase II study of autologous filgrastim (G-CSF)-mobilized peripheral blood progenitor cells to restore hemopoiesis after high-dose chemotherapy for lymphoid malignancies. *Bone Marrow Transplant.* **14,** 105–111.

Shinomiya, H., Shinomiya, M., Stevenson, G. W., *et al.* (1989). Activated killer monocytes: Preclinical model systems. *Immunol. Ser.* **48,** 101–126.

Shivdasani, R. A., Mayer, E. L., and Orkin, S. H. (1995). Absence of blood formation in mice lacking the T-cell leukaemia oncoprotein tal-1/SCL. *Nature* **373,** 432–434.

Simmons, P. J., Lévesque, J. P., and Zannettino, A. C. W. (1997). Cell adhesion molecules and their role in regulating hemopoiesis. *Baillère's Clin. Hematol.* **10,** 485–505.

Simmons, P. J., Zannettino, A., Gronthos, A., *et al.* (1994). Potential adhesion mechanisms for localisation of haematopoietic progenitors to bone marrow stroma. *Leuk. Lymphoma* **12,** 353–363.

Slepushkin, V., Salem, I., Andreev, S., *et al.* (1996). Targeting of liposomes to HIV-1-infected cells by peptides derived from the CD4 receptor. *Biochem. Biophys. Res. Commun.* **227,** 827–833.

Slovick, F. T., Abboud, C. N., Brennan, J. K., *et al.* (1984). Survival of granulocytic progenitors in the nonadherent and adherent compartments of human long-term marrow cultures. *Exp. Hematol.* **12,** 327–338.

Springer, T. (1994). Traffic signals for lymphocyte recirculation and leukocyte emigration: The multistep paradigm. *Cell* **76,** 301–314.

Stahl, N., Farruggella, T. J., Boulton, T. G., *et al.* (1995). Choice of STATs and other substrates specified by modular tyrosine-based motifs in cytokine receptors. *Science* **267,** 1349–1353.

Starr, R., Willson, T. A., Viney, E. M., *et al.* (1997). A family of cytokine-inducible inhibitors of signalling. *Nature* **387,** 917–921.

Sun, L. Q., Pyati, J., Wang, L., *et al.* (1995). Resistance to HIV-1 infection conferred by transduction of human peripheral blood lymphocytes with ribozyme, antisense or polyTAR constructs. *Proc. Natl. Acad. Sci. USA 92,* 7272–7276.

Takahira, H., Gotoh, A., Ritchie, A., *et al.* (1997). Steel factor enhances integrin-mediated tyrosine phosphorylation of focal adhesion kinase (pp 125FAK) and paxillin. *Blood* **89,** 1574–1584.

To, L. B., Roberts, M. M., Haylock, D. N., *et al.* (1992). Comparison of hematological recovery times and supportive care requirements of autologous recovery phase peripheral blood stem cell transplants, autologous bone marrow transplants and allogeneic bone marrow transplants. *Bone Marrow Transplant.* **9,** 277–284.

Torre-Amione, G., Beauchamp, R. D., Koeppen, H., *et al.* (1990). A highly immunogenic tumor transfected with a murine transforming growth factor type β_1 cDNA escapes immune surveillance. *Proc. Natl. Acad. Sci. USA* **87,** 1486–1490.

Traycoff, C. M., Yoder, M. C., Hiatt, K., *et al.* (1997). Functional association between expression of adhesion molecules and marrow repopulating potential of primitive murine hematopoietic progenitor cells (HPC). *Exp. Hematol.* **25,** 736a.

Tsui, H. W., Siminovitch, K. A., de Souza, L., *et al.* (1993). Motheaten and viable motheaten mice have mutations in the haematopoietic cell phosphatase gene. *Nat. Genet.* **4,** 124–129.

Ugelstad, J., Berge, A., Ellingson, T., *et al.* (1992). Preparation and application of new monosized polymer particles. *Prog. Polym. Sci.* **17,** 87–161.

Ugelstad, J., Mork, P. C., Mfutakamba, H. R., *et al.* (1983). Thermodynamics of swelling of polymer, oligomer and polymer–oligomer particles. Preparation and application of monodispersed particles. *In* "Science and Technology of Polymer Colloids" (G. W. Poehlein, R. H. Ottewill, and J. M. Goodwin, Eds.), pp. 51–99. Nijhoff, Boston.

Ugelstad, J., Stenstad, P., Kilaas, L., *et al.* (1993). Monodisperse magnetic polymer particles. *Blood Purif.* **11,** 349–369.

van Corven, E. J., Hordijk, P. L., Medema, R. H., *et al.* (1993). Pertussis toxin-sensitive activation of p21ras by G protein-coupled receptor agonists in fibroblasts. *Proc. Natl. Acad. Sci. USA* **90,** 1257–1261.

Van den Engh, G., and Stokdijk, W. (1989). Parallel processing data acquisition system for multilaser flow cytometry and cell sorting. *Cytometry* **10,** 282–293.

van der Loo, J. C. M., Xiao, X. L., McMillin, D., *et al.* (1997). Functional VLA-5 on hematopoietic stem cells. *Exp. Hematol.* **25,** 743a.

Van Pel, A., van der Bruggen, P., Coulie, P. G., *et al.* (1995). Genes coding for tumor antigens recognized by cytolytic T lymphocytes. *Immunol. Rev.* **145,** 229–250.

Vignais, M. L., Sadowski, H. B., Watling, D., *et al.* (1996). Platelet-derived growth factor induces phosphorylation of multiple JAK family kinases and STAT proteins. *Mol. Cell. Biol.* **16,** 1759–1769.

Villunger, A., Egle, A., Marschitz, I., *et al.* (1997). Constitutive expression of Fas (Apo-1/CD95) ligand on multiple myeloma cells: A potential mechanism of tumor-induced suppression of immune surveillance. *Blood* **90,** 12–20.

Wagner, J. E. (1992). Isolation of primitive hematopoietic stem cells. *Semin. Hematol.* **29,** 6–9.

Wagner, J. E., Defor, T., Rubenstein, P., *et al.* (1997). Transplantation of unrelated donor umbilical cord blood (UCB): Outcomes and analysis of risk factors. *Blood* **90,** 1767.

Wagner, J. E., Kernan, N. A., Steinbuch, M., *et al.* (1995). Allogenic sibling umbilical-cord blood transplantation in children with malignant and non-malignant disease. *Lancet* **346,** 214–219.

Walter, E. A., Greenberg, P. D., Gilbert, M. J., *et al.* (1995). Reconstitution of cellular immunity against cytomegalovirus in recipients of allogeneic bone marrow by transfer of T-cell clones from the donor. *N. Engl. J. Med.* **333,** 1038–1044.

Wang, Q., Stacy, T., Binder, M., *et al.* (1996). Disruption of the Cbfa2 gene causes necrosis and hemorrhaging in the central nervous system and blocks definitive hematopoiesis. *Proc. Natl. Acad. Sci. USA* **93,** 3444–3449.

Wang, T. Y., and Wu, J. H. D. (1992). A continuous perfusion bioreactor for long-term bone marrow culture. *Ann. N.Y. Acad. Sci.* **665,** 274–284.

Williams, S. F., Lee, W. J., Bender, J. G., *et al.* (1996). Selection and expansion of peripheral blood $CD34^+$ cells in autologous stem cell transplantation for breast cancer. *Blood* **87,** 1687–1691.

Woffendin, C., Yang, Z. Y., Ranga, U., *et al.* (1994). Nonviral and viral delivery of a human immunodeficiency virus protective gene into human T cells. *Proc. Natl. Acad. Sci. USA* **91,** 11581–11585.

Wu, G. Y., Wilson, J. M., Shalaby, F., *et al.* (1991). Receptor-mediated gene delivery *in vivo. J. Biol. Chem.* **266,** 14338–14342.

Wysocki, L. J., and Sato, V. L. (1978). "Panning" for lymphocytes: A method of cell selection. *Proc. Natl. Acad. Sci. USA* **75,** 2844–2848.

Yang, N. S., Sun, W. H., Yannelli, J. R., *et al.* (1995). Gene gun and other non-viral approaches for cancer gene therapy. *Nat. Med.* **1,** 481–483.

Yee, C., Gilbert, M. J., Riddell, S. R., *et al.* (1996). Isolation of tyrosinase-specific $CD8^+$ and $CD4^+$ T cell clones from the peripheral blood of melanoma patients following *in vitro* stimulation with recombinant vaccinia virus. *J. Immunol.* **157,** 4079–4086.

Yewdell, J. W., and Bennink, J. R. (1992). Cell biology of antigen processing and presentation to major histocompatibility complex class I molecule-restricted T lymphocytes. *Adv. Immunol.* **52,** 1–123.

Zandstra, P. W., Conneally, E., Petzer, A. L., *et al.* (1997a). Cytokine manipulation of primitive human hematopoietic cell self-renewal. *Proc. Natl. Acad. Sci. USA* **94,** 4698–4703.

Zandstra, P. W., Eaves, C. J., and Piret, J. M. (1994). Expansion of hematopoietic progenitor cell populations in stirred suspension bioreactors of normal human bone marrow cells. *Biotechnology* **12,** 909–914.

Zandstra, P. W., Petzer, A. L., Eaves, C. J., *et al.* (1997b). Cellular determinants affecting the rate of cytokine depletion in cultures of human hematopoietic cells. *Biotechnol. Bioeng.* **54,** 58–66.

Zeff, R. A., Nakagawa, M., Mashimo, H., *et al.* (1990). Failure of cell surface expression of a class I major histocompatibility antigen caused by somatic point mutation. *Transplantation* **49,** 803–808.

Index